INTELLIGENT CONTROL: PRINCIPLES, TECHNIQUES AND APPLICATIONS

SERIES IN INTELLIGENT CONTROL AND INTELLIGENT AUTOMATION

Editor-in-Charge: Fei-Yue Wang
(*University of Arizona*)

Vol. 1: Reliable Plan Selection by Intelligent Machines
(*J E McInroy, J C Musto, and G N Saridis*)

Vol. 2: Design of Intelligent Control Systems Based on Hierachical Stochastic Automata (*P Lima and G N Saridis*)

Vol. 3: Intelligent Task Planning Using Fuzzy Petri Nets
(*T Cao and A C Sanderson*)

Forthcoming volumes:

Vol. 4: Advanced Studies in Flexible Robotic Manipulators: Modeling, Design, Control, and Applications (*F Y Wang*)

Vol. 5: Computational Foundations for Intelligent Systems (*S J Yakowitz*)

Vol. 6: Modeling, Simulation, and Control of Flexible Manufacturing Systems: A Petri Net Approach (*M Zhou and K Venkatesh*)

Vol. 8: Advanced Topics in Computer Vision and Pattern Recognition
(*E Sung, D Mital, E K Teoh, H Wang, and Z Li*)

Vol. 9: Petri Nets for Supervisory Control of Discrete Event Systems: A Structural Approach (*A Giua and F DiCesare*)

Series in
Intelligent Control and Intelligent Automation
Vol. 7

INTELLIGENT CONTROL: PRINCIPLES, TECHNIQUES AND APPLICATIONS

Zi-Xing Cai
Research Center for Intelligent Control
Central South University of Technology, Changsha, China

Published by

World Scientific Publishing Co. Pte. Ltd.

P O Box 128, Farrer Road, Singapore 912805

USA office: Suite 1B, 1060 Main Street, River Edge, NJ 07661

UK office: 57 Shelton Street, Covent Garden, London WC2H 9HE

British Library Cataloguing-in-Publication Data
A catalogue record for this book is available from the British Library.

INTELLIGENT CONTROL: PRINCIPLES, TECHNIQUES AND APPLICATIONS

ISBN 981-02-2564-4

This book is printed on acid-free paper.

Printed in Singapore by Uto-Print

To
Huan, Jing-Feng and Yu-Feng

PREFACE

With the rapid development and great progress in modern science and technology which have given a new and higher requirement to the system and control sciences, automatic control is faced with a new and strong challenge. The traditional control including the classical feedback control, modern control and large-scale system theories has encountered many difficulties in its applications. First of all, the design and analysis for the traditional control system is based on their precise mathematical models that are usually difficult to achieve, owing to the complexity, nonlinearity, uncertainty, time varying, and incomplete characteristics of the existing practical systems. Secondly, some critical hypotheses have to be put forward in studying and modeling the control systems. However, these hypotheses are not always practical. Thirdly, in order to increase the control performances, the complexity of the control system has also been increased. For these reasons, automatic control has been looking for a new way to overcome the difficulties it has being facing with in recent years. One of the more effective ways to solve the problems mentioned above is to use the technique of intelligent control to the control systems, or to apply a hybrid method of the traditional control and intelligent control to control systems. The core of the new way is the intellectualization of the controller.

Intelligent control is one of the newest interdisciplinary courses today. Intelligent control has closed interconnections with artificial intelligence, system theory, cybernetics, operation research, informatics, cognitive psychology, neural network, pattern recognition, robotics, linguistics, space technique, bionics, computer science and engineering, etc. This book is intended to bring together all these interconnected fields.

Intelligent control has been a great value both for practical and potential applications in a wide range of areas. People with different backgrounds have been interested in intelligent control in recent years. Many developed industrial countries have paid close attention to intelligent control for years, and have decided to encourage and develop the research and application of intelligent control. The advanced robots, automatic manufacturing systems, industrial production processes, and space systems are typical application areas of intelligent control.

The current book, based on a course given at Central South University of Technology, is written as a textbook for graduate students and senior students, and as a valuable reference book for scientists and engineers working in the areas of electrical engineering and computer engineering, especially in automatic control, industrial automation, artificial intelligence and system engineering. One objective of the current book is to provide an introductory knowledge of intelligent control for students, scientists, and engineers. Another objective of this book is to bridge the gap between the theory and the application of intelligent control.

This book includes the basic concepts, principles, methodologies, and techniques of intelligent control and their applications. The outlines are as follows:

1. Introduces the development process, structural theories, and the research areas of intelligent control.
2. Presents the knowledge representations, as well as the searching and reasoning mechanisms as the fundamental techniques of the intelligent control.
3. Studies the constructions and design techniques and the theoretical principles of various intelligent control systems.
4. Analyzes the paradigms of the practical representative applications of intelligent control.
5. Discusses the research and development trends of intelligent control.

The book presents up-to-date research results both in theory and application of intelligent control and which present the highest level of automatic control in the world today. It also shows the close connection between theory and practice, enabling readers to use the principles to their case study and practical projects. The comprehensive materials, including fundamentals, methodologies, techniques, and theories , allow readers to learn much easier.

The book contains 10 chapters. Chapter 1 introduces the background, development and definition, structural theories, and research fields of intelligent control. Chapter 2 and Chapter 3 present the basic methods and techniques of knowledge representation and inference used in intelligent control, i.e. , to represent the control knowledge, to search for a solution, and to make a decision and/or planning. Chapter 4 to Chapter 8 discuss the various theories and systems of intelligent control that are the hierarchical intelligent control systems, expert control system, learning control system, fuzzy control systems, and neurocontrol systems. Each chapter includes control mechanisms, system architectures, and typical examples. Readers will have a more complete understanding in the intelligent control and learn the major development of the theories of the intelligent control. Chapter 9 gives the representative paradigms of the applications for intelligent control in the world. This chapter combines together the theories, techniques, tools, and systems. Readers will be able to know where intelligent control could be used and how to use it. Chapter 10 studies the prospect of intelligent control.

Readers are assumed to have some knowledge of artificial intelligence and essential mathematical prerequisites. We only present a limited amount of the prerequisites while related topics are discussed. For more systematic and formal fundamentals, readers should consult the related references.

Many professors, colleagues, friends, and students greatly contributed to this book during the writing of the book. Researchers in the research group in Center for Intelligent Control, Central South University of Technology, have contributed to the book through the results of their research supported by Natural Science Foundation of China, Doctoral Research Foundation of State Education Commission of China, Basic Research Foundation of Non-Ferrous Metals Industry Cooperation of China, and Natural Science Foundation of Hunan Province. I would like to thank the researchers in my group for their continuous help: Zhi-Ming Jiang, Jian-Qin Liu, Qiao-Guang Liu, Shao-Xian Tang , Jing Wang, and Dun-Wei Wen. My special thanks go to late Professor K. S. Fu for his patient guidance and fruitful collaboration when I was in Purdue University, and to the late Professor Zhong-Jun Zhang for his great enthusiasm and encouragement. Professor Bao-Sheng Hu and Professor Liang-Qi Zhang gave many valuable comments and suggestions. Discussions with Professors G. C. S. Lee, G. N. Saridis, Yao-Nan Wang and Yi-Xin Zhong helped me to have a better establishment and definition for many concepts. Dr. J. Wang, J.-F. Cai, and Miss B.-J. Din have typed the book manuscript and generated the final version of the book. I am grateful to all of them for their support and assistance.

In particular, I wish to extend my appreciation to Professor Fei-Yue Wang, the Editor-in-Charge of *Series in Intelligent Control and Intelligent Automation*, for his suggestion and comment to the organization of the book. I also appreciate Professors Jian-Qin Liu, Min-Jie Wei, Dun-Wei Wen, and Li Ma for their careful proof reading and corrections of the manuscript. Additional thanks are in order for Mr. Yew Kee Chiang and Miss Lakshmi Narayanan, the Editors from World Scientific Publishing, for his/her cooperative and enthusiastic support during the preparation and editing of the book. Finally, I wish to acknowledge my wife Huan Weng and my sons Jing-Feng and Yu-Feng for their continuous enthusiasm, patience, encouragement, and understanding, as well as much material and moral support.

Zi-Xing Cai
June 1997

CONTENTS

CHAPTER 1

INTRODUCTION

We are living in a fast changing era — the so-called the "Times of Second Scientific and Industrial Revolution." The major objective of the "times" is to make a breakthrough on the limitation in the human being using machine so that the social productivity could be raised to a higher level. In other words, we intend to achieve the automation in both the physical labors and mental activities of human being. The appearance and progress of the intelligent machine will fundamentally change the fashion of the work and life of human being.

The emergence and development of the **Artificial Intelligence (AI)** has great significance. In the short term, AI could be seen as a part of computer science concerned with designing intelligent computer systems, i.e., systems that exhibit the characteristics we associate with intelligence in human behavior, such as understanding, learning, reasoning, problem-solving, and so on. People with widely varying backgrounds and professions have been influenced by AI and have benefited greatly from it. They are discovering new ideas and new tools in this young science. It has been shown that the research and development of AI has promoted automatic control to a higher level, i.e., *intelligent control* [8].

1.1 A Challenge to Automatic Control

Automatic control as a discipline has not only made great contribution to science and technology both in theory and practice, but also brought benefit to mankind both in their living and work. Control technology has played a key role in the development of science and technology. However, there exist some essential challenges in the control area. These challenges come from the interaction among techniques such as computers, super VLSI and AI, the demands for current and future applications, and the new concepts and ideas in science and engineering. It is clear that the rapid development and great progress in modern science and technology have resulted in new demands to the control and system sciences.

As control methods found their way into standard practice, they have opened the door to a wide spectrum of complex applications. Such complex systems are characterized by uncertain model, a high degree of nonlinearity, distributed sensors and actuators, high level of noise, abrupt changes in dynamics, hierarchical and distributed decision makers, multiple time scales, complex information patterns, large amounts of data, stringent performance requirements, and so on.

The traditional control, which includes the classical feedback control, modern control theory and large-scale control system theory, has encountered many difficulties in its applications. First of all, the design and analysis for the traditional control systems are based on their precise mathematical models that are usually difficult to achieve owing to the complexity, nonlinearity, uncertainty, time-varying, and incomplete characteristic of the existing practical systems. Secondly, some critical hypotheses have to be put forward in studying and modeling the control systems; however, these hypotheses are hard to match in practice. Thirdly, in order to increase the control performances, the complexity of the control systems has to be increased too. As a result, the reliability of the control systems would be decreased. The degree to which a control system deals successfully with the above difficulties depends on the level of **intelligence** in the system. For these reasons, automatic control has been looking for new ways to overcome the difficulties and challenges it was faced with in recent years. And one of the more effective ways to solve the problems mentioned above is to use the technique of **intelligent control** to the control systems, or to apply a hybrid methodology of the traditional and intelligent control techniques to the control systems. The core of the new ways is the **intellectualization** of controller.

Many control scientists, engineers, managers, and government leaders are now aware of the challenge to control. For instance, a significant meeting was co-organized by the National Science Foundation (NSF), USA and the Control System Society (CSS), IEEE, and was held at University of Santa Clara in September 1986 [42]. Over 50 control theoreticians were gathered there. They discussed and affirmed every challenge. In their opinions, these challenges to control are in the following areas: robust control with multi-variables, adaptive and fault-tolerance control, nonlinear and stochastic control, control systems with distributed parameters, control systems with discrete variables and events, information processing and communication, structure for distributed information processing and decision-making, as well as integration, experimentation, and implementation for systems, etc. Most of the problems in the above research areas would be solved by using new ideas and new tools related to certain degree of intelligence. It is believed that the performances of the control systems would be improved, the design technique and analysis method would be simplified, and the complexity and cost of the control systems would be reduced. Another event as example is that the NSF and the Electric Power Research Institute (EPRI), USA, announced a cooperative program on Intelligent Control in the early 1992 [43]. This program has provided the first funds up to amount of three million dollars to support analytical and experimental research on intelligent control systems.

1.2 Advance in Intelligent Control

As we have mentioned above that the advancement of AI has promoted the discipline of automatic control towards a higher level, — the intelligent control. Intelligent control is one of the newest and most important areas where the computers are used to simulate human intelligence. More people are now aware that intelligent control stands for the future of the automatic control, and is the new frontier in the the development of the science of automatic control.

1.2.1 Automation and Artificial Intelligence

To better understand what is meant by **artificial intelligence**, it is helpful to first look at terms such as **mechanization** and **automation**. To do this, we will try to synthesize the views of others who have approached this problem.

The first industrial revolution was based on mechanization. Mechanization is the use of machines to take over the manual jobs previously done by either animals or human beings.

When we apply ordinary production techniques — the application of leverage and power — to a process, we are mechanizing it. Automation involves a good deal more. Automated devices are truly automated when feedback information automatically causes the machinery to adjust itself in order to re-achieve the norm. The internal adjustments of the machine or system are made by servomechanisms. Automation is the achievement of self-directing productive activity as a result of the combination of mechanization and computation.

The classification of mechanization depends on whether machines or combinations of animals and people are responsible for the three fundamental elements that occur in every activity: **power, action, and control.** Simple mechanized devices need a human to control them. However if a mechanical device is responsible for control, we will have a self-acting or automatic device. Automatic devices are not the same as automated ones; automation equals to mechanization plus automatic control plus one (or more) of three extra control features: a systematic approach, programmability, or feedback.

With a systematic approach, factories make parts by passing them through successive stages of a manufacturing process without people intervening. Thus the transfer lines of car factories in the 1930s count as automated systems. With programmability, the second of the three "extras" that define automation, an automated system can do more than one kind of job. Hence, an industrial robot is an automated, but not an automatic device. The computer that controls a machine can be fed different software to make the machine do different things, e.g., spray paint or weld bits of metal together. Finally, external feedback makes an automatic machine alter its routine according to changes that take place around it. For instance, an automatic lathe with feedback, in which a sensor detects that the metal to be cut is wrongly shaped and so instructs the machine to stop, is thus an automated device. It is clearly more useful than a lathe without this feature.

Though "intelligent behavior" is difficult to define, and is currently understood differently by different people, there have been some convergence of views within the AI community as the technical requirements for the computer solution of certain classes of problems becomes better understood. To be sure, the human solution of a complex equation might be classified as intelligent behavior, while the corresponding action by a machine might not be so classified, even though both the machine and man had been programmed for learning the process. One possible requirement is that there should be something unstructured, something nondeterministic, for the solution process to be qualified as intelligent. Another is that it depends on the knowledge that must be used in obtaining the solution, or on the methods used.

Another important aspect is the use of heuristic rules of the kind that humans use to solve problems. Although, in general, such rules cannot be proved effective, they often lead to solutions. Some computer scientists argue that heuristic programming better describes the field now called AI.

In summary, automation is closely connected to AI, and AI has a close relation with intelligent behavior, primarily with nonnumeric processes that involve complexity, uncertainty and ambiguity, and for which known algorithmic solutions do not usually exist. Unlike conventional computer programming, it is knowledge based, almost invariably involves search, and uses heuristics to guide the process to a solution.

1.2.2 Development of Intelligent Control

Intelligent control, which can be seen as an important part and a research area of AI and control in some sense, may be considered to be the top layer of automatic control on the hierarchical road to autonomous machines [29]. Figure 1.1 illustrates the developing process of the automatic control and increasing sophistication on the path to intelligent control. As we can see from Figure 1.1, the furthermost point on the path is the intelligent control that deals with the advanced decision making and closely connects to artificial intelligence. Intelligent control system can solve problems, identify objects, or plan a strategy for a complicated function of a system.

Although AI has promoted automatic control to the top layer, it is interesting to note that for many years there was no mention of any interconnection of control theory and AI. As the control theory has primarily dealt with analog or numeric computation in relation to servomechanism, and AI has primarily dealt with symbolic manipulation, the lack of overlapping was not too surprising.

Intelligent control has gained wide recognition since it first emerged in 1960s. Several ideas and methods of it have been proposed and developed since then.

Learning control was actively studied and used 30 years ago. The concept of learning machine was put forward at the same time when cybernetics appeared in early 1960s [93]. The self-learning and adaptive methods were developed for solving problems with stochastic properties in control systems. Initially, learning systems were applied to the aircraft control, pattern classification, and communication, e.g., to the control of nuclear power plant.

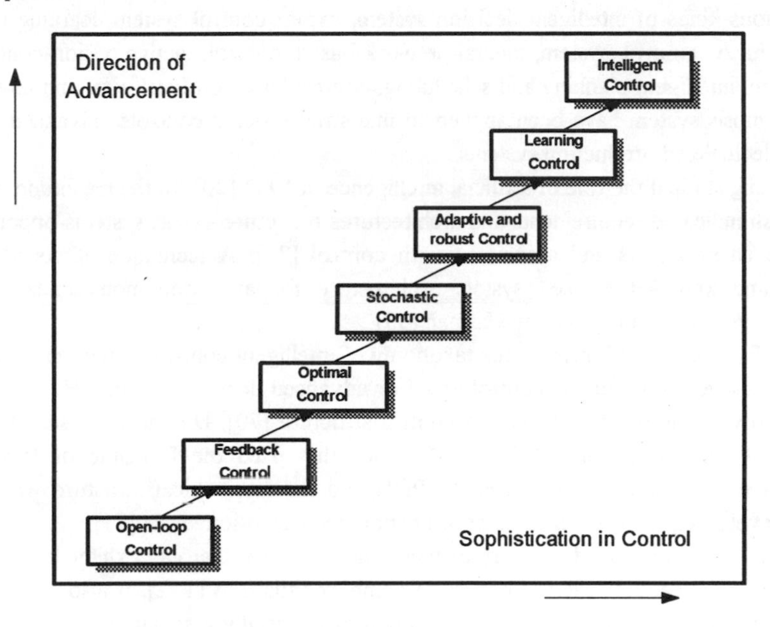

Figure 1.1. Development process of automatic control.

The automatic control began to overlap with artificial intelligence in the middle of 1960s. K. S. Fu applied heuristic inference rules to a learning control system in 1965 [24]; then he expounded the relationship between artificial intelligence and automatic control in his excellent paper entitled "Learning Control System and Intelligent Control System: An Intersection of Artificial Intelligence and Automatic Control [25]."

Fuzzy control is another active research area of intelligent control. Zadeh presented his famous paper entitled "fuzzy sets" which brought froth the era of fuzzy control in 1965 [95]. Since then, a great amount of research efforts have been carried out both in the therorectical investigations and practical applications [88]. It is noted that the application-intended research for fuzzy control has been made by many engineers and researchers, and has attained some interesting results since 1970s [49, 85].

The term **intelligent control** was originally and formally used in 1967 by Leondes. That was 47 years later than the term **robot**, and 11 years later than the term **artificial intelligence**. The early intelligent control systems applied some preliminary intelligent methods, e.g., pattern recognition and learning mechanism, and advanced very slowly.

In recent years, the intellectual current of the intelligent control has emerged again along with the rapid development of artificial intelligence, robotics, and space technology etc. Various kinds of intelligent decision system, expert control system, learning control system, fuzzy control system, neural-network-based control, active vision-control of perception, intelligent planning and scheduling, control based on Petri nets, and intelligent fault diagnosis system have been applied to industrial process controls, advanced robots and intellectualized production systems.

DeJong studied the role of artificial intelligence in 1983 [20]. In the meantime, Sauers and Walsh indicated requirements and architectures for future expert systems operating in real-time environments and associated with control [77]. A technique of coordination control and knowledge-based system architecture for an autonomous mobile robot guidance system was proposed by Harman [33].

Saridis made contribution to the taxonomy of intelligent control system. He indicated the highest level of automatic control as AI or advanced decision-making. He stated that AI can provide the topmost level in a control structure [70]. He and his research group established the Theory of Intelligent Machines that used the Principle of Increasing Precision with Decreasing Intelligence (IPDI) and the hierarchical structure with three control levels, i.e., the organization, coordination and execution levels [71].

Albus *et al.* developed a theory of hierarchical control that can exhibit learning and thereby provide learned reflex responses to complex situations [1, 2]. It also provides for the problem-solving and planning functions that are normally associated with the highest level of intelligent action that is considered to be the domain of AI, and incorporates expert system rules for error correction at the intermediate level of the control hierarchy. AI will become a central ingredient in advanced control systems.

Åström, de Silva, Zhou, Cai, Homem de Mello, and Sanderson *et al.* proposed and developed expert control [3, 4], knowledge-based control [79, 97], anthropomorphic control, expert planner [47], and hierarchical planning [40, 68] respectively in 1980s. Åström's paper entitled "expert control" has been well known and has prompted the research and development of expert control [3].

The study of **artificial neural networks (ANN)** has been going on for a long period of time since 1943 when McCulloch and Pitts established the brain model in their paper [51]. The initial motivation is to mimic the nerve systems of creatures [19, 37]. With the development of VLSI, photoelectronics, and computer technology, ANN has drawn more and more attention. The theory and mechanism of neural-network-based control has been developed and used in recent years. Although the capability of ANN-based control is still limited, many controllers based on neural networks have been proposed because of the important properties of the neural networks, e.g., capability of learning and memorizing through the interconnection weight, capability of forming and approaching arbitrary continuous nonlinear mappings, capability of generalization, ability to process information in parallel, capability of certain fault tolerance, and suitability for manufacturing by VLSI. Such controllers are called neural controllers. The neural controllers possess the properties of parallelism in information processing, rapidity in execution, good robustness in control, and strong adaptability for application. Therefore, these controllers have some special advantages and would be used widely.

The conditions for establishing the new discipline of **intelligent control** have matured gradually. The IEEE First Symposium on Intelligent Control was held in Troy, New York in 1985 [73]. In this meeting, the principles of the intelligent control as a discipline and the structure of an intelligent control system were discussed. After the first meeting, the IEEE Technical Committee on Intelligent Control was organized in the IEEE Control System Society. An interesting discussion about the definition of intelligent control and the possible syllabus of a graduate course on intelligent control has been in the progress. The IEEE International Symposium on Intelligent Control (ISIC) was established as a joint activity between the Control System Society and the Computer Society of IEEE, and was held in Philadelphia, Pennsylvania, USA in 1987 [55]. The second meeting was the first international conference on Intelligent Control, and 150 participants who came from USA, Europe, Japan, China, and other developing countries attended this symposium. At the second meeting, the mentioned discussions continued which demonstrated a growth of the interest of this new emerging discipline. The papers submitted and the discussion made in the meeting have shown that great progress has been made in intelligent control, and the need for reconsidering the situation of the control area and the adjacent disciplines has been determined by a host of new technological problems, as well as by the advances in the theories and the technologies. This meeting and related events which happened in the late 1980s were the milestones on the development of intelligent control, and there were signs that intelligent control as a relatively independent discipline, has become established at the international stage of science and technology.

A. Meystel, G. N. Saridis, J. Y. S. Luh, H. E. Stephanou and A. H. Levis, etc., have done excellent works for the ISIC and the development of intelligent control. Thousands of researchers including professors, graduate students, engineers, and research scientists from different professions and backgrounds have engaged in the research work of intelligent control all over the world in recent years. They have made great progress in their research on intelligent control.

Many contributors have published papers and reference notes on intelligent control during the past 10 years. A book entitled *Intelligent Control*, written by Cai, was published in 1990, and has been used as a textbook for graduates and senior undergraduates in China[13]. Another book, *Intelligent Control: Aspects of Fuzzy Logic and Neural Nets*, by

Harris *et al.* was published by World Scientific in 1993 [34]. Several conferences, symposia, or workshops on intelligent control have been held by IEEE, IFAC, IFIP, IMACS, IASTED, and other international, regional, or national academic institutions since 1987. A new international academic society, the International Society of Intelligent Automation, ISIA, has been set up since 1994, and a new journal, *International Journal Intelligent Control and Systems*, was published in January 1996. The world is expected to enter a new era of intelligent control.

1.3 What Is Intelligent Control

1.3.1 Definition of Intelligent Control

Like many advanced scientific disciplines or sciences such as artificial intelligence, knowledge engineering, and robotics, the discipline of intelligent control has also not been defined with an unified view until now. Although a suitable and general definition on intelligent control is difficult to make, it is necessary to give a definition for intelligent control, so that people could define characteristics, develop technology, and compare research results of intelligent control in the different countries. The following definitions are not perfect, but could be revised and improved through discussion and debate.

Definition 1.1. Intelligent Machine

Intelligent machine can perform anthropomorphic tasks autonomously or with the least human interaction in structured or unstructured, familiar or unfamiliar environments. In general, intelligent machines might replace human efforts in hazardous, remote, tedious, harmful, poisonous, or high precision jobs, where their higher efficiency proves to be more cost-effective in terms of human and humanistic values. Current flexible manufacturing systems are typical examples of interactive intelligent systems while advanced mobile robots are typical examples of autonomous system.

Definition 1.2. Automatic Control

Automatic control is a process in which machines or plants can be operated or controlled automatically according to the defined programs or procedures. Simply speaking, automatic control is a control without human interference. For instance, if a plant can receive measured physical variables of process, make computations, and regulate the process automatically, then this plant would be an automatic control plant. Feedback control, optimal control, stochastic control, adaptive control, learning control, and robust control are examples of automatic control.

Definition 1.3. Intelligent Control

Intelligent control is the process that drives an intelligent machine to attain its goal automatically. In other words, intelligent control is a kind of automatic control that can drive autonomous intelligent machines to reach their goals without any interaction or with the least human interaction. The control of an autonomous robot would be an example of intelligent control.

Definition 1.4. Intelligent Control Systems

Intelligent control systems are systems that can perform tasks of intelligent control. More precisely speaking, intelligent control systems are used to drive autonomous intelligent machines to reach their goals without any interaction with human operator. In doing so, the systems must be equipped with intelligence, scheduling, and execution capabilities. Therefore, the theoretical foundations of such systems should be found at the intersection of disciplines like cybernetics, artificial intelligencc, informatics, and operation research, which will be discussed in more details later.

1.3.2 Features of Intelligent Control

Intelligent control was advanced with regard to uncertainty of the controlled environment and task of plants or processes, and has a stronger adaptability to complexity in environment and task. An intelligent control system should possess the capabilities of learning, memory, reasoning, adaptability, and self-organization within a wide range of variation in environment and task. In other words, an intelligent control system should have the anthropomorphic ability of learning and reasoning, it can adapt the changing environment, deal with observed or sensory information, reduce uncertainty of systems, make planning, produce and execute control actions, and reach the preset objective and optimal or satisfactory performance.

Intelligent control possesses the following features [96]:

(1) Intelligent control is usually a hybrid control process that is represented by the knowledge of both the non-mathematical model and the mathematical model, i.e., it is a control process with generalized model. This control usually possesses complexity, incompleteness, ambiguity or uncertainty, and the known algorithm does not exist. In solving an intelligent control problem, the inference is made with knowledge and heuristic information. Therefore, in studying and designing the intelligent control system, the description task and world model, the recognition of symbol and environment, as well as the design of knowledge base and reasoning machine, rather than the expression, computation and processing of mathematical formulate, are emphasized. In other words, the key for designing an intelligent control system lies in the design of the prototype of intelligent machine rather than the traditional controller.

(2) The core of the intelligent control is in the higher level that would organize the

practical environment of process, make decision and planning, and realize generalized problem-solving. In order to realize these tasks, it is necessary to adopt the related technologies of the symbolic information processing, heuristic programming, knowledge representation, fuzzy logic, automatic similarities between the above solving process, and the thinking process of the human brain, i.e., there exists intelligence in the artificial problem-solving process.

(3) Intelligent control is an interdisciplinary discipline that has close connections with artificial intelligence, cybernetics, system theory, operation research, informatics, cognitive psychology, neural network, pattern recognition, robotics, bionics, linguistics, space technique, and computer science, etc. The further progress of intelligent control require the coordination with and assistance from the related disciplines and sciences.

(4) Intelligent control is one of the recently developing research areas and is still immature; it would experience faster and better development and should a better theory be found!

1.4 Structural Theories of Intelligent Control

Since K. S. Fu coined the name Intelligent Control as the field of intersection of artificial intelligence and automatic control system in 1971 [25], many researchers have attempted to formalize it into a scientific discipline. Several structural theories about intelligent control system have been proposed and discussed. These prompted a further understanding for the discipline of intelligent control [8, 71].

Intelligent control possesses the explicit interdisciplinary feature as we have stated in subsection 1.3.2. In this section we summarize and discuss three structural theories of intelligent control: **Two-element Intersection Structure, Three-element Intersection Structure, and Four-element Intersection Structure.** They could be called Two-element, Three-element, and Four-element Structure respectively, and are represented by the following intersets [8, 11]:

$$IC = AI \cap AC \tag{1.1}$$

$$IC = AI \cap AC \cap OR \tag{1.2}$$

$$IC = AI \cap AC \cap IN \cap OR . \tag{1.3}$$

The above intersection structures could also be represented by the conjunctions of the predicate calculus in discrete mathematics and artificial intelligent as follows:

$$IC = AI \wedge AC \tag{1.4}$$

$$IC = AI \wedge AC \wedge OR \tag{1.5}$$

$$IC = AI \wedge AC \wedge IN \wedge OR \tag{1.6}$$

where the abbreviation symbols for every subset and conjunctive term stand for the related

disciplines:
AI — Artificial Intelligent
AC — Automatic Control, or Cybernetics
OR — Operation Research
IN — Information Theory, or Informatics
IC — Intelligent Control
∩ — Symbol for intersection set
∧ — Conjunctive symbol

1.4.1 Two-Element Intersection Structure

K. S. Fu studied and discussed several areas related to learning control [25]. These areas include the followings:

(1) control system with human controller.
(2) control system with man-machine controller.
(3) automatic robotics control system.

In order to emphasize the problem-solving or high-level decision-making capability, he used a more general term, intelligent control systems, to cover all these areas of interest. He noted that the field of intelligent control systems describes the activities in the intersection of automatic control system and artificial intelligence, and many research activities in artificial intelligence are related to control problems. On the other hand, control engineers interested in adaptive and learning control have attempted to design more and more human-like controllers since 1970. We represent this intersection of the two areas by Formulae (1.1) and (1.4), and to show it in Figure 1.2. We call it a two-element intersection structure.

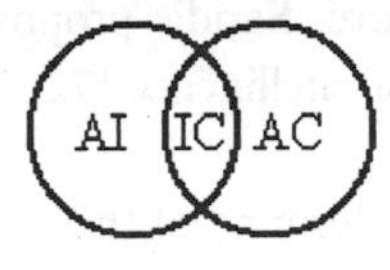

Figure 1.2. Two-element intersection structure of intelligent control.

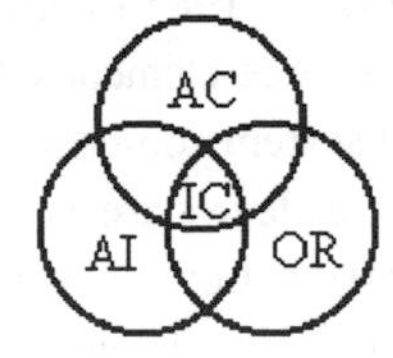

Figure 1.3. Three-element intersection structure of intelligent control.

The research on learning control systems is only a preliminary step toward a more general concept of intelligent control systems. In the use of a man-machine combination as a controller, decisions requiring higher intelligence can be performed by the human controller. For example, recognition of complex environmental situations, setting subgoals for the computer controller, correcting improper decisions made by the computer controller, and so on. The other activities requiring lower intelligence such as data collection, routine decision, and on-line computations, can usually be carried out by the traditional computer controller. In designing an intelligent control system, the purpose is to try to transfer, as much as possible, the designer's intelligence which is relevant to the specified tasks to the machine controller.

1.4.2 Three-Element Intersection Structure

G. N. Saridis thought that, in most cases one of the two subsets in Fu's structure dominated the other with a result of non-conducive to efficient and successful application. In order to drive autonomous intelligent machines to reach their goals without any interaction with human operator, the intelligent control system must be equipped with capabilities of intelligence, scheduling, and execution. Therefore, the theoretical foundations of such system should be found at the intersection of disciplines like artificial intelligence, operation research and automatic control, and should be structured according to a principle — The Principle of Increasing Precision with Decreasing Intelligence (IPDI) [76].

Saridis' three-element intersection structure for intelligent control was proposed in 1977 [70]. This theory expanded the field of intelligent control to include three areas represented in Formulae (1.2) and (1.4), and shown in Figure 1.3.

The three-element structural theory was discussed and debated at the IEEE First Symposium on Intelligent Control in 1985. This discussion is helpful for establishing the discipline of intelligent control. After this meeting, the debate on intelligent control was continued among the members of Intelligent Control Committee, IEEE, and was brought to a climax in the IEEE 1987 International Symposium on Intelligent Control [55].

Based on the three-element intersection structure, Saridis proposed a hierarchically intelligent control system composed of three levels of intelligence [72, 74], and depicted in Figure 1.4, The three levels are as follows.

(1) Organization Level. It represents the mastermind of the system with functions dominated by artificial intelligence. In general, there is only one organizer in this level.

(2) Coordination Level. It is the interface between higher and lower levels of intelligence with functions dominated by artificial intelligence and operation research. There may be multiple coordinators in the coordination level. The communication between the different coordinators is performed through a dispatcher [76].

(3) Execution Level. This level is the lowest one with high requirement in precision with functions dominated by traditional theory of automatic control system. Usually, the numbers of the traditional controllers is not less than the numbers of the coordinators.

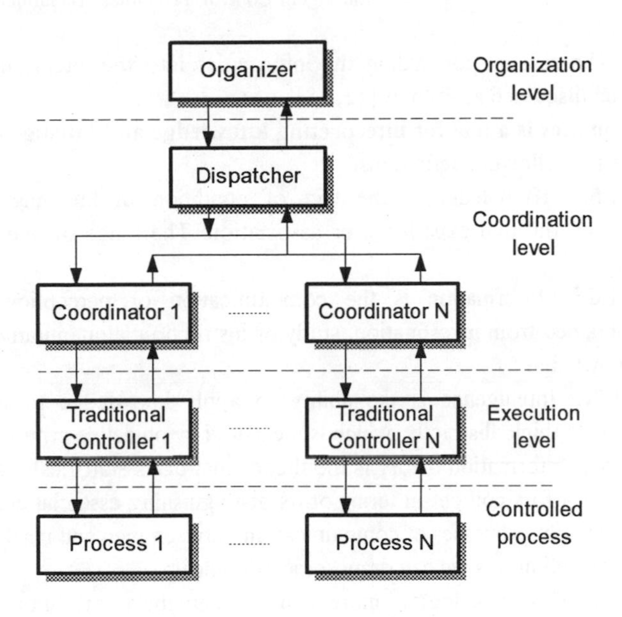

Figure 1.4. Saridis' hierarchical intelligent control system.

1.4.3 Four-Element Intersection Structure

After studying the previous intersection structural theories of intelligent control and the relationships among the disciplines related to intelligent control, Z.-X. Cai proposed the Four-Element Intersection Structure in 1986. He expanded the intelligent control to include four major areas as seen in Figure 1.5(a) and Formulae (1.3) and (1.6). A simplified diagram for representing the four-element intersection structure of intelligent control is depicted in Figure 1.5(b) [14, 96].

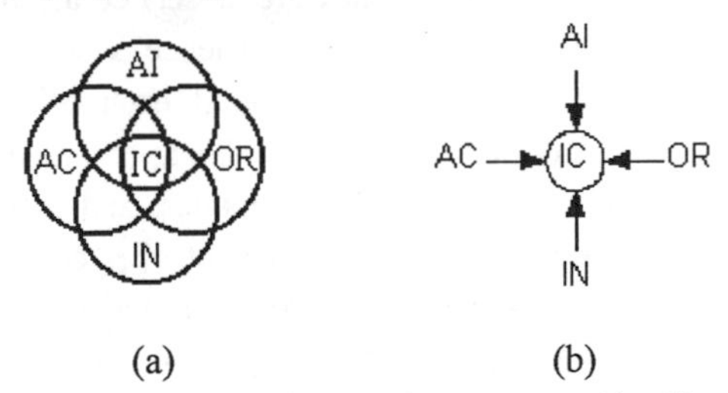

Figure 1.5. Four-element intersection structure of intelligent control.

We have found from Figure 1.5 that informatics becomes a new component of the intersection. The reasons for adding the informatics into the intersection structure are interpreted and discussed as follows [13, 18].

(i) Informatics is a tool for interpreting knowledge and intelligence.

We have the following definitions.

Definition 1.5. Knowledge is the fact of condition of knowing something with familiarity gained through experience or association. The range of one's information or understanding.

Definition 1.6. Information is the communication or perception of knowledge; knowledge obtained from investigation, study or instruction; a quantitative measure of the content of knowledge.

Definition 1.7. Intelligence is the ability to apply knowledge to manipulate one's environment or to think abstractly ,which is measured by objective criteria.

Definition 1.8. Information theory is the theory that deals statistically with information, the measurement of its content in terms of its distinguishing essential characteristics, and with the efficiency of processes of communication between men and machine.

From these definitions we can draw some inferences:

(1) The term "knowledge" is more general than the term "information", in other words, generally, $\{\text{information}\} \in \{\text{knowledge}\}$.

(2) Intelligence is the capacity to acquire and apply knowledge.

(3) Information theory may be utilized to interpret the machine knowledge and intelligence mathematically.

These tell us that informatics has become a tool for interpreting the knowledge and intelligence containing in the intelligent control systems.

(ii) Cybernetics, Systematology and Informatics intersect closely.

No matter what is artificial intelligence (including knowledge engineering), cybernetics (including engineering cybernetics and biological cybernetics), or systematology (including operation research), they are closely connected with informatics. As an outstanding disciplinary group in the frontier of science, systematology, informatics, and cybernetics have always interacted and intersected each other. For instance, an advanced robot with capability of autonomous guidance, i.e., an intelligent control system would need the participation of systematology, informatics, and cybernetics in order to control and communicate the system among the host controller component, to store and process signals, as well as to make optimization, decision, and motion. The information view has become a necessary resource for many disciplinary fields in the modern science and technology [66, 75].

Let us look at a scientific system diagram, shown in Figure 1.6, that was proposed by X. S. Qian, the founder of engineering cybernetics. For simplicity, a local part of the diagram is simply depicted. It could be seen from Figure 1.6 that like systematology, cybernetics, and operation research, informatics is an important component of the systematic.

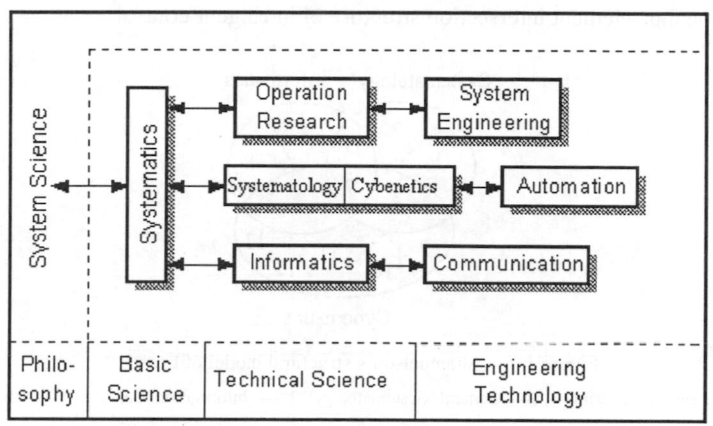

Figure 1.6. Qian's science system diagram (partly).

Cybernetics, systematology and informatics are intersected highly and closely. No matter what is AI, cybernetics, or systematology, they are closely connected with informatics — an outstanding disciplinary group in the frontier of science. Systematology, informatics and cybernetics have always interacted and intersected each other.

(iii) Informatics has become a tool for controlling intelligent machine.

The knowledge and experience within intelligent control come from the information, and could be processed to form new information such as command, decision strategy, and plan that are used to manipulate the activities of the controlled systems or plants.

The development of Informatics has generalized the concept of information to control area and has become a means to control the machine, organisms, and society — a tool to control bionics machines and anthropomorphic machine.

Intelligent control systems, either knowledge-based systems or neural-network-based systems, or others, would attempt to imitate the human activity, especially the thinking and decision-making process of the human brain. Whether the structure of the human organs is functionally a reflection to the relationship and function of the systematology, cybernetics and informatics? Shamuelson, the President of the 1976 International Symposium on General Systematology and Professor in University of Stockholm, Sweden, exhibited a slide of his general report at the symposium. This slide is shown in Figure 1.7. It presents a structural model of a human heart. One the one hand, this diagram describes the core relationships among the systematology, informatics, and cybernetics. If the governing function of the heart by brain and other central nerves could be considered in this structural model, i.e., the **intelligence** has been introduced in, then the model would really be a live model of the four-element intersection structure of intelligent control.

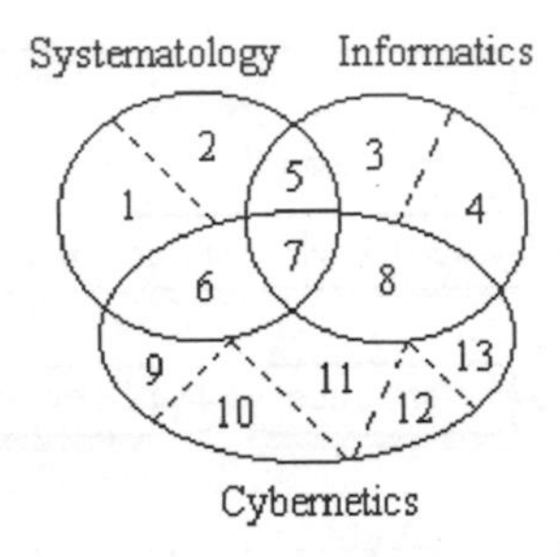

Figure 1.7. Shamuelson's structural model of heart.

1 — System Engineering, 2 — General Systematology, 3 — Information Science, 4 — Information Technology, 5 — Information System, 6 — Control System, 7 — Intersection of Systematology, Informatics and Cybernetics, 8 — Electronic Computer, 9 — Control, 10 — Communication, 11 — Human Communication, 12 — Human and Animal, 13 — Machines

(iv) Intelligent Control can be Measured by Information Entropy.

The hierarchical intelligent control theory is one of the more matured intelligent control theories that will be discussed in detail in Chapter 4. According to this control theory, the functions and costs of every control level are measured by **entropy**. Originally, the entropy is a measure to the total useless energy lost during the process of thermal transformation in thermos-mechanics. In information science, we define it as follows:

Definition 1.9. Entropy is the mean information contained in an information source. Let $P_1, P_2, \ldots, P_n$ be the probabilities of every event occurred in an information source, H be the mean information contained in the same information source, then we have

$$H = -K\sum_{i=1}^{n} P_i \log P_i \tag{1.7}$$

where K is a constant and is dependent upon the selected unit.

At the organization level, entropy has an information theoretic connotation and deals with knowledge representation and processing. Therefore, it is natural to consider Shannon's Entropy as the measure of lack of knowledge. At the coordination level, the entropy is used as the measure of uncertainty of coordination. At the execution level, the cost of execution is equivalent to the expended energy of the system which is expressed by an entropy in the sense of Boltzmann function. All these entropies may be added for every alternate complete action of the intelligent control system to represent its own cost, and could get a **total entropy**. The total entropy represents the total cost of actions. A criterion for building the intelligent control system would be to minimize its total entropy [71].

Entropy function is one of the fundamentals for modern information theory. The entropy function and information rate have been introduced into intelligent control systems

(v) Informatics takes part in the whole process of intelligent control, and a core role of Informatics is in the execution level of intelligent control system.

Generally speaking, informatics takes part in the whole process of intelligent control including information transfer, information conversion, knowledge reasoning, knowledge processing, knowledge representation, knowledge acquisition, knowledge retrieval, decision-making, and man-machine communication, etc. However, a core role of the informatics is in the execution level where control hardwares receive, transform, process, and output various information. For example, in our developed real-time expert processor that is used to process data and send the processed information to knowledge base and inference engine of the expert controller [90]. Another example deals with an intelligent robotics control system that consists of two components, i.e., the knowledge-based intelligent decision-making subsystem and the information (signal) identification and processing subsystem [9]. The former involves intelligent database and reasoning machine, and the latter may be measures and processors for different signals. In the two examples, the information processing or pre-processing is carried out by the information processors that are in the execution level.

All of the five reasons have given powerful evidences that informatics is a necessary subset for constructing the discipline of intelligent control.

The relationships among the four subsets in the four-element intersection structure, i.e., AI, automatic control, operation research and informatics, can be depicted in Figure 1.8 where every subset is connected with the other three subsets. The relationships of intersection can be clearly seen from the figure.

We can divide the all numbers into groups where the numbers from 1 to 6 show the relationships between AI and IN, 7 - 12 — relationships between AI and AC, 13 - 18 — relational between AI and OR, 19 - 23 — AC and IN, 24 - 28 — AC and OR, 29 - 33 — IN and OR.

The research on the structural theories has further revealed the essence of intelligent control, and is helpful to have a better understanding for intelligent control. The results of this research would propose some new research topics and issues for the study of theory and application of intelligent control.

1.4.4 General Structure of Intelligent Controller

Control engineers designed the future control systems beginning from task formalization to actuator operation. The control process design had been thought to be a product of system parameters for long period of time. Afterward, it became clear that the control process design is a product of system model and practical structure. It is confirmed that the task formalization has to be included in this synthesis. In addition, the coordination with task sets is also a component of this design.

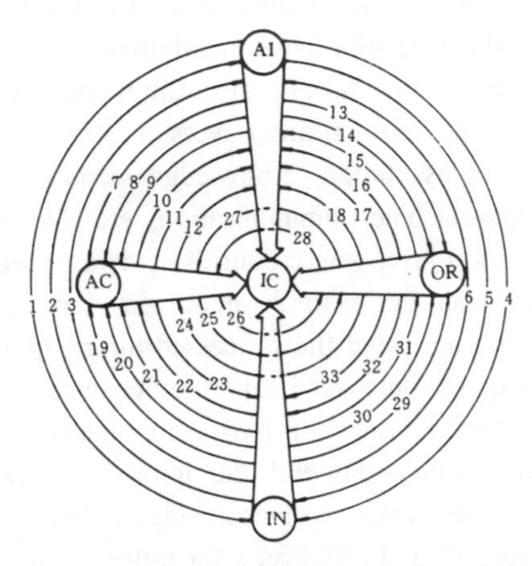

Figure 1.8. Relationships among subsets in four-element structure.

1 — formal language, 2 — knowledge acquisition, 3 — knowledge representation, 4 — knowledge retrieval, 5 — knowledge processing, 6 — man-machine communication, 7 — automata, 8 — learning, 9 — memory, 10 — optimization, 11 — adaptation, 12 — backtracking, 13 — intutitive inference, 14 — logic reasoning, 15 — decision-making, 16 — planning, 17 — scheduling, 18 — management, 19 — identification, 20 — estimation and prediction, 21 — modeling, 22 — sensing, 23 — feedback, 24 — organization, 25 — heuristic search, 26 — dynamic planning, 27 — coordination, 28 — gaming, 29 — information conversion, 30 — information transfer, 31 — dynamics, 32 — hierarchy, 33 — entirety

Design of intelligent control system deals with the combination of the following features [54].

(i) Hybrid system methodology with representation of traditional mathematical model and technical language description.

(ii) Hierarchical model with imprecise and incomplete plant.

(iii) Hierarchical and incomplete knowledge of the exosystem delivered by multiple sensors and being re-organized and refreshed constantly during the process of learning.

(iv) Task negotiation as a part of control system as well as control process.

Considering all of the above factors and features, we can picturize a typical structure for intelligent controller as shown in Figure 1.9.

A number of theories and techniques has been devised and developed for particular paradigms of the intelligent control systems: theory of hierarchical control, entropy methods for hierarchical controller design, principle of increasing precision with decreasing intelligence, design methods for expert controller, fuzzy controller and anthropomorphic controller, and the neural-network-based control methodology etc. Many research results are potentially applicable within these paradigms: theory of team control, theory of fuzzy sets and systems, theory of Petri Nets, combinative theory of fuzzy-Petri Nets and fuzzy neural network, and so on. Lots of the researches and developments of intelligent control theory are directed toward applications in various intelligent control systems: learning and self-organizing systems, neural-network-based systems, knowledge-based systems, linguistic and cognitive controllers, etc.

Based on the typical structure in Figure 1.9, a lot of schemes for intelligent controller have been proposed. We will go further into discussion in Chapter 4 to Chapter 8, and also in Chapter 9.

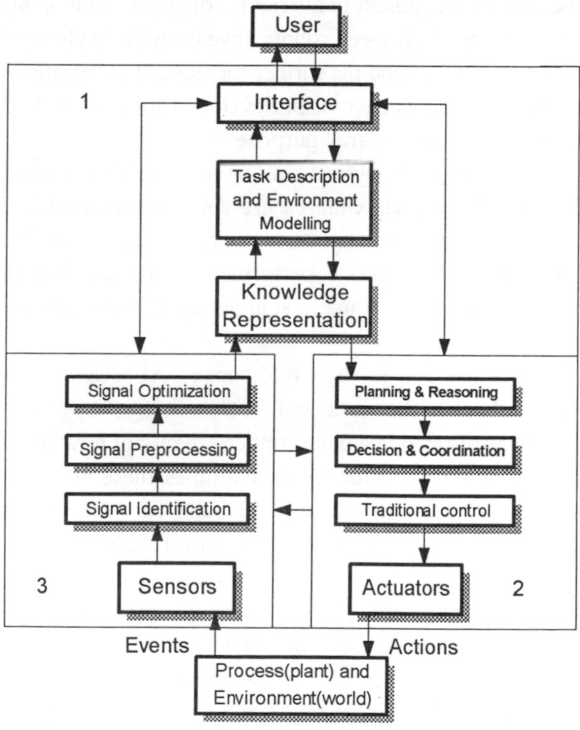

Figure 1.9. Typical structure of intelligent controller.

1 — intelligent control system, 2 — multilevel controller, 3 — multisensory system

1.5 Research and Application Fields of Intelligent Control

There is a wide range of research and application fields for intelligent control, and every field has its particular interesting topics. To help motivate our subsequent discussions, next we describe some of these research and application fields such as control of advanced robots, intelligent process control, intelligent scheduling and planning, expert control systems, voice control, control of restorative robots, and intellectualized instruments, etc. It has to be noted that these subfields of intelligent control are not completely independent. Most of the research fields deal with multiple intellectual areas. We will introduce and analyze some application examples of intelligent control in Chapter 9 and some paradigms of application in Chapter 4 to Chapter 8.

1.5.1 Advanced Robots

Although over 90 percent of the present industrial robots have none of intelligence, higher demands for a wide spectrum of advanced robots have been made along with the vigorous development of robotics technique and the further increase of automation level since early 1980s [35]. These advanced robots have been developed for land mobile vehicles, space and underwater explorations, and industrial purpose.

The problem of controlling the physical actions of a mobile robot might not seem to require much intelligence. Even small children are able to navigate successfully through their environment and to manipulate items such as light switches, toy blocks, and eating utensils, etc. However, these same tasks — performed almost unconsciously by human — when performed by a machine require many abilities used in solving more intellectually demanding problems.

Research on robotics has helped to develop many AI ideas. It has led to several techniques for modeling states of the world and for describing the process of change from one state to another. It has resulted in a better understanding of how to generate plans for operation sequences and how to monitor the execution of these plans. Complex robotics control problems have forced us to develop methods for planning at high-levels of abstraction, ignoring details, and the planning at lower levels where details become important.

Some sensor-based robots that possess sensory capabilities of vision, tactile, proximity, audio and force senses [7], and several interactive robots that possess interactive capabilities with environment [81], and several autonomous robot have been developed in recent years [33]. These research results enable robots to acquire information from environment more powerfully. The sensor-based robots have been successfully used in automatic operation, precise assembly, and product inspection [8, 26, 41].

Advance in research on mobile robots is conspicuous. They are single, two, four, six, and even more legs in mobile robots. They can imitate the walking ability of humans and animals on convex-concave ground, and can go up and down stairs interactively or autonomously.

Advanced robots have been widely applied to autonomous systems and flexible manufacturing systems. Autonomous robot could set its own goal, plan and execute its own actions, and change its own state to adopt the changing environment.

The control of the advanced robots is one of the most significant research and application fields. Almost every new idea and technique has been tested and used in this field.

1.5.2 Intelligent Process Planning and Control

The creation of manufacturing process plans is a key element in the design of the whole production cycle. It is the point at which "product designs" are translated into "manufacturing plans." The process planner interprets design requirements from engineering drawings. Based on experimental knowledge with respect to manufacturing process and equipment capabilities, the process planner must specify the sequence of manufacturing processes required to produce the part. This task involves determining the optimum equipment and tooling to be used and the definition of specific instructions concerning how each process is to be performed. If this kind of process planning is carried out by computers automatically, then we call this planning as **intelligent process planning** or **computer-aided process planning** [67]. The quality and consistency of the process plans have a direct impact on product cost as well as the overall factory performance.

Over the past 15 years, several approaches to intelligent process planning have been proposed and implemented. The first one was called **Variant Process Planning** which was based on the principles of Group Technology. Another one, called **Generative Process Planning**, was the concept generating a plan based on rules correlating part features to manufacturing methods. Some more sophisticated approaches to knowledge-based process planning have been developed in a prototyping environment for years. Two of the more advanced prototyping efforts have been conducted by Boeing Company and the Automated Manufacturing Research Facility at the National Bureau of Standards (NBS), USA. Boeing's prototype process planning system is driven directly from a product definition stored in a CAD system and generates routine sheets as well as machine control data for simple assembly. A prototype for high-level robotics planning based on the concept and techniques of expert system was proposed and implemented by Cai *et al.* during 1984-1987 [9, 14]. Another prototype process planning for machine assembly, which is based on the concept of topology and rotation mapping, was developed in China in 1991. Homem de Mello, Sanderson, Mathur, Will *et al.* have forwarded an automated assembly system and hierarchical planning system for space structure in 1990-1992 [40, 50, 68].

Intelligent process planning has been widely used in manufacturing system including Flexible Manufacturing Systems (FMS), Information Management Systems (IMS) and Computer Integrated Manufacturing Systems (CIMS) in recent years.

In many continuous production lines, for instance, in metallurgy, chemical engineering, oil refining, textile dyeing, paper production, steel and aluminum products processing, nuclear reaction as well as pharmaceutical production, the production processes require control with higher level of difficulties and performances. In order to maintain physical parameters which can be changed within a certain range of precision, and to guarantee high product quality and productivity in these continuous industrial plants or production lines, efficient control modes have to be applied to their process control. The traditional mathematical-model-based controllers has much limitation and can only suit a certain range of applications. The following are new trends for process control: to apply the technique of intelligent control into production processes, to imitate human experience by computer, to build the generalized and knowledge-based models, and to implement automatic reasoning, decision and control.

Many researches on intelligent process control have been carried out for years, and great achievements have been made. The early first research and application are in the process control of nuclear reactor [52]. The industrial process control for chemical metallurgy and flexible manufacturing have been emphasized during the past 10 years [53]. In designing and implementing the intelligent process controllers, the technique of pattern recognition and theory of fuzzy sets as well as Petri nets have been used.

1.5.3 Expert Control and Fuzzy Control Systems

Expert system is one of the systems that can handle real world, complex problem requiring an expert's interpretation and solve these problems by use of computer model of human expert reasoning to reach the same conclusions that the human expert would reach if he or she faces a comparable problem.

Expert system is one of the most active application research areas in artificial intelligence [3]. It has been widely used in problem-solving, consultation and control [36, 91].

The functions of expert system depend on its massive knowledge contained. An expert system seeks to capture enough of the human specialist's knowledge so that it would solve problems expertly. The key problem in the developing and using the expert system is how to present and use the expertise. The essential difference between expert system and traditional computer program lies in that the problems to be solved by expert system do not usually possess known algorithmic solutions, and the conclusions are drawn according to the information with uncertainty, incompleteness, and imprecision. These problems are similar to the ones to be solved by intelligent control.

Expert control system is one member of the family of expert system, and its task is to govern the overall behavior of an object or process. Expert control system can interpret the current situation of the control systems, predict the future behavior of the process, diagnose the probable problems which happened, revise and execute the control plans continuously. In other words, the expert system has the functions of interpretation, prediction, diagnosis, planning, and execution, etc.[53, 89].

Expert control systems have the structure of hierarchical control, and have two basic structural forms. The **expert control system** and the **expert controller**. The former is very complex; the latter is more simple. Therefore most of the expert control systems in operation have the structure of the expert controller [13, 44].

Expert control system can be applied to space traffic control, autonomous robotics control, fighting management, business management, fault detection and diagnosis, and task control (e.g., quality control) and so on. One example is the underwater autonomous robot [38]. We want this robot (vehicle) to recognize obstacles encountered in the underwater unfamiliar environment. This robot equips with expert control system and can infer what are the obstacles and how to handle them, then drives itself to the goal. Another example is the expert control system used to guide airplanes flying safely high above the sky in large airports or in the space "crossroads." [30].

Fuzzy control is a kind of control approach that uses the **fuzzy sets theory**. With its particular feature of fuzziness, fuzzy control can give rise to unconventional results which may invigorate the development of automatic control engineering. The usefulness of fuzzy control can be considered in two respects. On the one hand, fuzzy control offers a novel mechanism to implement such control laws that are often knowledge-based or even in linguistic descriptions. On the other hand, fuzzy control provides an alternative methodology to facilitate the design of nonlinear controllers for such plants being controlled that art usually uncertain and very difficult to cope with using conventional nonlinear control theory [31, 59, 62].

Expert control systems and fuzzy control systems have one thing in common: both of them want to model human experience and human decision-making behavior. Especially, almost all fuzzy control systems are rule-based control systems. If the rules are extracted from the human experts, then this fuzzy control system is called **fuzzy expert control system**.

Fuzzy control has been widely used in various industries [94], especially in industrial process control such as to control the temperature, flow, pressure, and other operating parameters. Fuzzy control has also been used to control the robot, steel mill, aircraft flight, helicopter flight, subway, elevator, and even camera focusing stepper, television focusing, and washing machine.

1.5.4 Neural-Network-Based Control

Recently, research in artificial neural networks (ANN, or NN) has been active as a new means of information processing. ANN try to mimic the biological brain neural networks into mathematical model. In ANN, a model of the brain connects many linear or nonlinear neuron models and processes information in a parallel distributed manner [48]. Because of their massively parallel nature, NN can perform computations at much higher speed. In addition, the neural network has many interesting and attractive features including nonlinearity, learning and adaptation, self-organization, function approximation, and massively parallel processing capabilities. Therefore, NN can adapt to changes in data, learning the characteristics of input signal, and learning a mapping between an input and an output space and synthesizing an associative memory [23].

The NN techniques has been widely used to information processing, such as pattern recognition, image identification, modeling, planning, and control of systems. Recently, in control field, many attempts have been made to apply the NN to control systems where the neural network is used to deal with nonlinearities and uncertainty of the control system and to approximate functions such as system identification [27, 45, 57]. Research in applications of neural network to control can be classified into some major methods depending on structures of the control systems [92], such as supervised control, inverse control, neural adaptive control, back-propagation of utility, and adaptive critics.

In recent years, much research has been trying to hybridize the fuzzy set theory and NN [21, 32, 58, 84], and it is often called as fuzzy neural network (FNN). Another hybrid on neurocontrol is to hybridize the expert system and the neural network [28], and it is called expert neural network (ENN) controller.

Research in theory and applications of neurocontrol is now in the ascendant. Many difficulties have to be solved in the future study. We expect that the research results of neural computation and neurocontrollers would bring intelligent control to a new stage for control science.

1.5.5 Intelligent Scheduling

An interesting class of problems is concerned with specifying optimal schedules or combinations. A classical example is a traveling salesman's problem, where the problem is to find a minimum distance tour, starting the journey from one of several cities, visiting each city precisely once, and returning to the starting one [60]. The problem can be generalized to be one of finding a minimum cost path over the edges of a graph containing nodes such that the path visits each of the nodes precisely once. In most problems of this type, the domain of possible combinations or sequences from which to choose an answer is very large.

Usually, Many combination and scheduling problems in real life and production activity are very complex. In order to find an optimal scheduling scheme from the possible combinations or sequences, a large searching space is needed, and a conbinational explosion of possibilities may soon be generated. As a result, the capabilities of large computers would be exhausted. The objective for studying this class of problems is to decrease or avoid this combination explosion. To do so, it is necessary to use the knowledge of problem domain which is the key to more efficient solution methods. Sometimes, we call the **intelligent scheduling** as **knowledge-base scheduling**.

Transportation scheduling such as that of trains and buses is a representative scheduling problems. For instance, a bus is stopped at its initial point A, and there are n passengers who have called for service. They have assigned their own stops for boarding and alighting. The problem is how to arrange an optimal traveling line that has minimum total traveling path and provides passengers with services as convenient as possible. Finally, the bus returns to its starting point, Point A. The operation scheduling management of trains in marshaling (classification) yard is another example of combination scheduling. In the last 10 years some results in solving this problem have been gotten [46, 50, 81, 86].

1.5.6 Voice Control

As humans communicate with one another using language, they employ, almost effortlessly, extremely complex processes that are not well understood. It has been very difficult to develop computer systems capable of generating and understanding even fragments of natural language. Language has evolved as a communication medium between intelligent beings. Its primary use is to transmit a bit of "mental structure" from one brain to another under circumstances in which each brain possesses largely and highly similar surrounding mental structures that serve as common context. Furthermore, part of these similar contextual mental structures allows each participant to know that the other can and will perform certain processes by using it during communication "acts." The evolution of language has apparently exploited the opportunity for participants to use their considerable computational resources and shared knowledge to generate and understand highly condensed and streamlined messages [29].

Generating and understanding language is an encoding and decoding problem of fantastic complexity. A computer system capable of understanding a message in natural language would seem to require (no less than would a human) both the contextual knowledge and the processes for making the inferences from this contextual knowledge and the message. Some results have been made toward natural language processing and understanding during past 10 years [61, 90].

The research results of the natural language understanding to automatic control is based on the voice control. Research on voice control has made a breakthrough. Some voice control systems have been used to control the robots and cars. A simple example for voice control is to control a robot to move up and down, go forward and backward, turn left and right, etc. A more complex example is to command a robotic manipulator to execute certain operation on blocks. For instance, the manipulator is commanded, by microphone, to "pick up the red and large block" among some blocks. As the instruction is received by the voice computer control system, the robotic manipulator will execute a series of actions, and complete the assigned task finally.

The key to build a voice control system is to build a world model shared by computer and programmer or human operator. The speech deals with some physical objects, actions and their relationships that are clearly defined and represented as an inner model in computer [16].

1.5.7 Control of Restorative Artificial Limbs

It is meaningful work to develop various restorative robots which would relieve wounded and disabled people from their suffering, improve their independent living abilities, and even help them return to work.

Several techniques related to restorative robots have been designed and developed. These techniques include artificial limbs (artificial hands and artificial legs), orthopedics (to assemble the maneuverable mechanism around the limbs of the paralytics so that their movement abilities could be trained, and their skeletal deformities could be corrected), telecontrol (to control the paralytics, and so to help them in doing some activities), as well as blind guidance (to guide blindman by mobile robot), and so on [5, 81].

Intelligent control for car driven by paralytic has also been developed during the past 10 years.

The control for restorative robots deals with even more related disciplines. Apart from the disciplines mentioned in four-element intersection structure, the biomedical engineering, physiology and micromechnanism are included.

1.5.8 Intellectualized Instruments

The function and versatility of the instruments would be expanded in a large extent through combination between the technique of electronic measurement and the technique of artificial intelligence. In fact, the current intellectualized instrument is one with general interface bus, an intellectualized digital controller.

The intellectualized digital controller rids of traditional design ideas for the analog instruments and digital instruments with multiple purposes. Its functional implementation depends upon the designer's knowledge and experience that are used to make dexterous control and management programs. These programs are solidified into ROM for dispatch by users. The users could exert their own intelligence and experience, and utilize the present control scheme. The more complex control scheme is implemented through interaction between the user (operator) and controller. As a result, the time for programming could be obviously reduced.

The advanced intellectualized instrument could not only measure multiple parameters adaptively and precisely, but also make record, display, memory, and print automatically, or transfer the changing situation of parameters during the process of real-time testing. Besides these, many intellectualized instruments also possess the functions of automatic operation, automatic correction, and fault self-diagnosis.

It is expected that the intelligent level of the intellectualized instrument would be increased, and the intellectualized instrument with higher capability of decision and control would be widely used in science and technology as well as in industry.

1.5.9 Others

Other research and application areas of intelligent control include intelligent fault detection and diagnosis, intelligent management and monitoring, intelligent weapon control as well as intelligent investigating and cracking crime, etc.

The research topics that intelligent control and its systems deal with will be further discussed in the last chapter of this book, Chapter 10, where we will talk about the prospect of intelligent control.

1.6 Outline of the Book

This book includes the basic concepts, principles, methodologies, and techniques of intelligent control and its applications. These are as follows.

(i) Introducing the development process, structural theories and research fields of intelligent control.

(ii) Presenting the knowledge representations, searching and reasoning mechanisms as the fundamental techniques of intelligent control.

(iii) Studying the theoretical principles and architectures of various intelligent control systems.

(iv) Analyzing the paradigms of the practical representative applications of intelligent control systems.

(v) Discussing the research and development trends of the intelligent control as a discipline.

This book contains 10 chapters. Chapter 1 introduces the background, development, and definitions. Chapters 2 and 3 briefly present the human cognitive process, the basic methods and techniques of knowledge representation and inference used in intelligent control, i.e., how to represent the control knowledge, how to search for a solution, and how to make a decision and/or planning. This part is especially necessary for the control engineers who have not had a systematic study in computer science and mathematics. We believe that this part will play an important role in studying the discipline of intelligent control.

Chapters 4 to 8 discuss the various theories and systems of intelligent control that are the hierarchical intelligent control systems, expert control systems, learning control systems, fuzzy control, and neural-network-based control. Each chapter is introduced with control mechanisms, system architectures, and typical examples. Readers will have a more complete understanding of intelligent control and learn the major development of the theories of intelligent control.

Chapter 9 gives the representative paradigms of the application for intelligent control in the world. In this chapter the theories, fundamentals, tools, and systems are integrated. This will enable readers to know where intelligent control could be used and how to use it.

Chapter 10 studies the prospect of intelligent control.

Over 400 references have been adopted and distributed in the end of each chapter of this book. These references reflect the up-to-date research achievements and will be helpful for further study.

1.7 Summary

In the introductory chapter, the difficulties, challenges, and opportunities to automatic control have been introduced, and it has been drawn that one of the more effective ways to solve the problems and difficulties encountered is to use the technique of intelligent control to the control systems, and the core of the new ways is intellectualization of the controllers.

After the proposed definitions for intelligent machine and intelligent control, the features of intelligent control have been summarized. The intelligent control usually possesses complexity, incompleteness, ambiguity, or uncertainty. The key for designing an intelligent control system is in the design of prototype of intelligent machine rather than the traditional controllers, and the key of the intelligent controller is in the higher level.

Discussion on the structural theories of intelligent control is beneficial for establishing this new discipline. Three essential structural theories of intelligent control were proposed by Fu, Saridis, and Cai in 1971, 1977 and 1986 respectively. They are two-element, three-element, and four-element intersection structural theories, and are represented by Formulae (1.1) to (1.6) and shown in Figures 1.2, 1.3, and 1.5.

The most recent structural theory is the four-element intersection structural theory of intelligent control. This structural theory has expanded the fields of intelligent control to four areas: artificial intelligence, automatic control, operation research, and informatics. The reason for adding the informatics into the structure of intelligent control has been discussed in details in Subsection 1.4.3. The relationships among the four subsets have been depicted in Figure 1.8. It has been clarified that informatics is a necessary component for constructing the new discipline of intelligent control.

A general structure of intelligent controller has also been shown in Figure 1.9, and its features have been set forth in Subsection 1.4.4.

Intelligent control has a wide range of research and application fields that include the advanced robot, intelligent process planning and process control, expert control, fuzzy control, neurocontrol, intelligent scheduling, control of restorative artificial lambs, voice control, intellectualized instruments, and others. The research objectives, situation, achievements as well as the practical and potential application areas of the above fields have been introduced systematically in Section 1.5. There will be more and more research and application fields of intelligent control in the coming decade and the next century.

References

1. J. S. Albus, "Theory and practice of hierarchical control," *Proc. 23th IEEE Computer Society Int. Conf.*, 1981.
2. J. S. Albus, "Hierarchical control for robots in an automatic factory," *Proc. 13th Int. Symp. on Industrial Robots and Robot 7*, 1983.
3. K. J. Åström, J. J. Anton, and K. E. Arzen, "Expert control," *Automatica*, **22** (1986) 227.
4. K. J. Åström, J. J. Mcavoy, "Intelligent control," *J. Process Control*, **2** (1992) 115.
5. G.Beni, and S. Hackwood, eds., *Recent Advances in Robots*, (John Wiley & Sons, New York, 1985).
6. Z.-X. Cai, and K. S. Fu, "Robot planning expert systems," *Proc. IEEE Int. Conf. Robotics and Automation, Vol.3*, IEEE Computer Society Press, 1986, 1973-1978.
7. Z.-X. Cai, and Z.-J. Zhang, "Development of Robots," *Robot*, **9** (1987).
8. Z.-X. Cai, Z.-J. Zhang, "Some issues on intelligent control," *Pattern Recognition and Artificial Intelligence*, **1** (1988) 45.
9. Z.-X. Cai, "An expert system for robotic transfer planning," *J. Computer Science and Technology*, **3** (1988)153.
10. Z.-X. Cai, *Robotics: Principles and Applications* (Central South University of Technology Press, Changsha, China, 1988).
11. Z.-X. Cai, "Robot pathfinding with collision-avoidance using expert system," *J. Computer Science and Technology*, **4** (1989) 229-235.

12. Z.-X. Cai, "Structural theories of intelligent control," *Proc. CAAI First Symp. Computer Vision and Intelligent Control* (CVIC, CAAI, Chongqing, 1989) 29.
13. Z.-X. Cai, *Intelligent Control* (Electronic Industry Press, Beijing, 1990).
14. Z.-X. Cai, "Four-element structure of intelligent control," *Proc. CAAI 2nd Symp. on Computer Vision and Intelligent Control* (CVIC, CAAI, Wuhan, 1991) 299.
15. Z.-X. Cai, "A Knowledge-based flexible assembly planner," *IFIP Transaction*, B-1 (1992) 365.
16. Z.-X. Cai, and G.-Y. Xu, *Artificial Intelligence: Principles and Applications*, Second Edition (Tsinghua University Press, Beijing, 1996).
17. Z.-X. Cai, "A multirobotic planning based on expert system," *High Technology Letters*, **1** (1995) 76.
18. Z.-X. Cai, "A new structural theory on intelligent control," *High Technology Letters*, **2** (1996) 45.
19. B. A. Curtis, S. Jacobson and E. M. Marcus, *An Introduction to the Neurosciences* (W. B. Saunder Company, NewYork, 1972).
20. K. Dojong, "Intelligent control: integrating AI and control theory," *Proc. IEEE Trends and Applications*, 1983.
21. Y. Dote, and K. Kano, "DPS-based neuro-fuzzy position controllers for servomotor," *Proc. IEEE IECON*, 1992, 986-989.
22. H. F. Durrant-Whyte, ed., *Integration, Coordination and Control of Multi-Sensor Robot Systems* (Kluwer Academic Publishers, Boston, MA, 1990).
23. F. H. Eeckman, ed., *Analyses and Modeling of Neural Systems* (Kluwer Academic Publishers, Boston, MA, 1992).
24. K. S .Fu *et al.*, "A heuristic approach to reinforcement learning control system," *IEEE Trans.* **AC-10** (1965) 390-398.
25. K. S. Fu, "Learning control systems and intelligent control systems: a intersection of artificial intelligence and automatic control," *IEEE Trans.* **AC-16** (1971) 70-72.
26. K. S. Fu, "Computer vision for automatic inspection," *Proc. of Robotic Intelligenc and Productivity Conference*, 1983.
27. T. Fukuda and T. Shibata, "Research trends in neuromorphic control," *J. Robotics Mechation,* **2** (1991) 418.
28. T. Fukuda and T. Shibat, "Theory and applications of neural networks for industrial control systems," *IEEE Trans. on Industrial Electronics*, **9** (1992) 478.
29. W. B. Gevarter, *Artificial Intelligence, Expert Systems, Computer Vision and Natural Language Processing* (NOYES, NJ, 1984).
30. D. D. Grossman, "Trafic control of multiple robot vehicles," *IEEE J. of Robotics and Automation*, **4** (1988) 491.
31. M. M. Gupta, *Fuzzy Logic in Knowledge-Based Systems: Decision and Control*

(North-Holland, 1988).
32. M. M. Gupta and J. Qi, "On fuzzy neuron models," in: *Proc. Int. Joint Conf. of Neural Networks, Vol.2*, Seattle, WA, 1991, 431-436.
33. S. Y. Harmon, "Coordination between control and knowledge-based systems for autonomous vehicle guidance," *Proc. IEEE Trends and Applications*, 1983.
34. C. J. Harris, C. G. Moore and M. Brown, *Intelligent Control: Aspects of Fuzzy Logic and Neural Nets* (World Scientific, Singapore, 1993).
35. J. Hartley, *Robots at Work, A Pratical Guide for Engineers and Managers* (IFS Publications, England, 1983)
36. F. Hages-Roth, D. Waterman and D. Lenat, eds., *Building Expert Systems* (Addison Wesley, Readings, MA, 1983).
37. D. O. Hebb, *Organization of Behavior: A Neuropsychological Theory* (John Wiley and Sons, New York, 1949).
38. M. Herman, J. S. Albus and T. H. Hong, "Intelligent control for multiple autonomous undersea vehicle," in: *Neural Network for Control*, ed. W.T.Miller *et.al.*(MIT Press, Combridge, MA, 1990) 42.
39. R. Herrrod and J. Rickel, "The industrial automation-AI connection," *TI Tech. J.* Winter, 1987, 2-13.
40. L. S. Homem de Mello and A. C. Sanderson, "An AND/OR graph representation of assemebly plans," *AAAI Proc. of the 5th National Conf. on AI*, 1986: 1113-1119.
41. S. H. Hopkin and P. J. Drazan, "Semiautonomous systems in automatic assembly," *IEE Control Theory and Application, Part D*, **132** (1985) 174.
42. IEEE Editorial, "Challenges to control: a collective view," *IEEE Trans. Automation Control*, **32** (1987).
43. *Intelligent Control: Annoucement of Program Initiative*, NSF and EPRI, USA, 1992.
44. D. Ionescu *et al.*, "Expert system for computer process control design," *Proc. IEEE ISIC* (IEEE Computer Society Press, 1987) 185.
45. B. Kosko, ed., *Neural Network and Fuzzy Systems: A Dynamical System Approach to Machine Intelligence* (Prentice-Hall, Englewood, Cliffs, 1992).
46. J. Latomee, *Robot Motion Planning* (Kluwer Academic Publishers, Norwell, MA, 1991).
47. Z. Li, M. Xu and Q. Zhou, "A novel simulating of human intelligent controller," *ACTA Automation Sinica*, **16** (1990) 503.
48. R. P. Lippmann, "An introduction to computing with neural nets," *IEEE ASSP Magazine*, **4** (1987) 4.
49. J. Maiers and Y. S. Sherif, "Applications of fuzzy set theory," *IEEE Trans. System, Man and Cybernetics*, **15** (1985) 175.
50. R. K. Mathur, R. Münger and A. C. Sanderson, "Hierarchical planning for space-truss

assembly," in: *Intelligent Robotic Systems for Space Exploration*, ed. A.A.Desrochers (Kluwer Academic, Boston, MA, 1992) 141.

51. W. S. Meculloch and W. Pitts, "A logical calculus of the ideas immanent in nervois activity," *Bulletin of Machematical Biophysics*, **5** (1943) 115.
52. J. E. Mcdonall and B. V. Koen, "Application of AI techniques to digital computer control of nuclear reactors," *Nuclear Science and Engineering*, 56, 1975.
53. J. McGhee, "Holistic approaches in knowledge-based process control," in: *Knowledge Based Systems for Industrial Control*, (Peter Peregrinus Ltd, UK, 1990).
54. A. Meystel, "Intelligent control: highlights and shadows," *Proc. IEEE ISIC* (IEEE Computer Society Press, 1987) 2.
55. A. Meystal and J. Y. S. Luh, eds., *Proceedings IEEE International Symposium of Intelligent Control* (IEEE Computer Society Press, Philadelphia, Pennsylvania, 1987).
56. A. Meystel, J. Herath and S. Gray, eds. *Proc. IEEE ISIC*, Philadelphia, Pennsylvania, 1990
57. W. T. Miller III, R. S. Sutton and P. J. Werbos, *Neural Network for Control* (MIT Press, Cambridge, MA, 1990).
58. H. Narqzaki and A. L. Ralescu, "A synthesis method for multilayered neural network using fuzzy sets," *Proc. IJCNN*, 1991, 54-66.
59. C. V. Negoita, *Expert Systems and Fuzzy Systems* (Benjammin/Cummings Publishing Co. 1985).
60. N. J. Nilsson, *Principle of Aritifical Intelligence* (Tioga Publishing Co., Palo Alto, CA, 1980).
61. G. Novak, Understanding natural language with diagram, *Proc. AAAI*, 1990.
62. W. Pedrycz, *Fuzzy Control and Fuzzy Systems* (Research Studies Press, England, 1993).
63. *Proc. IEEE ISIC*, IEEE Computer Society Press, 1989.
64. *Proc. IEEE ISIC*, IEEE Control System Society, 1991.
65. *Proc. IEEE ISIC*, IEEE Control System Society, 1992.
66. X.-S. Qian, *On System Engineering* (Hunan S&T Press, Changsha, China, 1982).
67. J. Richardson, "AI applied to process planning," *TI Techniques*, July-Aug, 1987, 13-18.
68. A. C. Sanderson *et al.*, "Assembly sequence planning, *AI Magazine*, Special Issue on Assembly planning," **11** (1990) 62.
69. R. Sanz *et al.*, "Intelligent process control: the CONEX architecture," in: *Engineering Systems with Intelligence: Concepts, Tools and Applications*, (Kluwer Academic Publishers, Boston, 1991).
70. G. N. Saridis, "Toward the realization of intelligent controls," *Proc. IEEE,* **67** (1979).
71. G. N. Saridis, "Intelligent robotic control," *IEEE Trans,* **AC-28** (1983) 547.

72. G. N. Saridis, "Architectures for intelligent machines," *CIRSSE Technical Report, No. 96*, RPI, 1991.
73. G. N. Saridis and A. Meystel eds. *Proceedings IEEE Workshop on Intelligent Control* (IEEE Computer Society Pross, Troy, New York, 1985).
74. G. N. Saridis and H. E. Stephanou, "A hierarchical approach to the control of a prosthetic arm," *IEEE Trans.* **SMC-7** (1977) 407.
75. G. N. Saridis and K. P. Valavanis, "Information theoretic approach for knowledge engineering and intelligent machines," *Proc. America Control Conf.* 1985.
76. G. N. Saridis and K. P. Valavanis, "Analytical design of intelligent machines," *Automatica* **24** (1988) 123.
77. R. Sauers and R. Walsh, "On requirements of future expert systems," *Proc. 8th IJCAI*, 1983
78. R. J. Schilling, *Fundamentals of Robotics: Analysis and Control* (Prentice-Hall, Englewood, Cliff, NJ, 1990).
79. C. W. de Silva, A. G. J. MacFarlane, *Knowledge-Based Control with Application to Robots* (Springer-Verlay, Berlin, 1989).
80. H. A. Simon, *Cognition of Human Being: Thinking Theory of Information Processing* (Science Press, Beijing, 1986).
81. M. W. Spong, F. L. Lewis and C. T. Abdallah, eds., *Robot Control: Dynamics, Motion Planning and Analysis* (IEEE Press, 1993).
82. L. Stark, *Neurological Control Systems, Studies in Bioengineering* (Plenum Press, New York, 1968).
83. H. E. Stephanou, A. Meystel and J. Y. S. Luh, eds. *Proc. IEEE ISIC* (IEEE Computer Society Press, 1988).
84. M. Strefezza and Y. Dote, "Fuzzy and neural network controller," *Proc.IEEE IECON*, Los Alamitos, 1991, 1437-1442.
85. M. Sugeno, *Industrial Application of Fuzzy Logic* (North Holland, 1985).
86. A. Tate, "Applications of knowledge-based planning systems," in: *Knowledge-Based Expert Systems in Industry*. ed. J.Kriz (Ellis Horwood Ltd, England, 1987) 130.
87. M. Togai, "Japan's next generation of robots," *IEEE Computer*, March 1984, 19-25.
88. R. M. Tong, "A control engineering review of fuzzy system," *Automatica*, **13** (1977) 559.
89. S. G. Tzafestas, ed.,*Engineering Systems with Intelligence: Concepts, Tools, and Applications* (Kluwer Academic, Boston, MA, 1991).
90. M.-H. Wang *et al.*, "A neural network for Chinese formation processing," *Proc. IJCNN*, Vol.2, Beijing, 978-981.
91. S. M. Weiss and C. A. Kulikowski, *A Pratical Guide to Designing Expert Systems* (Rowmand and Allenkeld Publishers, New Jersey, 1984).

92. P. J. Werbos, "Neurocontrol and related techniques," in: *Handbook of Neural Computing Applications* (Academic Press, New York, 1990) 345.
93. N. Wiener, *Cybernetics, or Control and Communication in the Animal and the Machine* (MIT Press, Cambridge, MA, 1948).
94. R. R. Yager and L. A. Zadeh, eds., *An Introduction to Fuzzy Logic Applications in Intelligent Systems* (Kluwer Academic, Boston, 1992).
95. L. A. Zadeh, "Fuzzy sets," *Information and Control*, **8** (1965) 338.
96. Z.-J. Zhang, and Z-X.Cai, "Intelligent control and intelligent control systems," *Information and Control*, **18** (1989) 30.
97. Q. J. Zhou and J. K. Bai, "An intelligent controller of novel design," *Instrument Society of America*, 1983.

CHAPTER 2

METHODOLOGIES OF KNOWLEDGE REPRESENTATION

When we discussed the structures and features of intelligent control in Chapter 1, it was mentioned that the intelligent control system possesses a hierarchical structure, and the highest level of this system, the organization level, is based on knowledge. Knowledge is necessary either for task description of the system or for the representation of the problems, and even for reasoning and decision making. A fundamental step is to make an investigation into the methodology of knowledge representation in the study of intelligent control. Any problem-solving deals with two key techniques: representation and searching.

There may be various methods of knowledge representation for a problem. All of these representation methods possess different representation space. In this chapter, we are going to discuss some knowledge representation methods commonly used, i.e., state space, problem reduction, predicate logic, and semantic network, etc. Some more advanced problem-solving techniques such as expert systems, neural network and fuzzy sets method systems will be explained in related sections of the follow-up chapters [2, 15, 25, 26].

Before discussing the knowledge representation and reasoning, let us briefly introduce the process of human cognition.

2.1 Process of Human Cognition and AI

The cognitive process of a human being is a very complex behavior, and it could not be explained completely until recently. People with different backgrounds and professions have studied the cognitive process from various points of view, which led to the discovery of related disciplines such as cognitive psychology, cognitive physiology, and cognitive engineering, etc.[4, 5, 20, 23]. Some theories about the cognitive process are representative. One of them is based on the physical symbol system, or symbol operation system. People usually call this theory and its researchers symbolism (or logicism) and symbolists (or logicists) respectively. Another theory is established on the basis of neural networks of the human brain, and this theory and theorists are named as connectionism and connectionists (or bionicsts). Yet another one declares that an intelligent system can be decomposed into independent and parallel activity procedures which all interface directly to the world through perception and action, and the cognitive process is based on this principle. This theory and theorists are called actionism and actionists (or evolutionists). Discussion on these disciplines and theories is beyond the scope of this book. We would only talk about several essential issues related to intelligent control, and will briefly introduce the process of human cognition in the view of computer engineering, and this discussion is basically based on the opinion of symbolism [13, 19, 21].

2.1.1 Tasks for Studying Cognitive Process

There are several different levels in the psychological activity of the human cognition. These levels can be compared with the levels in a computer, which are shown in Figure 2.1. From Figure 2.1(a) we can see that the highest level of the psychological activity for a human being is thinking strategy; the next one is preliminary information processing; and the lowest level is physiological process including the activities of central nervous system, neuron, and brain [18, 22, 27]. The corresponding parts in the computer are the hardware, the programming language and the program of computer, see Figure 2.1(b).

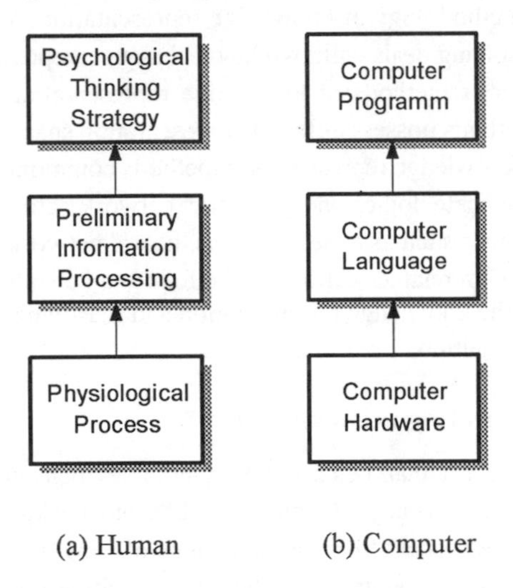

(a) Human (b) Computer

Figure 2.1. A comparison of human cognitive process with computer.

Definition 2.1. Let T be a time variable, x be a cognitive operation, Δx be the variation of x , and Δx is a function of both the organism state S (the physiological and psychological state of organism as well as the memory in brain) and the stimulus R of outside world at that time. As the outside stimulus R is applied to the organism with a specified state S, a variation Δx will appear, i.e.,

$$\begin{aligned} &T \to T+1 \\ &x \to x+\Delta x \\ &\Delta x = f(S,R) \end{aligned} \tag{2.1}$$

The computer works with a similar mechanism. In a certain period of time, the memory of the computer corresponds to the state of organism, and the input of the computer corresponds to some stimulus exerted to the organism. The computer operates after receiving the input, and then the internal state of the computer changes with time. We could study this computer system in different levels. This system carries out intelligent information processing based on a model of the human cognition. Obviously, this system is an intelligent computer system. A specific and important objective for studying the cognitive process is to design this high-level intelligent information processing system available to a special field. An automatic control system that has the capability of intelligent information processing is an intelligent control system, and it could be either an expert system or an intelligent decision-making system etc.

2.1.2 Hypotheses of Intelligent Information Processing

The human could be seen as an information processing system.

Sometimes, we call the information processing system the symbol operation system or physical symbol system. A symbol is a pattern. Any pattern would be a symbol if the pattern could be simply distinguished from the other one. For instance, the different Chinese spelling characters or English characters are different symbols. An operation of symbols is equivalent to a comparison of the symbols, then the same symbols and different symbols can be found. The basic task and function of the physical symbol system are to identify the same symbols and to distinguish the different symbols. Therefore, it is necessary for this system to be able to distinguish the essential difference among the different symbols. The symbols may be physical symbols or physical actions, abstract symbols or micromotion of neuron in the brain, and motion mode of an electron in an electronic computer, etc.

Definition 2.2. A complete symbol system should possess the following six basic functions:

(1) Input: to input symbols;
(2) Output: to output symbols;
(3) Store: to store symbols;
(4) Copy: to copy symbols;
(5) Set up symbol structure: to find the relations among symbols, and to form the symbol structure in a symbol system;
(6) Condition transfer: to continue and complete a process of activity according to the present symbols.

If a physical symbol system possesses all the six functions above and can accomplish the overall process, then this system will become a complete physical symbol system. The human can input signal, for example, to see through the eye, to hear by the ear, and to touch by the hand, etc. The computer can also input symbol through card, punching belt, keyboard, tape, or video tape, etc. Therefore, both the human and the computer possess the six mentioned functions of the physical symbol system.

Hypothesis 2.1. If any system can exhibit the intelligence, then it must be able to carry out the six functions. In contrast, if any system possesses the six functions, then it can exhibit the intelligence. This intelligence is a kind of intelligence that the human has. This hypothesis is called the Hypothesis of Physical Symbol System.

Deduction 2.1. Since the human possesses the intelligence, he or she should be a physical symbol system.

Deduction 2.2. Since the computer is a physical symbol system, it should exhibit the intelligence.

Deduction 2.3. Since both the human and the computer are the physical symbol system and exhibit intelligence, the human activity can be simulated by computer.

We can make a computer program according to the cognitive process of the human being. This work is exactly the research content of the artificial intelligence and intelligent control. If we could do so, then we would be able to describe the cognitive activity process of a human being formally by a computer, or could find a theory describing the intellectual activity process of the human being.

It is noted that the computer is not always meant to simulate the activity of the human being; the computer can make some complex programs to solve equations and compute a complex problem. This kind of computation would not precisely be cognitive process of the human being.

2.1.3 Computer Simulation of Human Intelligence

The intellectual currents of the times directly help scientists to study certain phenomena. For the evolution of artificial intelligence, the two most important forces in the intellectual environment of the 1930s and 1940s were mathematics logic and new ideas about computation. A. Turing, who has been called the father of AI, not only invented a simple; universal and nonnumerical model of computation, but also argued dirctly for the possibility that computational mechanisms could behave in a way that would be perceived as intelligence.

There were other strong intellectual currents from several directions that converged in the middle of this century to the people who founded the science of AI. The concepts of cybernetics and self-organizing systems of Wiener, McCulloch, and others focused on the microscopic behavior of locally simple system [12, 24]. The cybernetics influenced many fields because its thinking spanned many fields, linking ideas about the integration of the neural network and nervous system with information theory, control theory, as well as symbol, logic, computation and simulation. The ideas of cyberneticists were part of the intellectual currents of the times, and in many cases they influenced the early workers in AI directly as their mentors. Meanwhile, their ideas carried within itself the seeds of intelligent control to some extent.

There has been a long-standing connection between the idea of complex mechanical devices and intelligence. Starting with the fabulously intricate clocks and mechanical automata of past centuries, people had made an intuitive link between the complexity of a machine's operation and some aspects of their own mental life [11]. Over the past few centuries, new technologies have resulted in a dramatic increase in the complexity that we can achieve in the things we build. Modern computers are orders of magnitude more complex than anything that man has ever built before.

The first task of computers in this century is focused on the numerical computations that had previously been performed cooperatively by teams of hundreds of clerks. However, most of the intellectual activities of the human being are not in the area of the numerical computations but in the areas of the logic inference and thinking in terms of images. As we have learnt, the **Deduction 2.1** of the **Hypothesis 2.1** of physical symbol system stated that the human possesses intelligence, and so he or she could be a physical symbol system. **Deduction 2.3** of the hypothesis declared that the cognitive activities of human could be simulated by means of computer programs. That is, both the humans and the computers are physical symbol systems, and use the same physical symbols. Thereby, the computer could simulate the intellectual activity process of the human being.

Not long after the dramatic success demonstrated by the first digital computers with these elaborate calculations, people began to explore more generally the possibility of intelligent mechanical behavior: could machines play chess, or prove theorems? They could. The computer can really perform many intellectual functions such as playing chess, proving theorems, translating languages, solving puzzles, making decisions and so on. All of these functions are accomplished by writing and executing the computer programs that imitate the human intelligence. However, recently machine intelligence is not as good, in most cases, as compared with human intelligence. The intellectual level of the machine intelligence simulated by computer programs is very limited. It is difficult at least today to write a computer program that exhibits exactly the same behavior as a human.

As an example, let us consider the programs that play chess. Current programs are quite proficient; the best experimental systems play at the human "expert" level, but not as well as human chess "masters". The programs work by searching through a space of possible moves, i.e., considering the alternative moves and their consequences several steps ahead in the game, just as human players do. Computers can search through thousands of moves at the same time, and techniques for efficient searching constitute some of the core of AI. The reason that computers cannot beat human players is that looking ahead is not all there is to chess, since there are too many possible moves to search exhaustively, alternatives must be evaluated without knowing for sure which will lead to a winning game, and this is one of the abilities that human experts cannot explicate.

Psychological studies have shown that chess masters have learnt to "see" thousands of meaningful configurations of pieces when they look at a chess position, which presumably help them decide on the best move, but no one has yet designed a computer program that can identify these configurations until now.

Research on the intelligent computer has made several great advances in recent years, and research on the neural computer is a new paradigm of them. The neural computer can make "thinking" similar to the human thinking and intend to rebuild the image of human brain [4, 6]. It was reported from MITI, Japan, that a research on the feasibility of the neural computer system was completed in Japan in the late April of 1989, and a long-term and detailed research program about this system was proposed. Many research groups and researchers in the USA, Europe, Japan, China, and other countries have engaged into the research work of the artificial neural network (ANN), and a great upsurge in studying the ANN is spreading around the world. A lot of research results including ANN-based control have been obtained in late 1980s and early 1990s [8, 9, 10].

It is predicted that the neural computer will be put into practical use by the end of this century, and the products of the neural computer will be available commercially at the same time.

The human brain is a mystical and wonderful organ. It can duplicate interaction and process information rapidly and in quantity, and execute multiple tasks at almost the same time. Up to now, all kinds of computers are still based on the structure of Von Neumann machine, and can only "solve" single problems one by one in some order. Although the parallel and distributed processing (PDP) computer can speed up the data computation and information processing greatly in the 1980s, its performances are still very limited, and some essential problems remain to be solved. It is expected that research on the neural computing could build the neural computer whose capability of information processing would be much stronger and with the level of real artificial intelligence. Hopefully, the time when the neural computer replaces the traditional electronic computer would come.

2.1.4 *Basic Functions of Human Brain*

The process of information processing of a human can be studied through psychological experiment, in order to understand human thinking. We can study how the human acquires the ability of information processing from the viewpoint of human evolution. During the evolution process over millions of years the human being possesses the ability to acquire new information from the environment and the ability to make effective activity through adaptation with the environment. Now the question is how to design a system that could simulate the cognitive activity of human efficiently.

In order to simulate the cognitive activity of human beings for a system, the following four conditions should be satisfied, in other words, four functions should be provided:

(1) This system should act step by step and in series since the human could simply think or do one thing exactly at the same time.

(2) This system could simply make limited computation since the computational ability of the human being is limited.

(3) This system could develop various desires because the human being has various desires both in material living and mental (cultural) living.

(4) This system could process the unforeseen events.

The human brain has developed three skills for its cognitive activity during the evolution process, and these skills enable it to complete the four functions mentioned above.

(1) To solve a problem by searching

Searching means proposing a strategy that is used to solve the encountered problem. Because the searching process is carried out in series and the computational ability of the human is limited, the solution could only be found through trials. The human being could not simultaneously consider various possibilities to solve a problem and to compare all the possibilities in their searching processes. For example, there can be a lot of different ways of placing pieces on a chess board, and people have a limited ability to search in his or her game. There are 18×18 checkers and 19×19 intersections, and there are three possible states for every intersection, i.e., black, white or empty. Therefore, there exist about 10^{190} possibilities for the distributed pattern of the chess pieces! It is not difficult to understand even if the best human player of chess could not consider all the possibilities during the process of a game. Human could not consider all the possibilities in his or her brain, and simply uses the searching methods that are efficient in human life.

(2) To look for satisfactory solution(s) rather than optimal one(s)

Humans usually look for satisfactory solution(s) rather than optimal solution(s) when they solve problems. It is hard to look for an optimal solution because much time and more memory space for the computer have to be used. It is much easier to look for a satisfactory solution since it is not necessary to search for all the faces of the problem, and the objective of solving the problem could be achieved.

(3) There is an aspiration level

The character of a human is the ability to regulate the satisfactory level of his or her own desires which is called the aspiration level by psychologists. The aspiration level of a human could be self-regulated according to the change of the condition in the outside world. This character could be applied to both the life of a lone person and the activity of the human community.

Apart from the above three skills for cognitive activity of the human being, a cognitive system is demanded to have some mechanisms for information processing such as the distribution, memory and movement of the information.

2.2 State Space Representation

Problem-solving is a big topic that deals with many core concepts such as reduction, inference, planning, common-knowledge reasoning, theorem-proving and related processes, etc. After analyzing the problem-solving methods used in AI research and application, we have found that many problem-solving methods applied some trial-and-error search mechanism. In other words, these methods solved the problems by searching for a solution in a possible solution space. A method for problem representation and problem-solving based on the solution space is called **state space method** that is based on the **state** and **operator** [7, 14, 16, 17].

2.2.1 State Description of Problem

First of all, let us define the followings.

Definition 2.3. State Q is a sequential set with minimum variables $q_0, q_1, q_2, \ldots, q_n$ for describing the difference among different things or events of some class. The vector form of state is:

$$Q = [q_0, q_1, \ldots, q_n]^T \tag{2.2}$$

where every element q_i, $i = 0, 1, \ldots, n$, is the component of the set and is called state variable. A special state can be got by assigning a set of values to every component or element. For example, a state k has the following form

$$Q_k = [q_{0k}, q_{1k}, \ldots, q_{nk}]^T. \tag{2.3}$$

Definition 2.4. A means that is used to transform a problem from one state to another is called an operator. An operator may be a movement, process, rule, mathematical operator, computational symbol or logical symbol and so on.

Definition 2.5. State space of problem is a graph that represents all possible states and their relationships with the problem, and includes three explanatory sets, i.e., set S of all possible initial states of problem, set F of operators, and set G of goal states. Therefore, a state space could be a triple state $\{S, F, G\}$.

Example 2.1. 15-puzzle problem

Let us use the 15-puzzle problem to explain the concept of state space representation. The 15-puzzle problem consists of 15 tiles and are moved on the board either horizontally or vertically. There always exists one empty square cell on the board, so it is possible to move an adjacent numbered tile into the empty cell. We can trace moving tiles from the moving empty cell. The 15-puzzle problem is shown in Figure 2.2. There are two situations — initial situation and goal situation — of which correspond to the initial state and goal state for the problem.

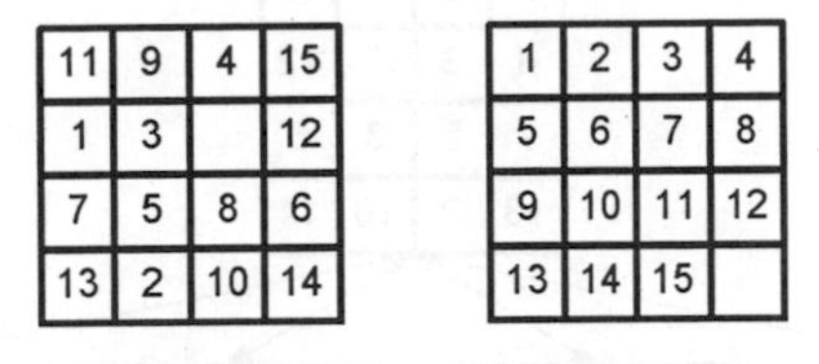

(a) initial state (b) goal state

Figure 2.2. 15-puzzle problem.

How to transform the initial situation into the goal situation? A solution for the problem is the suitable movement sequence of the tiles, e.g., move tile 12 left, tile 15 down, tile 4 right, and so on.

The most direct method of solving the 15-puzzle problem is to try every different move until the goal situation is achieved accidentally. Naturally, this deals with some trial-and-error search. Beginning with the initial situation, try every legal move to get to a new situation, then assess the next situation which results from it. Continue this process until the goal situation is reached. We consider the **space** that consists of states reachable by the initial state as a **graph** that consists of all the nodes corresponding to the states. The graph is called state graph. The state graph for the 15-puzzle problem is partly demonstrated in Figure 2.3 where every node is labeled with its represented situation. First of all, apply suitable operator(s) to the new state(s) and so on, until the goal state is generated.

We usually use the term of state space method to represent the following method: the beginning with initial state, add one operator each time, set up experimental sequence incrementally, until the goal state is reached.

The whole process for searching the state space includes generating new state descriptions from old state descriptions, then examining these new state descriptions to see whether the goal state has been reached. In general, this examination simply checks whether some state description is matched with the given one. Sometimes, however, a more complex goal test has to be carried out. For some optimization problem, finding a path to reach the goal is not enough. An optimal path corresponding to some criterion has to be found. For instance, the least number of moves for a puzzle or game problem needs to be found.

In summary, in order to complete a state description for some problem, we have to define three things: (a) the form of the state description, especially the initial state description; (b) the set of operators and their actions to the state description; (c) the features of the target state description.

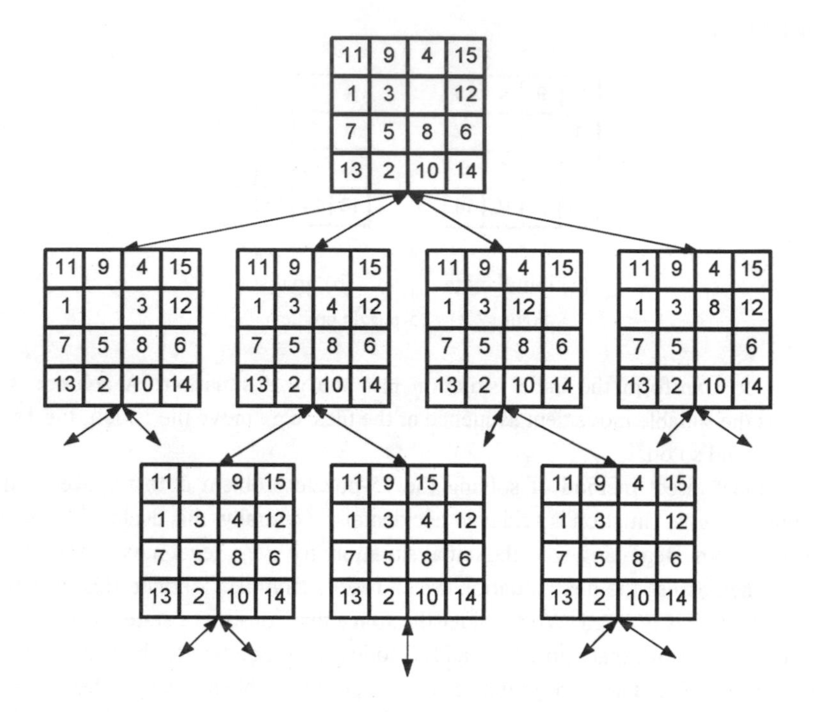

Figure 2.3. Local state graph of 15-puzzle problem.

2.2.2 Graph Notion of State

We have used a graph to illustrate the state space for the 15-puzzle problem in Figure 2.3. In order to have a better understanding of the state space graph, let us introduce some terms and formal graph notions in graph theory.

Definition 2.5. A graph consists of a (not necessarily finite) set of nodes. Certain pairs of nodes are connected by arcs directed from one node to another node. Such a graph is called a directed graph. If an arc is directed from node n_i to node n_j, then node n_j is said to be a descendant node or successor of node n_i, while the node n_i is said to be a parent node or ancestor of node n_j. For our interested graph, only one node has finite descendant nodes. A pair of nodes may be descendant nodes of each other. When we use a graph to represent a state space, every node in the graph is labeled by related state description, and every arc is labeled by related operator.

Definition 2.6. A sequence of nodes $(n_{i1}, n_{i2}, \ldots, n_{ik})$, if there always exists a descendant node n_{ij} with every node $n_{i,j-1}$ for j=2,3, ... , k, is called as a path of length k from node n_{i1} to node n_{ik}. If there exists a path beginning from node n_i and ends at node n_j, then the node n_j is said to be accessible from node n_i or a successor of node n_i, and the node n_i is an ancestor of node n_j.

It can be found that looking for a sequence of operators used to transform a state into another state is equivalent to looking for a path of graph.

It is convenient to assign the cost to arc as a representation of the cost of applying the corresponding operator. Let $c(n_i, n_j)$ be the cost of an arc directed from node n_i to node n_j. The cost of a path between two nodes is equal to the sum of the costs of all arcs that connect every node, one by one, on the path. For optimization problems, a path with minimum cost between two nodes has to be found.

For the simplest type of problems, a path, there may exist a minimum cost in the path, between a specified node s (representing the initial state) and another specified node t (representing the goal state) is desired to be found. Two representative problems of this simplest type are:

(1) Find a path between node s and any node in set of node $\{t_i\}$.

(2) Find a path between any node in set of nodes $\{s_i\}$ and any node in set of nodes $\{t_i\}$.

The set $\{t_i\}$ is called the **goal set** that is not necessary to be given explicitly and might be implicitly defined by the properties possessed by the corresponding goal state descriptions. Every node in set $\{t_i\}$ is a goal node.

A graph may be specified explicitly or implicitly. For the explicit specification, the nodes and arcs associated with costs are listed in a table explicitly. This table might list every node and its descendant nodes as well as the costs of the connected arcs in the graph. Obviously, an explicit specification is impractical for large-scale graphs and impossible for those that have an infinite set of nodes.

For an implicit specification, a finite set $\{s_i\}$ of nodes is given as start nodes. In addition, it is convenient to have a successor operator Γ that can be applied to any node and to all the successors of that node and the costs of the associated arcs. (In our state space terminology, the successor operator is defined in terms of the set of operators applicable to a given state description.) The application of Γ to the members of $\{s_i\}$, to their successors, and so on infinitely, then makes explicit the graph implicitly defined by Γ and $\{s_i\}$. The process of applying descendant operators to a node is a process of expanding the node. Therefore, the process of searching for a solution sequence of operators through a state space corresponds to making explicit a sufficient portion of an implicit graph to include a goal node. Searching graphs in this manner is therefore a central element of state-space problem-solving.

The effort solving a problem is greatly dependent on the method of problem representation. Of course, a less state space representation is expected. Many problems that seem to be very difficult might possess a lesser and simpler representation if a suitable method of problem representation could be used.

2.3 Problem Reduction Representation

Problem reduction is another approach to problem description and problem-solving. Given a problem description, transform the original problem into a set of subproblems through a series of transformations; and the solution of these subproblems can be obtained directly, so that the original problem is solved.

A representation applying problem reduction involves three components:

(1) An initial problem description

(2) A set of operators transforming the initial problem into subproblems and sub-subproblems.

(3) A set of description of primitive problems.

Problem reduction is to use the backward reasoning from the problem to be solved, establishing subproblems and sub-subproblems until the original problem is finally reduced to a set of trivial primitive problems.

2.3.1 Problem Reduction Description

Example 2.2. Tower of Hanoi Puzzle

In order to illustrate how a problem might be solved by the problem reduction approach, let us consider another puzzle, the **Tower-of-Hanoi Puzzle** that can be stated as follows:

There are three pegs — 1, 2 and 3 and three disks of different sizes A, B and C. The disks have holes in their centers so that they can be stacked on the pegs. Initially, the disks are all on peg 1; the largest disk C at the bottom, and the smallest disk A is at the top. The goal is to transfer the disks to peg 3 within the constrains that:

(1) only one disk may be moved at a time;

(2) only the top disk on a peg can be moved;

(3) larger disk may never be placed on top of a smaller one.

The initial and goal configurations are given in Figure 2.4.

If we would apply state space approach to this problem, then the graph of state space should have twenty seven nodes, each representing one of the legal configurations of disk on pegs. However, we can also solve this problem by a simple problem reduction approach. The original problem can be reduced to a set of simpler problems that involves the following chain of reasoning:

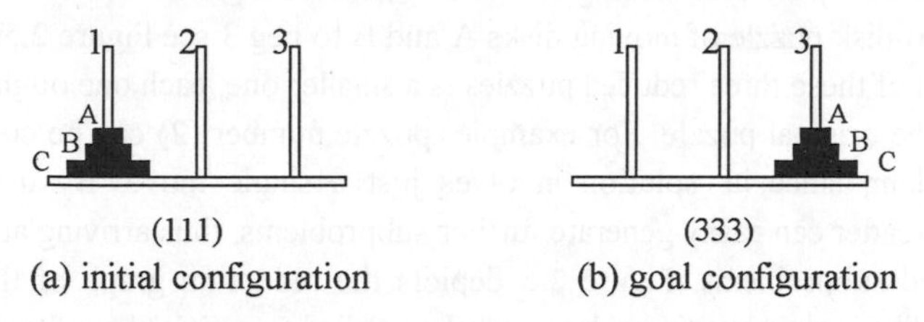

(a) initial configuration (b) goal configuration

Figure 2.4. The Tower-of-Hanoi puzzle.

(1) We must move disk C to peg 3 in order to move all of the disks to peg 3, and peg 3 must be empty just prior to moving disk C to it.

(2) We can not move disk C anywhere until disks A and B are moved first, and disk A and B had better not be moved to peg 3 since then we would not be able to move disk C there. Therefore, we should move disks A and B to peg 2 first.

(3) Then the key step of moving disk C from peg 1 to peg 3 can be completed, and go on to solve the rest of the puzzle.

From the above discussion we see that the original puzzle can be reduced to the following three sub-puzzles:

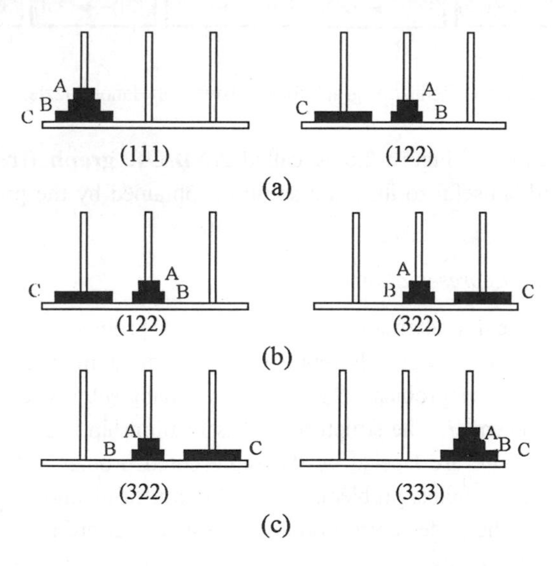

Figure 2.5. Reduction for Tower-of-Hanoi Puzzle.

(1) the two-disk puzzle of moving disks A and B to peg 2 see Figure 2.5(a).

(2) the one-disk puzzle of moving disk C to peg 3 see Figure 2.5(b).

(3) the two-disk puzzle of moving disks A and B to peg 3 see Figure 2.5(c).

Since each of these three reduced puzzles is a smaller one, each one ought to be easier to solve than the original puzzle. For example, puzzle number (2) can be considered as a primitive problem, since its solution involves just a single move. By using a similar reasoning, the reader can easily generate further subproblems, thus arriving at solutions for all the unsolved subproblems. Figure 2.6 depicts the reduction graph of the Tower-of-Hanoi Puzzle. The same reduction scheme can be applied to an initial configuration with an arbitrary number of disks. As exercise, the reader may try to solve the Tower-of-Hanoi Puzzle with four or five disks.

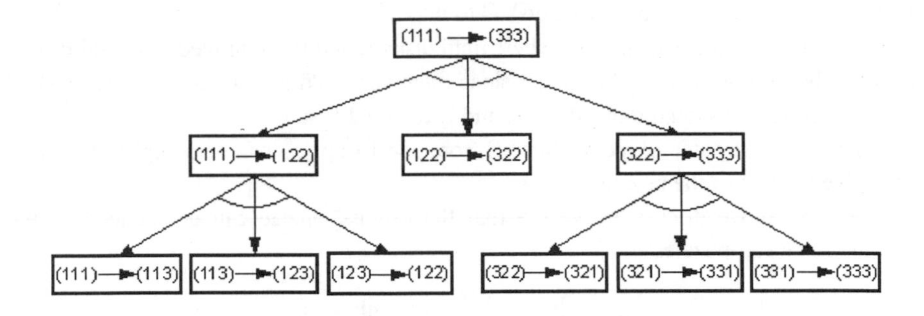

Figure 2.6. Reduction graph for the Tower-of-Hanoi Puzzle.

The graph structure of Figure 2.6 is called **AND/OR graph (tree)** or **problem reduction graph**, and is useful to illustrate solutions obtained by the problem-reduction approach.

2.3.2 AND/OR Graph Representation

The reduction of a problem to alternative sets of successor problems or subproblems by a graph-like structure can be conveniently represented by a diagram. Thus, problem A may be solved either by solving subproblems B and C or by solving subproblems D, E and F, or by solving a single problem H. The structure of this relationship is shown in Figure 2.7. The nodes of this structure are labeled by the represented problems. Problem B and C constitute one set of subproblems, problems D,E and F constitute another, and problem H constitutes a third one. The nodes corresponding to a given set are indicated by a special mark linking their incoming arcs.

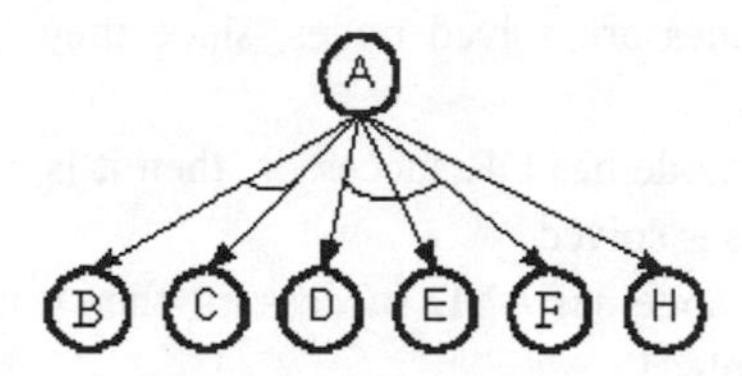

Figure 2.7. Structure for alternative sets of subproblems.

It is usual to introduce some extra nodes into the structure so that each set, containing more than one descendant problem, can be grouped below its own parent node. With this convention, the structure of Figure 2.7 becomes one in Figure 2.8 where the added nodes labeled N and M serve as exclusive parents for set $\{B, C\}$ and set $\{D, E, F\}$ respectively. If we think of N and M as playing the role of problem descriptions, then the problem A is reduced to single alternative subproblem N, M or H. For this reason, the nodes labeled N, M and H are called OR nodes. However, problem N is reduced to a single alternative set of subproblems B and C, all of which must be solved in order to solve N. For this reason, the nodes labeled B and C are called AND nodes. For the same reason, the nodes labeled D, E and F are AND nodes too. AND nodes are indicated by marks on their in coming arcs. Structures like that shown in Figure 2.8 are called AND/OR graphs. One of the nodes in this graph, called the start node, corresponds to the original problem description. Those nodes in the graph corresponding to primitive problem descriptions are called terminal nodes.

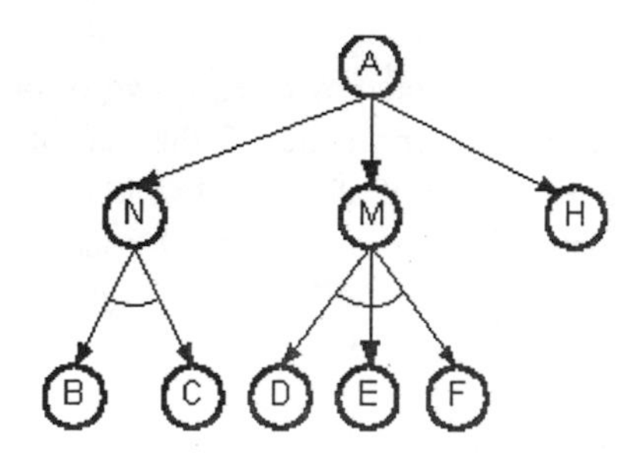

Figure 2.8. Structure of an AND/OR graph.

The object of the search process carried out in an AND/OR graph is to show that the start node is solved.

Definition 2.7. A solved node in an AND/OR graph can be given recursively as follows:

(1) The terminal nodes are solved nodes, since they are associated with primitive problems.

(2) If a nonterminal node has OR successor, then it is a solved node if and only if at least one of its successors is solved.

(3) If a nonterminal node has AND successors, then it is a solved node if and only if all of its successors are solved.

Definition 2.8. A solution graph is a subgraph of solved nodes that demonstrates how the start node is solved according to Definition 2.7.

Definition 2.9. An unsolvable node in AND/OR graph can be defined recursively as follows:

(1) Nonterminal nodes with no successors are unsolvable nodes.

(2) If a nonterminal node has OR successors, then it is unsolvable if and only if all of its successors are unsolvable.

(3) If a nonterminal node has AND successors, then it is unsolvable if and only if at least one of its successor is unsolvable.

Unsolvable nodes are indicated by small circles.

We can summarize the constructing rules for AND/OR graphs as follows:

(1) Every node in AND/OR graphs represents a single problem or a set of problems to be solved. The **start node** contained in AND/OR graph corresponds to the original problem.

(2) The nodes corresponding to primitive problems are called **terminal nodes** that do not have any successor.

(3) For any possible case of applying operators to a problem, the problem would be transformed into a set of subproblems.

(4) For any node that is represented by a set of two or more than two subproblems, the directed links are towards every node of the set of subproblems. The set of subproblems are solvable if and only if all elements within the set are solvable. For this reason, these nodes of subproblems are called AND nodes. In order to distinguish OR nodes with AND nodes, all links of the AND successors nodes with the same parent node are connected by a small portion of an arc.

2.4 Predicate Logic Representation

Although prepositional logic can represent various facts in the real world, its applicability is limited since it is not suitable to represent more complex problems. However, the predicate logic allows us to represent such things and problems that can not be represented by the prepositional logic. More precisely speaking, the first-order predicate calculus is one of formal language, and its basic objective is to make the logic demonstrated symbolically. If we can apply the form of mathematical deduction to prove new statements from a set of given correct statements, then we can assert that the new statements are also correct.

2.4.1 Predicate Calculus

As a formal language, the predicate calculus is defined by its syntax. In order to specify a syntax we must specify the alphabet of symbols to be used in the language and how these symbols are to be put together into legitimate expressions in the language. An important class of expressions of the predicate calculus are called the well-formed formulas (wffs).

The relationship between language and the domain of discourse is specified by the semantics of the language.

1. Syntax and semantics

The basic components of the predicate logic symbols involve predicates, variables, functions, constants, punctuation marks, connectives and quantifiers. The punctuation marks including comma(,), brackets( ) and sometimes { } and [ ] are used to separate the related components to represent the relationships among the domains of discourse. In our discussion, we only use the first-order predicate symbols. From these symbols we can construct various expressions. The classes of interesting expressions involve terms, atomic formulas and well-formed formulas. For example, if we want to represent "There is a robot in Room 1", then a simple atomic formula can be written as follows:

INROOM (Robot, r1)

where Robot and r1 are constant symbols. INROOM is a predicate symbol. In general, atomic formulas consist of predicate symbols and terms. The constant symbols are the simplest terms and are used to represent objects and entities in the domains of discourse. The constant symbols may be either objects, persons, concepts or other things with names. The variable symbols are also terms that allow us to represent something regardless of specific entities. For example,

INROOM (x, y)

where x and y are variable symbols.

Function symbols represent relationships over a domain of discourse. For instance, function symbol "mother" can be used to express a mapping between someone and his (or her) mother. The following atomic formula expresses the relationship of "Lee's mother married to his father"

MARRIED (father(LEE), mother(LEE)).

In predicate calculus, a well-formed formula can be interpreted by specifying relationships among the elements of language over a domain of definition. For every predicate symbol, a relationship over a domain must be specified. For every constant symbol, a related entity over a domain of definition must be specified. For every functional symbol, a related function over a domain of definition must be specified. All these specifications define the syntax of predicate calculus language. In our applications, the predicate calculus is used to represent defined statements over a domain. A defined atomic formula with some interpretation possesses True value if and only if the statement corresponding to this atomic formula is T (true) over a domain of definition. The defined atomic formula possesses False value if and only if the corresponding statement is F (false) over a domain of definition. For some settings of variables, the atomic formula is with T; for other settings of variables, the atomic formula is with F.

2. Connectives and quantifiers

Atomic formulas are basic construction blocks of predicate calculus. We can combine multiple atomic formulas to construct a more complex well-formed formula by use of connectives $\wedge$ (AND), $\vee$(OR), and $\rightarrow$ (implication, in some literature the symbol $\Rightarrow$ is used).

Connective $\wedge$ is used to represent a compound sentence with AND relationship. For example, the sentence "I like music and painting" can be written as

$$\text{LIKE (I, MUSIC)} \wedge \text{LIKE (I, PAINTING)}.$$

A simple sentence can also be written into compound form. For example, the sentence "Mr. Lee lives in a house which is yellow in color" can be represented as

$$\text{LIVES (LEE, HOUSE-1)} \wedge \text{COLOR (HOUSE-1, YELLOW)}$$

where predicate LIVES represents a relationship between the person and the object, while predicate COLOR represents a relationship between the object and its color.

Several formulas connected by connective $\wedge$ form a new formula that is called conjunction. Every component of the conjunction is called conjunctive term. Any conjunction composed of several wffs is a well-formed formula too.

Connective $\vee$ is used to represent a compound sentence with OR relationship. For instance, the sentence "Mr. Smith plays basketball or football" could construct a wff of the form

$$\text{PLAYS (SMITH, BASKETBALL)} \vee \text{PLAYS (SMITH, FOOTBALL)}.$$

Several wffs connected by connective $\vee$ form a new formula called disjunction. Every component of the disjunction is called disjunctive term. Any disjunction composed of several wffs is a wff, too.

The truth values of conjunction and disjunction depend on the truth values of their components. If every conjunctive term has the value T, then its conjunction has the value T also, otherwise, its conjunction has the value F. If there exists at least one value T in disjunctive terms, then its disjunction has the value T also, otherwise, its disjunction has the value F.

Connective $\rightarrow$ is used to represent a compound sentence with relationship "IF-THEN". For example, the sentence "If the book is owned by Ms. Paul, then the color of this book is blue" might form the wff

$$\text{OWNS (PAUL, BOOK-1)} \rightarrow \text{COLOR (BOOK-1, BLUE)}.$$

Another example says "If Bill runs with the fastest speed, then he wins the championship". We would form the wff

$$\text{RUNS (BILL, FASTEST)} \rightarrow \text{WINS (BILL, CHAMPIONSHIP)}.$$

A formula constructed from two formulas by connective $\rightarrow$ is called implication, and the symbol $\rightarrow$ is read implies. The left term of an implication is called an antecedent, and the right term is called a consequent. If both the antecedent and the consequent of an implication are wffs, then the implication is also wff. If the consequent has the value T (regardless of the value of the antecedent), or the antecedent has the value F (regardless of the value of the consequent), then the implication has the value T, otherwise has the value F.

Symbol ~ (in some literature, a symbol Γ) is used to negate the value of a formula, and is read NOT. In other words, the symbol ~ transforms the value from T to F of wff, or vice versa. For example, the sentence "Robot is not in Room 2" can be expressed as

$$\sim \text{INROOM (ROBOT, r2)}.$$

If a formula is with symbol ~ in front of it, then it is called negation. The negation of a wff is also a wff.

Sometimes, an atomic formula, say $P(x)$, has the value T for all variables of x. We can express this property by adding a symbol $(\forall x)$ to the front of $P(x)$, i.e., $(\forall x)P(x)$. If there exists at least one value of x that makes $P(x)$ have the value T, then this property can be expressed by adding a symbol $(\exists x)$ to the front of $P(x)$, i.e., $(\exists x)P(x)$. $\forall$ (meaning for all) sign is called universal quantifier, and $\exists$ (meaning there exists one) sign is called existential quantifier. With these assignments, we can represent the sentence "Every robot has its color of gray" as follows

$$(\forall x)\ [\text{ROBOT}(x) \rightarrow \text{COLOR}(x, \text{GRAY})]$$

and the sentence "There is an object in Room 1" can have the form

$$(\exists x)\ \text{INROOM}\ (x, r1)$$

where x is a quantified variable. If a variable in a wff is quantified, then the quantified wff is also a wff, and this variable is called bound variable, otherwise, it is called a free variable. We are interested when all variables in wff are bound ones. Such a wff is called a sentence.

We have noted that all predicate calculus in this book are the first-order predicate calculus. It is inadmissible to quantify the predicate and function symbols. For instance, in the first-order predicate calculus, a formula such as $(\forall P)\ P(A)$ would not be a wff.

2.4.2 Predicate Formulae

1. Definition of predicate formulae

Definition 2.10. If $P(X_1, X_2, \ldots, X_n)$ is a n-element predicate formula, where P is the n-element predicate, $X_1, X_2, \ldots, X_n$ are objective variables, then the formula is called an atomic formula or atomic predicate formula. A compound predicate formula consisting of atomic formulae and connected by connectives is called molecular formula. Therefore, the predicate formula can be defined recursively as follows:

(1) An atomic formula is a wff.

(2) If A is a wff, so is $(\sim A)$.

(3) If A and B are wffs, then so are $(A \wedge B)$, $(A \vee B)$, $(A \rightarrow B)$ and $(A \leftrightarrow B)$.

(4) If A is a wff, x is any variable, then $(\forall x)$ and $(\exists x)$ are also wffs.

(5) Only such a predicate formula that is derived according to the above four rules is a wff.

2. Properties of wffs

If P and Q are any well-formed formulas, then the compound expressions constructed from P and Q have the truth values given in Table 2.1.

Table 2.1. Truth table of compound expressions

P	$\sim P$	Q	$\sim Q$	$P \vee Q$	$\sim P \vee Q$	$P \wedge Q$	$P \rightarrow Q$	$Q \rightarrow P$	$(P \rightarrow Q) \wedge (Q \rightarrow P)$
T	F	T	F	T	T	T	T	T	T
F	T	T	F	T	T	F	T	F	F
T	F	F	T	T	F	F	F	T	F
F	T	F	T	F	T	F	T	T	T

Definition 2.11. Two wffs are equivalent if they have the same truth value under every interpretation.

The equivalent expressions of wffs can be list in Table 2.2. In this table, P, Q and R denote the wffs not containing the variables, and $P(x)$, $Q(x)$, $P(y)$ and $Q(y)$ denote the wffs containing variables x or y. The equivalent laws or rules can be easily verified with the truth table / Table 2.1 / and simple arguments for the expressions containing quantifiers.

The last two expressions in Table 2.2 show that the bound variable in a quantified expression is a class of virtual element and can be substituted by any other variable that has not been applied in the expression until now.

2.4.3 Substitution and Unification

1. Substitution

In the predicate logic, some reference rules can be applied to certain wffs and sets of wffs to generate new wffs. One important reference rule is called modus ponens that uses the wffs W_1 and $W_1 \rightarrow W_2$ to generate wff W_2. Another reference rule called universal specialization uses the wff $(\forall x)W(x)$ to generate wff $W(A)$, where x is a bound variable, A is a constant. We can apply the combination of modus ponens and universal specialization to produce $W_2(A)$ from the wffs $(\forall X)[W_1(X) \rightarrow W_2(X)]$ and $W_1(A)$. It is necessary to find the substitution of "A for x" that makes $W_1(A)$ and $W_1(X)$ identical.

Table 2.2. Equivalent expressions of wffs

No	Name of rule	Equivalent expression
1	Double Negation	$\sim(\sim P) = P$
2	Conditional Elimination	$P \rightarrow Q = \sim P \vee Q$
3	Bi-Conditional Elimination	$P \leftrightarrow Q = (\sim P \vee Q) \wedge (\sim Q \vee P)$
4	Demorgan's Laws	$\sim(P \vee Q) = (\sim P \wedge \sim Q)$ $\sim(P \wedge Q) = (\sim P \vee \sim Q)$
5	Distributivity	$P \wedge (Q \vee R) = (P \wedge Q) \vee (P \wedge R)$ $P \vee (Q \wedge R) = (P \vee Q) \wedge (P \vee R)$
6	Commutativity	$P \wedge Q = Q \wedge P$; $P \vee Q = Q \vee P$
7	Associativity	$(P \wedge Q) \wedge R = P \wedge (Q \wedge R)$ $(P \vee Q) \vee R = P \vee (Q \vee R)$
8	Reverse Negation	$P \rightarrow Q = (\sim Q \rightarrow \sim P)$
9	Commutativity with Quantifiers	$(\forall x)[P(x) \wedge Q(x)] = (\forall x) P(x) \wedge (\forall x)Q(x)$ $(\exists x)[P(x) \vee Q(x)] = (\exists x) P(x) \vee (\exists x)Q(x)$
10	Negation with Quantifiers	$\sim(\forall x)P(x) = (\exists x) [\sim P(x)]$ $\sim(\exists x)P(x) = (\forall x) [\sim P(x)]$
11	Variable Substitution	$(\forall x)P(x) = (\forall y)P(y)$ $(\exists x)P(x) = (\exists y)P(y)$

The terms of an expression can be variable symbols, constant symbols or functional expressions, the later consisting of function symbols and terms. A substitution instance of an expression is obtained by substituting terms for variables in that expression.

The composition of substitutions is associative, i.e.,

$$\begin{aligned}(Ls1)s2 &= L(s1s2)\\(s1s2)s3 &= s1(s2s3)\end{aligned} \tag{2.4}$$

where L is an expression, and $s1$-$s3$ are instances of expression.

However, substitutions are not commutative in general, i.e.,

$$s1s2 \neq s2s1 \tag{2.5}$$

2. Unification

Finding substitutions of terms for variables to make expressions identical is an important process in artificial intelligence, and is called unification. If a substitution s is applied to every member of a set $\{E_i\}$ of expressions, then the set of substitution instances can be denoted by $\{E_i\}$s. A set $\{E_i\}$ of expressions is unificable if there exists a substitution s such that

$$E_1 s = E_2 s = \ldots = E_n s$$

where s is said to be an unifier of $\{E_i\}$ since its use collapses the set to a singleton.

2.5 Semantic Network Representation

In the previous section, we discussed a binary-predicate version of predicate calculus that can lead to a graphical representation. A collection of predicate calculus expressions of this type can be represented by a graph structure that is called semantic network. Generally speaking, semantic network is a graphical representation of knowledge and consists of nodes and arcs or links (arcs with arrows). The nodes are used to represent entities, concepts and situations, and are connected by arcs that represent the relationships between two nodes.

2.5.1 Representation of Two-element Semantic Network

First of all, let us use semantic network to represent some simpler facts, for example, " All seagulls are birds " . We set two nodes to represent SEAGULL and BIRD respectively, and they are connected by a link ISA (meaning a) as shown in Figure 2.9 (a). If we want to represent a young seagull by its name SEABIRGER, then we simply add a new node SEABIRGER and a link ISA to the original semantic network, see Figure 2.9 (b). People working in the areas of AI and Intelligent Control require the knowledge of taxonomy and the knowledge of properties of related objects. For instance, we can use the semantic network shown in Figure 2.9 (c) to represent the fact that a bird has wings.

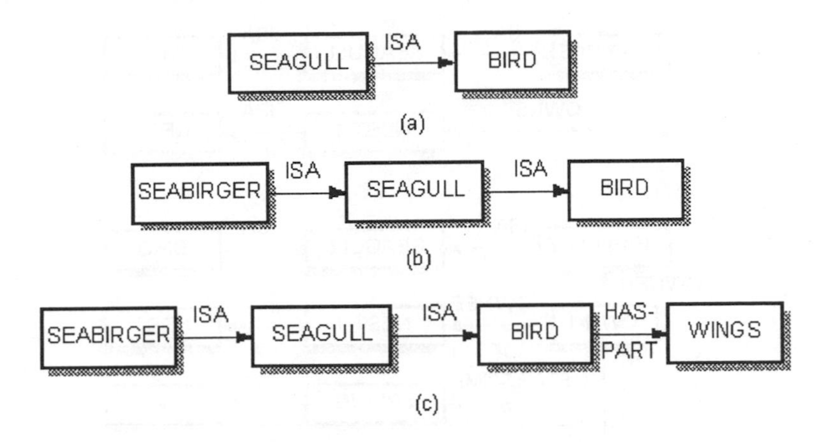

Figure 2.9. Simple examples of semantic network.

Suppose we want to represent a fact "Seabirger owns a nest", then a link of ownership OWNS can be linked to the node of the nest of seabirger, NEST-1 / see Figure 2.10(a) /. NEST-1 is only one of the nests , that is , node NEST represents an example of this kind of an object. If we would like to add the information " Seabirger owns a nest from spring to fall " into the semantic network , it is impossible to realize this representation by current semantic network since ownership is a link in semantic network and can simply represent the relationship between two elements. If the approach of predicate calculus is applied to this example, then it would seem that a four-element predicate calculus is desired . We require a semantic network that is equivalent to the four-element predicate calculus and can represent the relationships among the starting time , ending time , owner and ownee of this occupation .

An approach proposed by Simmons and Slocum can be used to represent both an object (or a set of objects) and a situation (or action) [21]. Every situation node has a set of outcoming arcs (so called instance arc), and is used to explain the situation and the action. For example, the fact "Seabirger owns a nest from spring to fall" can be represented by a semantic network with instance arcs and situation nodes shown in Figure 2.10 (b), where node OWN-1 is a set and shows that seabirger has its own nest. Of course, seabirger can own other things, and OWN-1 is only an instance of ownership. Ownership is a specific " situation ". SEABIRGER is a specific owner of OWN-1, and the NEST-1 is a specific ownee of OWN-1. The period of time owned by seabirger is from spring to fall, and the time instances are SPRING and FALL.

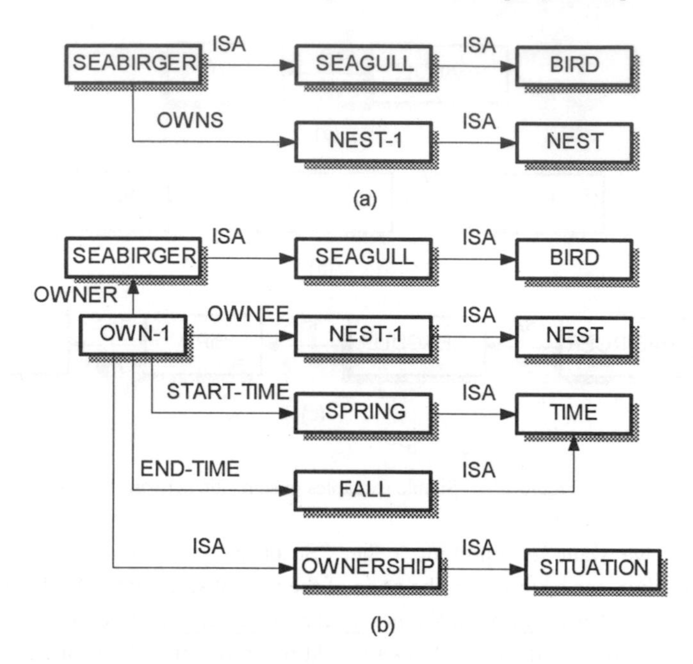

Figure 2.10. Example of semantic network with instance node.

In selecting nodes, the first thing to do is to know what are represented by the nodes: basic objects, concepts or multiple purpose. Otherwise, if the semantic network is only used to represent a specific object or concept, then more semantic networks are desired for representing more instances. This will make the problem complex.

Usually, a related object or concept, or the knowledge for a set of objects or concepts is represented by a semantic network. Otherwise, more number of nets will make the problem more complex. The related issue is to look for basic concepts and some basic arcs. This issue is called the problem of selecting semantic primitive. In doing so, a set of primitives is intended to represent knowledge. These primitives describe basic knowledge. For example, we intend to define a semantic network to represent the concept of a car. We establish node MY CAR to explain the fact " this car is mine ". To explain "My car is colored brown ", a node BROWN is added and linked with node MY CAR by link COLOR. To explain "The covering of my car is steel", a new node STEEL is introduced and is connected by link COVERING. To explain "Car is a traffic tool", node T-TOOL is introduced. To explain "Seat is a part of car", node SEAT is added. In order to explain the occupation of the car's owner, node *X* is set and connected by link

OWNER, then we set node ENGINEER and PERSON to represent the occupation of the owner. As a result, a semantic network related to the car is established in Figure 2.11.

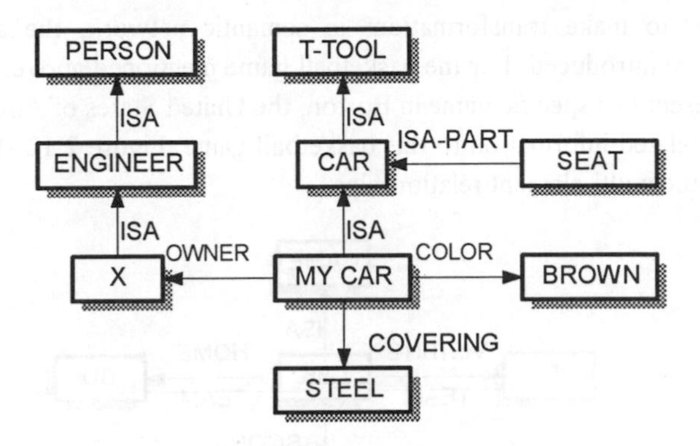

Figure 2.11. Semantic network for a car.

2.5.2 *Representation of Multi-element Semantic Network*

Semantic network is a network structure. Its nodes are connected to each other by links. Essentially, the connection between nodes is a relation of two elements. For example, we can represent "Tom is a worker" as WORKER (TOM) by using predicate calculus. By using semantic network, it can be represented as

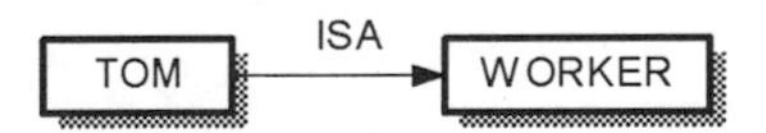

that is equivalent to ISA (TOM, WORKER) in the predicate calculus.

If a represented problem or fact is with multiple relations, for instance, to represent a score of 85:89 in a basketball game between two teams: Boston University (home team) and Tsinghua University (visiting team), the representation is SCORE (BU, TU, (85:89)). There are three terms in the expression. However, semantic network can only represent relationship as a combination of two-element relationships or conjunction of two-element relationships. Specially, the multi-element relationships $(X_1, X_2, \ldots, X_n)$ can always be transformed into

$$R_1(X_{11}, X_{12}) \wedge R_2(X_{21}, X_{22}) \wedge \ldots \wedge R_n(X_{n1}, X_{n2}).$$

For example, the sentence "Three lines a, b and c construct a triangle" can be represented as TRIANGLE (*a*, *b*, *c*), and this three-element relationship can be transformed into a set conjunction of two-element relationships

$$CAT(a,b) \wedge CAT(a,b) \wedge CAT(c,a)$$

where CAT stands for catenate.

In order to make transformations in semantic network, the additional node is necessary to be introduced. For the basketball game mentioned above, we can set a node G96 to represent this specific game in Boston, the United States of America in 1996, then connect the related information to this basketball game. Figure 2.12 shows the semantic network for the multi-element relationships.

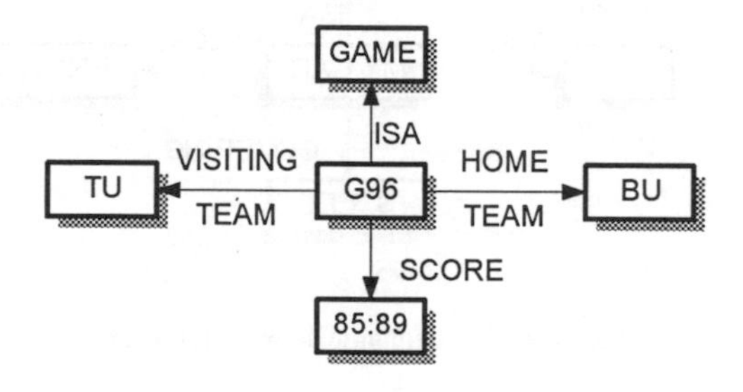

Figure 2.12. Semantic network representation with multi-element relationships.

The connectives and quantification in predicate logic can be represented by the semantic networks [15, 26].

2.6 Summary

Starting with this chapter, we have a discussion in the cognitive process of a human being. This cognitive process is a very complex behavior. There are three theories about the cognitive process, i.e., Symbolism(Logicism), Connectionism and Actionism. Arguments on the cognitive process would continue for several decades or even centuries.

The Symbolism is a more mature cognitive theory. According to this theory, the human form can be seen as a physical symbol system. A hypothesis and three deductions can be drawn. Therefore, we could describe the cognitive activity process of a human being by computer programs. In other words, a computer can imitate the human intelligence.

Knowledge representation is the major topic in this chapter. We have a study of some basic methodologies for knowledge representation in artificial intelligence. These basic methodologies for knowledge representation have played an important role in the development of AI and are still useful for intelligent control.

State space representation can be seen as a basis for the other approaches. Problem reduction representation is more general than the state space one. In other words, the state space representation is a special case of problem reduction when only OR nodes have occurred in its AND/OR graph (tree). The above two approaches and the semantic network representation are all graphic structures while the predicate logic representation is based on the first-order predicate calculus.

State space representation uses a state to represent a problem or knowledge, then applies a suitable operator to the state, and generates a new state. We apply the state space graph to search until a goal state is reached. The sequence of operators that corresponds to a path from the initial state to the goal state is a solution of the related problem.

For problem reduction (AND/OR graph) method, a node is used to a problem, then it decomposes (reduces) the original problem into several subproblems, and into sub-subproblems. We can get the solution by means of searching the AND/OR graph (tree) while transferring the problem into primitive problems. The solution is a solution tree rather than a path from start node to goal node.

In predicate logic representation, we apply the predicate calculus to AI and Intelligent Control Systems, and the wffs are used to represent a problem and the goal. We will learn in the next chapter that the wffs can be transformed into a set of clauses. With the clause set and the negation of the goal set, we can get an empty clause NIL at the root of the proof tree.

The semantic network that we have discussed uses node to represent the object, concept and situation, and uses arc to stand for the relationship between nodes. The answer to the question is in the semantic network. It can be applied to represent multi-element relations, and can be easily extended to represent new and more complex problems.

According to the discussion, the four methods of knowledge representation have closely similar ideas as shown in Table 2.3.

Table 2.3. Relations among four approaches of representation

Method	Problem	Operation	Goal	Solution
State Space	state	operator	goal state	solution path
Problem Reduction	node	arc	terminal nodes	solution tree
Predicate Logic	wffs	rule	root of tree	answer statement
Semantic Network	node	link	network	network

There are some more methods for knowledge representation, such as, frame, script, unit, procedure, blackboard, Petri nets, and some heuristic algorithms etc. The readers who are interested in the mentioned methods might refer to related materials.

References

1. M. A. Arbib, *Brains, Machines and Mathematics* (McGraw-Hill, New York, 1964).
2. A. Barr and E. A. Felgenbaum, *Handbook of AI, Vol.1* (William Kaufmann, New York, 1981).
3. A. Barr and E. A. Felgenbaum, *Handbook of AI, Vol.2* (William Kaufmann, New York, 1982).
4. R. M. Cotterill, ed. *Computer Simulation of Brain Science* (Cambridge University Press, Cambridge, England, 1988).
5. T. C. Eccles, *The Understanding of the Brain* (McGraw-Hill, New York, 1973).
6. R. Hecht-Nielson, "Neurocomputer applications," in: *Neural Computers*, eds. R. Eckmiller and Ch. v. d. Malsburg (Springer-Verlay, Berlin, 1988) 445-453.
7. E. B. Hunt, *Artificial Intelligence* (Academic Press, New York, 1975).
8. K. J. Hunt, D. Sbarbaro, R. Zbikowski and P. J. Gawthrop, "Neural networks for control systems — a survey," *Automatica*, **28**(1992) 1083-1112.
9. IEEE special issue on neural networks, *IEEE Control System Magazine*, **9,10,11** (1988, 1989, 1990).
10. *Int. J. Control*, Special issue on intelligent control, **56** (1992).
11. P. McCorduck, *Machines Who Think* (Freeman, San Francisco, 1979).
12. W. S. McCulloch and W. Pitt, "A logical calculus of ideas immanent in nervous activity," *Bulletin of Mathematical Biophysics*, **9** (1943) 127.
13. R. Morell *et al.*, *Minds, Brains, and Computers: Perspectives in Cognitive Science and Artificial Intelligence* (Ablex, New York, 1992).
14. N. J. Nilsson, *Problem Solving Methods in Artificial Intelligence* (McGraw-Hill, New York, 1971).
15. N. J. Nilsson, *Principles of Artificial Intelligence* (Tioga, Palo Alto, CA, 1980).
16. E. Rich, *Artificial Intelligence* (McGraw-Hill, New York, 1983).
17. E. Rich and K. Knight, *Artificial Intelligence*, Second Edition (McGraw-Hill, New York, 1991).
18. A. C. Sanderson and R. J. Peterka, "Neural modeling and model identification," *CRC Critical Reviews in Biomedical Engineering*, **12** (1985) 237.
19. R. C. Schank and P. G. Childers, *The Cognitive Computer on Language, Learning and Artificial Intelligence* (Addison-Wesley, Mass., 1984).
20. R. Seera and G. Zanarini, *Complex Systems and Cognitive Processes* (Springer-

Verlag, New York, 1990).

21. R. E. Simmons and J. Slocwm, "Generating English discourse from semantic networks," *CACM*, **15** (1972) 891-905.
22. H. A. Simon, *Human Cognition, Information Processing Theory of Thinking* (Science Press, Beijing, 1986).
23. L. Stark, *Neurological Control Systems, Studies in Bioengineering* (Plenum Press, New York, 1968).
24. R. W. Thathcer and E. R. John, *Functional Neuroscience, Vol.1, Foundations of Cognitive Process* (LEA, Hillsdale, NJ, 1977).
25. N. Wiener, *Cybernetics, or Control and Communication in the Animal and the Machine* (MIT Press, Cambridge, MA, 1948).
26. P. H. Winston, *Artificial Intelligence*, Second Edition (Addison-Wesley, Reading, Mass. 1984).
27. P. H. Winston, *Artificial Intelligence*, Third Edition (Addison-Wesley, Reading, Mass., 1992).
28. D. E. Wooldridge, *The Machinery of the Brain* (McGraw-Hill, New York, 1963).

CHAPTER 3
GENERAL INFERENCE PRINCIPLES

We have given an introduction to the knowledge representation for problem-solving in Chapter 2. Our goal is to solve the problem using problem representation. A problem solving procedure, which leads problem representation to problem solution, is called search process or searching. Inference is the base of knowledge application and a typical multistep procedure which deduces a conclusion from premises. Moreover, each step in the procedure must satisfy an acceptable rule which is called inference rule. In an intelligent system, the inference, which is closely related to the methods of knowledge representation, deals with applied specific control strategies to solve the existing problems in terms of acquired knowledge and proven facts. During the inference procedure, not only the conclusion of the solved problem can be deduced, but also new knowledge might come into being.

3.1 Methods and Strategies for Inference

There are various methods and strategies taken during the inference procedure.

3.1.1 Methods of Inference

If the knowledge which the inference is dependent on possesses a certainty factor (CF), the inference procedure leading premises to conclusions is also a procedure of CF transfer. In this sense, the inference modes can be classified as following.

1. Inference by deduction

Assuming that a domain knowledge is represented as a certain causality, the conclusion deduced by inference procedure with logical relations is also positive. That is, the conclusion deduced is acceptable if the premises are firmly believable. If the CF of a premise is P, then the CF of the conclusion is P, too.

The deduction inference can be classified as forward deduction inference, backward deduction inference and hybrid deduction inference which combines the forward and backward deduction together [1, 10]. The forward deduction inference is a condition-driven inference mode in which all the available knowledge (rules) is searched one by one from a set of facts after it starts, and new knowledge is added until terminal condition involving the goal wff is achieved. To the contrary, the backward deduction inference is a goal driven inference mode in which evidences from supposed goal are searched to support the goal. The implications used as rules operate on a global database of goals until a terminal condition which contains facts is achieved.

Unfortunately, there are some limitations to both forward and backward deduction systems. It is a fact that the forward deduction inference is restricted to goal expressions consisting of literal disjunction although it could handle fact expressions with arbitrary form. On the other hand, the backward deduction inference could only handle the goal expressions with arbitrary form but is restricted to fact expressions consisting of literal conjunction. It is possible to combine forward and backward inferences into one, which is called hybrid inference, so that we could take the advantages from each one without any limitations of either mode.

2. Inference by induction

Induction inference can get a conclusion with a lower CF than the CF of premise when it starts from a premise with a certain CF. It means that the CF would change in the course of induction, and only a part of CF of the premise can be transferred to conclusion. Therefore, the CF of conclusion is lower than the CF of premise. The induction inference can infer universal laws from individual instances or phenomena.

There are two methods of induction inference in common use: the simple trial-and-error method and the analog method. In the course of the trial-and-error method, the subclasses of a class of instances and their attributes are searched, and then such a conclusion in which the set of instances that contain this attribute could be reached, if the opposite instance is not found. This method can be written in the following implication form:

$$[p(x_1), p(x_2), \ldots, p(x_n)] \to (\forall x)p(x) \tag{3.1}$$

where $p(x_i)$, i=1, 2, ..., n, refers to the CF of the individual subclasses, $p(x)$ is the CF of the set of instances.

Analog inference, another method of the induction inference, is based on the principle of analog. That is, if two instances or two class of instances have the same condition in many attributes, they should have the same attributes. Let A and B refer to different classes of instances; a_i and b_j are the different attributes of instances, i=1, 2, ... ,n, j=1, 2, ... , m; and $A(a_i)$, $B(a_i)$, $A(b_j)$ and $B(b_j)$ refer to the attributes of A and B respectively. The induction inference can be expressed as follows:

$$[A(a_i) \wedge B(a_i) \vee A(b_j)] \to B(b_j) \tag{3.2}$$

$$i=1, 2, \ldots ,n, \quad j=1, 2, \ldots, m$$

3. Inference under uncertainty

Inference under uncertainty draws a conclusion under uncertain facts, insufficient evidences and incomplete knowledge. Based on condition retrieval and inference, the uncertain knowledge is processed. There are many proposed inference methods under uncertainty in common use: the method of certain factor [2], probability-based subjective Bayes method, Dempster-Shafer evidence theory-based inference [3, 15] and fuzzy subset method, etc. The latter one will be introduced in Section 3.5.

4. Nonmonotonic reasoning

The axioms and/or the rules of inference in nonmonotonic reasoning are extended to make it possible to reason under incomplete information, might change some previous referred knowledge to false during the inference procedure. In other words, nonmonotonic reasoning retains the property that a statement is either considered to be true, false or neither of them at any given moment [13, 14].

The procedure of nonmonotonic reasoning is more complex than the one of monotonic reasoning. As soon as new knowledge is added to the knowledge base, some previously built and stored statements in the knowledge base have to be deleted. Moreover, all the statements which relied on the deleted knowledge have to be deleted or refine themselves by using new evidences. This knowledge deleting process is performed repeatedly until no further knowledge deletion is needed. With regard to the changes in the real world or object, basic properties such as the knowledge in knowledge base and the data in the database will be modified all the time. In this case, the nonmonotonic reasoning is supposed to be applicable.

3.1.2 Control Strategies for Inference

The control strategies of inference have a direct impact on the inference efficiency. As we know, there are some common control strategies including forward reasoning, backward reasoning, bi-directional reasoning, pattern matching and backtracking, metarule-based reasoning and various searching strategies. The first three are described briefly as following.

1. Control strategy with forward reasoning

The forward reasoning is a fact-driven searching in which, starting with all possible initial evidences (facts) of a problem, all the applicable knowledge can be recognized and an applicable set of knowledge called a conflict set is formed through matching premises of every piece of knowledge (rules). As a result, new facts would be reached which would generate a new knowledge to match with the original facts in the knowledge base. This problem-solving process will go on until a goal state is reached. If the conclusion of the problem has been included in the generated facts, this problem is solved. If the conflict solving cannot find the new knowledge available, it can be concluded that the problem is unsolved or there is no new solution. Therefore, forward reasoning is a "matching-conflict-solving-executing" recurrent process.

2. Control strategy with backward reasoning

Backward reasoning is a goal-driven reasoning where it starts with proposed goal hypotheses, then an available knowledge set can be extracted by searching all the knowledge whose conclusions can match their hypotheses. If the available knowledge set is an empty one, the reasoning ends with failure; otherwise, if the knowledge set is a non-empty one, a piece of knowledge can be chosen from the knowledge set to verify the premise. When the verification has succeeded, the conclusion can be found in the set of goals or hypothesis and a certainty factor or a specific value can be computed according to the knowledge and it can be put into context. If a premise condition of a piece of knowledge is invalidated by user or context, its conclusion could not be verified, then the reasoning system returns to the available knowledge set and reselects a new available knowledge.

3. Control strategy with bi-directional reasoning

In order to avoid the high cost of matching procedure and the blind knowledge selection in forward reasoning and to overcome the blindness of goal selection in backward reasoning, the two reasoning strategies are combined and a hybrid control strategy with bi-directional reasoning is proposed. In bi-directional reasoning, the forward reasoning starts with original evidences and the backward reasoning begins from possible conclusion. It is expected that the two procedures could connect with each other in some middle node(s). The connection of the two reasoning procedures shows that the mid-results deduced from forward reasoning could satisfy the middle hypotheses of some conclusions generated from backward reasoning, and the bi-directional reasoning is completed.

3.1.3 Graph Search Strategies

There are various search strategies including blind search and heuristic search. All of these searches are based on a graph-search strategy. We can consider a graph-search control strategy as a means of finding a path in a graph from a node which represents an initial database to another node that satisfies the terminal condition of the production system. The algorithms of graph searching are of special interest. A graph-search control strategy can be viewed as a process where an explicit portion of an implicit graph sufficient to contain a goal node is generated.

The graph-search procedure could be defined and an illustration is given as follows:

Procedure GRAPHSEARCH.

(1) Create a search graph G consisting solely of the starting node s. Put s in a list called OPEN.

(2) Create a list called CLOSED that is initially empty.

(3) LOOP: if OPEN is empty, exit with failure.

(4) Select the first node on OPEN, remove it from OPEN, and put it on CLOSED.

Call this node n.

(5) If n is a goal node, exit successfully with the solution obtained by tracing a path along the pointers (see step 7) from n to s in G.

(6) Expand node n, generating the set M of its successors that are not ancestors of n. Install these members of M as successors of n in G.

(7) Establish a pointer to n from those members of M that were not already in G. Add these members of M to OPEN. For each member of M that was already on OPEN or CLOSED, decide whether or not to redirect its pointer to n. For each member of M already on CLOSED, decide for each of its descendants in G whether or not to redirect its pointer.

(8) Reorder the list OPEN, either according to some arbitrary scheme (for blind search) or according to heuristic merit (for heuristic search).

(9) Go Loop.

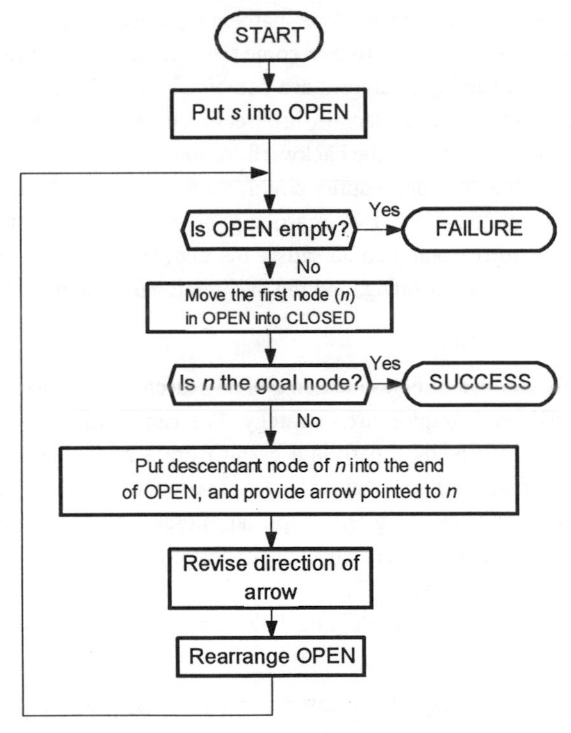

Figure 3.1. Flowchart for procedure GRAPHSEARCH.

3.2 Blind Search

In order to set the nodes on OPEN, if no heuristic information from the problem domain is applied, some arbitrary scheme must be used in step 8 of the algorithm. We call this search procedure uninformed search or blind search. For purposes of comparison, we will give a brief description about the two types of blind search, breadth-first search and depth-first search.

3.2.1 Breadth-first Search

Breadth-first search explores the space in a level-by-level fashion, that is, node expansion in the search tree proceeds along "contours" of equal depth. Only when there are no more nodes to be explored at a given level does the algorithm move onto the next level. The search tree generated by a breadth-first search in the 8-puzzle problem is illustrated in Figure 3.2 [10]. The numbers next to each node indicate the expanding order during the search. The dark path gives a solution.

It is shown that breadth-first search is guaranteed to find a shorter path to a goal node, provided a path exists at all. If no solution path exists, the search exits with failure on finite graphs or will never terminate on infinite graphs.

3.2.2 Depth-first Search

In depth-first search, when a node is examined, all of its off-springs and their descendants are examined before any of its siblings. Depth-first search goes deeper into the search space whenever this is possible. Only when no further descendants of a node can be found are its siblings considered. The nodes in the same depth are examined arbitrarily. This search is chosen for depth-first search because the deepest node in the search tree is put in the first place for expansion. In order to prevent the search from running along a fruitless path, a depth bound is generated. A depth-first search of the graph of Figure 3.3 describes the 8-puzzle problem [10]. The nodes labeled with their corresponding database (states or situations) are numbered in the order of search procedure. In this example, we assume a depth bound five and we can see from Figure 3.3 that a depth-first search goes deeper along one path until it meets the depth bound, then it begins to examine an alternative path in the same depth (level), or less, that differs only in the last step; then those that differ in the last two steps; and so on.

3.3 Heuristic Search

The blind search methods to find a path approaching a goal node are time-consuming and exhaustive. Although the blind search does provide a solution to a path-finding problem, it is often infeasible to apply the search in AI systems because there are too many expanded nodes before a path is found. As far as practical lengths of time and computer storage are concerned, it is vital to find more efficient alternatives to blind search.

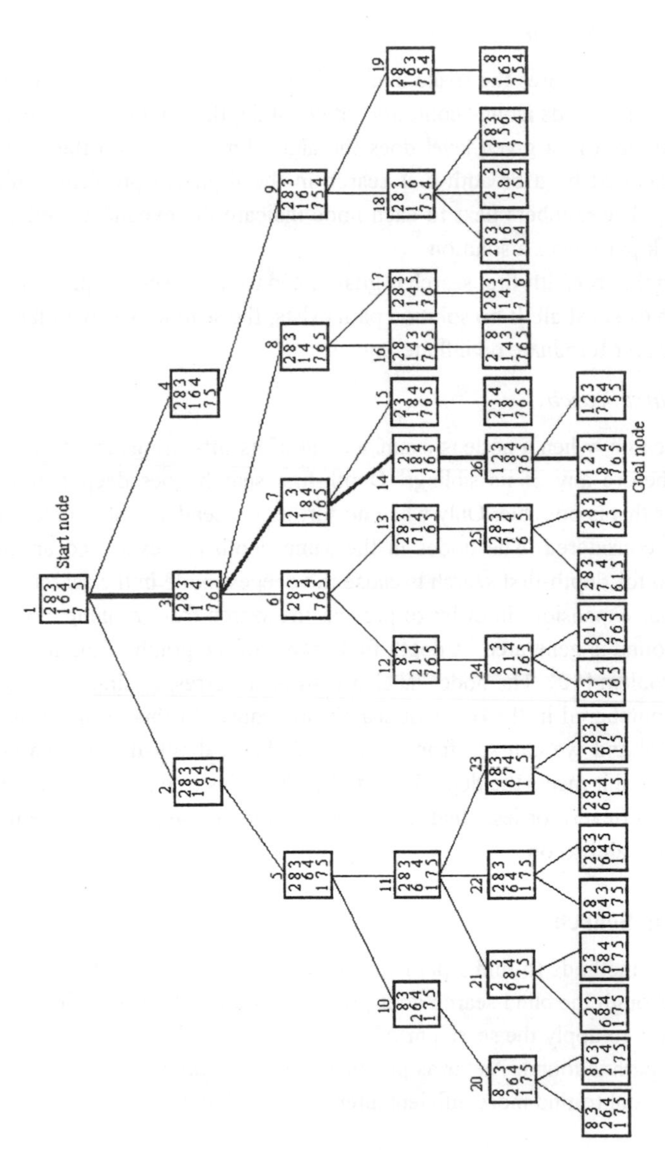

Figure 3.2. A search tree produced by breadth-first search in 8-puzzle problem.

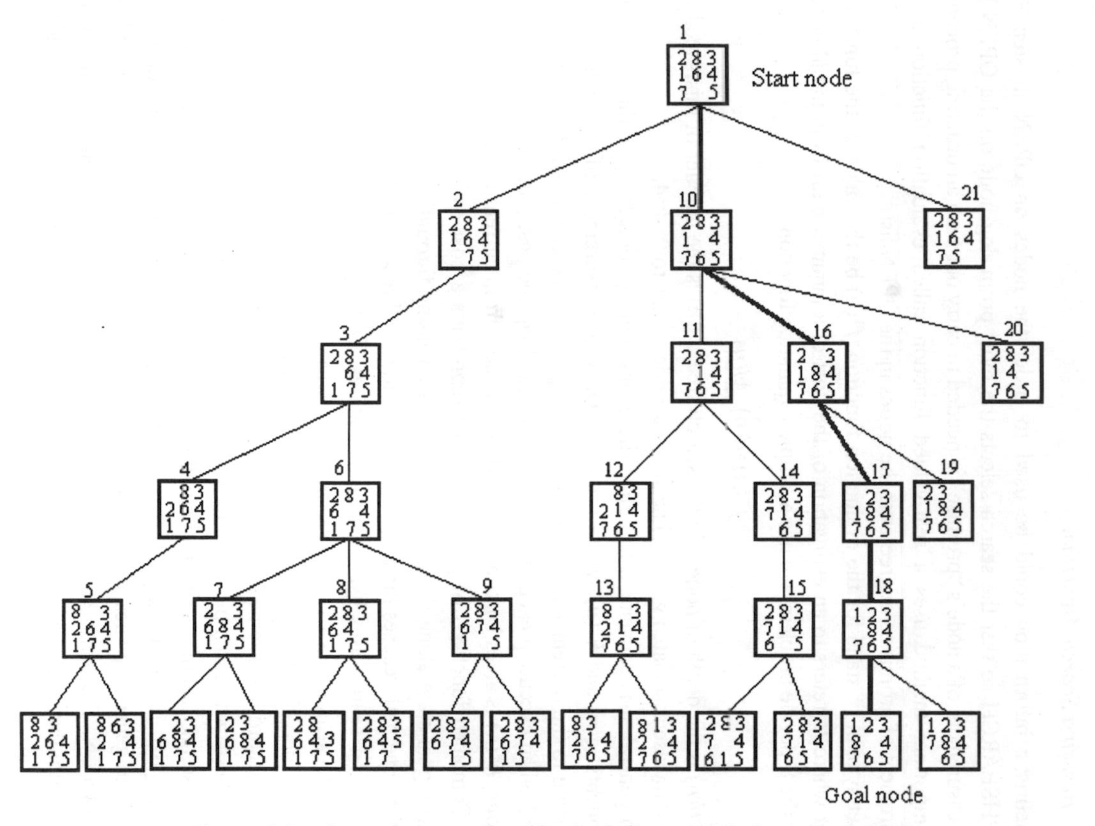

Figure 3.3. A search tree produced by depth-first search in 8-puzzle problem.

For problem-solving, it is possible to use task-dependent information to speed up the search process. This sort of information is called heuristic information. In other words, a heuristic is only an informed guess of the next step to be specified heuristics that reduces the search effort without sacrificing the guarantee of finding the shortest path. For most practical problems, we are interested in minimizing the cost combinations of path and search, that is, control strategy cost and rule application cost.

3.3.1 Best-first Search Algorithm

The heuristic information could be used to order the nodes on OPEN in step 8 of GRAPHSEARCH so that the search selects the most promising node on the OPEN list. The measurement of a node's "promise" is needed to carry out such an ordering procedure. One important method uses a real-valued function called evaluation function as the measurement of promising degree over the nodes in the OPEN list.

Let symbol f^* stand for the evaluation function, $f^*(n)$ be the value of the function at node n where f^* refers to an estimation of the cost of a minimal path from initial node n. As for the 8-puzzle problem, we use a simple evaluation function:

$$f^*(n)=d(n)+M(n)$$

where $d(n)$ is the depth of node n in the search tree, $M(n)$ is the number of misplaced tiles in that database associated with n. The starting node equals to 0+4=4.

In general, it is convenient to define the evaluation function $f^*(n)$ as the sum of two components: $g(n)$ and $h^*(n)$ where function $g(n)$ is a measurement of the cost from the initial state to the current state n. In other words, it is the exact sum of those costs which are paid when executing each of the applied rules along the best path to the node n. The latter one, $h^*(n)$ is an estimation of the additional cost from current node n to a goal state $f^*(n)$. Combined functions of $g(n)$ and $h^*(n)$ represents a cost estimate from the initial state to a goal state along the path while steadily passing through the current node n. If more than one path reaches the node n, the algorithm will record the best one.

Best-first search uses lists to maintain states: OPEN to keep track of the current fringe of the search and CLOSED to record states already visited. An added step in the algorithm orders the states already visited. An added step in the algorithm orders the states on OPEN according to some heuristic estimate of their "closeness" to a goal. Thus, at each iteration, best-first search picks the most promising node that has been generated but not expanded so far. If it meets the goal condition, the algorithm returns the solution path that led to the goal. If it is not a goal, the algorithm applies all matching production rules or operators to generate its descendants and applies the evaluation function to them and adds them to OPEN. The algorithm checks if any of its descendants has been generated earlier to make sure that each node only appears once in the graph. The algorithm expands one node at each step until it generates a goal. The algorithm is given by Nilsson [9]. The evaluation function $f^*(n)$ is defined so that the more promising a node is, the lower the value of the function is. The node selection for expansion is the one at which f^* is minimum.

The Algorithm of Best-first search is as follows:

Best-First-Search algorithm

(1) Start with OPEN containing just the initial state.

(2) Until a goal node is found or there are no nodes left on OPEN do the following:

(i) Pick the best node on OPEN.

(ii) Generate its successors.

(iii) For each successors do the following:

(a) If it has not been generated before, evaluate it, add it to OPEN, and record its parent.

(b) If it has been generated before, change the parent if this new path is better than the previous one. In that case, update the cost of getting to this node and to any successors that this node may already have.

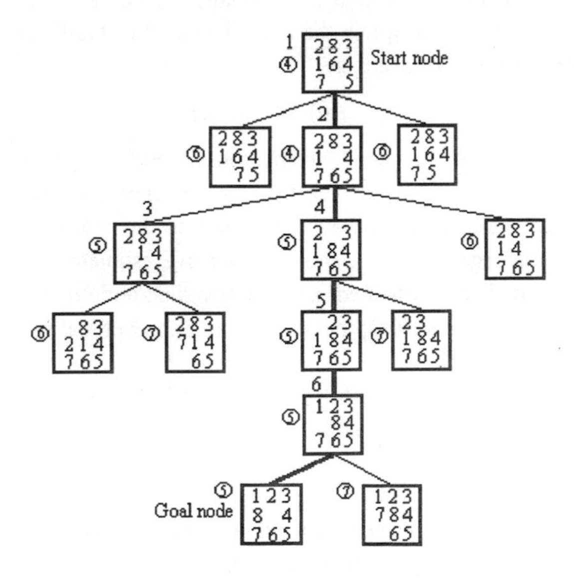

Figure 3.4. A search tree using algorithm of Best-First-Search.

Figure 3.4 shows the results of using evaluation function and the algorithm of Best-first-Search on the 8-puzzle problem. The value of f^* for each node is circled; the uncircled numbers refer to expanding order during the search. It can be seen that the solution path is the same as that generated by uninformed search method even though expanded nodes are reduced sharply by means of evaluation function. It is vital that the choice of evaluation function has great impact on search results.

*3.3.2 An Ordered State-Space Search —Algorithm A**

The A^* algorithm was first presented by Hart, Nilsson, and Raphael [8]. It addressed the problem of finding a minimal cost path joining the start node and a goal node. The A^* algorithm is an ordered State-Space Search whose distinctive feature is its evaluation function f^*. In the usual ordered search, the node chosen for expansion is always the one at which f^* is minimum. f^* is defined as:

$$f^*(n)=f(n)+h^*(n) \tag{3.4}$$

$$g(n)=c(s,n) \tag{3.5}$$

where $g(n)$ is the minimal cost of the path from start node to node n, and $h^*(n)$, estimate of the minimal cost from node n to a goal node. Thus, the value $f^*(n)$ is the estimate of minimal cost of a solution path in which the node is passed through. The actual costs are denoted by $f(n)$, $g(n)$ and $h(n)$ respectively:

$$f(n)=g(n)+h(n). \tag{3.6}$$

The function $g(n)$, applied to a node n being considered for expansion, is calculated as the actual cost from the start node s to n along the cheapest path found so far by information and can be defined in any way appropriate to the problem domain. However, the $h^*(n)$ should be nonnegative, and it should never overestimate the cost of reaching a goal node from the node being evaluated; i.e., for any such node n, it should always hold that $h^*(n)$ is less than or equal to $h(n)$, the actual cost of an optimal path from n to a goal node.

The best-first-search algorithm is a simplification of the algorithm A^*. The latter uses the same f^*, g and h^* functions, as well as the lists OPEN and CLOSED, that we have already described.

Algorithm A^*

(1) Start with OPEN containing only the initial node. Set that node's g value to 0, its h^* value to whatever it is, and its f^* value to $h^* + 0$, or h^*. Set CLOSED to the empty list.

(2) Until a goal node is found, repeat the following procedure: If there are no nodes on OPEN, report failure. Otherwise, pick the node on OPEN with the lowest f^* value. Call it BESTNODE. Remove it from OPEN. Place it on CLOSED. See if BESTNODE is a goal node. If so, exit and report a solution(either BESTNODE if all we want is the node or

the path that has been created between the initial state and BESTNODE if we are interested in the path). Otherwise, generate the successors of BESTNODE but do not set BESTNODE to point to them yet. (First we need to see if any of them has already been generated.) For each such SUCCESSOR, do the following:

(a) Set SUCCESSOR to point back to BESTNODE. These backwards links will make it possible to recover the path once a solution is found.

(b) Compute *g*(SUCCESSOR)=*g*(BESTNODE) + the cost of getting from BESTNODE to SUCCESSOR.

(c) See if SUCCESSOR is the same as any node on OPEN (i.e. it has already been generated but not processed). If so, call that node OLD. Since this node already exists in the graph, we can throw SUCCESSOR away and add OLD to the list of BESTNODE's successors. Now we must decide whether OLD's parent link should be reset to point to BESTNODE. It should be if the path we have just found to SUCCESSOR is cheaper than the current best path to OLD (since SUCCESSOR and OLD are really the same node). So note whether it is cheaper to get OLD via its current parent or to get to SUCCESSOR via BESTNODE by comparing their *g* values. If OLD is cheaper (or just as cheap), then we need do nothing. If SUCCESSOR is cheaper, then reset OLD's parent link to point to BESTNODE, record the new cheaper path in *g*(OLD), and update *f**(OLD).

(d) If SUCCESSOR was not on OPEN, see if it is on CLOSED. If so, call the node on CLOSED OLD and add OLD to the list of BESTNODE's successors. Check to see if the new path or the old path is better just as in step 2(c), and set the parent link and *g* and *f** values appropriately. If we have just found a better path to OLD, we must propagate the improvement to OLD's successors. This is a bit tricky. OLD points to its successors. Each successor in turn points to its successors, and so forth, until each branch terminates with a node that either is still on OPEN or has no successors. So to propagate the new cost downward, do a depth-first traversal of the tree starting at OLD, changing each node's *g* value (and thus also its *f** value), terminating each branch when you reach either a node with no successors or a node to which an equivalent or better path has already been found. This condition is easy to check for. Each node's parent link points back to its best known parent. As we propagate down to a node, see if its parent points to the node we are coming from. If so, continue the propagation. If not, then its *g* value already reflects the better path of which it is a part. So the propagation may stop here. But it is possible that with the new value of *g* being propagated downward, the path we are following may become better than the path through the current parent. So compare the two. If the path through the current parent is better, stop the propagation. If the path we are propagating through now is better, reset the parent and continue propagation.

(e) If SUCCESSOR was not already on either OPEN or CLOSED, then put it on OPEN, and add it to the list of BESTNODE's successors. Compute f^*(SUCCESSOR)=g(SUCCESSOR) + h^*(SUCCESSOR).

3.3.3 *A Heuristic AND/OR Graph Search —Algorithm AO**

The A^* algorithm can only be used to State-Space search. For more complex graphs, AND/OR graphs, a more efficient search method must be established.

We might define a breadth-first search algorithm for searching implicit AND/OR graphs to find a solution using an evaluation-function with a heuristic function component. We can now describe a search procedure using a heuristic function $h^*(n)$ that is an estimate of $h(n)$, the actual cost of an optimal solution graph from node n to a set of terminal nodes.

Just as with GRAPHSEARCH procedure, simplifications in the statement of the procedure are possible if $h^*(n)$ satisfies certain restrictions. A monotone restriction is imposed onto $h^*(n)$, that is, for every connector in the implicit graph directed from node n to successors $n_1, \cdots, n_k$, we assume:

$$h^*(n) \leq c+h(n_1)+, \cdots, + h(n_k) \tag{3.7}$$

where c is the cost of the connector. This restriction is analogous to the monotone restriction on heuristic function of ordinary graphs. If $h^*(n)=0$ for n in the set of terminal nodes, then the monotone restriction implies that h^* is a lower bound on h, i.e., $h^*(n) \leq h(n)$ for all nodes n.

The heuristic search procedure for AND/OR graphs can be started as follows [7]:

Algorithm AO*

(1) Create a search graph G consisting only of the start node s. Associate with node s a cost $q^*(s)=h^*(s)$. If s is a terminal node, label s SOLVED.

(2) Until s is labeled SOLVED, do the following:

(3) Begin.

(4) Compute a partial solution graph G'.

(5) Select any nonterminal leaf node n from G'.

(6) Expand node n generating all of its successors and putting these in G as successors of n. For each successor n_j, not already occurring in G, associate the cost $q^*(n_j)=h^*(n_j)$. Label SOLVED any of these successors that are terminal nodes.

(7) Create a singleton set of nodes, S, containing just node n.

(8) Until S is empty, do the following:

(9) Begin.

(10) Remove a node m from S such that m has no descendants in G occurring in S.

(11) Revise the cost $q^*(m)$for m as follows:

For each connector directed from m to a set of nodes $\{n_{1i},\cdots,n_{ki}\}$, compute $q_i*(m)=c_i+q^*(n_{1i})+,\cdots,+q^*(n_{ki})$. Set $q^*(m)$ to be the minimum over all outgoing connectors of $q_i*(m)$ and mark the connector through which this minimum is achieved, erasing the previous mark if deferent. If all the successor nodes through this connector are labeled SOLVED, then label node m SOLVED.

(12) If m has been marked SOLVED or if the revised cost of m is different from its previous cost, then add all those parents of m to S such that m is one of their successors through a marked connecter.

(13) End.

(14) End.

Algorithm AO* can be summarized as a repetition of the following two major operations. First, a top-down graph-growing operation (steps 4-6) finds the best partial solution graph by tracing down through the marked connectors. These previous computed marks indicate the current best partial solution graph from each node in the search graph. One of the nonterminal leaf nodes of this best partial solution graph is expanded, and a cost is assigned to its successors. The second major operation in AO* is a bottom-up cost-revising, connector-marking, and SOLVED-labeling procedure (step 7-12). Starting with the node just expanded, the procedure revises its cost (using the newly computed costs of its successors) and marks the outgoing connector on the estimated best paths to terminal nodes. This revised cost estimate is propagated upwards in the graph. The revised cost is an updated estimate of the cost of an optimal solution graph from n to a set of terminal nodes. When the AND/OR graph is an AND/OR tree, the bottom-up operation can be simplified somewhat because each node has only one parent.

3.4 Resolution Principle

Another important rule of inference is resolution that can be applied to a certain class of wffs called clauses. A clause is defined as a wff consisting of a disjunction of literal. The resolution process is applied to a pair of parent clauses to produce a derived clause.

3.4.1 Conversion Procedure to Clause Form

Before explaining the resolution process, let us show that any predicate calculus wff can be converted to a set of clauses. We illustrate this conversion process by the following example.

Let us suppose that all Americans who know Jackson either like Lincoln or think that anyone who likes anyone is funny. This statement can be represented in the following wff:

$$(\forall x)[\text{American}(x) \wedge \text{know}(x, \text{Jackson})] \rightarrow$$
$$[\text{like}(x, \text{Lincoln}) \vee (\forall y)(\exists z)(\text{like}(y, z) \rightarrow \text{thinkfunny}(x, y)] \quad (3.8)$$

The conversion process consists of the following steps:

(1) Eliminate implication symbols by making the substation $P \vee Q$ for $\sim P \rightarrow Q$ throughout the wff. For the above example, we have

$$(\forall x) \sim [\text{American}(x) \wedge \text{know}(x, \text{Jackson})] \vee$$
$$[\text{like}(x, \text{Lincoln}) \vee (\forall y) (\sim (\exists z)(\text{like}(y, z)) \vee \text{thinkfunny}(x, y)]$$

(2) Reduce the scopes of negation symbols by repeatedly making use of deMorgan's laws and the equivalencies of quantifier negation. We change the wff in Step 1 to:

$$(\forall x) [\sim \text{American}(x) \vee \sim \text{know}(x, \text{Jackson})] \vee$$
$$\text{like}(x, \text{Lincoln}) \vee ((\forall y)(\forall z)(\sim \text{like}(y, z)) \vee \text{thinkfunny}(x, y))]$$

(3) Standardize variables so that each quantifier binds a unique variable. The variables bound by quantifiers are just dummy variables, therefore, this standardizing process cannot change the true value of the wff. In our example, the variables have been standardized as we represent the statement in wff (3.8), each quantifier has its own unique dummy variable x, y or z.

This step is in preparation for the next.

(4) Move all quantifiers to the left of wff without changing their relative order, convert the wff to prenex normal form. Performing this operation on the formula of Step 2, we have the following:

$$(\forall x)(\forall y)(\forall z) [\sim \text{American}(x) \vee \sim \text{know}(x, \text{Jackson})] \vee$$
$$[\text{like}(x, \text{Lincoln}) \vee (\sim \text{like}(y, z) \vee \text{thinkfunny}(x, y))]$$

A wff in prenex form consists of a string of quantifiers called prefix followed by a quantifier-free formula called a matrix.

(5) Eliminate existential quantifiers. A formula consisting of an existentially quantified variable asserts that there is a value that can be substituted for the variable that makes the formula true. The quantifier can be eliminated by substituting for the variable a reference to a function that produces the desired value. A new function name called Skolem function is used for each such replacement. Sometimes a Skolem function without arguments is called Skolem constant.

For our example, an existential quantifier,($\exists z$), has been converted to universal quantifier by deMorgan laws. Therefore, this step for the example is not necessary. However, for some other problems, this step might be needed to operate.

(6) Remove the prefix and eliminate universal quantifiers. At this point, all remaining variables are universally quantified, and we can assume that any variable it sees is universally quantified.

The formula in Step 4 can be converted into the form:

$$[\sim \text{American}(x) \vee \sim \text{know}(x, \text{Jackson})] \vee$$
$$[\text{like}(x, \text{Lincoln}) \vee (\sim \text{like}(y, z) \vee \text{thinkfunny}(x, y))]$$

(7) Convert the matrix into a conjunction of disjunctions. In our example, since there are no conjunction symbols, it is only necessary to exploit the equivalent laws of wff including the associative and distributive properties and simply remove the parentheses, getting as a result:

$$\sim \text{American}(x) \vee \sim \text{know}(x, \text{Jackson}) \vee$$
$$\text{like}(x, \text{Lincoln}) \vee \sim \text{like}(y, z) \vee \text{thinkfunny}(x, y)$$

(8) Eliminate conjunction symbols and create a separate clause corresponding to each conjunction by replacing the wff $[P(x) \wedge Q(y)]$ with the set of wff's $\{P(x), Q(y)\}$. Any wff consisting solely of a disjunction of literals is called a clause. Again, in our example, the step is not needed.

(9) Rename variables and standardize the variables apart in the set of clauses if necessary. This step means that the variables are renamed so that no two clauses make references to the same variable.

After this entire procedure is applied to a set of wff's, we will get a set of clauses, each of which is a disjunction of literal. For our special example, a single clause consisting of five atomic formulae has been attained. These clauses can generate proofs by the rules of resolution inference that will be discussed in the next subsection.

It is noted that the conversion procedure to clause form consisting of nine steps is a general procedure for problems. However, for special problems, some steps in the procedure might be not necessary, and the procedure can be simplified.

Next we give another example that converts a wff into clause form in nine steps.

$$(\forall x)\{P(x) \to \{(\forall y)[P(y) \to P(f(x, y))]$$
$$\wedge \sim (\forall y)[Q(x, y) \to P(y)]\}\}$$

(1) $$(\forall x)\{\sim P(x) \vee \{(\forall y)[\sim P(y) \vee P(f(x, y))]$$
$$\wedge \sim (\forall y)[\sim Q(x, y) \vee P(y)]\}\}$$

(2) $(\forall x)\{\sim P(x) \vee \{(\forall y)[\sim P(y) \vee P(f(x,y))]$
$\wedge (\exists y)\{\sim [\sim Q(x,y) \vee P(y)]\}\}\}$
$(\forall x)\{\sim P(x) \vee \{(\forall y)[\sim P(y) \vee P(f(x,y))]$
$\wedge (\exists y)[Q(x,y) \wedge \sim P(y)]\}\}$

(3) $(\forall x)\{\sim P(x) \vee \{(\forall y)[\sim P(y) \vee P(f(x,y))]$
$\wedge (\exists w)[Q(x,w) \wedge \sim P(w)]\}\}$

(4) $(\forall x)\{\sim P(x) \vee \{(\forall y)[\sim P(y) \vee P(f(x,y))]$
$\wedge [Q(x,g(x)) \wedge \sim P(g(x))]\}\}$
where $w=g(x)$ is a skolem function.

(5) $$\underbrace{(\forall x)(\forall y)}_{\text{prefix}}\underbrace{\{\sim P(x) \vee \{[\sim P(y) \vee P(f(x,y))] \wedge [Q(x,g(x)) \wedge \sim P(g(x))]\}\}}_{\text{matrix}}$$

(6) $(\forall x)(\forall y)\{[\sim P(x) \vee \sim P(y) \vee P(f(x,y))]$
$\wedge [\sim P(x) \vee Q(x,g(x))] \wedge [\sim P(x) \vee \sim P(g(x))]\}$

(7) $\{[\sim P(x) \vee \sim P(y) \vee P(f(x,y))]$
$\wedge [\sim P(x) \vee Q(x,g(x))] \wedge [\sim P(x) \vee \sim P(g(x))]\}$

(8) $\sim P(x) \vee \sim P(y) \vee P(f(x,y))$
$\sim P(x) \vee Q(x,g(x))$
$\sim P(x) \vee \sim P(g(x))$

(9) Using new variables $x1$, $x2$ and $x3$ to substitute for x in the above three clauses respectively, the variables can be renamed and standardized, then we get the following final set of clauses:

$\sim P(x1) \vee \sim P(y) \vee P(f(x1,y))$
$\sim P(x2) \vee Q(x2,g(x2))$
$\sim P(x3) \vee \sim P(g(x3))$

3.4.2 Rules of Resolution Inference

In a theorem-proving system, resolution is used as a rule, and the set of wffs from which a theorem could be proved is converted first into clauses. It is shown that if the wff P logically follows a set of wffs, S, then it also logically follows the set of clauses obtained by converting the wffs in S to clause form. Therefore, clauses and sets of clauses are completely general forms of expressing wffs.

Let us suppose two ground clauses, $P1 \vee P2 \vee \cdots \vee PN$ and $\sim P1 \vee Q2 \vee \cdots \vee QM$, where a literal, $P1$, in one clause is the negation of one of the literal, $\sim P1$, in the other clause. From the two parent clauses a new clause called the resolvant of the two can be inferred. The resolvant is obtained by taking the disjunction of the two clauses and then eliminating the complementary pair, $P1$ and $\sim P1$. The following are some special examples of derived resolvants from their parent clauses.

(1) Modus Ponens

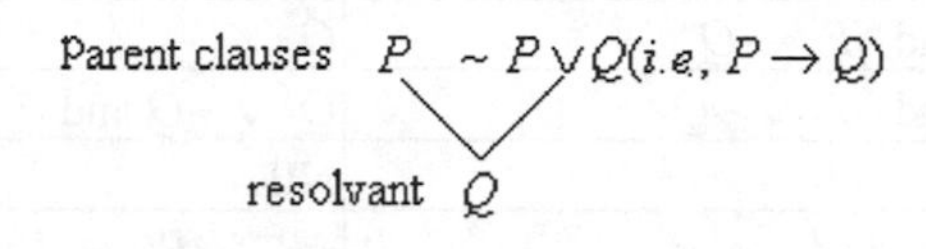

(2) Merge

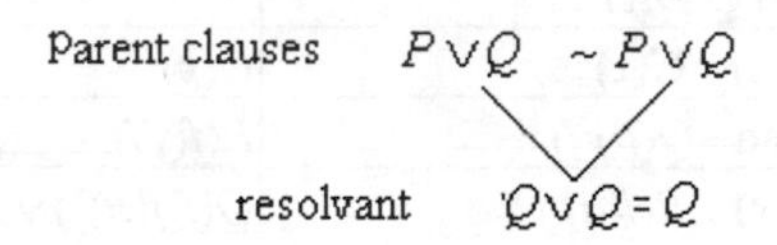

(3) Tautologies

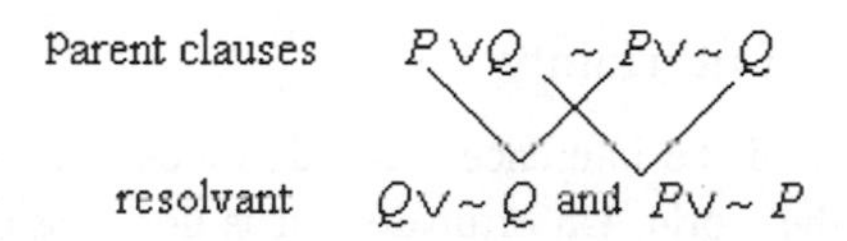

(4) Empty (contradiction)

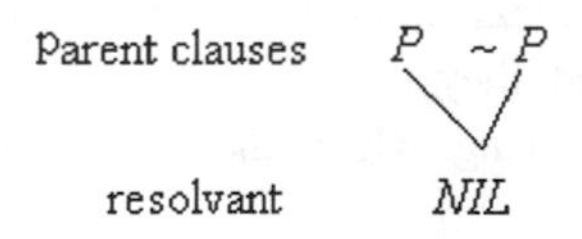

(5) Chaining

Parent clauses $\sim P \vee Q (i.e., P \rightarrow Q)$ $\sim Q \vee R (i.e., Q \rightarrow R)$

resolvant $\sim P \vee R (i.e., P \rightarrow R)$

From the above examples we can see that resolution allows the incorporation of several operations into one simple inference rule.

3.4.3 Resolution for Clauses with Variables

For clauses containing variables, the use of resolution needs to find a substitution that can be applied to the parent clauses so that they contain complementary literals. It is helpful to represent a clause by a set of literals.

The parent clauses and their resolvants of the examples discussed and the same ones with variables can be listed in Table 3.1, where w, x, y and z are bound variables, a is a constant, $f(\_)$ are Skilem functions, σ are substitutions.

Table 3.1. Parent clauses and resolvants

No.	Parent clauses	Resolvants
1	P and $\sim P \vee Q$ (i.e., $P \rightarrow Q$)	Q
2	$P \vee Q$ and $\sim P \vee Q$	Q
3	$P \vee$ Q and $\sim P \vee \sim Q$	$Q \vee \sim Q$ and $P \vee \sim P$
4	$\sim P \vee P$	NIL
5	$\sim P \vee Q$ and $\sim Q \vee R$ (i.e. $P \rightarrow Q$ and $Q \rightarrow R$)	$\sim P \vee R$(i.e., $P \rightarrow R$)
6	$B(x)$ and $\sim B(x) \vee C(x)$	$C(x)$
7	$P(x) \vee Q(x)$ and $\sim Q(f(y))$	$P(f(y))$, $\sigma=\{f(y)/x\}$
8	$P(x,f(y)) \vee Q(x) \vee R(f(a),y)$ and $\sim P(f(f(a)),z) \vee R(z,w)$	$Q(f(f(a))) \vee R(f(a),y) \vee R(f(a),w)$ $\sigma=\{f(f(a))/x$ and $f(y)/z\}$

3.5 Inference under Uncertainty

So far, we have discussed the inference methodologies with a complete, consistent and unchanging model of the world. Unfortunately, it is not possible to build such models in many problem domains. We have to explore new techniques for solving problems with uncertain, incomplete and fuzzy models. A variety of techniques for handling these problems have been proposed, they are:

(1) Nonmonotonic reasoning, which allows statements to be deleted from or added to the database.

(2) Probabilistic reasoning, which makes it possible to represent likely but uncertain inferences.

(3) Fuzzy reasoning, which provides a way of representing fuzzy or continuous properties of objects.

In the following, we are going to describe two approaches to reasoning under uncertainty, nonmonotonic reasoning and statistical reasoning.

3.5.1 Nonmonotonic Reasoning

Conventional reasoning systems, such as first-order predicate logic, are worked with information that has three important properties:

(1) Completeness. All necessary facts are presented or can be derived by the conventional first-order logic rules.

(2) Consistent. The model of the problem world is unchanging.

(3) Monotonicity. If new facts added as and when available are consistent with all the

other facts asserted, then nothing will ever be retracted from the set of facts that are known to be true.

However, if any of these properties is not satisfied, conventional logic-based reasoning systems become inadequate. Nonmonotonic reasoning systems can solve problems without these properties. In nonmonotonic reasoning, the axioms and/or the rules of inference are extended to make it possible to reason with incomplete information, and a statement is either believed to true, to be false, or to be neither.

In order to do nonmonotonic reasoning, several key issues must be addressed.

(1) Extension of the knowledge base. We need an extended system that allows reasoning to be based both on the first-order and second-order predicate logic, and call any inference that depends on the lack of some piece of knowledge as a nonmonotonic inference. Allowing such reasoning has a significant impact on a knowledge base. Because nonmonotonic reasoning systems derive their names from facts that are incomplete knowledge, the knowledge base may not grow monotonically as new assertions are made. Adding a new assertion may invalidate an inference that is dependent on the absence of that assertion. On the other hand, the first-order predicate logic system is monotonic in this respect. As new axioms are asserted, new wffs may become provable, but no old proofs ever become invalid. A knowledge base that can allow inferences based on incomplete knowledge has to be designed.

(2) Updating of the knowledge base.

When a new fact is added to or an old fact is removed from a system, the knowledge base needs to be updated properly. In nonmonotonic reasoning, the addition of a fact can cause previously discovered proofs to be invalid. These proofs and all the conclusions that depend on these proofs have to be found. The usual solution to a problem is to keep track of the proofs, which are often called justifications. This makes it possible to find all the justifications that depend on the absence of the new fact, and those proofs can be marked as invalid. It is also possible to support conventional, monotonic reasoning in the case where axioms must occasionally be retracted to reflect changes in the modeled world.

(3) Methods for resolving conflicts.

As the inferences are based on the lack of knowledge and on its presence, contradictions are likely to occur. Particularly, in nonmonotonic systems, portions of the knowledge base are often locally consistent but globally inconsistent. We require additional methods for solving such conflicts in ways that are most appropriate for the particular problem being solved [7].

There are many logical approaches to nonmonotonic reasoning such as default reasoning, minimalist reasoning, and circumscription reasoning [4, 6, 11, 12].

For understanding how nonmonotonic reasoning works, let us see Figure 3.5 [14]. In Figure 3.5 , the box A corresponds to an original set of wffs, and the large circle contains all the models of A. When some nonmonotonic reasoning capabilities are added to A, a new set of wffs represented by the box B is attained. Wffs B usually contains more information than A does. As a result, fewer models satisfy B than A. The set of models corresponding to B is shown at the lower right of the large circle. Now suppose some new wffs representing new information is added to A, and they are represented by the box C. However, if the set of models corresponding to C is shown in the smaller interior circle, then a difficulty may arise. In order to find a new set of models that satisfy C, we need to accept models that were previously rejected, and the wffs that were responsible for those models being thrown away need to be eliminated. This is the essence of nonmonotonic reasoning.

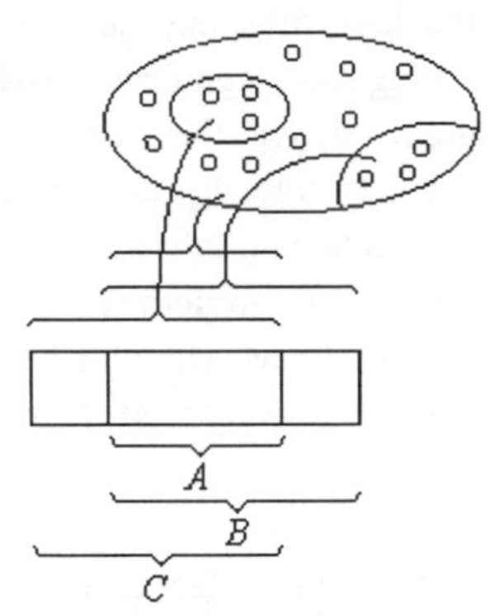

Figure 3.5. Mechanism of nonmonotonic reasoning.

3.5.2 Statistical Reasoning

Up to now, we have discussed several representation methods that can be used to model belief systems in which, at any given point, a particular fact is believed to be true, or false, or not considered one way or the other. However, for some kinds of problem solving, it is useful to be able to describe beliefs that are not certain but for which there is some supporting evidence. Consider one class of such problems in which there is genuine randomness in the world. Playing card games such as bridge and blackjack are good examples of this class. Although in these problems it is not possible to predict the world with certainty, some knowledge about the likelihood of various outcomes is available, and we would like to be able to exploit it.

Another class of such problems could be modeled by using the technique of nonmonotonic reasoning in principle. In these problems, the relevant world is not random, and it behaves "normally" unless there is some kind of exception. Many common sense tasks fall into this category. In this case, statistical reasoning may serve a very useful function as summaries of the world, rather than enumerating all the possible exceptions.

We can use a numerical summary that tells us how often an exception of some sort can be expected to occur.

For statistical reasoning, the statistical measures can be used to describe levels of evidence and belief that augment knowledge representation. In describing this technique, the probability and Bayes' Theorem have to be introduced first, followed by discussing certainty factors (CF) and rule-based systems, then using Bayesian networks as an alternative approach to reduce the complexity of a Bayesian reasoning system and to approximate to the formalism [14]. Readers who are interested in this technique can refer to related papers and books.

3.6 Summary

We have described several principles and techniques for searching and solving problems in this chapter.

There exists various search strategies including blind search, heuristic search, rule-based resolution, and inference under uncertainty etc. Most of these searches are based on a graph-search control strategy.

Blind search, or uninformed search, does not use heuristic information. Breadth-first search and the depth-first search are exhaustive searching methods to find paths to a goal node. In general, blind search provides a solution to the path-finding problem; however, it is often infeasible to use it for controlling intelligent systems because the search expands to too many nodes before a path (solution) is found. When we use heuristic information in searching process, the search cost can be minimized, and the search becomes more efficient. In heuristic search, we have introduced the best-first search algorithm and ordered searches — algorithm A* and algorithm AO* for state-space search and AND/OR graph search respectively.

Resolution is another important search method, a rule-based inference and can be used to a certain class of wffs, the clauses. The resolution process is applied to a pair of parent clauses to produce a derived clause that shows that a problem is solved or a theorem is proved.

As for inference under uncertainty, we have simply discussed its general idea. Two approaches to this reasoning are introduced, they are nonmonotonic reasoning and statistical reasoning.

References

1. A. Barr, and E. A. Feigenbaum eds., *The Handbook of Artificial Intelligence*, Vol.1 (William Kaufmann, Los Altos, CA, 1981).

2. P. R. Cohen, *Heuristic Reasoning About Uncertainty: An Artificial Intelligence Approach* (Pitman Advanced Published Program, 1985).
3. A. P. Dempster, "A generalization of Bayesian inference," *J. the Royal Statistical Society, Series B*, 30 (1968) 205.
4. D. W. Etherington, *Reasoning with Incomplete Information* (Morgan Kaufmann, Los Altos, CA, 1988).
5. K. S. Fu, Z.-X. Cai, and G.-Y. Xu, *Artificial Intelligence and Its Applications* (Tsinghua University Press, Beijing, 1987).
6. M. Genesereth, M. L. Ginsberg, and J. S. Rosenschein, "Cooperation without communication," In *Proc. AAAI-87* (Reprinted in *Readings in Distributed AI*). Eds. A. H. Bond and L. Gasser (Morgan Kaufmann, San Mateo, CA, 1988).
7. M. L. Ginsberg *et al.*, *Readings in Nonmonotonic Reasoning* (Morgan Kaufmann, Los Altos, CA, 1987).
8. P. E. Hart, N. J. Nilsson, and B. Raphael, "A formal basis for the heuristic determination of minimum cost paths," *IEEE Trans. Systems, Science and Cybernetics*, SSC-4 (1968) 100.
9. N. J. Nilsson, *Problem-Solving Methods in Artificial Intelligent* (McGraw-Hill, New York, 1971).
10. N. J. Nilsson, *Principles of Artificial Intelligence* (Tioga Publishing Co., Palo Alto, 1980)
11. R. Reiter, "Nonmonotonic reasoning," *Annual Review of Computer Science*, 1987, 147-186.
12. R. Reiter, "A theory of diagnosis from first principles," *Aritificial Intelligence*, 32 (1987) 57-95.
13. E. Rich, *Artificial Intelligence* (McGraw-Hill, New York, 1983) 173-184.
14. E. Rich, and K. Knight, *Artificial Intelligence, Second Edition* (McGraw-Hill, New York, 1991) 195-205.
15. G. Shafer, *A Mathematical Theory of Evidence* (Princeton University Press, Princeton, NJ, 1976).
16. P. H. Winston, *Artificial Intelligence, Third Edition* (Addison-Wesley, Reading, MA, 1992).

CHAPTER 4
HIERARCHICAL CONTROL SYSTEMS

We have briefly described some structural theories of intelligent control system in Section 4 of Chapter 1. From this chapter, we are going to discuss the theories and applications of intelligent control systems in some detail. The systems involved are the hierarchical intelligent control systems, expert control systems, learning control systems, fuzzy control systems, and neural network based control systems. In each chapter control mechanisms, system architecture and typical examples will be discussed. As a matter of fact, several approaches and mechanisms are usually combined and applied to one practical intelligent control system or plant, and hybrid control system or systems can be established. For convenience of explanation and study, we try to introduce them separately in the following chapters.

Hierarchical intelligent control is one of the earlier and more mature intelligent control theories. The formation of this theory is based on the study of early learning control and the summary of the relationships among artificial intelligence, adaptive control, learning control and self-organizing systems from the point of view of engineering cybernetics [8-10, 31, 34, 46].

Two versions of hierarchical intelligent control will be introduced in this chapter, they are the hierarchical intelligent control based on three control levels and the principle of increasing precision with decreasing intelligence [32-35, 39], and the hybrid multi-layer intelligent control based on knowledge description and mathematical analysis [46, 47]. These two intelligent control theories are closely connected in some aspects. For the convenience of description, we also discuss them in this chapter separately but with different details.

4.1 Hierarchical Intelligent Control System

As an unified methodology of cognitive and control systems, the control intelligence of hierarchical intelligent control approach proposed by Saradis and Meystal, etc. is hierarchically distributed according to the principle of increasing precision with decreasing intelligence (IPDI) that is a very evident principle in all hierarchical management systems [32, 33].

During the past twenty years, several contributors have made considerable efforts to develop the theory of hierarchical intelligent machines and build working models to implement such a theory [1-3, 20, 23, 24, 33-35, 51]. Such intelligent machines have been designed to perform anthropomorphic tasks with minimum interaction with human in potential applications to robotics systems. The theoretical results of such efforts have been obtained in two distinct directions, i.e., the logical based approach and the analytic approach. The former has been described by Nilsson and Fikes [26-28] and its available technology is still in research and development. The latter has reached a more mature level both theoretically and experimentally.

Several new methodologies and technologies, such as Boltzmann machines, neural network and Petri nets etc, have provided new tools for analytic formulation of the theory of intelligent machines. As a result, modifications and refinements of the theory of intelligent machines have been done in the past five years.

The three levels of t he intelligent machine with introduction of neural network to the organization level, Petri nets to the coordination level and entropy measures to the execution level have been presented in detail [38] in this section. Furthermore, a general view of the theory of intelligent machines is given, which is followed by a discussion on the architecture of the hierarchical intelligent control, and then an explanation of the development of the mathematical model for intelligent robotic system.

4.1.1 General Theory of Hierarchical Intelligent Machines

1. General structure

As we have learnt in subsection 1.4.2 and Figure 1.4 the resulting structure of the hierarchical intelligent control system is composed of three basic levels of control. The three interactive levels are shown in Figure 4.1, where f_E^c is the on-line feedback from the execution level to the coordination level; f_c^o is the off-line feedback from the coordination level to the organization level; the input commands are $C = \{c_1, c_2, \ldots, c_m\}$; the classified input commands are $U = \{u_1, u_2, \ldots, u_m\}$.

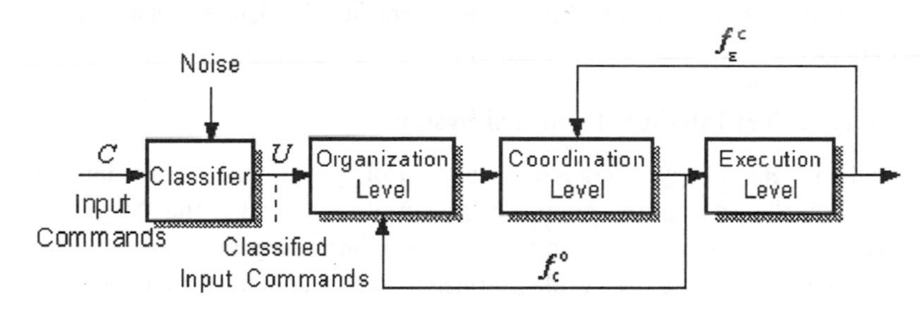

Figure 4.1. Interactive structure of an intelligent machine.

The system is viewed as an entity of transforming a (qualitative) user command to a sequence of mechanical actions. The output of the system is realized via specific commands to a set of actuators. Once the initial user command is received, the system will be brought into action depending on input information from a set of sensors interacted with the environment. These external and internal sensors provide information to monitor the workspace environment (external) and the status of each subsystems (internal), e.g., position, velocity and acceleration in a robot paradigm. The intelligent machine fuses the information and chooses between alternative actions.

The functions and structures of the three level are as follows.

(1) The organization level represents the mastermind of the control system and is designed to organize a sequence of abstract actions, general tasks and rules from a set of primitives stored in a long term memory regardless of the present world model. In other words, it serves as the generator of the rules of an inference engine by processing high levels of information for machine reasoning, planning, decision making, learning (feedback) and memory operations as shown in Figure 4.2 that can be viewed as an architecture of Boltzmann Machine in the organization level. The Boltzmann architecture may be interpreted as a machine that searches for the optimal interconnection of several nodes (neurons) which represent different primitive events to produce a string defining an optimal task [25].

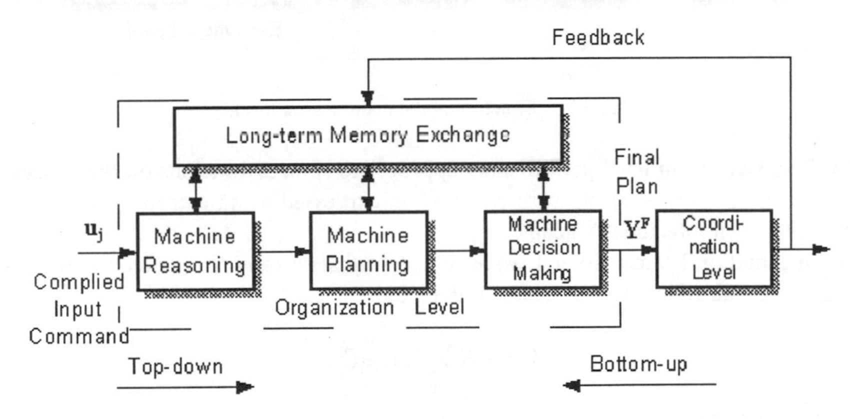

Figure 4.2. Block diagram of the organization level.

(2) The coordination level is an intermediate structure serving as an interface between the organization level and the execution level. It deals with real-time information of the world by generating a proper sequence of subtasks pertinent to the execution of the original command. It involves coordination of decision making and learning on a short term memory, e.g., a buffer. It utilizes linguistic decision schemata with learning capabilities [11, 12], and is assigned subjective probabilities for each action. The respective entropies may be obtained directly from these subjective probabilities. Petri net transducers have been investigated to implement such decision schemata. In addition, Petri nets provide the necessary protocols to communicate among the various coordinators in order to integrate the activities of the machine [41, 49]. The coordination level is composed of a specific number of coordinators of fixed structure, each performing specific functions.

Communication between the coordinators is accomplished through a dispatcher who's variable structure is dictated by the organizer as shown in Figure 4.3.

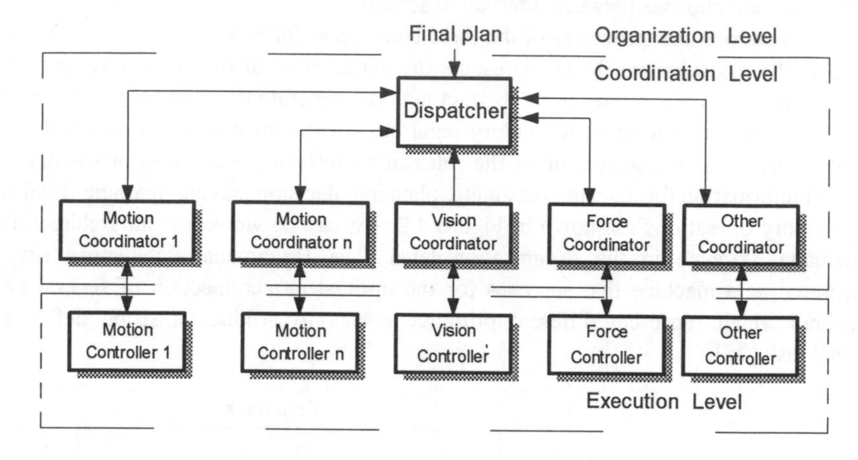

Figure 4.3. Structure of the coordination level.

(3) The Execution level performs the appropriate control functions on the processes involved. Their performance measure can also be expressed as an entropy, thus unifying the functions of an Intelligent Machine.

From definition 1.9 in chapter 1 we have known that entropy H is defined as the mean information contained in an information source, and

$$H = -K\sum_{i=1}^{n} P_i \log P_i \tag{1.7}$$

where H is called **Shannon's negative entropy**.

We have also learnt that entropy function has been introduced into hierarchical intelligent control systems. We can transform equation (1.7) into

$$H = -\int_{\Omega_s} P(s) \log P(s) ds \tag{4.1}$$

where Ω_s is the space of information signals transmitted, while the base of the logarithm may be defined accordingly. The negative entropy is a measure of uncertainty of transmission of information, i.e., uncertainty about the state of the system, obtained as a function of a probability density exponential function of the involved system entropy.

The three level structure of Figure 4.1 has the potential of top-down and bottom-up knowledge (information) processing. The top-down flow of knowledge is related to task generation and decomposition at finer levels of detail. The bottom-up flow of knowledge is related to aggregation of selective information from simple feedback at the lowest layer of the execution level (as defined in classical control theory) to accumulate knowledge at the highest level. This feedback information is essential for learning. Learning in intelligent machines is required and is necessary to choose alternative actions.

A more detailed description of the analytic functions of each level is given in [38].

The functions involved in the upper levels of the intelligent machine are imitating functions of human behavior and may be treated as elements of knowledge based systems. Actually, the functions of planning, decision making, learning, data storage and retrieval, and task coordination may be considered as knowledge handling and management. On the other hand, the problem of a control system can be reformulated by using entropy as a control measure to integrate all the hardware activities associated with the machine at the higher levels. Thus, vision coordination, motion control, path planning and force sensing in a robot paradigm may be integrated into the pertinent functions. Therefore, the flow of knowledge may be considered as the key variable of such a system. Knowledge flow in an intelligent machine represents the following aspects respectively:

(a) Data handling and management.
(b) Planning and decision performed by the central processing unit.
(c) Sensing and data acquisition obtained through peripheral devices.
(d) Formal languages which define the software.

Subjective probabilistic models are assigned to the individual functions whose entropies may be evaluated for every task executed. This provides an analytical measure of the total activity. Since all levels of a hierarchical intelligent control system can be measured by entropies and their rates, the optimal operation of an intelligent machine may be obtained through the solution of mathematical programming problems.

Another development of this theory is the structure of the nested hierarchical systems [21]. Even the hierarchy is not tree-like, still using hierarchy is beneficial since the hierarchy of resolutions (errors per level) helps to increase the effectiveness of the system under limited computing power which is important to mobile systems.

The various aspects of the theory of hierarchically intelligent controls may be summarized as follows: the theory of intelligent controls may be postulated as the mathematical problem of finding the right sequence of decision and control for a system structured according to the principle of increasing precision with decreasing intelligence (constraint) such that it minimizes its total entropy.

2. Definitions

In order to establish an theoretic information framework for the system model, it is necessary to clarify the actual meaning of the terms such as knowledge, intelligence and information, and then, give appropriate definitions for machine knowledge, machine intelligence, machine precision and machine imprecision.

The definitions of knowledge, information and intelligence can be quoted from *Webster's New Collegiate Dictionary* as follows:

Definition 4.1. Knowledge is the fact or condition of knowing something with familiarity gained through experience or association; the range of one's information or understanding.

Definition 4.2. Information is the communication or perception of knowledge; knowledge obtained from investigation, study or instruction; a quantitative measure of the contents of information.

Definition 4.3. Intelligence is the ability to apply knowledge to manipulate one's environment or to think abstractly as measured by objective criteria. Or quoting from *The American Heritage Dictionary of the English Language:* **Intelligence** is the capacity to acquire and apply knowledge.

Definition 4.4. Information theory deals statistically with information, the measurement of its contents in terms of its distinguishing essential characteristics, and with the efficiency of processes of communication between men and machines.

According to the above definitions, it is clear that the term **knowledge** is more general than **information**. Information theory may be utilized for the mathematical interpretation of machine knowledge and intelligence.

The definitions of the corresponding terms for intelligent machines are defined as follows:

Definition 4.5. Machine Knowledge (K) is defined to be the structured information acquired and applied to remove ignorance or uncertainty about a specific task pertaining to the intelligent machine.

Machine knowledge within an intelligent machine includes both a priori and a posterior knowledge. A priori knowledge is acquired through any initial information given by the designer and/or user. A posterior knowledge is acquired and accumulated through learning and experience. Machine knowledge is represented within the Intelligent Machine explicitly and implicitly. Explicit knowledge representation requires data, data bases and knowledge bases, semantic networks, etc. Implicit knowledge representation is embedded within the system control and learning algorithms.

The term machine knowledge indicates the total amount of information that has been accrued in the long term memory of the organizer. It represents memory operations like storage and retrieval of information. Retrieval of information is needed for the internal decision making procedures of the organizer, while storage of information is necessary to upgrade previously stored information.

Definition 4.6. The Rate of Machine Knowledge (R) is the flow of knowledge through an intelligent machine.

Definition 4.7. Machine Intelligence (MI) is the process of analyzing, organizing and converting data into knowledge. Machine intelligence is the set of actions or rules which operates on a data base (DB) of events or activities to produce a flow of knowledge (R).

Definition 4.8. Machine Imprecision is the uncertainty of execution of the various tasks of the intelligent machine.

Definition 4.9. Machine Precision is the complement of machine imprecision and represents the complexity of the process.

We may also derive definition 4.5 to definition 4.7 and the relations analytically to provide an analytical interpretation of the system concepts.

Machine knowledge (K) representing a type of accrued information may be represented as

$$K = -\alpha - \ln p(K) \tag{4.2}$$

where $p(K)$ is the probability density of knowledge and α is an appropriately chosen constant. From equation (4.2), the probability density function $p(K)$ satisfies the expression in agreement with Jaynes principle of maximum entropy [15, 16]:

$$p(K) = e^{-\alpha - K} \; ; \alpha = \ln \int_{\Omega_s} e^{-K} ds \tag{4.3}$$

where Ω_s is the state space of knowledge, and all the other terms have been defined previously.

The rate of knowledge R which is the main variable with an intelligent machine with discrete states, is defined over a fixed interval of time T:

$$R = K/T \, . \tag{4.4}$$

It was intuitively thought that the rate of knowledge (RK) must satisfy the following relation characteristic of the principle of IPDI [36]:

$$(MI) : (DB) \rightarrow (R) \tag{4.5}$$

which is proven as evidenced in definition 4.7.

The relations among the above variables in the forms of **Ohmic Law** are as follows:

$$(MI) \times (DB) = (R) \tag{4.6}$$

$$P = (MI) \times (R) \tag{4.7}$$

$$E = \int (MI) \times (R) dt \tag{4.8}$$

$$K = \int (R) dt \tag{4.9}$$

where P stands for power, E for energy, and K is for knowledge.

An analytic formulation of the above principle derived from simple probabilistic relations among the rate of knowledge, machine intelligence and the data base of knowledge may now be derived and justified. The entropies of the various functions come naturally into the picture as a measure of their activities.

3. Analytic formulation of principle IPDI

In order to mathematically formulate the concepts of knowledge based systems, one must consider the state space of knowledge Ω_s with state s_i, i=1, 2, ... , n. They represent the state of events at the nodes of a network defining the stages of a task to be executed. Then the intelligent machine knowledge between two states is considered as the association of the state s_i with another state s_j and is expressed as:

$$K_{ij} = (1/2) w_{ij} s_i s_j \tag{4.10}$$

where w_{ij} are state transition coefficients, which are zero in case of inactive transmission.

Knowledge at the state s_i is the association of that state with all the other active states s_j and is expressed as:

$$K_i = 1/2 \sum_j w_{ij} s_i s_j \; . \tag{4.11}$$

Finally, the total knowledge of a system is considered as:

$$K = 1/2 \sum_i \sum_j w_{ij} s_i s_j \tag{4.12}$$

and has the form of energy of the underlying events. The rate (flow) of knowledge is the derivative of knowledge and the discrete state space Ω_s is defined respectively as:

$$R_{ij} = K_{ij}/T, \quad R_i = K_i/T, \quad R = K/T \tag{4.13}$$

where T is a fixed time interval.

We have defined the machine knowledge as structured information, so it can be expressed by a probabilistic relation similar to the one given by Shannon, and expressed for each level according to equation 4.2:

$$\ln p(K_i) = -\alpha_i - K_i \tag{4.14}$$

which yields a probability distribution satisfying Jayne's principle of maximum entropy [15, 16].

For $E\{K\}$ = const:

$$p(K_i) = e^{-\alpha_i - K_i} \; , \; e^{\alpha_i} = \sum_i e^{K_i} \tag{4.15}$$

The rate of knowledge is also related probabilistically by considering $K_i = R_i T$:

$$p(K_i) = p(R_i T) = e^{-\alpha_i - TR_i} = e^{\dot{\alpha}_i - \mu_i R_i} \tag{4.16}$$

where α_i and μ_i are appropriate constants, i=2, 3.

The principle of IPDI is expressed probabilistically by

$$PR(MI, DB) = PR(R) \tag{4.17}$$

where PR is the probability, MI is the machine intelligence, and DB is the data base associated with the task to be executed and represents the complexity of the task which is also proportional to the precision of execution. The following relation is obtained by conditioning and taking the natural logarithms:

$$\ln p(MI/DB) + \ln p(DB) = \ln p(R). \tag{4.18}$$

Taking the expected value on both sides one obtains the entropy equation:

$$H(MI/DB) + H(DB) = H(R) \tag{4.19}$$

where $H(x)$ is the entropy associated with x, as previously defined. For a constant rate of knowledge which is expected during the conception and execution of a task, increase of the entropy of DB requires a decrease of the entropy of MI for the particular data base, which manifests the principle. If MI is independent of DB then:

$$H(MI) + H(DB) = H(R). \tag{4.20}$$

This principle is applicable across one level of the hierarchy as well as through the other levels, in which case the flow R is represented throughout the system in an information theoretic manner. The partition law of information rates can be applied naturally to such a system as shown in [43]. A case study demonstrating the validity of the above is given in [19, 38].

4.1.2 Architecture of Hierarchical Intelligent Control Systems

The proposed system architecture partitions the overall system into subsystems and derives a computational module for each subsystem based on the principle of increasing precision with decreasing intelligence. It interconnects all subsystems into a tree like structure producing multilayer hierarchies.

Some special simple architectures are desired to implement the hierarchically intelligent control algorithms. Specific hardware units are built within each of the higher two levels for the upgrade phase of the system. This phase involves the upgrade of the individual and accrued costs, probability distribution functions and knowledge base associated with a particular plan. Upgrade follows plan execution (completion of the decision phase) and is performed in a bottoms-up way: costs and probabilities associated with the lower level are upgraded first, followed by the ones with the higher level. A complete upgrade of the knowledge base is also performed.

The architectural models for the three interactive levels describe the hardware units including special memories to store permanent and temporary information, memories to store probabilities, costs and terminal sets, and other special hardware units for each level separately.

In this subsection, we would like to present the decision phase architectures associated with the two models of the organization level first, followed by the models for the coordination and execution levels.

1. Organization level

Before describing the hardware architecture of the organization level, let us define the functions of this level as follows. Functions are internal operations on the activities and ordered activities. They are defined in their right order within the organizer.

Function 4.1. Machine Reasoning (MR) is the association of the compiled input command, $u_j \in U$, with a set of pertinent activities A_{jm} and production rules and procedures constituting the inference engine (reasoning machine) of the system.

Function 4.2. Machine Planning (MP) is the formulation of complete and compatible ordered activities to execute the requested job. Machine planning contains ordering of the active non repetitive primitive events within the activities, rejection of incompatible ordered activities, insertion of valid sequences of repetitive events in the strings of compatible ordered non repetitive events, checks for completeness and organization of complete plans.

Function 4.3. Machine Decision Making (MDM) is the selection of the complete and compatible ordered activity with the highest associated probability of success.

Function 4.4. Machine Learning and Feedback (MFB) is calculation of the different individual and accrued cost functions associated with the execution of a requested job, and upgrading of the respective probabilities through learning algorithms. The machine feedback function is performed after the completion of the requested job and communication of selective feedback from the lower to the higher levels.

Function 4.5. Machine Memory Exchange (MME) is retrieval, storage and upgrading of information from the long-term memory of the organizer. Retrieval is performed during machine reasoning and machine planning, storage and upgrading after machine decision making and actual execution of the requested job.

The first three functions are associated with top-down local goals while the last two functions are with bottom-up local goals.

The organization level performs general knowledge (information) processing tasks in association with a long term memory. The architectural model of the organizer accommodates fast and reliable operation for each individual function of the organization level. The mechanisms of the memory exchange function associated with the machine reasoning, planning and decision making functions are analyzed. When describing the architectural units, it has been assumed that every string is compatible, therefore, the necessary memory required to store all activities, ordered activities, etc., indicates the worst case scenario.

1A. Probabilistic based architectural model

The architectural models for the machine reasoning function, machine planning function and machine decision making function are shown in Figures 4.4, 4.5 and 4.6 respectively. The functions of the organization level such as machine reasoning, machine planning, and machine decision making will be further explained in the next subsection.

The input to the machine reasoning model is the compiled input command u_j and the corresponding output Z_j^R, the set of the maximum$(2^N - 1)$ pertinent activities A_{jm} (strings X_{jm}) with the corresponding addresses which store the activity probability distribution functions $P(X_{jm}/u_j)$. The reasoning block *RB* contains a maximum of $(2^N - 1)$ strings of binary valued random variables $X_{jm} = (x_1, x_2, \ldots, x_{N-L}, \ldots, x_N)$, $m = 1, 2, \ldots, (2^N - 1)$, stored at a particular order, which represents the activities (A_{jm}) associated with any compiled input command. The corresponding addresses which store the probability distribution functions related to these activities (strings) are transferred from the memory D^R, a part of the long term memory of the organizer. The memory D^R consists of M different memory blocks $D_1, D_2, \ldots, D_M$. One memory block, D_j, is associated with each compiled input command u_j. Each block D_j contains the specific addresses which store the elements of the probability vector S_j that corresponds to u_j. Once the compiled input command u_j has been recognized, realization of the switch S_1 activates (enables) the memory block D_j. Transfer of the data contained in D_j is accomplished via the realization of the switch S_2. The switches are coupled with each other. The contents of the *RB* are transferred to the right most positions of the *PRB* (probabilistic reasoning block) while the corresponding addresses occupy the left most positions. The information stored in D^R is not modifiable by or during the machine reasoning function. Therefore, D^R is considered as permanent memory whose values can not be changed during an iteration cycle, i.e., from the time the user has requested a job until its actual execution. The values of the probability distribution functions associated with the set of pertinent activities stored in the addresses are upgraded only after the completion of the requested job through a specific hardware unit.

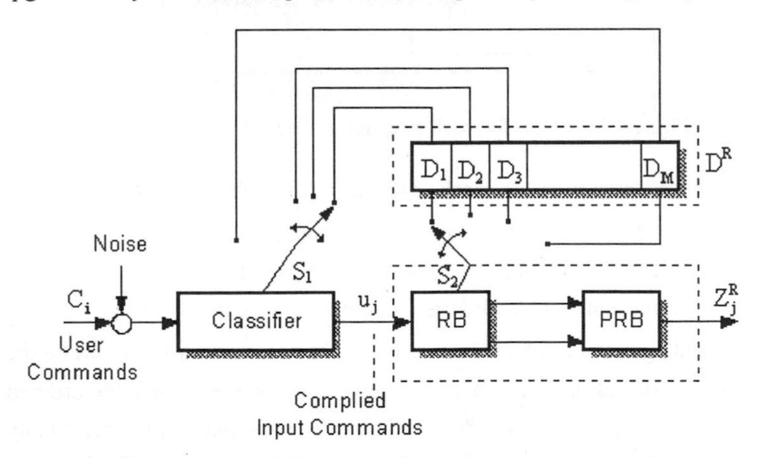

Figure 4.4. Model for machine reasoning function.

In the model of machine planning function, the input is Z_j^R, or equivalently, the corresponding activated memory block. The output from the machine planning model is $Z_j^P = Z_{jmv}$, the set of all complete and compatible plans capable of completing a requested job. This set is a subset of the set of all compatible augmented ordered strings of primitive events formulated during the machine planning function. All compatible ordered activities (strings, valid permutations of primitive events) are stored in the first planning box *PB1*. Every compatible argumented ordered activity (strings which includes the valid sequences of repetitive events) is stored in the second box *PB2*. The compatibility tests are performed within the boxes *CPT1* and *CPT2*. The specific hardware unit for the compatibility tests is omitted here.

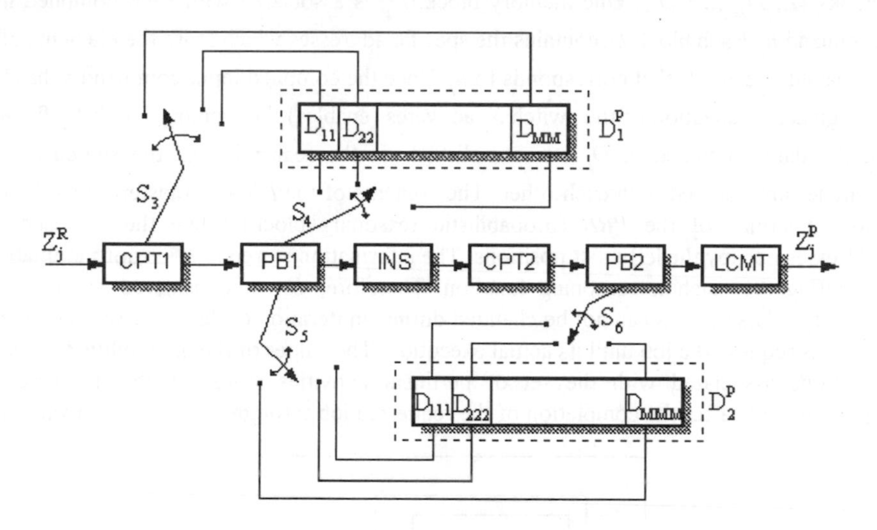

Figure 4.5. Model for machine planning function.

The corresponding addresses of the mask matrices with probabilities $P(M_{jmr}/u_j)$ are transferred from the memory D_1^P via the realization of the coupled switches S_3 and S_4 and are multiplied by the probability distribution functions $P(X_{jm}/u_j)$ to yield the probability distribution functions of the compatible ordered activities (strings) Y_{jmr}. The switch S_3 activates the corresponding memory block D_{jj} (once Z_j^R has been recognized), while the switch S_4 permits the transfer of data. The box *PB1* now contains the compatible ordered activities (strings) with their corresponding addresses containing the probability distribution functions $P(Y_{jmr}/u_j)$. The insertion of the valid sequences of repetitive primitive events is performed in the box *INS* as previously explained while the compatible augmented ordered activities are stored in the second planning box *PB2*. The corresponding probabilities are transferred from D_2^P. Activation and transfer of data from the appropriate memory block D_{iij} is accomplished via the realization of the two coupled switches S_5 and S_6 in a way similar to the one described before. The box *PB2* contains now the augmented compatible ordered activities (strings) with their corresponding addresses containing the pertinent probability distribution functions. The information stored in D_1^P and D_2^P is not modifiable also during the machine planning or by machine planning function. This information is considered permanent within an iteration cycle. The values of the probability distribution functions stored in the addresses are upgraded after the completion of the requested job. The output from the machine planning function Z_j^P is the set of all complete and compatible ordered activities (strings) that may execute the requested job. The completeness test is performed within *LCMT*. This test accepts every meaningful syntactically correct plan as explained through the operational procedures. Every compatible augmented ordered activity (string) is checked. If a compatible augmented ordered activity (string) does not begin with a repetitive primitive event and end with a non repetitive one and does not follow the limitations imposed by the operational procedures, then it is rejected as incomplete. Hence, every complete plan is compatible but every compatible plan is not necessarily complete.

In the model of machine decision making function, all complete and compatible plans are stored in *MDMB* (machine decision making box) and checked by pairs to find the most probable one. The most probable plan is stored in *RR*. If during the check, a complete plan with higher probability than the one already stored in *RR* is found, it is transferred to *RR* while the already stored one is discarded. Once the check is over, the contents of *RR* indicate the most probable complete and compatible plan to execute the requested job.

Every complete and compatible plan is also stored in a particular part of the long term memory of the organizer D_{ss}. This memory contains every complete and compatible plan related to every compiled input command as shown in Figure 4.6. The idea of this particular memory is very important: it represents the situation of a well trained system (under the assumption that no unpredictable events may occur). An intelligent robotic system which has reached this mode of operation associates immediately after the recognition of the compiled input command the most probable of the plans (more than one available) stored in D_{ss}, without going through every single individual function.

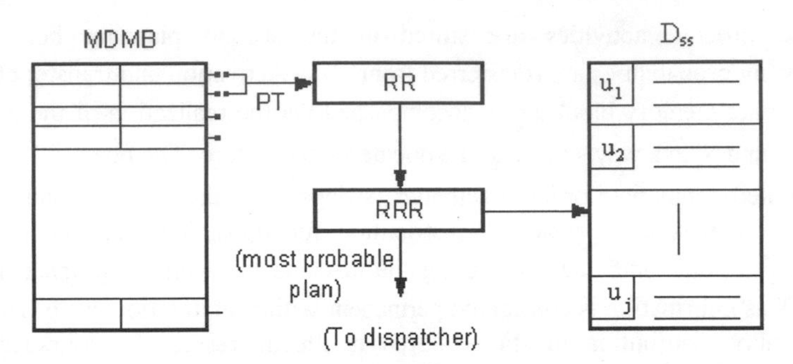

Figure 4.6. Model for machine decision making function.

1B. Expert system based architectural model

The architectural model for the expert system cell configuration of the organization level follows, in principle, steps similar to the probabilistic model. This model represents a valid hardware alternate, suitable to the machine intelligent functions.

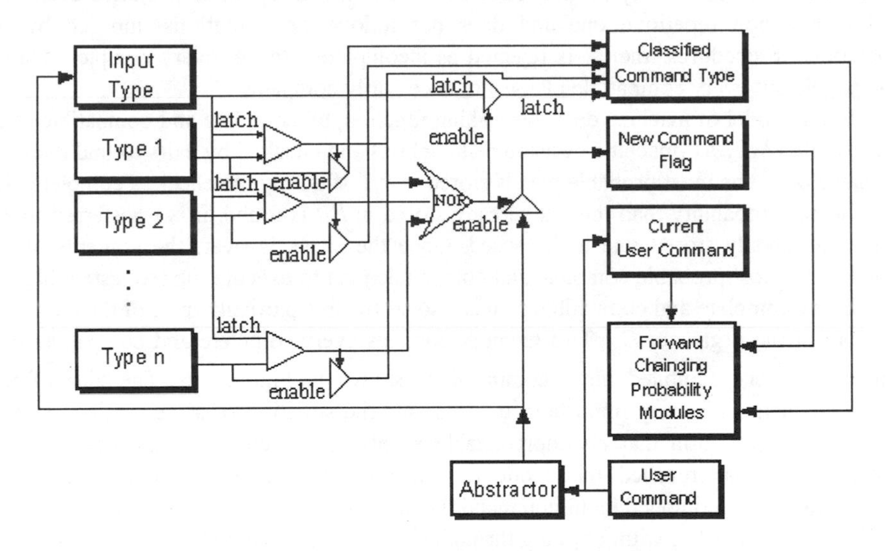

Figure 4.7. Hareware implementation of classifier model.
(From Valavanis and Saradis, 1992, Figure 5.4)

The classification technique, which is the same for both models, may be implemented in hardware using an abstractor, a NOR gate, a few latches and resisters as shown in Figure 4.7. The user command gets filtered through the abstractor and is classified into a specific command type which is then simultaneously compared with all existing command

types residing in the knowledge base. When determined whether or not the user command is of a new command type, a flag is set accordingly and relevant resisters are loaded for the next processing stage.

For further understanding of both the probabilistic based and the expert system cell based architectural models, the reader may read [44], where some hardware implementations for the forward chaining method, the probabilistic method and the permutator hardware with biasing and feedback have been shown and explained.

2. Coordination level

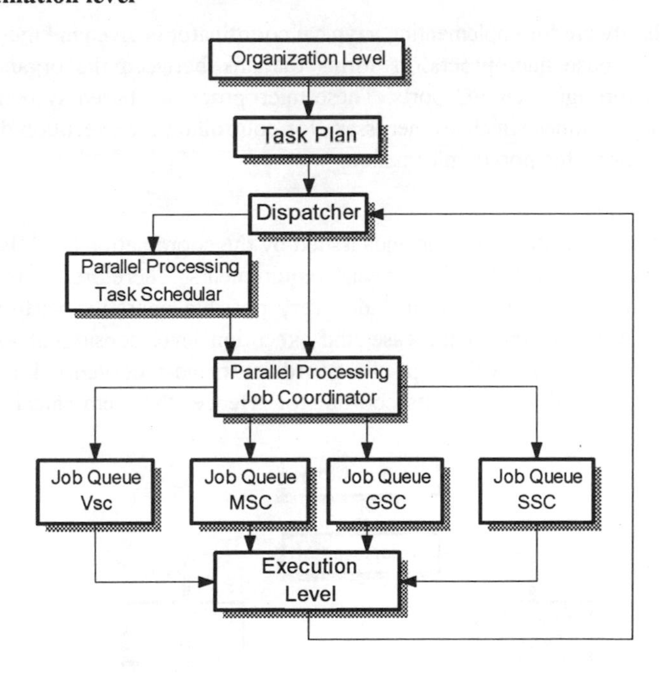

Figure 4.8. Block diagram of coordination level.

The coordination level consists of different coordinators, with dedicated microprocessors for each of them. Once the best task sequence (complete plan) generated and chosen by the organizer, is sent to the coordination level, all the necessary details are provided, and the chosen plan is successfully executed. Parallel tasks are performed where possible to achieve minimum time performance. An alternative block diagram for the coordination level is shown in Figure 4.8. The architecture is also inherently horizontal in which data is shared between the coordinators through the dispatcher. However, this does not violate the overall tree like structure. In order to execute tasks simultaneously, the parallel processing task scheduler (PPTS) assigns time stamps to each process and synchronizes these operations with the parallel processing job coordinator (PPJC).

The purpose of the PPJC is to control the flow of the task execution based upon the time stamp given to each of these subprocesses and the availability of the different execution devices. The assignment of a time stamp is based on the sequence of the execution of task and the dependability of one task to the next. To determine whether two tasks in sequence could be executed in parallel, The PPTS searches the parallel execution table (PET) for that particular ordered pair. If the pair exists in this table, then these tasks are assigned with the same time stamp indicating that they could be executed simultaneously.

The main hardware for implementing a typical coordinator is given in Figure 4.9. Each of the special purpose microprocrssors forms the link between the organization and execution levels through their I/O ports. These microprocessor based systems use local ROMs for storing routines which are necessary for controlling the execution devices, and also RAMs for storing temporary information.

3. Execution level

The execution level executes the commands issued by the coordination level. Each control problem is analyzed in the light of its special requirements. Therefore, there is not one general architectural model that can include every possible operation performed at the execution level. Although this is the case, the execution level consists of a number of devices that each associates with a specific coordinator and vice versa. Each device is accessed via a command issued by its coordinator. Hence, the hierarchical structure is preserved.

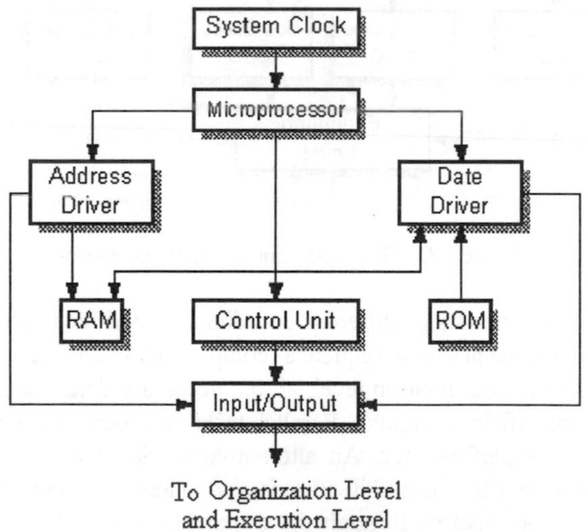

Figure 4.9. Hardware configuration of a coordinator.

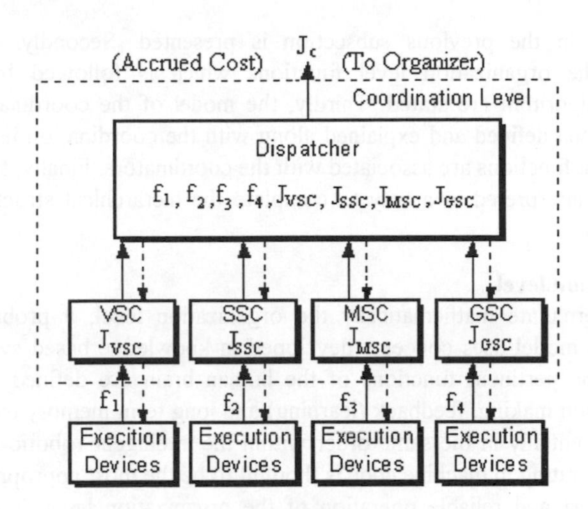

Figure 4.10. Architectural model for coordinators and executors.
(From Figure 5.13, Valavanis and Saridis, 1992)

In the case of an intelligent robotic system the execution devices at the execution level are the components of the vision system, the different sensors and the manipulators with their corresponding gripers. Consider the example for motion system coordinator. When a command is issued for a specific motion, the robotic arm controller applies direct inputs to each of the joint actuators to move the arm to the desired final position.

The on-line feedback mechanism (from the execution level to the coordination level) is activated during the execution of the requested job. A block diagram of this mechanism is shown in Figure 4.10. Solid lines represent the online feedback information from the execution devices to the different coordinators. Individual and accrued costs are calculated within each coordinator and are transferred to the dispatcher of the coordination level where the overall accrued cost associated with the coordination level is calculated and communicated back to the organizer after the execution of the job. The dotted lines illustrate how the information from the dispatcher is used by the different coordinators and their execution devices for the completion of the job.

4.1.3 Mathematical Model for Intelligent Robotic Systems

In this subsection, we would like to derive the mathematical models suitable to the hierarchical structure of intelligent robotic systems operating under the constraints of hierarchically intelligent control systems, define the levels' individual functions and explain the flow of knowledge (information) within complex machines and systems.

The subsection is arranged as follows: first, the probabilistic model of the organization level in which the organizer's individual functions have been defined and interpreted mathematically in the previous subsection is presented. Secondly, operational rules imposed on the organization level functions which is followed by a step-by-step mathematical algorithm are stated. Thirdly, the model of the coordination level and its functions are then defined and explained along with the coordination level algorithm and appropriate cost functions are associated with the coordinators. Finally, the execution level is defined and interpreted in a way to complete the hierarchical structure of intelligent robotic systems.

1. Organization level

In order to formulate mathematically the organization level, a probabilistic model is proposed. This model uses concepts developed in knowledge based systems of artificial intelligence. The pertinent functions of the human brain are defined to be: reasoning, planning, decision making, feedback (learning) and long term memory exchange. They are performed sequentially in the same order within the intelligent robotic system. A special architecture operating in machine code is thought to be the most appropriate structure that can generate fast and reliable operation of the organization level. This is obtained by defining the various operations in a mathematical way and then assigning a probabilistic structure to organize the appropriate events and actions for execution.

Some prior information (knowledge) including information related to availability of resources, tools, materials, timing requirement, etc., are needed and provided in terms of : (1) a set of definitions establishing the organizer workspace; (2) a set of operational procedures governing the organizer functions.

1A. Definitions

For every system with the proposed three level structure, the following are defined.

Definition 4.10. The set of typed in user commands:

$$C = \{c_1, c_2, \dots, c_M\} \tag{4.21}$$

with *a priori* probabilities $p(c_n)$, $n = 1, 2, \dots, M$, M is fixed and finite, sent to the system through any remote or noncommunication channel.

Definition 4.11. The set of classified compiled input commands:

$$U = \{u_1, u_2, \dots, u_M\} \tag{4.22}$$

with associated probabilities $p(u_j/c_n)$, $j = 1,2, \dots, M$, M is fixed and finite, which are the actual inputs to the organization level of the system.

Definition 4.12. The task domain of the system is defined with the set

$$E_t = \{e_1, e_2, \dots, e_N\} \tag{4.23}$$

of primitive events (actions). These primitives are basically divided into the disjoint subsets of non repetitive and repetitive events, defined as E_{nr} and E_r respectively. Without loss of generality, consider $(N - L)$ non repetitive events and L repetitive ones within E_t. Given the task domain and the specific application problem, all subsets shown in Table 4.1 must be defined.

Table 4.1. Notations of set of events

Notation	Meaning
E_t	set of primitive events in the task domain
E_{nr}	set of non repetitive events
E_r	set of repetitive primitive events
E_c	set of active non repetitive events associated with a u_j
E_{snr}	set of non repetitive events allowed to start an activity associated with a u_j
E_{sr}	set of repetitive events allowed to start a plan
E_{end}	set of non repetitive primitive events allowed to end a plan
E_{cr}	list of sets of critical non repetitive events associated with a u_j
E_{comp}	list of compatible pairs of primitive events
E_{unw}	list of unwanted precedence pairs of primitive events
E_{ord}	list of valid repetitive orderings

The **set theoretic relations** are as follows:

$$E_{snr} \subset E_c \subset E_{nr} \tag{4.24}$$

$$E_{end} \subset E_c \subset E_{nr} \tag{4.25}$$

$$E_{sr} \subset E_r \tag{4.26}$$

$$E_{nr} \cup E_r \cup E_c \cup E_{snr} \cup E_{sr} \cup E_{end} = E_t \tag{4.27}$$

$$E_{comp} \subset (E_t \times E_t) \tag{4.28}$$

$$E_{unw} \subset (E_t \times E_t) \tag{4.29}$$

Definition 4.13. The binary valued random variable x_i, $i = 1,2, \ldots, N$, is associated with each e_i indicating whether e_i is active ($x_i = 1$) or inactive ($x_i = 0$) within a plan given a specific u_j, with corresponding probabilities $p(x_i = 1/u_j) = p_{ij}$ and $p(x_i = 0/u_j) = 1 - p_{ij}$, respectively.

Definition 4.14. The set of activities A_{jm} (groups of primitive events concatenated together to form a complex task) associated with a specific u_j, and they are represented in terms of the binary string X_{jm} indicating which events/actions are active and/or inactive within an activity A_{jm}. The indices indicate the mth activity-string associated with the jth compiled command. Since the x_i's are considered binary, the initial maximum number of activities A_{jm} (strings X_{jm}) associated with one u_j is $(2^N - 1)$. The corresponding probability of an activity-string is defined as

$$p(X_{jm} / U_j) = \{\prod_{i=1}^{N} p(x_j / u_j)\}_m$$

$$m = 1, 2, \ldots, (2^N - 1). \tag{4.30}$$

Definition 4.15. The set of complete plans B_{jmr} (augmented strings Y_{jmr}) associated with each u_j obtained by inserting strings of valid active repetitive events within appropriate positions of the valid activities A_{jm} (strings X_{jm}) where r indicates the rth string of valid active repetitive events. The corresponding probability of an augmented activity-string is defined as

$$P(Y_{jmr}/u_j) = p(M_{jmr}/X_{jm})P(X_{jm}/u_j) \tag{4.31}$$

$$\sum_m \sum_r p(M_{jmr} / X_{jm}) = 1 \text{ ,for all } j \tag{4.32}$$

where M_{jmr} is the rth permutation (mask) matrix, or augmented mask matrix, associated with X_{jm}, $p(M_{jmr}/X_{jm})$ denotes the probability of the rth string of valid active repetitive events associated with X_{jm} and defined via the permutation matrix M_{jmr} (or augmented permutation matrix), and r is defined accordingly based on the specific application problem.

Definition 4.16. The set of complete plans Z_{jmr} is associated with each u_j which is a subset of B_{jmr} and whose members are augmented activities satisfying some criteria for completion.

1B. Procedures

Any hierarchical system is viewed as an entity of transforming user requested jobs into a sequence of specific actions.

Once a user command is issued, it is preprocessed for classification and translation in machine code that is easily understood by the organization level of the system. Since these commands are general and sometimes fuzzy, it is expected that the system itself will develop suitable plan scenarios from its task domain to execute the user requested job. The formulation and organization of these plans requires the previously defined five functions to be performed within the organizer.

Definition 4.17. A plan is a sequence of relevant and compatible non repetitive and repetitive active events-strings from the task domain E_t, appropriately arranged for sequential execution. The non repetitive events provide the framework for the plan(s) while the repetitive ones are necessary for completeness.

The operational procedures of the organizer are described as follows:

Rule 4.1. The user commands c_n, $n = 1,2, \ldots, M$, are considered to be independent of each other. The classified compiled input commands to the organization level u_j, $j = 1, 2, \ldots, M$, are also considered to be independent of each other. However, due to classification, u_j depends on c_n. Since u_j and the primitive events e_i are assumed to be probabilistically independent, different activities are independent of each other.

Rule 4.2. A string of events satisfies the compatibility test if for every event e_i in the string, e_i can follow the event immediately to its left in the string, and e_i can come before the event immediately to its right. A exception to this rule is the first and the last events of the string, in which case additional information has to be supplied as to know whether a particular repetitive event can start the string or a non repetitive event can end a string.

Rule 4.3. A string of non repetitive events satisfies the precedence test if for every event e_i in the string, e_i can precede all other events that follow it, though not necessarily immediately. Event e_i can precede event e_z if the pair $(e_i e_z)$ is not in the list of unwanted precedence pairs. An unwanted precedence pair $(e_z e_i)$ implies that e_z cannot precede e_i.

Rule 4.4. The relative order of non repetitive events is crucial for the formulation of activities. If there is a maximum number of $(N - L)$ non repetitive events associated with an input command and q $(q \leq N - L)$ is the minimum number of critical non repetitive events related to the input command and determined during the initialization of the system, then the activities associated with the particular command may be generated by considering all the valid permutations of these events taken l at a time, $l = 1, 2, \ldots, (N - L)$. It follows an activity that has at least q but not more than $(N - L)$ non repetitive events. To specify q, assume there are several basic operations to execute a command, with critical (active non repetitive) events:

$$\{e_1, e_2, \ldots, e_{s1}\}, \{e_3, e_5, \ldots, e_{s2}\}, \ldots, \{e_7, e_9, \ldots, e_{sv}\} \tag{4.33}$$

corresponding to each basic operation. Then $q = \min\{s1, s2, \ldots, sv\}$.

Rule 4.5. Let L be the number of repetitive events in the task domain. A valid ordering of repetitive events meets the following conditions:

(1) It has at least one but not more than L repetitive events, each event in the string being unique.

(2) The first event of the string can follow at least one non repetitive event in the task domain or it can start the plan.

(3) The last event of that same string can come before at least one non repetitive event, and the ordering satisfies the compatibility test.

Rule 4.6. Formulation of augmented activities is performed by inserting valid strings of repetitive events in all different positions between non repetitive in each activity. In general, a complete plan represented as an ordered list of events will have the form $(*e_i*e_j* \ldots *e_z)$ where the e's are the non repetitive events (with the exception of the first *). Considering that the original activity used to derive the complete plan is $(e_ie_j \ldots e_z)$, an augmented activity will have any of the following forms:

$$
\begin{aligned}
&(*e_ie_j \ldots e_z)\\
&(*e_i*e_j \ldots e_z)\\
&\quad\cdot\\
&\quad\cdot\\
&\quad\cdot\\
&(*e_i*e_j* \ldots *e_z).
\end{aligned}
\tag{4.34}
$$

Rule 4.7. A complete plan is an augmented activity that:

(1) begins with a repetitive event which can start the plan and ends with a non repetitive event;

(2) contains at least one repetitive event between non repetitive events;

(3) satisfies all compatibility tests.

The simple plan is one that starts with a repetitive event followed by a non repetitive event.

Rule 4.8. The generated plans are coupled with the classified compiled input commands u_j, because their corresponding probabilities are conditioned on the received commands u_j. This coupling is used to provide further information about the formulation of meaningful strings to complete the requested job.

All pertinent information is stored in the long term memory of the organizer at specific addresses. This information is accessed when a specific user command is issued.

1C. Algorithm

The algorithm of the organization level is briefly introduced as follows.

Algorithm 4.1. Algorithm of organization level

Step 1. System input

An *a priori* probability distribution function $p(c_n)$ is assumed for each user command. The compiled input command to the organization level u_j has a probability distribution function $p(u_j)$. Both $p(c_n)$ and $p(u_j)$ can be expressed by certain equations. Each u_j depends on the user command c_n.

The uncertainties of c_n and u_j are given in terms of the entropy functions. Then the total entropy associated with all user and compiled commands can be gotten and represented.

Step 2. Machine reasoning

Once u_j is known, an initial set of $(2^N - 1)$ pertinent activities (strings) of active/inactive primitives is associated with it. Each activity A_{jm} is represented by a string of binary valued random variables $X_{jm} = \{x_1, x_2, \ldots, x_N\}$ corresponding to the *m*th activity and *j*th compiled command. Each string is assigned a probability, and it is updated by the learning algorithm in Step 5.

The entropy associated with a particular activity A_{jm}, the total entropy of a set of pertinent activities associated with one u_j, and the entropy of all activities (strings) associated with all compiled commands can be shown.

Step 3. Machine planning

Once the set of activities A_{jm} has been associated with the compiled command, based on the operational procedures, eight substeps from substep 1 to substep 8 occur which actually simplify the algorithm complexity and reduce memory storage requirements.

Step 4. Machine decision making

Decision making is defined as the plan corresponding to a given command with the highest probability of success. After the selection of the most probable plan associated with a specific command u_j, this plan is communicated to the coordination level where a specific execution scenario is developed by that level. After the plan execution, feedback information is communicated from the coordination level to the organization level.

Step 5. Machine feedback

Define the overall accrued cost J_c associated with the execution of specific plans where the particular form of J_c depends on the specific application problem. Upgrade all pertinent probabilities.

Step 6. Machine memory exchange

This function involves retrieval, storage, and upgrading of information in the long term memory of the organizer. Once a compiled command has been recognized, a memory package is associated with it that contains relevant information. After the formulation of the plan and execution of the requested job, the upgraded information is stored in the organizer memory.

The above algorithm has been described in detail. In addition, an alternative model, the expert system cell, for the organization level has been presented. This alternative model is basically an expert system cell and it is suitable and works better for situations where the system has a good understanding of the different problems it is faced with. However, the expert system cell model does include a "probabilistic mode" for cases of very incomplete knowledge related to the user requested job, which is essentially the model already described in this subsection. The reader interested in the mentioned algorithm and model might study them for further learning [44].

2. Coordination level

The best possible plan to execute the requested job is stored in a buffer accessible by the dispatcher of the coordination level. Due to the hierarchical control structure, planning is hierarchical, too. Each task in the organization level is decomposed into subtasks at the coordination level, all the details associated with plan execution are defined at the coordination level. Thus, it is the objective of the coordination level to formulate the actual control problem and decide how the plan will be executed.

2A. Structure

As we have introduced early, the coordination level is an intermediate structure serving as an interface between the organization level and the execution level. It is essential for dispatching organizational information to the execution level. Its objective is the actual formulation of the control problem associated with the most probable complete and compatible plan formulated by the organization level that will execute the requested job in real time. This includes selection of one among alternative plan scripts that accomplish the same job in different ways according to the constraints imposed by the workspace model and timing requirements. The coordination level is composed of a specified number of coordinators of fixed structure. Specific hardware (execution devices) from the execution level is associated with each coordinators. These execution devices lead to well defined tasks when a command is issued to them by their corresponding coordinator.

The individual functions of each coordinator may be defined *a priori* (during the design phase of the system) because they are considered to be unmodifiable with time. Thus, they are assumed to be deterministic functions since the number of parameters involved in each one of them is also prespecified. This structure implies that the coordination level does not have any reasoning capabilities like the organizer. Its intelligence is related to its ability on how to execute the organizer plan in the best possible way based on past experience and real time constraints associated with the workspace environment.

The coordination level involves decision making associated with specific knowledge (information) processing based on the already formulated plan. The flexibility of the coordination level and its ability to formulate alternate real time scenarios is influenced by the variable structure of the dispatcher which is dictated by the organization level.

All tasks generated at the organization level are decomposed into subtasks within it. All information related to task and subtasks completion and execution is stored in the dispatcher memory and passed to the different coordinators when necessary.

The block diagram of the coordination level for an intelligent robotic system is shown in Figure 4.11. There are four different coordinators and a dispatcher in the coordination level as seen from Figure 4.11. A Petri Nets structure of the coordination level for the intelligent robotic system is shown in Figure 4.12 [17].

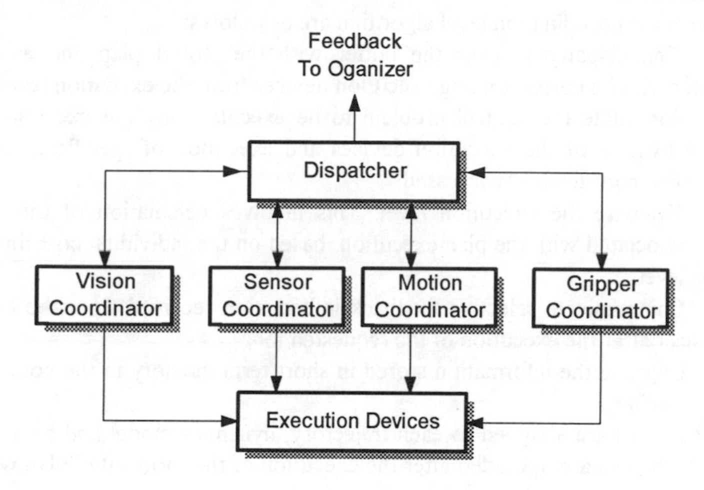

Figure 4.11. Structure of coordination level for an intelligent robotic system.

2B. Functions

Each coordinator when accessed, performs a prespecified number of different functions. A cost is assigned to each individual function. An accrued cost is associated with the operation of each coordinator. An overall accrued cost is calculated in terms of the weighted sum of the accrued costs of the coordinators after the execution of the requested job. This cost is communicated to the organizer after the completion of the requested job and is used to upgrade the information stored in the long term memory of the organization level. This feedback information (sent from the coordination level to the organization level after the completion of the requested job) is off-line feedback information. On the other hand, feedback information is communicated to the coordination level from the execution level during the execution of the requested job. Each coordinator when accessed issues a number of commands to its associated execution devices (at the execution level). During the execution and upon completion of the issued commands feedback information is received by the coordinator and is stored in the dispatcher. This information is used by other coordinators if necessary, and also to calculate the individual, accrued and overall accrued costs related to the coordination level. Therefore, the feedback information from the execution to the coordination level is real-time, on line feedback information.

The functions of different coordinators such as the vision system coordinator, the sensor system coordinator, the motion system coordinator, and the gripper system coordinator for the intelligent robotic system, can be explained in detail. Due to the limited volume of the book, they have to be omitted.

2C. Algorithm

The steps of the coordination level algorithm are as follows:

Step 1. The dispatcher scans the buffer with the stored plan and associates each coordinator with its corresponding execution devices from the execution level.

Step 2. Formulate the control problem to be executed by the execution level. This involves activation of the execution devices and execution of specific tasks when their corresponding coordinator is accessed.

Step 3. Evaluate the execution level. This involves calculation of the accrued cost functions associated with the plan execution, based on the individual cost functions of the execution level.

Step 4. Communicate selective feedback information received from the lower level to the organizer after the execution of the requested job.

Step 5. Upgrade the information stored in short term memory in the coordination level (in the dispatcher).

Probabilities are assigned to each trajectory, dynamics model and control algorithm. These probabilities are upgraded after the execution of the formulated plan with the aid of learning algorithms.

3. Execution level

The execution level is composed of a number of execution devices related to the coordinators of the coordination level. Its main objective is to execute the specific tasks issued by the different coordinators as accurate as possible. Individual cost functions are assigned to each operation of the execution devices, the form of which depends on the specific application problem. This requires system theoretic methods to be suitably implemented to control the hardware of the execution level.

The entropy measures have been established and are equivalent to the performance criteria of the optimal control problem to which a physical meaning is provided. This is done by expressing the problem of the control system design probabilistically and assigning a distribution function representing the uncertainty of selection of the optimum solution over the space of admissible controls. When the distribution satisfies Jaynes' maximum entropy criterion [15, 16], the performance criterion of the control problem is associated with the entropy of selecting a certain control. Minimization of the differential entropy which is equivalent to the average performance of the system yields the optimal control solution. The adaptive control formulation of the problem is derived as a special case of its optimal formulation [36].

The cost of control at the execution level can be expressed as an entropy which measures the uncertainty of selecting an appropriate control to execute a task. By selecting an optimal control, one minimizes the entropy, i.e., the uncertainty of execution.

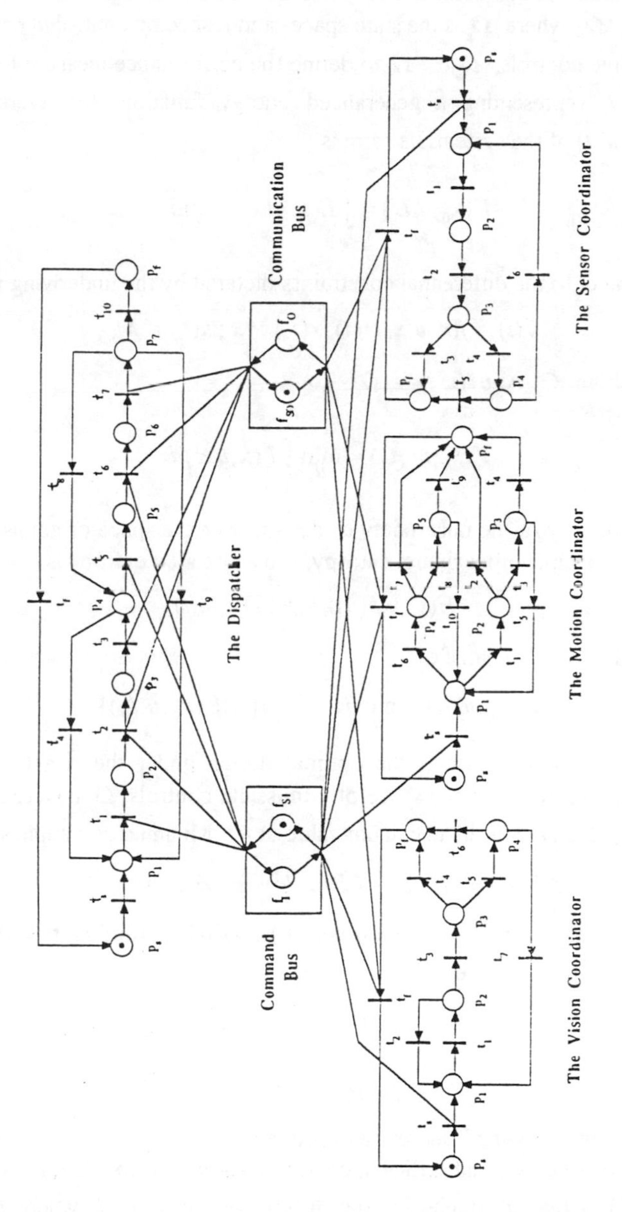

Figure 4.12. Petri nets structure of coordination level for an intelligent robotic system.

It is known that optimal control theory has utilized a non negative functional state of the system $x(t) \in \Omega_x$, where Ω_x is the state space, and a specific control $u(x,t) \in \Omega_u$ is the set of all admissible controls, $\Omega_u \subset \Omega_x$, to define the performance measure for some initial conditions x_0, t_0 representing a generalized energy function, the average value of Lagrangian $L(x, u, t)$ of the system. Its form is

$$V(x_0,u,t_0) = \int_{t_0}^{t} L(x,u(x,\tau),\tau)d\tau \tag{4.35}$$

where $L > 0$, subject to the differential constraints dictated by the underlying process:

$$\dot{x}(t) = f(x, u(x, t), t), x(t_0) = x_s; x(t_f) \in M_f \tag{4.36}$$

with M a manifold in Ω_x. At $u^*(x, t) \in \Omega_u$:

$$V(x_0,u^*,t_0) = \min_u \int_{t_0}^{t} L(x,u,\tau)d\tau \tag{4.37}$$

by selecting the density of the uncertainty of design over the space of admissible controls to satisfy Jaynes principle of maximum entropy, the associated entropy is:

$$H(x_0,u,p(u)) = E\{V(x_0,u,t)\} \tag{4.38}$$

Then the optimal control u^* satisfies

$$H(x_0,u^*,t) = \min_u H = \min_u \{E\{V(x_0,u,t_0)\}\} \tag{4.39}$$

The problem then is recast as the optimal design under the worst uncertainty of selecting the best control from the space of admissible controls Ω_u covered by a density function $p(x_0, u)$ which attains its maximum value at u^*. Minimization implies:

$$dH / du = \partial H / \partial u + (dH / dp)(\partial p / \partial u) = 0\,. \tag{4.40}$$

Since the probability density function $p(x_0,u)$ maximizes (differential) the entropy H according to Jaynes principle, i.e.:

$$dH/dp = 0 \tag{4.41}$$

then

$$dH / du = \partial H / \partial u\,. \tag{4.42}$$

Based on (4.38), the following Theorem was proven [36].

Theorem 4.1. A necessary and sufficient condition for $u^*(x,t)$ to minimize $V(u(x,t), x_0, t_0)$ subject to (4.36) is that u^* minimizes the differential entropy H, where $p(x_0,u)$ is the maximum entropy density function according to Jaynes' maximum entropy principle.

Entropy satisfies the additive property and any system composed of a combination of subsystems will be optimal by minimizing its total entropy.

4.2 Multilayer Hybrid Intelligent Control System

The multi-layer hybrid knowledge based/analytical control was proposed by Villa [46, 48]. This control theory can be used to the control design problem for a complex discrete-event driven system. The main difficulties of the design problem appear to be the uncertainty in describing the complex system and the driving events, and the complexity of the control structure.

4.2.1 Complex Event Driven System

The practical experience on control design of complex industrial systems has proven that the main features of such systems are as follows [7, 21]:

(1) **The complexity**. It is shown by the representability in terms of composition of a number of interconnected basic subsystems or sections.

(2) **The event-driven nature of each component section**. Let us consider a multi-stage and multi-product manufacturing system that is composed by a number of work shops, connected serially, and decoupled by inter-shop storage space partially. Each shop can be affected by failures (a prior unknown endogenous events) and the connecting material handling system. In addition, the whole manufacturing system is driven by the arrival of new orders of products (exogenous events) [48]. From the production manager's point of view, the key task is to plan a new execution order over some future horizon and to control the part flows by up-dating production plans. One event (either endogenous or exogenous) occurs simply at one instance and is then detected.

An efficient control architecture of complex event-driven systems should possess the following basic properties [14, 46]:

(1) The capability to recognize and classify any occurred event. This ability is capable of identifying the model of the event-affected sections.

(2) The capability to coordinate a set of local (decentralized) control strategies that is capable of making the strategies consistent to a given overall control objective.

In order to implement such a control architecture, a problem, so called **large-scale uncertain event-driven control problem**, should be solved first. An optimal solution for this problem has not been found until now.

The objective of the multi-layer hybrid knowledge based and analytical control theory is to obtain a guaranteed approximation solution on the basis of the following two ideas.

The first idea is to approach the problem in terms of an optimization of adaptive control strategy. An occurred event is recognized and up-dated each time, and its effects on the model parameters are identified. This means that a large-scale open-loop feedback control problem must be solved before an event recognition is solved.

Owing to the complexity of the system, we decompose the problem into two subproblems as follows:

(1) A set of lower-layer control optimizations, one for each section, corresponds to a higher-layer goal coordination and faces consistency with local control strategies. This set of control design problems has to be solved in such a way that each time an event simply affects one individual section.

(2) A higher-layer reference-control optimization is based on a reduced-order model of the entire system and defines a low-frequency reference behavior.

This problem together with the above mentioned ones must be solved in such a way that each time an event simply affects a group of sections or their interactive variables.

In order to make such a decomposition, an approximated separation between the frequency bands of the two optimization problems is adopted. On the other hand, each time an unknown event occurs, an event recognition and classification procedure have to be applied. To do so, an approximated separation between the event-and-model learning phase and the control design phase must be assured at each decision layer.

The approach mentioned above points out the essential difficulty in solving an event-driven control problem, i.e., the impossibility of defining a robust control strategy during the initial transient after an event occurrence. Due to the high uncertainty, few data is available, and event recognition is affected.

In order to overcome this difficulty the second idea is applied. This idea uses a knowledge based event recognition procedure to improve the learning capability. This implies that an event classification is performed by comparing the detected system evolution with a number of previously occurred failed situations, thus implementing a diagnosis procedure. Since a control action has been associated with an analyzed situation, the diagnosis procedure also allows a control strategy to be chosen and to drive the initial transients. Intuitively, this approach corresponds to the design of a knowledge based control strategy to be used during initial transient after the event occurs. This approach must be associated with the subsequent analytically stated control strategy to reduce uncertainty of the event by use of further measurements. Hence, a jointly knowledge-based/analytical control architecture is proposed.

To assure an efficient performance, both the control design approaches must be proved to be coordinable and consistent. For this purpose, a common mathematical formulation will be introduced, and both the analytical identification and control design problem, and the knowledge based event recognition and control choice problems can be stated in terms of the standard projection theorem [18]. The event diagnosis and the related control choice are performed by projecting current behavior on previously detected failed situations. For this reason, an integrated set of criteria for the design of a multilayer control architecture will be described for event-driven complex systems.

4.2.2 Mechanisms of Multilayer Control Architecture

The control design problem for a complex event-driven dynamic system can be formally described on the basis of the following assumptions:

Assumption 4.1. The related complex system is composed of N, N = 1,2, ..., n, sections and is obtained by recognizing the interactive variables of the N sections.

Assumption 4.2. Each section is described by an event-driven dynamic model in terms of a set of state equations for which parameters depend on the occurred and *a priori* unknown event.

Assumption 4.3. The local identification and its control objective are associated with each other for every section, and the global objectives will be obtained by a linear combination of the local objective.

Assumption 4.4. Only a finite number of events can occur for every section and every set of interactive variables, and any *a priori* definable law of occurrence is unknown. This implies that the model parameters of the sections or the interactive variables can only assume a finite number of values that directly influence the occurred events.

Based on the above four Assumptions, a mathematical representation of each basic section j, j = 1, 2, ... , n, can be described as follows:

$$x_j(t) = A_j(p_j(t))x_j(t) + B_j(q_j(t))u_j(t) + z_j(t) \tag{4.43}$$

$$x_j(t_0) = x_j^0$$

$$z_j(t) = \sum_{n=1}^{N} C_{jn}(r_j(t))x_j(t) \tag{4.44}$$

where t_0 is the occurring time for the last detected event that affects section j; $x_j(\_)$ is the n_j-dimensional state vector: $u_j(\_)$ is the m_j-dimensional control vector: $z_j(\_)$ is the n_j-dimensional vector of interactive variables: and

$$p_j(\_) \in P; \quad q_j(\_) \in Q; \quad r_j(\_) \in \Re$$

are the vectors of parameters subject to abrupt changes due to assurance of an event, where $p_j(\_)$ and $q_j(\_)$ are the types of local events (or small scale events), and $r_j(\_)$ is the type of global events (or large scale events).

1. Local identification problem

If the occurrence of an event of type $p(\_)$ or $q(\_)$ is detected, then the following local identification problem must be solved:

$$\min_{A_j(t_0),B_j(t_0)} \int_{t_0}^{t+t_0} h_j(x_j^m(\tau), x(\tau), u_j^m(\tau))d\tau \tag{4.45}$$

subject to:

$$\dot{x}_j = A_j(t_0)x_j^m + B_j(t_0)u_j^m + z_j^m$$

where $x_j^m(\_)$, $u_j^m(\_)$ and z_j^m denote the measured section state, control and interactive variable evolutions respectively, and t bounds the increasing identification horizon.

2. Local control design problem

Before each step of local identification is made, the following local control design problem should be solved:

$$\min_{x_j(\_),u_j(\_)} \int_{t_0}^{t_0+T_j} g_j(x_j^*(\tau),x_j(\tau),u_j(\tau))d\tau \tag{4.46}$$

subject to:

$$\dot{x}_j = \hat{A}_j x_j + \hat{B}_j u_j + z_j$$

$$z_j = \sum_{n=1}^{N} \hat{C}_{jn} x_j$$

where $\hat{A}_j, \hat{B}_j$ and $\hat{C}_{jn}$ denote the currently estimated parameters; T_j denotes the control replanning horizon; and $\dot{x}_j$ defines the reference state trajectory for the jth section of the desired evolution.

In order to solve a higher-layer identification problem without increasing its complexity too much, a reduced-dimension overall model must be defined by applying an aggregation operator to the set of section models. To achieve this aim, a model approximation will be used which allows to separate the slow and fast models of each section.

Let $x_j(\_)$ be the reduced vector of the jth section state, and

$$\begin{cases} \dot{\overline{x}}_j = \overline{A}_j x_j + \overline{B}_j u_j + \overline{z}_j \\ \overline{z}_j = \sum_{n=1}^{N} \overline{C}_{jn}(x_j(t))\overline{x}_j \end{cases} \tag{4.47}$$

be the related approximated model equations. An approximated overall model can be deduced by summarizing all models of sections into an unique representation:

$$\dot{\overline{x}} = \overline{A}(r(t))\overline{x} + \overline{B}u \tag{4.48}$$

3. Global identification problem

While this overall model is being used, if the occurrence of an event of type $r(\_)$ is detected, then the following problem has to be solved:

$$\min_{\overline{A}(t_0)} \int_{t_0}^{t_0+t} h(x^m(\tau),x(\tau),u^m(\tau))d\tau \tag{4.49}$$

subject to Equation (4.48).

4. Global control design problem

The following global control design problem has also to be solved:

$$\min_{x(\_),u(\_)} \int_{t_0}^{t_0+T} g(x^*(\tau), x(\tau), u(\tau))d\tau \tag{4.50}$$

subject to equation (4.48).

The procedure to link global optimization and local ones is suggested by the aggregation procedure introduced above, and the global optimization must define the reference state trajectories of low frequency for the set of control design problems of lower layer under the assumption that no local event $p_j(\_)$ or $q_j(\_)$ will occur. For this reason, the separation between the frequency bands is associated with the control design problems of the two decisional layers. In addition, both concepts of the event classification and the effective tasks to be performed at the higher decisional layer can be clarified. For the former, it is suggested to classify the events according to their types, i.e., their effects either on individual sections (small scale events) or groups of sections (large scale events). For the latter, it can be formally stated in terms of the following.

5. Goal coordination problem

The goal coordination problem can be described by

$$\begin{aligned}\min_{\lambda(\_)} \sum_{j=1}^{N} \{\max_{p_j(\_)}[\min_{x_j(\_),u_j(\_)} \int_{t_0}^{t_0+T} (g_j(x_j^*(\tau), x_j(\tau), u_j(\tau)) \\ + p_j^T(\tau)(\hat{A}_j x_j(\tau) + \hat{B}_j u_j(\tau) + z_j(\tau) - \dot{x}_j(\tau)) \\ + \lambda_j^T(\tau)(\sum_{n=1}^{N} C_{jn} x_n(\tau) - z_j(\tau))d\tau]\}\end{aligned} \tag{4.51}$$

connecting the global control design problem and the goal coordination problem as depicted in Figure 4.13. This shows the necessity of defining three types of control design phases:

(1) If a small scale event (or local event) p, q is occurred, then the local control optimizations and the global goal coordination have to be solved.

(2) If a large scale event (or global event) r is occurred, then the global control optimization has to be solved along with the solution of the chain of problems mentioned above.

The main feature of the two-layer hierarchical control architecture is the ability to define an event-driven control strategy of the open-loop-feedback type, and only one occurred event is detected and updated each time. Nevertheless, with regard to its practical applicability, some more problems have to be solved:

(1) How to recognize the event occurrence and classify the event type.

(2) How to choose an aggregation operator and the related separation among the frequency bands of different layers.

(3) How to perform identification during the initial transient after the event occurs and then choose a sufficiently reliable control action.

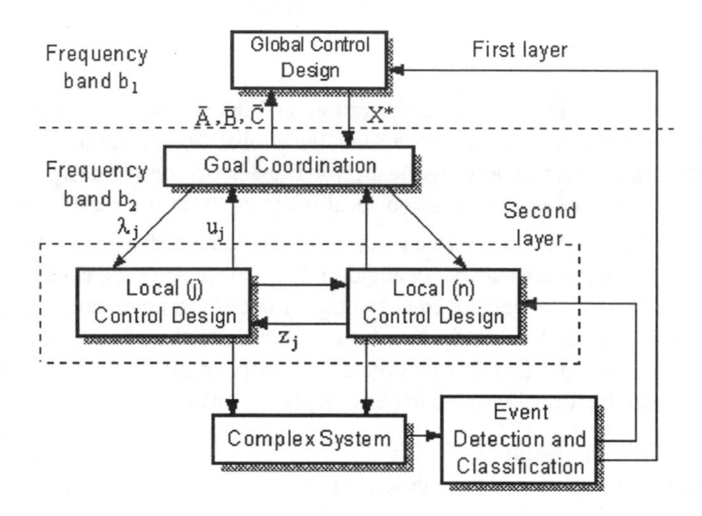

Figure 4.13. Two-layer hierarchical structure of intelligent control.

A reasonable solution to such problems should apply to the natural knowledge of a human expert. This is a right answer, but is not sufficient if an automated learning and control system is desired. In this case, the following problems are desired:

(1) To describe a formal representation of problem-solving procedures based on human expertise of the two types:

- Ability of recognizing a current situation similar to a previously occurred one:
- Ability of classifying a new situation according to some event models (matrices).

(2) To state a mathematical formulation of such problem-solving procedures coherent with the formulation adopted for similar analytical procedures discussed above. Therefore, a representation of knowledge-based and analytical control system is obtained.

(3) To define the conditions in which the complete mathematical formulation must be satisfied in order to give a robust solution to the overall event-driven control problem. Therefore, a representation of a hybrid knowledge-based and analytical control system is given.

The first problem has already been analyzed and studied widely in recent years [36].

In essence, the procedures of knowledge-based control problem-solving for event-driven systems can be implemented in terms of :

- A pattern recognition technique for event classification.
- An inferential reasoning technique for control choice.

On the contrary, the second and third problems have not yet been solved, and some new approaches are required.

4.2.3 Model of Hybrid Knowledge-based/Analytical Control System

In order to formulate both the knowledge-based identification and control procedures and the analytical procedures mathematically in a common way, let us refer to the theoretical basis outlined in the past subsection. According to the expression (4.43), the model of each section is described by a dynamic operator $D_j(t_0)$

$$D_j(t_0) : \{x_j(t_0), z_j(\tau), u_j(\tau), e_j\} \rightarrow X_j(t) \qquad t_0 \leq \tau < t \tag{4.52}$$

where $e_j \in E_j$ denotes the set of event-driven parameters. For every time-dependent function, a projection operator is assumed as:

$$p_{jk} = \langle x_j(\_) / x_k(\_) \rangle$$

and a norm operator

$$p_i^2 = \langle x_i(\_) / x_i(\_) \rangle$$

are assumed [18].

For the local identification and control problems, the norm operators are defined as:

$$I_j^2 = \langle (x_j^m(\_) - x_j(\_)) / (x_j^m(\_) - x_j(\_)) \rangle$$
$$N_j^2 = \langle x_j^*(\_) - x_j(\_)) / (x_j^*(\_) - x_j(\_)) \rangle$$

Consequently, every local identification problem and control design problem can be rewritten as minimum norm projection problems

$$\min_{e_j} I_j^2 \qquad \text{and} \qquad \min_{x_j, u_j} N_j^2$$

(1) For event identification problem

The event classification and related model identification problem can be stated as a projection over a set of admissible event-affected parameter :

$$\min_{E_j} I_j^2, \quad \text{for } e_j \in E_j, \ \text{subject to } D_j(t_0)$$

(2) For local control design problem

The control choice problem by inferential reasoning can be stated as a projection over a set of admissible event-driven control strategies:

$$\min_{x_j,u_j} N_j^2, \quad \text{for } u_j \in U_j, \text{ subject to } D_j(t_0)$$

Every low-layer problem to be solved in the desired hybrid control system can be described in terms of the following tasks.

(1) **Task 1**

Task 1.1. Event recognition and model identification phase

To minimize I_j^2 referred to e_j either for $e_j \in E_j$ if initial transient after small scale event is occurred, or subject to $D_j(t_0)$ if the set of measures is sufficient and/or the event has not been classified previously.

Task 1.2. Control design phase

To minimize N_j^2 referred to (x_j,u_j) either for $u_j \in U_j$ or subject to $D_j(t_0)$ under the same conditions as above.

Let us consider the global identification and control problems. The global identification problem can be reformulated as the local identification one exactly both in knowledge-based and analytical procedures. The global control problem requires to be rewritten in terms of a new projection problem. Let

$$N^2 = \sum_j N_j^2$$

be the overall control cost, and

$$D_j(t_0), H_j(t_0), j=1,2,\ldots,N$$

$$H_j(t_0): \{x_n(t)\} \rightarrow z_j(t), \quad n=1,2,\ldots,N$$

be the set of the jth dynamic operators and the linear operators modeling the interactive variables described to account for dynamic and interactive constraints:

$$\max_{\lambda} \sum_j \{\max_{p_j}[\min_{x_j,u_j}(N_j^2 + P_j^T D_j(t_0) + \lambda_j^T H_j(t_0))]\} \tag{4.53}$$

The suggested separation approach of frequency can be used to approximate such a complex problem by a chain of subproblems that are parts of the following task to be solved at the higher decisional layer.

(2) **Task 2**

Task 2.1. Global optimization phase of low-frequency

To maximize N^2 referred to (x,u) either for $u \in U$ with initial transient after a large scale event has occurred and the event is recognizable, or subject to $D(t_0) = A\{D_j(t_0)\}$,

j=1, 2, ..., N, where A is the aggregation operator adopted.

Task 2.2. Goal coordination phase of high-frequency

To maximize

$$\sum_j \{\min(N_j^2 + \lambda_j^T H_j(t_0))\}$$

referred to $\lambda_j(\_)$, subject to $D_j(t_0)$, j=1,2,...,N.

Task 2.3. Local control design of high-frequency

This task corresponds to task 1.2 to be solved for each section.

The task structure discussed clarifies the operation mode for the low-layer hybrid control system and gives evidence of the link between the knowledge-based solution and the analytical one for the higher-layer identification and control problems. Intuitively, a knowledge-based procedure for large-scale event classification and the related control choice must be included in task 2.1. On the contrary, the goal coordination action can only be based on analytical procedures. This result is particularly important since it remarks that multi-layer control hierarchies can only define robust strategies in case of a reduced level of the event and model uncertainty.

It is noted that the proposed hybrid control system can satisfy the following two necessary and sufficient conditions:

(1) The control architecture can be represented as a decision tree to assure the coordinability of the defined hybrid control system.

(2) The higher-layer module can perform an active coordination to assure the consistency of all local control strategies with the adopted overall control objective.

The proposed hybrid multi-layer knowledge-based/analytical control system allows to clarify some misunderstandings about the possible use of control design procedures obtained either by analytical rules or by knowledge-based problem-solving procedures.

4.3 Paradigms of Hierarchically Intelligent Control Systems

The hierarchically intelligent control has been applied to various devices, plants and systems both in case study and in industrial applications since late 1970s. The early case studies were in the control for robotic manipulators in laboratories, mobile robotic systems for emergency operation in nuclear power plants, prosthetic arm control for disabled personnels, and the motor control for urban and freeway traffic [4, 5, 29, 31, 32, 34, 40]. Some hierarchically intelligent control systems have been used to industrial process control, robotic assembly and in robotic operations in hospitals since 1980s [13, 42, 45, 50]. Based on the research and application works, some International Workshop such as on Nuclear Robotics, etc. have been held since middle 1980s.

In this section, two examples will be introduced and illustrated , they are:

1. A robotic assembly system designed to manufacture a mechanical assembly out of four parts coming at random on a conveyor belt.

2. A hierarchical control system for a nuclear reactor using uncertain dynamic

techniques subject to multiple demands.

In addition, some more examples will be presented in the following chapters where fuzzy control and neural network-based control will be introduced. For instances, a hybrid symbolic and neuromophic control for hierarchical intelligent control of robotic manipulator, a hierarchically intelligent control for industrial boilers in a factory self-supply power plant, and so on.

4.3.1 Hierarchically Robotic Assembly System

Figure 4.14 shows a robotic work station (cell) that was designed to assemble a product from four different parts, say A, B, C and D, as described in Figure 4.15 [44].

1. Assumptions

The product assembly is based on the following assumptions:

(1) Parts A, B, C and D are fed from a conveyor belt at random order, but they always come together. When a part is at location F, the conveyor stops until the part is buffered or fetched; the next part is then brought to F.

(2) The manipulator can execute the following actions:

• **buffering** — temporally store parts located in the conveyor

• **mating** — joining of subassemblies (parts) one of which must already be held by the manipulator

• **fetching** — retrieving a part in the part buffer or directly from designated locations in the workspace.

(3) Parts A, B, C and D are buffered in locations B_1, B_2, B_3 and B_4 respectively. Part A always has to be buffered, while parts B, C and D may or may not be buffered.

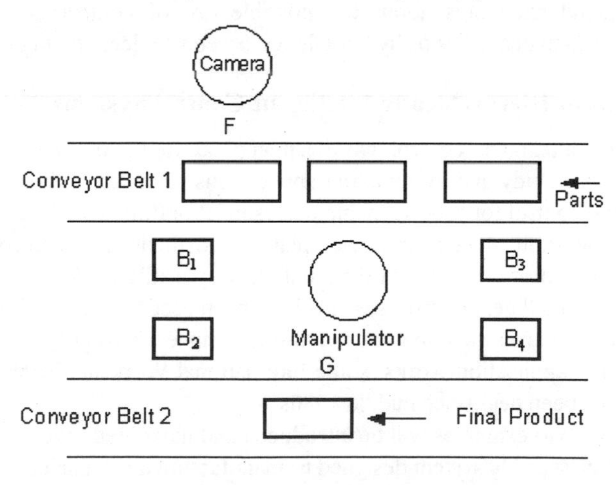

Figure 4.14. A robotic workcell.

(4) There are two basic assembly sequences:

a. Assuming A is already in buffer B_1, B is fetched either from buffer B_2 or directly from the conveyor belt and the mating operation involving B and A (resulting in subassembly S_1) is done in B_1. In a similar manner, if D is already in B_4, C is fetched either from the conveyor belt or from B_3, and the mating operation involving C and D (resulting in subassembly S_2) is done in B_4. Once S_1 and S_2 are available, S_2 is fetched from buffer B_4 and moved to buffer B_1 to be mated with S_1. S_1 and S_2 when mated together result in the final product S_4 which is then moved to G.

b. This sequence differs from the above in that after S_1 is assembled, C is fetched and moved to B_1 to be mated with S_1 (resulting in subassembly S_3, and then D is fetched and again moved to B_1 to be mated with S_3 to produce the final product S_4. In the first sequence D has to be buffered, while in the second case, it could be fetched directly from the conveyor belt.

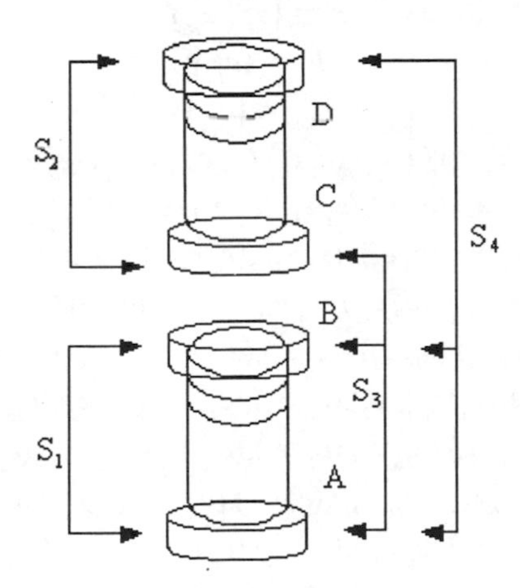

Figure 4.15. Description for a product assembly.

(5) Assembly of S_1 may be done before or after assembly of S_2.

(6) It is desired that if A is already in B_1, once B is seen in F, B is fetched right away and the mating of B to A follows next. In a similar manner, if D is already in B_4, once C is seen in F, C is fetched right away and mated with D. This is to eliminate the unnecessary bufferings of B and C.

The efficiency of this assembly operation depends on the ability to handle incoming parts at random order.

Based on the assumptions made, there will be four buffering operations, five mating operations, four fetching operations, and one operation to move the finishing product to the second conveyor belt.

2. Algorithm

The user command is : $c_1 = u_1$ = assemble part.

The task domain is given below ($N=16$, $N\text{-}L=14$, $L=2$), along with the corresponding subsets from Table 4.1.

$$E_t = \{e_1,e_2,e_3,e_4,e_5,e_6,e_7,e_8,e_9,e_{10},e_{11},e_{12},e_{13},e_{14},e_{15},e_{16}\} \tag{4.54}$$

where e_1 = buffer A, e_2 = buffer B, e_3 = Buffer C, e_4 = buffer D, e_5 = fetch B, e_6 = fetch C, e_7 = fetch D, e_8 = fetch S_2, e_9 = mate C to D, e_{10} = mate B to A, e_{11} = mate C to S_1, e_{12} = mate D to S_3, e_{13} = mate S_2 to S_1, e_{14} = move S_4, e_{15} = see, e_{16} = move.

$$E_{nr} = \{e_1,e_2,e_3,e_4,e_5,e_6,e_7,e_8,e_9,e_{10},e_{11},e_{12},e_{13},e_{14}\}$$

$$E_r = \{e_{15}, e_{16}\}$$

$$E_{sr} = \{e_{15}\}$$

$$E_{end} = \{e_{14}\}$$

$$\begin{aligned}E_{comp} = \{&(e_1\, e_3),(e_1\, e_4),(e_1\, e_5),(e_1\, e_6),(e_2\, e_1),(e_2\, e_3),\\ &(e_2\, e_4),(e_2\, e_6),(e_3\, e_1),(e_3\, e_2),(e_3\, e_4),(e_3\, e_5),\\ &(e_4\, e_1),(e_4\, e_2),(e_4\, e_5),(e_4\, e_6),(e_5\, e_{10}),(e_6\, e_9),\\ &(e_6\, e_{11}),(e_7\, e_{12}),(e_8\, e_{13}),(e_9\, e_1),(e_9\, e_2),(e_9\, e_5),\\ &(e_9\, e_8),(e_{10}\, e_3),(e_{10}\, e_4),(e_{10}\, e_6),(e_{10}\, e_8),(e_{11}\, e_7),\\ &(e_{16}\, e_1),(e_{16}\, e_2),(e_{16}\, e_3),(e_{16}\, e_4),(e_{16}\, e_5),(e_{16}\, e_6),\\ &(e_{16}\, e_7),(e_{16}\, e_8),(e_{16}\, e_9),(e_{16}\, e_{10}),(e_{16}\, e_{11}),(e_{16}\, e_{12}),\\ &(e_{16}\, e_{13}),(e_{16}\, e_{14}),(e_1\, e_{15}),(e_2\, e_{15}),(e_3\, e_{15}),(e_4\, e_{15}),\\ &(e_5\, e_{15}),(e_6\, e_{15}),(e_7\, e_{15}),(e_8\, e_{15}),(e_9\, e_{15}),(e_{10}\, e_{15}),\\ &(e_{11}\, e_{15}),(e_{12}\, e_{15}),(e_{13}\, e_{15}),(e_{12}\, e_{14}),(e_{13}\, e_{14})\}\end{aligned} \tag{4.55}$$

The algorithm of organization level is as follows:

The initial set of pertinent activities associated with the input command is $(2^{16} - 1)$. According to the operational procedures, the following justifiable simplifications are considered:

It can be verified that $(e_{15}\, e_{16})$ is the only valid repetitive ordering. Store $E_{ord} = \{(e_{15}\, e_{16})\}$ in the long term memory of the organizer. Then, given the command c_1, we have:

$$E_c = \{e_1,e_2,e_3,e_4,e_5,e_6,e_7,e_8,e_9,e_{10},e_{11},e_{12},e_{13},e_{14}\}$$

$$E_{cr} = \{\{e_1,e_4,e_5,e_6,e_8,e_9,e_{10},e_{13},e_{14}\},$$

$$\{e_1, e_5, e_6, e_7, e_{10}, e_{11}, e_{12}, e_{14}\}\}, \; q = 8$$
$$E_{snr} = \{e_1, e_2, e_3, e_4\}$$
$$E_{ord} = \{(e_{15}\, e_{16})\}$$
$$E_{sr} = \{e_{15}\} \tag{4.56}$$
$$(e_i\, e_{15}) \in E_{comp}, \;\; \forall i \in \{1, \ldots\ldots, 13\}$$
$$(e_{16}\, e_j) \in E_{comp}, \;\; \forall j \in \{1, \ldots\ldots, 14\}$$

To demonstrate the feasibility of the algorithm and the necessity of the operational procedures, several different cases were considered, related to the complete presence/absence of the precedence constraints, incompletely specified precedence constraints, and incompletely specified critical events. All cases demonstrated how the precedence constraints make the algorithm robust by preventing the generation of absurd activities and the generation of meaningful activities that do not satisfy the imposed assumptions.

3. Assembly plan

It has been shown that a set of 76 activities represents all possible valid sequences of active non repetitive events which also satisfy all assumptions made. It follows that for each activity (with non repetitive primitives) of the form $(e_i, e_j, \ldots, e_k)$, the corresponding plan is $(e_{15}\, e_{16}\, e_i\, e_{15}\, e_{16}\, e_j\, e_{15}\, e_{16} \ldots e_{15}\, e_{16}\, e_k)$ [42].

To demonstrate the convergence of the algorithm, a tabular look at the cost of execution of primitive tasks has been considered, along with the following assumptions: the cost of execution of the repetitive events is negligible compared with the cost of execution of the non repetitive ones. Therefore, the cost of execution of each plan is proportional to the cost of the corresponding activity of critical non repetitive events. The total cost is assumed to be the sum of the individual costs. Moreover, it is considered that the order parts appeared in F are possible to be identified.

Based on the above assumptions and from the fact that for this example, there is only one plan associated with any activity it follows that plans and activities may be interchangeable without loss of generality.

Table 4.2 lists the estimated and actual costs for executing the non repetitive events. Normally, the actual costs of plans are determined from the data fed back from the coordinators. From the assumption that parts come at random order, it follows that there are 24 ways (4!) in which parts A, B, C and D could come and be seen at location F.

Table 4.2. Actual and estimated costs of events

Event	e_1	e_2	e_3	e_4	e_5	e_6	e_7	e_8	e_9	e_{10}	e_{11}	e_{12}	e_{13}	e_{14}
Estimated Cost	2	2	2	2	3	3	3	4	6	5	7	8	9	5
Actual Cost	6	3	2	4	5	4	5	7	8	9	8	9	11	7

Given a part sequence, the plan chosen is the one with the highest probability among the plans that are relevant to that sequence as previously explained. After execution of the plan, the costs of the plans are recomputed in the following manner. Initially, the costs of the plans are determined from the cost estimated alone; subsequently, the actual costs are used for events that have been used in previous executions. The plan which has the minimum cost among those which are relevant to the part order is rewarded, resulting in an increase of its probability. The rest are penalized with consequent decreases in corresponding probabilities.

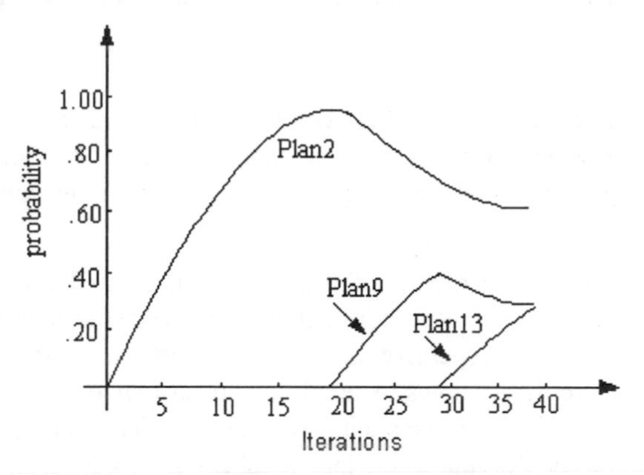

Figure 4.16. Probability of success for best plans.
[From Valvanis and Carelo, 1990, J.of Intelligent and Robotic Systems, 3(4)]

Three plans were chosen to demonstrate the convergence of the learning algorithm. There are forty training sessions (iterations). For the first twenty training sessions, parts are presented in order A-C-D-B (plan 2); during the next ten, A-D-C-B (plan 9) and during the last ten, A-B-C-D (plan 13). Figure 4.16 shows the plots of probabilities for 40 iterations. Initially, the plans are assigned equal probabilities. During the first 20 iterations, the probability of Plan 2 is increasing because it is constantly rewarded. Plan 2 has the lowest cost of execution of three potential candidates. During the next 10 iterations, the probability of Plan 9 is increasing while that of Plan 2 is decreasing. This is because only this plan is relevant to parts order A-D-B-C. Hence, it is constantly rewarded. During the last 10 iterations, it has been shown that Plan 13 is the one with the lowest cost. Since Plan

13 has the lowest cost of the two, it is constantly being rewarded, hence a resulting increase in its probability. The probabilities of all the plans except plans 2, 9 and 13 are monotonically decreasing and are negligible [42].

4.3.2 Hierarchical Control of a Nuclear Reactor

Recent advances in the nonlinear optimal control area are opening new possibilities towards its implementation in process control. In this example, a good agreement is shown between a centralized and a hierarchical implementation of a controller for a nuclear power plant subject to multiple demands.

1. Schematic diagram

Figure 4.17 shows a schematic diagram of a nuclear power plant with the main energy flow loops, sensors, actuators and other major components. The plant is partitioned into four control subsystems labeled from SS1 to SS4. The decoupling points between subsystems were selected on the basis of physical boundaries that are present in the real systems, such as heat exchangers, in which energy but no mass transfers are allowed, minimizing in this way the number of variables contributing to the coupling terms that have a physical meaning such as feedback reactivity and power transfer [30].

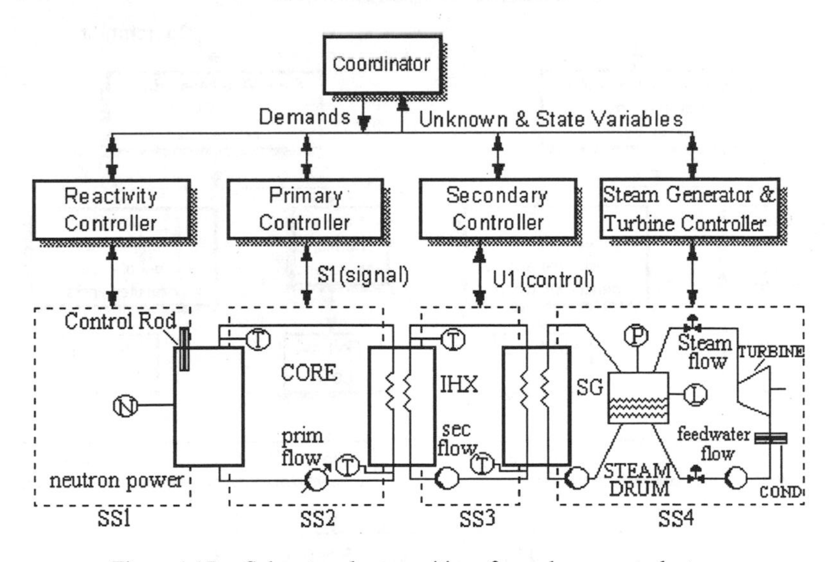

Figure 4.17. Subsystem decomposition of a nuclear power plant.

The role of the coordinator module is to provide a set of consistent demands for each of the subsystems. The local controller of each subsystems generates the optimal control actions to be sent to the actuators to fulfill the supervisor's demands. In the process, the generated unknown terms and the computed state variables are set up to the coordinator for plant performance evaluation, but this does not require any iterative computation with the subsystems.

By observing the time dependency of the unknown terms calculated by the different subsystems, the coordinator monitors the status of the plant and the local controllers. As new conditions arise, the coordinator recognizes the perturbance, identifies the failed subsystem, and changes the distribution of demands according to pre-established control strategies for the particular subsystem and for the entire plant.

In this two-layer control system, a coordinator module with a small set of anomaly-detection rules has been implemented to test the viability of the behavior of the approach. One set of rules is based on the analysis of the behavior of coupling terms only, and the other is based on the tracking of state variables only. Figure 6.18 gives a schematic diagram of the coordinator with the two sets of rules used in sample cases described here. These rules apply to the detection of specific anomalies in the intermediate heat transport lock (pump failure) and can take corrective actions.

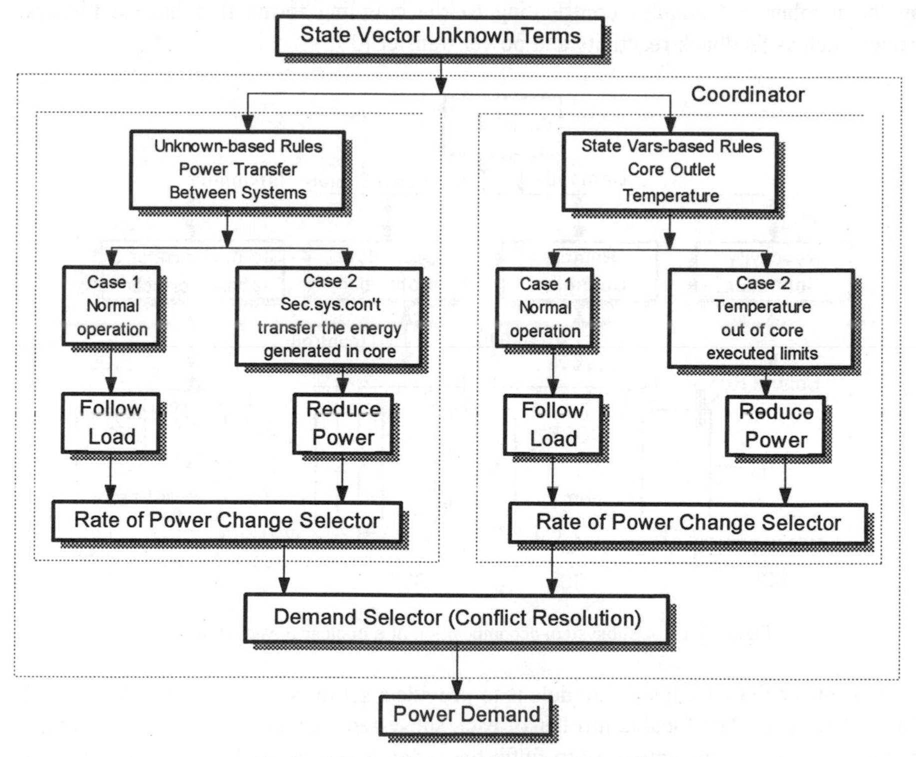

Figure 4.18. Schematic diagram of coordinator.

2. Mathematical models

Algorithms for multivariate control with parameter tracking, actuator saturation effects and physical limits to state variable capabilities can be developed on the basis of a consistent mathematical formulation.

Once a system decomposition is accomplished and a hierarchical structure is established, the coordinator module controls the operation of one or more subsystems and can be controlled by other modules higher in the hierarchy.

Let a large-scale system be represented by the following state vector equation:

$$dX/dt = F(X, U) \tag{4.57}$$

where X represents the state vector and U is the control vector. This system can be targeted for decomposition into n subsystems governed by a set of state equations of the form:

$$dX_i/dt = F_i(X_i, U_i) + P_i \tag{4.58}$$

where X_i and U_i represent the state and control vectors of the ith subsystem, and P_i is an unknown term vector that accounts for both the effects of those state variables not explicitly modeled in the ith subsystem and the effects of its modeling inaccuracies.

We define that, for each subsystem i, two Hamiltonian functions: one to be used to generate optimal controls for demand following, and the other to generate the unknown terms for optimal matching of the controller's state predictions and the measured signals. These Hamiltonians take the forms:

$$H_{Ci} = 0.5(U_i\text{-}U_{oi})_T Q_i(U_i\text{-}U_{oi}) + 0.5(D_i\text{-}X_i)_T R_i(D_i\text{-}X_i) + W_{Ti}(F_i\text{+}P_i) \tag{4.59}$$

$$H_{Si} = 0.5(P_i\text{-}P_{oi})_T M_i(P_i\text{-}P_{oi}) + 0.5(Y_i\text{-}S_i)_T N_i(Y_i\text{-}S_i) + Z_{Ti}(F_i\text{+}P_i) + E_{Ti}G_i \tag{4.60}$$

where W_i, Z_i, and E_i are adjoint state vectors; Q_i, R_i, M_i and N_i are weight matrices; D_i is demanded state vector; Y_i and S_i are the estimated and actual sensor reading vectors respectively; and G_i is the sensor transfer function.

By using Pontryagin's Maximum Principle (PMP) to the above Hamiltonian functions, a set of differential and algebraic equations can be computed from these equations.

Detailed mathematical models of nuclear power plants may involve more than 500 state variables to represent the dynamics of mass, energy, momentum and neutron processes across the plant. For this study a simplified model consisting of 29 state equations and 4 controls has been implemented on a VAX 11/780 computer using ACSL simulation language.

For illustration, let us consider the following system of equations representing energy balance for the 3-node model of the heat exchange between subsystems SS2 and SS3:

$$dT_{op}/dt = W_p/M_p(T_{ip}-T_{op}) - (HA)_{pm}/(MC)_p(T_{op}-T_m) \tag{4.61}$$

$$dT_m/dt = (HA)_{pm}/(MC)_m(T_{op}-T_m) - (HA)_{ms}/(MC)_s(T_m-T_{os}) \tag{4.62}$$

$$dT_{os}/dt = W_s/M_s(T_{is}-T_{os}) + (HA)_{ms}/(MC)_s(T_m-T_{os}) \tag{4.63}$$

where T stands for temperature; M for mass; W for mass flow rate; HA for heat transfer coefficient; i, o for inlet and outlet; p, s, m for primary, secondary and metal.

Since the temperatures of the coolant at the outlet nodes are measurable it is possible to decouple the subsystems by replacing the metal equations and coupling terms with P_1 and P_2 as follows:

$$dT_{op}/dt = W_p/M_p(T_{ip}-T_{op}) + P_1 \tag{4.64}$$

$$dT_{os}/dt = W_s/M_s(T_{is}-T_{os}) + P_2 \tag{4.65}$$

Equations (4.64) and (4.65) are then made part of the controller's model for subsystems SS2 and SS3 respectively. By virtue of the Hamiltonian and the application of the PMP, P_1 and P_2 will be synthesized by their respective controllers while optimally matching their internal model's computations to the measured plant data [6].

3. Performance analysis

The performances of both a centralized optimal controller and a corresponding uncertain-dynamics-based hierarchical controller in response to a set of demands have shown that there is perfect agreement between the demand, plant and controller's model of the neutron power and steam drum pressure for both implementations.

The control model can track the plant exactly and the demand is maintained with an error of about 0.1 % during the transient [30].

Figure 4.19 and Figure 4.20 show the transient of state variables and the corresponding control actions with different triggered rules, rule 1 or rule 2 respectively, where the supervisor detects that the core exit temperature has reduced a high limit and imposes a runback demand on subsystem's SS1 neutron power. The core outlet temperature does not attain the demanded value since the capabilities of the flow controller are exceeded due to saturation.

The approach to hierarchical control of large systems described here eliminates the need for the typical iterative computations to account for the coupling effects between subsystems. In addition, the computational independence between the unknown terms generated by each subsystem's controller facilitates both its implementation in a distributed network of CPUs, and the isolation and diagnoses of sensor and system component failures.

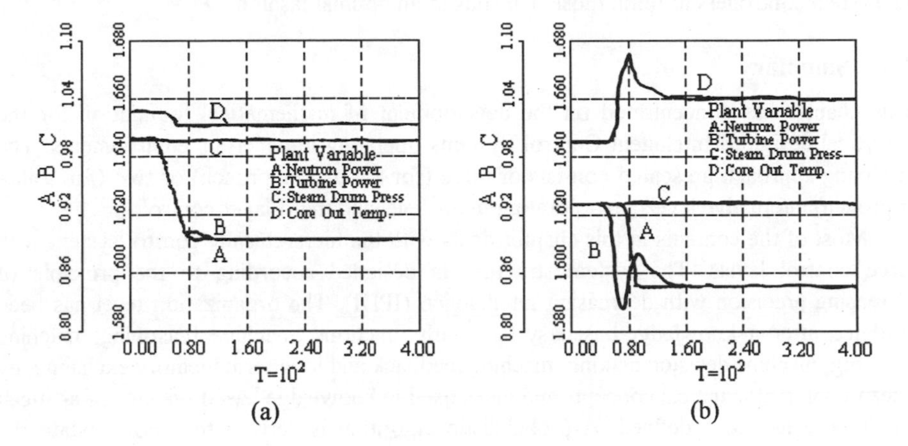

Figure 4.19. State variables (pump failure).
(a) Rule 1 triggered; (b) Rule 2 triggered
(with Rovers *et al.*, 1988)

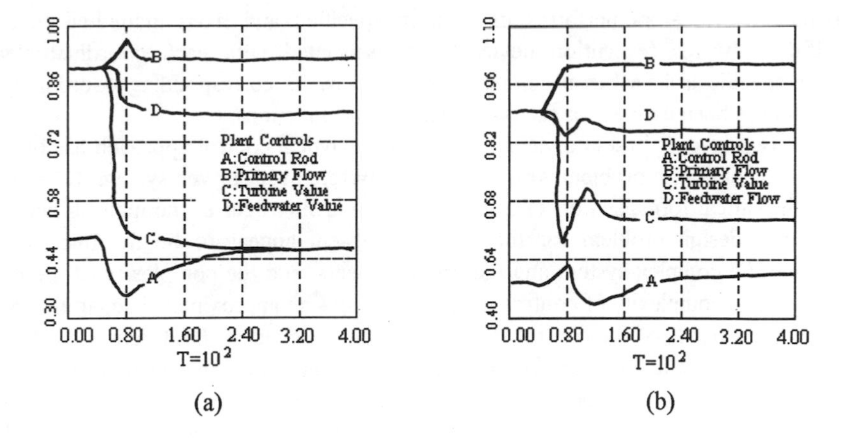

Figure 4.20. Plant controls (pump failure).
(a) Rule 1 triggered; (b) Rule 2 triggered
(with Rovers *et al.*, 1988)

The role of the coordinator is transformed from that of an inflexible black-box-like numerical procedure to that of an intelligent supervisor of subsystem performance. By incorporating symbolic and numerical recipes, the supervisor issues the appropriate set of demands for each subsystem required for the specific goal. It is the task of the individual subsystem controllers to fulfill those demands in an optimal fashion.

4.4 Summary

This chapter has concentrated on the development of mathematical foundation for the analytical design of Intelligent Control Systems operating in uncertain environments. The modeling approach presented consists of three (for Saridis's approach) or two (for Villa's approach) interactive levels incorporating multi-level and multi-layer control.

Most of the contents in this chapter deals with the hierarchically control system with three control levels. The system structure is designed according to the principle of increasing precision with decreasing intelligence (IPDI). The organization level has been modeled after a knowledge-based system with functions: machine reasoning, machine planning, machine decision making, machine feedback and long term memory exchange, by interpreting mathematical concepts and ideas used in knowledge based systems of artificial intelligence has been defined. A probabilistic algorithm is derived to accommodate the organizer functions. Furthermore, an expert system cell is mentioned.

The model of coordination level is composed of a number of coordinators. The coordinators have fixed structures and do not communicate with each other. A dispatcher of variable structure serves as the communicator between the different coordinators. Each individual coordinators performs its own prespecified and fixed individual functions. Specific hardwares (execution devices) are associated with each coordinator which execute specific tasks when a command is issued by its corresponding coordinator. An algorithm has been derived for the coordination level functions.

The two-layer hybrid knowledge based and analytical control approach has proposed for the control design problem with a complex discrete event-driven system. By analyzing the mathematical formalization of a multi-layer control structure, the main difficulties of the control design problem for the complex system appear to be the uncertainty in describing the complex system and the driving events, and the complexity of the control structure. The complexity of control structure requires an approximated separation among the frequency bands of the control strategies to be designed at the different layers of the decisional architecture. For the problem of event complexity, the techniques of pattern recognition are used to classify the events before an unknown event has occurred. In addition, due to the initial high uncertainty about the event model, the knowledge based control choice procedure has also been used to drive initial transient.

Based on four assumptions, a mathematical representation of this hybrid control is described, and five design problems, i.e., local identification, local control design, global identification, global control design, and goal coordination problems are analyzed. By connecting the global control design problem and the goal coordination problem, a two-layer hierarchical control architecture is derived, and three types of control design phases are defined, i.e., $p(\_)$, $q(\_)$, and $r(\_)$. All of the design problems are discussed, and every low-layer problem to be solved in the desired hybrid control system is described in terms of the related tasks. This task structure clarifies that the operation modes of two-layer hybrid control system gives evidence of the link between knowledge-based solution and the analytical solution.

Two application paradigms of hierarchically intelligent control systems are demonstrated in Section 4.3. One is robotic assembly system designed to manufacture a mechanical assembly out of four parts coming at random on a conveyor belt; another is hierarchical control system for a nuclear reactor plant using uncertain dynamic techniques subject to multiple demands. Usually, a combination of different techniques for knowledge representation and searching and reasoning, such as state space, AND/OR graph, predicate logic, semantic network, fuzzy sets, neural network, Petri nets and so on, is applied to a hierarchically intelligent control system. Therefore, the application paradigms introduced in this chapter are very limited, some more examples will be and can be only presented in the succeeding chapters.

References

1. J. S. Albus, "A new approach to manipulation control: The cerebella model articulation controller," *Trans of ASME J. Dynamics Systems, Measurement and Control*, **97** (1975) 220.
2. J. S. Albus, *Brains, Behavior, and Robotics* (Byte/McGraw-Hill, Peterborough, NH, 1981).
3. J. S. Albus, "Outline for a theory of intelligence," *IEEE Trans. SMC*, **21** (1991) 1094.
4. S. M. Babcock, W. R. Hamel, and H. C. Martin, "Real-time control of mobile robot with multiple manipulators," *ORNL CESAR/TRP-O5*, 1983.
5. J. Barhen, G. Saussure, and C. R. Weisbin, "Strategy planning by an intelligent machine," *ORNL CESAR/TRP-01*, 1983.
6. C. R. Brittain, P. I. Otaduy, R. B. Perez, "New approach to hierarchical decomposition of large scale systems," *Proc IEEE ISIC*, 1988, 108-112.
7. B. Conterno *et al.*, "A large scale system approach to the production and control problem," *Proc.24th IEEE Conference on Decision and Control*, 1985.
8. K. S. Fu, "Learning control systems — review and outlook," *IEEE Trans. on Automatic Control*, AC-**15** (1970) 210.
9. K. S. Fu, *et al.*, "A heuristic approach to reinforcement learning control system," *IEEE Trans.* AC-**10** (1965), 390-398.
10. R. M. Glorioso, *Engineering Cybernetics* (Prentice-Hall, Englewood Cliffs, NJ, 1975).
11. J. H. Graham, and G. N. Saridis, "Linguistic methods for hierarchically intelligent control," *TR-EE 80-34,* Purdue University, West Lafayette, IN, 1980

12. J. H. Graham, and G. N. Saridis, "Linguistic structures for hierarchical systems," *IEEE Trans. SMC*, **12** (1982) 325.
13. C. Guo, Z. Sun, S. Zhu, and G. Fu, "Hierarchical intelligent control for industrial boiler," *Proc. IEEE ISIC*, 1992, 116-121.
14. Y. C. Ho, and C. Cassandras, "A new approach to the analysis of discrete event dynamic system," *Automatica*, **19** (1983).
15. E .T. Jaynes, "Information theory and statistical mechanics," *Physical Review*, **106** (4), 1957.
16. E. T. Jaynes, "On the rationale of maximum entropy methods," *Proc.IEEE*, **70** (9) 1982.
17. D. R. Lefebvre, and G. N. Saridis, "A computer architecture for intelligent machines." *CIRSSE Technical Report*, No.108, RPI, 1991.
18. D. Luemberger, *Optimization by Vector Space Methods* (Wiley, New York, 1969).
19. J. E. McInroy, and G. N. Saridis, "Reliability based control and sensing design for intelligent machines," in: *Reliability Analysis*, H. H. Graham ed. (Elsevier, Horth Holland, 1991).
20. A. Meystel, "Intelligent control of a multiactuator system," *Proc. 4th IFAC-IFIP Symposium on Information Control Problems in Manufacturing Technology* (NBS, Gainthersburg, MD, 1982).
21. A. Meystel, "Cognitive controller for autonomous systems," *Proc. IEEE Workshop on Intelligent Control*, 1986, 222.
22. A. Meystel, "Knowledge-based controller for intelligent mobile robots," in: *AI and Man-Machine Interface* (Springer-Verlag, New York, 1986).
23. A. Meystel, "Intelligent control in robotics," *J. of Robotic Systems*, 5(1988).
24. A. Meystel ed., "Special Issue on Intelligent Control," *J. of Intelligent and Robotic Systems*, **2** (1989).
25. M. C. Moed, and G. N. Saridis, "A Boltzmann machine for the organization of intelligent machine," *IEEE Trans. SMC*, **20** (1990).
26. N. J. Nilsson, "A mobile automation: an application of AI techniques," *Proc. IJCAI*, Washington, D.C., 1969.
27. N. J. Nilsson, *Problem Solving Methods in AI* (McGraw-Hill, New York, 1971).
28. N. J. Nilsson, *Principles of Artificial Intelligence* (Tioga Publishing Co., Palo Alto, 1980).
29. E. M. Oblow, and J. Barhen, "Mathematical modeling and robot dynamics," *ORNL CESAR/TPR-03*, 1983.
30. L. A. Rovere, P. J. Otaduy, C. R. Brittain, and R. B. Perez, "Hierarchical control of a nuclear reactor using uncertain dynamics techniques," *Proc. IEEE ISIC*, 1988, 713-717.

31. G. N. Saridis, *Self-Organizing Control of Stochastic Systems* (Marcel Dekker, New York, 1977).
32. G. N. Saridis, and H. E. Stephanou, "A hierarchical approach to the control of a prosthetic arm," *IEEE Trans. Systems Man and Cybernetics*, SMC-7 (1977) 407.
33. G. N. Saridis, "Toward the realization of intelligent controls," *Proc. IEEE*, **67**, **8** (1979).
34. G. N. Saridis, "Intelligent robotic control," *IEEE Trans. Automatic Control*, 28 (1983) 547.
35. G. N. Saridis, "Foundations of intelligent controls," in: *Proc. IEEE Workshop on Intelligent Controls* (RPI, Troy, New York, 1985) 23.
36. G. N. Saridis, and K. P. Valavanis, "Analytical design of intelligent machine," *Automatica*, **24** (1988) 123.
37. G. N. Saridis, "Analytic formulation of the IPDI for intelligent machines," Automatica, **25** (1989) 461-467.
38. G. N. Saridis, "On the revised theory of intelligent machine," *CIRSSE Technical Report No.58*, RPI, 1990.
39. G. N. Saridis, "Architectures for intelligent machines," *CIRSSE Technical Report*, No.96, RPI, 1991.
40. A. D. Solomon, "World model,: *ORNL CESAR/TRP-02*, 1983.
41. K. P. Valavanis, "On the hierarchical modeling analysis and simulation of FMS with extended Petri nets," *IEEE Trans. SMC*, **20** (1990).
42. K. P. Valavanis, and S. J. Carelo, "An efficient planning technique for robotic assemblies and intelligent robotic systems," *J. Intelligent and Robotic Systems,* **3** (4), 1990.
43. K. P. Valavanis, and G. N. Saridis, "Information theoretic modeling of intelligent robotic systems," *IEEE Trans. SMC*, **18** (1988).
44. K. P. Valavanis, and G. N. Saridis, *Intelligent Robotic Systems: Theory, Design and Applications* (Kluwer Academic Publishers, Boston, 1992).
45. V. P. Valavanis, and P. H. Yuan, "Hardware and software for intelligent robotic systems," *J.Intelligent and Robotic Systems*, **1(4)** 1989.
46. A. Villa, "Hybrid knowledge-based/analytical control of uncertain systems," in: *Proc. IEEE ISIC*, 1987, 59-90.
47. A. Villa, "Expert control system: A rationale to handle system uncertainty and control structure complexity," *IEEE Trans. on SMC*, 1987.
48. A. Villa, *et al.*, "Expert control theory: a key for solving production planning and control problems in flexible manufacturing," *Proc IEEE Int. Conf. Robotics and Automation*, 1986.
49. F. Y. Wang, *et al.*, "A Petri net coordination model of intelligent mobile robots,"

IEEE Trans. SMC, **21** (1991) 777.

50. Y. N. Wang, Z.-X. Cai, "Real-time expert intelligent control system REICS," *Proc. IFAC-AARTC*, 1992, 307-312.
51. G. Zames, "On the metric complexity of causal linear systems, ε-entropy and ε-dimension for continuous time," *IEEE Trans. Automatic Control*, **24** (1979) 222.

CHAPTER 5
EXPERT CONTROL SYSTEMS

Expert systems are becoming more and more popular in all kinds of application fields where highly reliable, rapid decision-making and different functions are involved. These functions include explanation, prediction, analysis, diagnosis, debugging, design, planning, control, monitoring, instruction, testing, counsel, management, evaluation and decision support and so on. Expert control system is one of the important members in the family of expert systems. Expert control system possesses multiple functions such as analysis, diagnosis, monitoring, debugging and prediction, etc.

As we know from its name, the expert control system is a control system that is based on the technique of expert systems, and is a typical and widely used knowledge-based control system.

Hayes-Roth *et al.* introduced expert control systems in 1983 [6]. They described that an expert control system is a system whose overall behavior can be adaptively governed. To do this, the control system must repeatedly interpret the current situation, predict the future, diagnose the causes of anticipated problems, formulate a remedial plan, and monitor its execution to ensure success. The first application of the expert control system was reported in 1984 [13]. It is a distributed real-time process control system for an oil-refinery. Åström *et al.* published his paper entitled "Expert Control" in 1986 [1]. Since then, many kinds of expert control systems have been developed and applied. Expert systems and intelligent control are closely connected; they at least have one thing in common, i.e., both are based on imitating the human intelligence, and deal with the same uncertainty problems.

This chapter mainly concerns with three issues: fundamentals to expert systems, expert control systems, and real-time expert control systems. We will treat them as different sections from Section 5.1 to Section 5.3 which are as follows.

5.1 Expert Systems

Experts are people who are very familiar with solving specific types of problems. Their skill usually comes from extensive experience and detailed specialized knowledge of the problems they handle.

Expert systems have been one of the most emphasized areas in the field of artificial intelligence during the past twenty years. For this reason, many highly specialized systems containing the expertise needed to solve problems of medical diagnosis and treatment, chemical structure analysis, geological exploration, computer configuration selection, consultation, supervision, planning, control and fault diagnosis of industrial systems, and so on have been built by various research groups.

What is an expert system? Until now, no unified definition has been given. An expert system can handle real-world complex problems which need an expert's interpretation and solve problems by using a computer model of human expert reasoning to reach the same conclusions that the human expert would do if he or she faces with a comparable problem.

Feigenbaum, a pioneer in the field of expert systems, defined an expert system as an intelligent computer program that uses knowledge and inference procedures to solve problems that are difficult enough to require significant human expertise for their solutions. The knowledge needed to perform at such a level, plus the inference procedures used, can be regarded as a model of the expertise of the best practitioners of the field.

In short, an expert system is an intelligent computer program that can perform special and difficult task(s) in some field(s) at the level of human experts.

The fundamental function of the expert system depends upon its knowledge, therefore, the expert system is sometimes called knowledge-based system.

5.1.1 Feature of Expert Systems

Comparing with traditional programs, the expert systems possess the following features:

1. Heuristic. The problems to be solved by expert systems usually are of illegitimate structures whose problem-solving knowledge includes not only the theoretical knowledge and common-sense knowledge, but also the private heuristic knowledge of the experts. The heuristic knowledge might be incomplete and inaccurate; however, they can perform high-class analysis and inferences to solve complex or difficult problems in the case of incomplete and inaccurate information. During a problem-solving process the expert system utilizes and combines the heuristic knowledge, even multiple expertise, simulates the thinking and cognitive process of experts. Therefore, the expert systems possess the heuristic features and are able to make inference, judgment, decision and conclusion(s) with high effectiveness and accuracy.

2. Transparency. The expert system can interpret its own inference process and answer user's queries that make the user understand and heighten confidence in the expert system. When an expert system interprets user's queries, it utilizes knowledge applied and middle results generated during the problem-solving process in the knowledge base. This mechanism of interpretation provides a transparent interface for the expert system. Therefore the users could accept the expert system easily. Meanwhile, the reasonableness of the knowledge utilization in problem-solving process can be examined by checking the display for the inference path of this mechanism. When solved results of a problem are unsatisfactory, the knowledge engineer and user can find the failure of the inference.

3. Flexibility. The flexibility of an expert system is an ability to extend and enrich the knowledge base, and improve the system performance under nonprogramming states, i.e., a self-learning ability. Because the knowledge base and the reasoning machine in the expert system are independent relatively, and the knowledge representation in knowledge base is obvious, the extension and revision of the knowledge base of the expert system are more flexible and more convenient. Therefore the expert system can add new knowledge, revise and refresh original ones. After an expert system is built, the reasoning machine can select various relative knowledge from the knowledge base and construct problem-solving sequences according to the particulars of the specific solved problems. This ability of the reasoning machine can be realized as one side of the flexibility of the expert system. This feature has led the expert system to a very wide field of applications.

4. Symbolic operation. Different from the traditional programs that deal with data processing, the expert systems emphasize upon symbolic processing — symbolic operation and symbolic inference rather than numerical computation. The expert systems use symbols to represent knowledge, set of symbols to represent concept of problems. A symbol is a string programming and can be used to represent concepts in the realistic world.

5. Reasoning with uncertainty. The method for solving problems by domain experts is mostly experimental; the experimental knowledge is usually represented with inaccuracy and exists with a certain probability. In addition, the provided information concerning the problem is often uncertain. The expert systems can integrate and utilize the fuzzy and uncertain information and knowledge to make inference.

Owing to the features described above, the expert systems possess a lot of advantages [2, 19].

5.1.2 Architectures and Types of Expert Systems

Figure 5.1 gives a simplified block diagram of a rule-based expert system. As seen from Figure 5.1, knowledge base and a reasoning machine are the most essential components of the expert system.

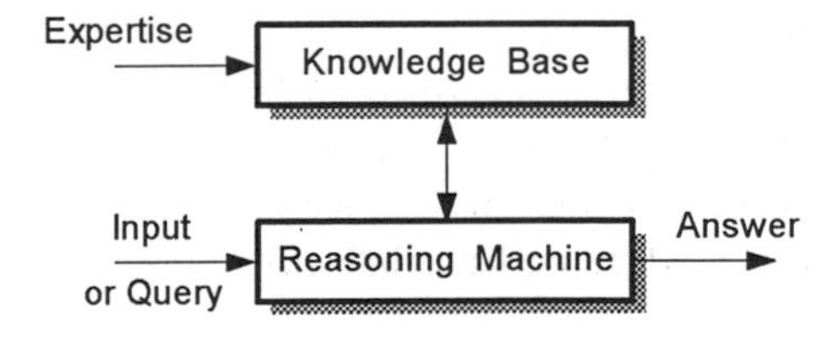

Figure 5.1. Simplified block diagram of expert system.

An ideal architecture of an expert system is shown in Figure 5.2. The system contains the following components.

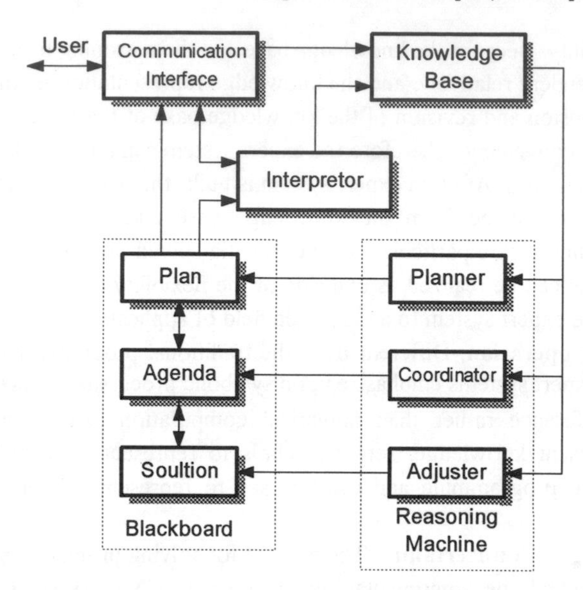

Figure 5.2. Architecture of an ideal expert system.

1. Knowledge Base. It is used to store knowledge from the experts of special field(s). It contains facts and feasible operators or rules for heuristic planning and problem solving. The other data is stored in a separate database called global database, or database simply.

2. Reasoning Machine. It is used to memorize the reasoning rules and the control strategies applied. According to the information from the knowledge base, the reasoning machine can coordinate the whole system in a logical manner, draw inference and make a decision. A planner applies the corresponding rule in the knowledge base to devise an ideal planning sequence for a given task specified. A coordinator is used to control the order of rule processing. An adjuster can adjust previous conclusions when new data (or knowledge) alter their bases of support.

3. User Interface. An interface is used to communicate between the user and the expert system, and is called communication interface or user interface. The user interacts with the expert system in problem-oriented language such as in restricted English, graphics or a structure editor. The interface mediates information exchanges between the expert system and the human user.

4. Interpreter. Through the user interface, interpreter explains user questions, commands and other information generated by the expert system, including answers to questions, explanations and justifications for its behavior, and requests for data. For example, it answers questions why some conclusions was reached or why some alternatives was rejected. Therefore, the user can input data, ask questions, realize the reasoning process and result, and get a better understanding of them.

5. Blackboard. The blackboard is applied to record intermediate hypotheses and decisions that the expert system manipulates. Three types of decisions are recorded in the blackboard, i.e., plan, agenda, and solution elements. The plan elements describe the overall or general attack that the system will pursue against the current problem, including current plans, problem states, and contexts. The agenda elements record the potential actions waiting for execution, which generally correspond to knowledge base rules that seem relevant to some decision placed on the blackboard previously. The solution elements represent the candidate hypotheses and decisions the system has generated thus far, along with the dependencies that relate decisions to one another.

It is noted that the almost no exiting expert system contains all the components shown above, but some components, especially the knowledge base and reasoning machine, occur in almost all expert systems. Moreover, many expert systems use global database in place of the blackboard. The global database contains information related to specific tasks and the current state.

There are ten types of expert systems, they are interpretation, prediction, diagnosis, design, planning, monitoring, design, debugging, repair, instruction and control systems which is listed in Table 5.1 [6]. All of them have some concrete applications.

Table 5.1. Types of expert systems

Category	Problem Addressed
Interpretation	inferring situation descriptions from sensor data
Prediction	inferring likely consequences of given situations
Diagnosis	inferring system malfunctions from observable
Design	Configuring objects under constrains
Planning	Designing actions
Monitoring	Comparing observations to plan vulnerabilities
Debugging	Prescribing remedies for malfunctions
Repair	Executing a plan to administer a prescribed remedy
Instruction	Diagnosing, debugging, and repairing student behavior
Control	Interpreting, predicting, repairing, and monitoring system behaviors

5.1.3 Step for Building Expert Systems

The key for successfully building an expert system is to begin it from a smaller one, and extend and test it step by step, make it into a larger-scale and more perfect system.

The general procedure for building expert systems is as follows.

1. Design of initial knowledge base.

Design of knowledge base is the most important and most difficult task. The design involves the following five stages which is shown in Figure 5.3.

(i) Problem identification. In the first stage, the essence of the studied problem, such as what is the task(s) to be solved, how to define the task(s), can the task(s) be decomposed into sub-tasks or sub-problems, which typical data are contained in the problem, etc., are identified, and key concepts and relations are found.

(ii) Knowledge conceptualization. In the second stage, the key concepts and relations mentioned above are made explicit. The knowledge engineer may find it useful to diagram these concepts and relations to make the conceptual base permanent for the prototype system. The concepts and relations include data types, given and inferred states, for names subtask, for names control strategy, hypotheses, related objects in the domain, hierarchy diagram and relations, processes and constrains in problem-solving, information flow, and so on.

(iii) Concept formalization. The third stage includes mapping the key concepts, subproblems and information flow characteristics into more formal representations based on various knowledge-engineering tools or frameworks. Three essential factors in the formalization process are the hypothesis space, the underlying model of the process, and the characteristics of the data.

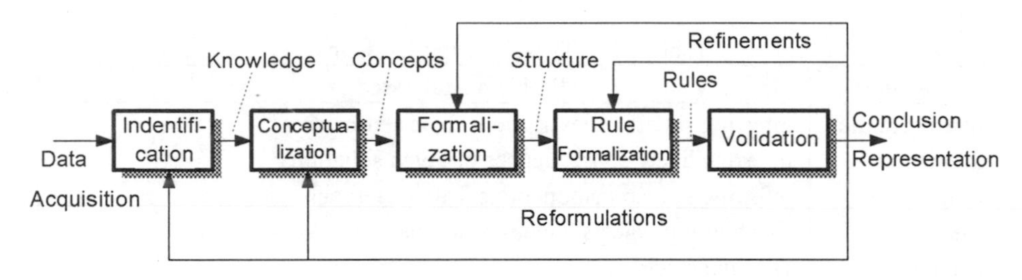

Figure 5.3. Stages for designing knowledge base.

(iv) Rule formulation. In the fourth stage, set up a prototype expert system, formulate rules, and implement the knowledge base. The implementation involves mapping the formalized knowledge from the previous stage into the representational framework associated with the chosen tool. The knowledge in the framework becomes an executable program or statements.

(v) Rule validation. This stage involves evaluating the prototype system and the representational forms used to implement it, affirming the reasonableness of the rule-formulating knowledge, and testing the effectiveness of the rules.

From the above discussion and Figure 5.3 we know that developing the expert system comprises two main phases. The first phase involves identifying and conceptualizing the problem. Identification means selecting and acquiring an expert, knowledge sources, and resources, and defining the problem clearly. Conceptualization means uncovering the key concepts and relations needed to characterize the problem. The second phase deals with the formalization, implementation, and validation of an appropriate architecture for the expert system, including constant reformulation of concepts, redesign of representations, and refinement of the implemented system. Revision results from the expert's criticisms and suggestions for improving the system's behavior and competence.

2. Development and test for prototype system

After selecting the methodologies of knowledge representation, the experimental subsets needed for the whole system can be built. These subsets involve typical knowledge of the whole model, and deal with tasks and inference process related to the test simply.

3. Improvement and induction for the knowledge base

In the last step for building expert systems, the knowledge base and inference rules need to be tested repeatedly, so more perfect results can be induced. For building expert systems, the knowledge engineer has a long way to go. After several months of interacting with the expert to define the basic concepts, extract rules, and modify the consultant's problem-solving strategy, the knowledge engineer will have constructed a knowledge base with a few hundred rules and parameters. Then he or she continues to refine and extend the prototype model for several more months. At this point the knowledge engineer decides whether the current prototype has become unwieldy as a result of ad hoc incremental improvement. If it is, the knowledge engineer discards it and quickly implements a new and more efficient system, which now must be extended and refined. Thus, after more than a year of concentrated efforts, the knowledge engineer will have completed the development of a credible expert system. Several more months will be needed to test the new system in the field and effectively change it from a prototype model to a working expert system.

5.1.4 Model-Based Expert Systems

When the advantages and shortages of the expert system are concerned, many AI researchers think that the idea of knowledge base developed from rule-based expert systems is very important. This idea has not only promoted the development of artificial intelligence but also effected the development of the whole computer science. However, they point out that the idea of rule-based knowledge base restricts the further development of the expert systems. Single rule-base representation is not perfect enough to solve problems faced in developing the expert system. One point of view for artificial intelligence is that AI deals with study on acquisition, representation and utilization of various qualitative models including physical, perceptive, cognitive, and social system models. According to this view point, a knowledge base in a knowledge-based system can be viewed as synthesis of various models, and these models are usually qualitative ones. Constructing a model is closely connected with knowledge, therefore, the acquisition, representation and utilization of knowledge. With this point of view, an expert system is an integrated system with various models different in mechanism and operation mode. This expert system sometimes is called model-based expert system.

In the model-based expert system, a deep model of domain is built. The deep model knows the problem-solving space for heuristic searching. As a result, this new system can make inference through two reasoning modes: (i) make inference by using heuristic knowledge as it does in rule-based expert system; (ii) make inference by generating and testing the searching space of the deep model.

The model-based expert system is one of a new generation of expert systems. The new generation of expert systems possesses the following features.

1. Parallel and distributed processing

Based on various algorithms, the new expert systems use parallel techniques for inference and implementation and is suited to work in the hardware environment with multiprocessors, so the systems possess the function of parallel processing. The multiprocessors in the system can make processing both in synchronous and asynchronous mode, and realize synchronization of communication among components of the expert system distributed in every processor. This function would increase the efficiency of processing and reliability of the system.

2. Synergetic work with multiple expert systems

Synergetic expert system is a new kind of expert system, in which there are multiple expert systems worked synergetically. In the synergetic expert system communication among every sub-expert system can be carried out. The output(s) of one or more sub-systems might be the input of another sub-system, which can even become a feedback information of itself or its parent system. As a result, the problem-solving capability of the multi-expert system would increase through the synergetic work of the multiple expert systems.

3. High-level descriptive languages

In order to build a new expert system, a high-level descriptive language of expert systems which describes the function, performance and interface of the system, and a knowledge representation language which describes domain knowledge are used. These two high-level languages make it possible to select or synthesize a suitable mode of knowledge representation automatically or semi-automatically, then the described knowledge can be transformed into a knowledge base followed by forming correspondent inference-execution mechanism, interpretation mechanism, user interface, and learning module, and so on.

4. Self-learning function

The new expert system should provide high-level functions in knowledge acquisition and learning, as well as in available knowledge acquisition tool. The system can make inference, acquire new knowledge, summarize new experiences, and extend knowledge base by means of the existing knowledge in the knowledge base and the dynamic answer of the system to the user's query.

5. New inference techniques

The most existing expert system can only make a deductive inference, but the new expert system can do more things including inductive inference (e.g., associative inference and analog inference), non-standard logic inference (e.g., monotonous reasoning and weighed logic reasoning) and incomplete knowledge-based inference as well as fuzzy inference, and so on.

6. Advanced intellectual man-machine interface

The new generation of expert systems is expected to understand natural language, implement multimedia input/output. This requirement needs support from hardware.

No existing expert system possesses all the features mentioned above, but some new expert systems possess several of them.

5.2 Expert Control Systems

There are some important differences between the expert systems and the expert control systems. They are:

(i) The expert systems simply complete consultative function for problems of special domains and aid users to work. Its inference is knowledge-based inference, and the inference results are knowledge items, new knowledge items or changed from the original one. However, the expert control systems need to make decisions to control action independently and automatically, and its inference results are either changed knowledge items or an activation of some analytic algorithms.

(ii) The expert systems usually work in off-line mode, but need to acquire dynamic information in on-line mode and make real-time control for the system. The requirements for real-time are faced with the following difficult issues: non-monotonic reasoning, asynchronous events, time-based reasoning, parallel reasoning, and other real-time problems.

Emerging from the automatic control fields, expert control is viewed as new paradigms for solving control problems, and has been developed and applied in various areas during the past ten years. People in different areas and with different professional backgrounds have exhibited great interest in the expert control systems.

There exist two main types of expert control: the expert control system and the expert controller. The former is with a more complex structure, higher cost, better performance, and used to plants or processes where higher technical requirements are needed; the latter is with a simpler structure, lower cost and has a performance that can meet the general requirements for the industrial process control. Therefore, the expert controller is more widely used in industry control. In this section, first of all, we present the control requirements and design principles for the expert controller, then introduce the structures and types of expert control systems. In the end we discuss and interpret some design paradigms of expert controllers.

5.2.1 Control Requirements and Design Principle of Expert Control Systems

1. Some control requirements to expert controllers

In the evolution of adaptive control, the expert controller is a new and significant milestone for the design of practical adaptive mechanisms. Up to now, the two major drawbacks of adaptive control have been the requirements for an accurate model of the plant and the inability to set meaningful goals for the adaptive mechanism. The expert controller does not suffer from such shortcomings, since it circumvents the need for a mathematical model of the plant and provides meaningful time-domain based goals for the adaptive design.

Generally speaking, there is no unified and fixed requirements for an expert control system. The requirements are dependent upon the specific applications. However, we may still consider some general requirements for the expert controllers (expert control systems) which are as follows.

(i) High reliability of operation

If the expert controller were not used to replace conventional controllers for some specific plants or systems, then the whole control system would become much more complex, especially in the structure of hardware; as a result, the reliability of the system would decrease. Therefore, a higher reliability is required for an expert controller which usually has a convenient monitoring capability.

(ii) High ability of decision-making

Decision-making is one of the key abilities for knowledge-based control systems. Most expert control systems are required to have different levels of ability of decision-making. Expert control systems can cope with problems such as uncertainty, incompleteness and imprecision that are hard to deal with by conventional control approaches.

(iii) Versatility of application

Versatility of application includes easy to be developed, paradigm diversity, hybrid knowledge representation facilities, flexible dimensionality of the global database, underlying hardware flexibility, multiple reasoning mechanisms such as hypothetical reasoning, non-monotonic reasoning, approximate reasoning, and open and extensive architecture.

(iv) Flexibilities

This principle involves flexibility in control strategy, flexibility in data management, flexibility in expressing the expertise, flexibility in explanation, flexibility in pattern matching and flexibility in procedural attachment, etc.

(v) Human-like capability

The control level of the expert control system has to reach the level of human experts.

According to the cases of application, the control requirements are specified. For example, in the case of plants with significant non-linearities of actuator, the objectives of the expert controller are two-fold. The first objective is to assess the non-linear characteristics of the actuator in terms of both the saturation levels and the step-up and step-down linearity. This information is then used by the knowledge base of the expert controller to set up the logic for the anti-vibration protection on the integral part of the controller. The second objective of this expert controller is to tune a PI control action so that the closed-loop step response lies within limits set by the commissioning engineer or plant operator. The limits within which the expert controller works are maximum overshoot, maximum undershoot, and damping ratio. These limits are then used by the expert controller to compare the untuned closed-loop transient step response with the desired response shape and to update the controller in accordance with this disparity [16].

As another example, let us look at the case of process control. In this case, the special requirements for expert controller might deal with: continuous operation, multiple expert operating at different shifts, changing propagation and amplifying through the system, inconsistency in the quality of input materials, the process which itself changes gradually over time, very complex structures of the plant, multiple sensors, different levels of description of the plant maybe appropriate for different control tasks, the models of the plant maybe in many different forms, and so on.

2. Design principles for expert controllers

According to the discussion of the requirements for the expert controllers, the design principles for expert controllers can be described as follows [2].

(i) Multiplicity in model description

The principle of multiplicity in model description deals with applying multiple description forms of models for the controlled object and controller rather than one form such as an analytical model in the process of design. The design technique of control systems in conventional control theories depends uniquely on the mathematical analytical model of the controlled object. However, in the design of an expert controller, since different kinds of quantitative and qualitative knowledge, accurate and fuzzy information can be processed by using the technique of expert system, the multifarious description forms of models would be allowed. The main description forms include:

(a) **Analytical model** This is a well known description form that involves differential equations, difference equations, transfer functions, expressions of state space and impulse transfer functions, etc., and is employed to describe continuous systems.

(b) **Discrete event-based model** As the name indicated, this model is used for a discrete system and has found even more applications in analysis and design of complex systems.
(c) **Fuzzy model** It is specially effective for describing qualitative knowledge. As the accurate mathematical model of an object is unknown, and only some qualitative knowledge about the controlled process is available, it is more convenient to establish the input and output fuzzy sets as well as their fuzzy relation in terms of fuzzy sets and fuzzy logic.
(d) **Rule-base model** The basic form of a production rule is

$$\langle \text{IF} \rangle \ldots \langle \text{THEN} \rangle \ldots \tag{5.1}$$

it provides organizations for large bodies of facts with mechanisms for inferencing access path development and comparison based on symbol manipulation. The rule-based symbolized model is specially suited to describe the cause-and-effect relations of process and the non-analytical mapping relations, etc. This description form is with higher flexibility, and is convenient to add and modify the rules.
(e) **Model-based model** For model-based expert control system, its knowledge base consists of various models that include physical and psychological models such as neural network model and visual knowledge model, etc., and are usually qualitative models. This approach can reduce on-line computation by off-line pre-computation to produce simplified models individually matched to the task which needs to be performed.

In addition, some other description of models can be used in terms of the different specific cases. For instances, a cause-and-effect model of a system can be set by predicate logic, an associative memory model by symbol matrix.

To sum up, in the design of an expert controller we should select one or more available description forms of the model in order to reflect the process performances better and enforce the capability of information processing of the control system.

The general model of an expert controller can be represented as follows:

$$U = f(E, K, I, G). \tag{5.2}$$

Its basic form is:

$$\text{IF} \quad E \quad \text{AND} \quad K \quad \text{THEN (IF} \quad I \quad \text{THEN } U) \tag{5.3}$$

where

f is intelligent operator;

$E = \{e_1, e_2, \ldots, e_m\}$ is the input set of the controller;

$K = \{k_1, k_2, \ldots, k_n\}$ is the set of experience data and facts in knowledge base;

$I = \{I_1, I_2, \ldots, i_p\}$ is the output set of inference engine;

$U = \{u_1, u_2, \ldots, u_q\}$ is the output set of the controller.

These can be seen in Figure 5.5.
The meaning of the intelligent operator is to make inference in terms of input information

(*E*) and experience data (*K*) and rules in knowledge base, and output the control behavior (*U*) according to the inferred result (*I*). The specific form of the intelligent operator can be implemented by the forms of models introduced above.

(ii) Dexterousness in on-line processing

An important feature of the intelligent control system is the capability of dividing and constructing information in useful forms. The attention on the information processing and utilizing of on-line process have to be especially paid in the design of expert controller. In information storage, the feature information that is meaningful for making control decisions should be memorized, and the out of date information should be forgotten. In information processing, the numerical computation and symbol operation should be combined. In information utilization, the feature information that reflects process performances should be extracted and utilized. Processing and utilizing on-line information dexterously would enhance the information processing capability and decision-making level of the control system.

(iii) Flexibility in control strategies

The flexibility of control strategy is another important principle that should be followed in the design of the expert controller. Owing to the time-variability and uncertainty of industrial objects and stochastic presence of disturbance in the factory fields, some different control strategies of open-loop and closed-loop are required. The control strategies and control parameters can be flexibly modified through on-line information acquired in order to ensure a super control behavior. In addition, the adaptive strategy for handing abnormal phenomena should be designed in the expert controller to enforce the adaptivity of the system.

(iv) Hierarchy in decision-making mechanisms

The neural system of a human being is a hierarchical decision-making system and consists of the cerebrum, cerebellum, brain stem and spinal cord, etc. The core of intelligent control is to anthropomorphize the human intelligence, hence, the design of intelligent controller should follow the principle of hierarchy, i.e., to construct the decision-making mechanisms with hierarchical structure according to the different levels of intelligence of the controller.

(v) Reasoning and decision-making in real-time

This principle is necessary for designing expert controller used to control industrial processes. In order to realize real-time reasoning and decision-making, the size of knowledge base should not be too large, and the inference mechanism should be as simple as possible.

Since expert controller applies multiple model descriptions, its implementation is multiple, too. There are two main methods for implementing the expert controllers. One keeps the structural features of the expert control system, but with a smaller size of knowledge base and simpler inference mechanism; another which combines with the technique of expert systems to increase the decision-making level of the original controller is based on some control algorithm, e.g., PID Algorithm.

5.2.2 Structures of Expert Control Systems

Although the structures of expert control systems may be different due to the different applications and their requirements, almost every expert control system/controller consists of knowledge base, inference engine (reasoning machine), control rules and/or algorithms and so on.

The general roles of the knowledge base and inference engine have been explained in Section 5.1 where we introduced expert systems. Their roles in an expert controller (control system) are similar to the ones in an expert system.

The basic structure of an expert control system is depicted in Figure 5.4. From the view point of the performance indices, the expert control system should provide the controlled object same or very similar performance indices as operated by a master or expert under the same circumstances. In addition, the expert control system should execute the following tasks.

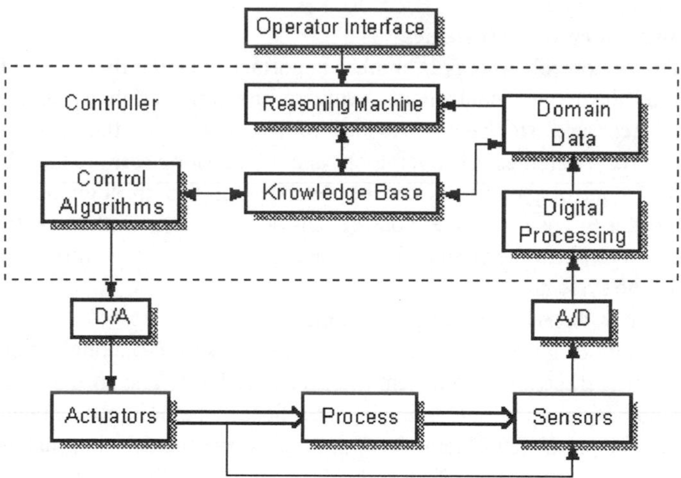

Figure 5.4. A typical structure of expert control system.

(i) Supervise the operation of the plant (process) and controller.
(ii) Examine possible failure or fault of the system components, replace these faulty components or revise control algorithms to keep the necessary performance of the system.
(iii) In special cases, select suitable control algorithm to adapt the variation of the system parameters and environment.

The followings are two specific structures of expert controllers.

1. Industrial expert controller

Figure 5.5 shows a structure of an industrial expert controller (EC) [2]. The knowledge base (KB) is a basic component of the expert controller, and is used to store the domain knowledge of industrial process control, experience of experts (expertise) and facts.

The key problem for building knowledge base is how to express the acquired knowledge. The KB in most industrial expert controllers is built by production rules that possess high flexibility. Every production rule can be added, deleted and modified independently so that the content of the knowledge base can be easily refreshed.

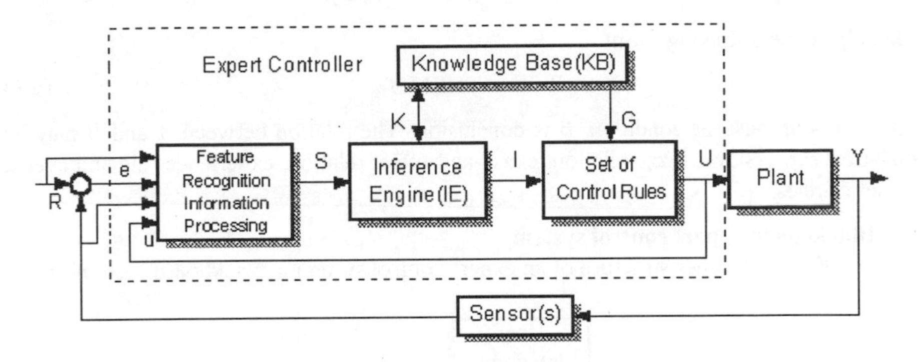

Figure 5.5. Simplified structure of industrial expert controller.

The set of control rule (SCR) sums up every control pattern and control experience of the controlled process. Since the number of rules is limited and the search space is very small, the mechanism of the inference engine (IE) is very simple. A forward chaining reasoning method is used to judge the conditions of every rule in the sequence. If the condition is satisfied, then the rule would be carried out; otherwise, the search will be continued.

The input set of the expert controller is

$$E = (R, e, Y, U) \tag{5.4}$$

see Figure 5.1 and Expression (5.2), where

R is the reference control input;

e is the error signal and

$$e = R - Y \tag{5.5}$$

Y is the controlled output;

S is the output set of the feature information;

G is the command for modifying rules;

I, U and K have been denoted above, and

$$U = f(E, K, I, \mathrm{G}) \tag{5.2}$$

where the intelligent operator f is a composite operation of several other operators:

$$f = g \cdot h \cdot p \tag{5.6}$$

where g, h, p are also intelligent operators, and

$$\left.\begin{aligned} g&: \ E \to S \\ h&: \ S \times K \to I \\ p&: \ I \times G \to U \end{aligned}\right\} \tag{5.7}$$

which have the following form:

$$\text{IF } A \quad \text{THEN } B \tag{5.8}$$

where A is premise or condition, B is conclusion. The relation between A and B may be analytical expressions, fuzzy relation, cause-and-effect relation, experience rule and even a subset of rules.

2. Blackboard expert control system

Figure 5.6 gives another structure of an expert control system, a blackboard system [4, 6].

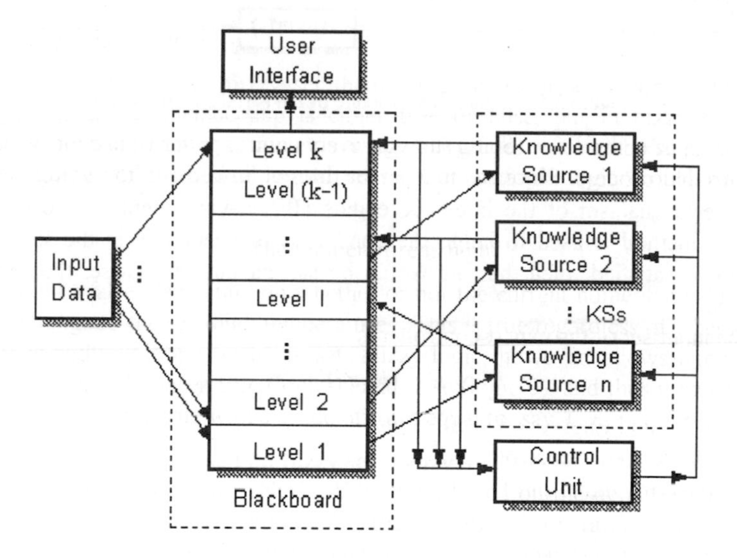

Figure 5.6. Architecture of blackboard expert control system.

Blackboard architecture is a powerful expert system architecture and problem-solving model. It can deal with large amounts of diverse, erroneous, and incomplete knowledge to solve problems. Basic blackboard architecture consists of a shared data region called the blackboard (BB), a set of independent knowledge sources (KSs), and a control unit called the scheduler. It provides a means of organizing the application of this knowledge and the cooperation needed between the sources of this knowledge.

The main advantages that a blackboard system can offer are its flexibility of control and its ability to integrate different kinds of knowledge representation and inferencing techniques. For example, a production rule system or a frame-based system can be part of a blackboard system.

The three components of the blackboard control system can be explained as follows.

(i) **Blackboard (BB)**

Blackboard is used to store knowledge accessible to all the knowledge sources. It is a global data structure used to organize the problem-solving data and to handle communications between the KSs. The objects that are placed on the BB could be input data, partial results, hypotheses, alternatives, and the final solution. Interaction among the KSs is carried out via the BB. A blackboard may be partitioned into an unlimited number of sub-blackboards. That is, a BB can be divided into several BB levels corresponding to different aspects of the solution process. Hence, the objects can be organized hierarchically into different levels of analysis.

Each entry on the BB can have an associated certainty factor. This is one way that the system handles uncertainty in the knowledge. The mechanism of the blackboard ensures that there is a uniform interface between each KS and the partial solutions found so far.

(ii) **Knowledge sources (KSs)**

Knowledge sources are self-selecting modules of domain knowledge. Each knowledge source can be viewed as an independent program specializing in processing a certain type of information or knowledge of a narrower domain. Each knowledge source should have the ability to assess itself on its contribution to the problem-solving process at any instance. The knowledge sources in a blackboard system are separated and independent. Each has its own set of working procedures or rules, and its own private data structure.It contains information necessary for a correct run of the knowledge source. The action part of a knowledge source performs the actual problem-solving and produces changes to the BB. It can follow different kinds of knowledge representation and different inference mechanisms. Hence, the action part of KS can be a production rule system with forward/backward chaining or a frame-based system with slot-filling procedures attached to some slots.

(iii) **Control unit**

The basic problem-solving mechanism in a blackboard system starts with the addition of information to the blackboard by a knowledge source. This event then triggers other knowledge sources interested in the newly posted information. Some testing procedures will be performed on these triggered knowledge sources to determine whether they are eligible for execution. Finally, a triggered knowledge source is chosen to be executed, which will add information to the blackboard, and the cycle starts again.

A control unit is a database with control data objects. Control data objects are used by the control unit to select which knowledge source to be executed from a set of potentially executable knowledge sources. High-level plans and strategies are pre-defined before program execution and chosen opportunistically in a manner that best suits the problem situation. A group of control knowledge sources can dynamically construct plans for achieving the system's behavior. They describe actions that solve the control problem. When executed, they add or modify information on the control blackboard. The control unit then applies whatever control heuristics are recorded on the control blackboard.

The blackboard control architecture enables the system to give higher priority to those knowledge sources that match with the currently chosen focus. The focus of attention can be changed on the control blackboard. Therefore, the system can explore various problem-solving strategies and concentrate on possible solutions which are the most promising.

The great interest for flexibility of control that a blackboard architecture offers is in the control of an autonomous mobile robot. A blackboard architecture for an expert system for controlling a mobile robot was proposed, and the implementation of this expert control system has been carried out.

5.2.3 *Types of Expert Control Systems*

We have sorted the expert control systems into the expert control system and the expert controller according to the complexity of the system structure. In this sub-section, we are going to discuss the type of expert control systems on the basis of the control mechanism of the system.

The expert controllers sometimes are called knowledge-based controllers. On the basis of the role of the knowledge-based controller in the whole control system, the expert control systems can be sorted into the direct expert control system and indirect expert control system. In the former, the controller provides a control signal to the system that effects the controlled process directly, as shown in Figure 5.7(a). The control system shown in Figure 5.5 is an example of the direct expert control system. In the latter, the controller effects the controlled process indirectly, as shown in Figure 5.7(b). The indirect expert control system can be called supervised expert control system or parameter-adaptive control system.

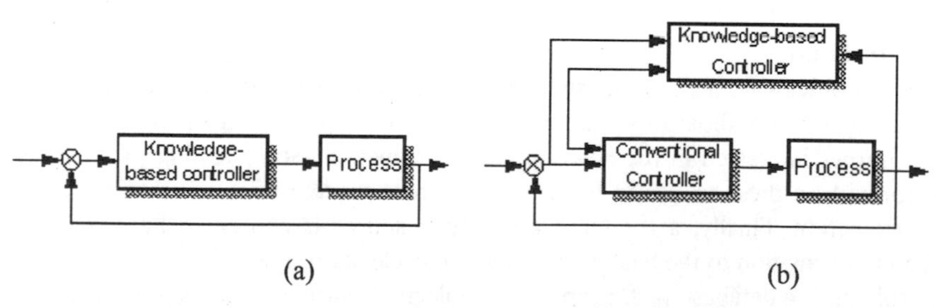

Figure 5.7. Two types of expert control systems.

The main difference between these two control systems is in the design objective of the knowledge base. In the direct expert control systems, the knowledge-based controller imitates the cognitive ability of the human expert or human being directly, and two kinds of rules are designed in the controller: the training rules and the machine rules. The training rules consist of a series of production rules, and map control error into actions of the controlled object directly. The machine rules are dynamic ones acquired by accumulating and learning the control experience of human experts/masters, and are used to implement the learning process of the machine.

In the indirect expert control system, the intelligent (knowledge-based) controller is used to regulate the parameters of the conventional controller, supervise some features of the controlled object such as overshoot, rising time and steady time, etc., then draw up rules for tuning PID parameters so that a steady and high-quality operation of the control system can be ensured. Figure 5.8 in next sub-section presents a paradigm of indirect expert control system.

5.2.4 Paradigm of Expert Controller

As an example, let us introduce an expert tuner for PI controller, which is simply called PI expert controller [11, 16].

1. Software architecture

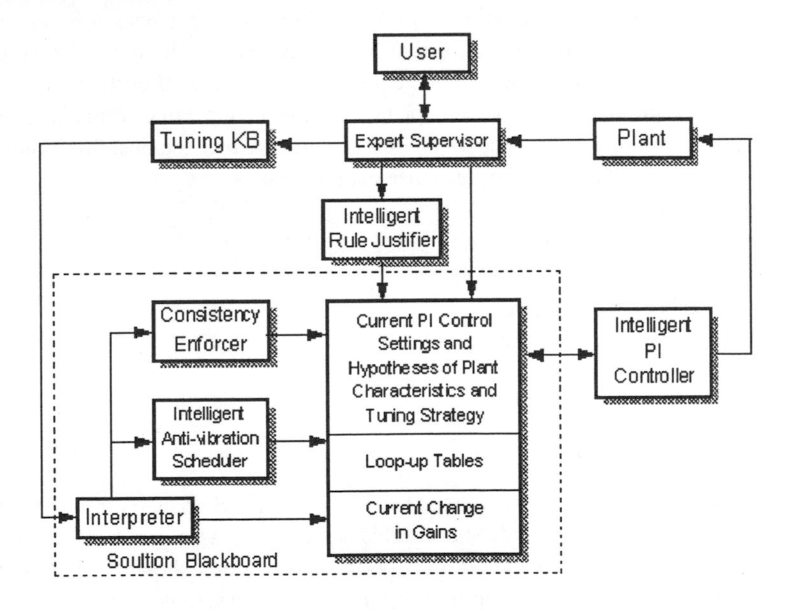

Figure 5.8. Software architecture of an expert tuner.

The software architecture of the control system is shown in Figure 5.8. This expert controller is fundamentally different from existing expert tuners because both outputs of the plant and the variables manipulated by the controller in response to set-point changes are used to adjust the gains of the controller. Also the expert tuner provides facilities for intelligent anti-vibration protection on the integral part of the controller for plants with non-linearities of the actuator.

The software architecture of the expert tuner consists of a hierarchy of tasks which can be decomposed into the fundamental subtasks, and these subtasks in turn yield the following subsystems:

- Expert Signal Conditioner
- Intelligent PI Controller
- Expert Supervisor
- Expert Tuner Inference Engine
- Intelligent Anti-Vibration Protection Scheduler
- Expert Tuner Performance Justifier

2. Knowledge encoding and knowledge deployment

(i) **Knowledge encoding**

This expert tuner uses meta-knowledge to deal with a variety of plants. The meta-knowledge enables previous experiences to be used in order to interpret new situations. Each meta-rule contains information with closed-loop transient response characteristics, open-loop plant characteristics, and actuator characteristics. The characterization of the closed-loop transient response employs the magnitude and time of both the first overshoot and undershoot of the output of the plant and the manipulated variable evaluated at strategic instants along the transient. These data are used to establish the following nine categories of closed-loop transient response characteristics:

- too low monotone
- too low oscillatory
- overshoot-undershoot
- no overshoot-undershoot
- no overshoot-no undershoot
- overshoot-no undershoot
- overshoot-monotone
- overshoot-oscillatory
- over safety limit

In these categories, the closed-loop transient response characteristics can be described as too low, overshoot and undershoot. This situation is as a result of the excursions outside the upper and lower closed-loop performance limits shown in Figure 5.9. This information concerning closed-loop transient response characteristics is coupled with that concerning open-loop plant characteristics and nonlinear actuator characteristics, which are described by the following eight categories:

- no delay monotone
- no delay oscillatory
- short delay monotone
- short delay oscillatory
- medium delay monotone
- medium delay oscillatory
- long delay monotone
- long delay oscillatory

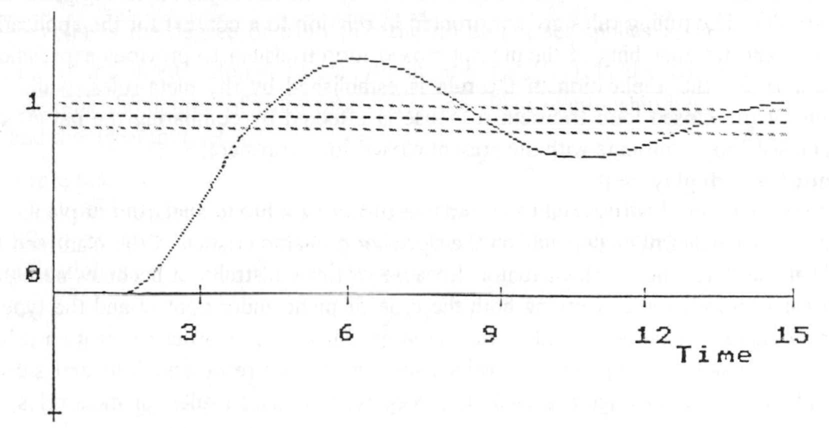

Figure 5.9. Upper and lower closed-loop performance limits.

(with Poter and Jones, 1987)

In these categories, the description of open-loop plants as short delay, medium delay and long delay arises from the consideration of the ratio of the pure time delay in the plant to the dominant time constant. The characteristics associated with the non-linearity of actuator are described by the following eight categories:

- no positive actuation
- no negative actuation
- positive actuator saturation
- negative actuator saturation
- small positive set-point change
- small negative set-point change
- large positive set-point change
- large negative set-point change

The meta-knowledge using this kind of information, organizes the transient responses into recognizable collections of objects, logically decides which anti-vibration rules to be applied to the integral controller and which tuning rules to be applied to the PI controller, and indicates the nature of the consequential improvement in the transient response. The anti-vibration rules and the tuning rules are based on the combined experience of many control engineers and constructed in such a manner that a quantitative judgment can be made concerning the setting up of the constraining logic for the integrator and the modification of the proportional-plus-integral gains for the controller. The rules for setting up the logic to control the vibration of integrator are constructed in relation to a set of open-loop tests performed on the plant for inputs of different magnitude. These broad characterizations of non-linearity are then assumed to be fixed during the tuning process of the controller. The tuning rules are constructed in relation to a context for the application of the rule and the matching of the present closed-loop transient to previous experiences. The context for the application of the rule is established by the meta-rules, whilst the matching of the closed-loop transient response is effected by comparing the patterns of stored closed-loop transients with the present closed-loop transient.

(ii) **Knowledge deployment**

Since the expert tuner is required to be capable of tuning a broad spectrum of plants, the application of certain rules depends on the open-loop characteristics of the plant and the non-linear characteristics of the actuator. Because of this constraint, it becomes necessary to construct hypotheses concerning both the type of plant under control and the type of non-linear characteristic possessed by the actuator. Therefore in order to create a robust expert tuner which can explain and justify itself, both deep representations and surface representations of knowledge are used. The deep representation rules, or meta-rules, are used not only by the consistency enforcer to create intermediate hypotheses but also by the inference engine to modify controller gains, whereas the surface representation rules are used only to change either the logic for the anti-vibration protection in the integral part of the controller or the gains of controller. The consistency enforcer attempts to maintain a constant representation of the emerging solution by using the meta-rules to construct hypotheses concerning the current plant characteristics and the current tuning strategy, and it possesses a facility to advance or retract these hypotheses. The hypotheses created by deep representation rules in the consistency enforcer are contained in the solution blackboard. This contains information on the current hypothesis concerning plant characteristics and non-linear actuator characteristics, the current tuning strategy, the current trend of changes of controller gains, and the anti-vibration in the integral controller.

The operation of the expert tuner is that the open-loop step response tests with different step sizes are performed, and initial hypotheses concerning the plant characteristics and nonlinearity characteristics of actuator are advanced. The interpreter then executes the anti-vibration protection agenda by setting up the logic tables for the anti-vibration protection in the integral controller, and then the inference mechanism switches its attention to tuning the PI controller. The interpreter then executes the tuning agenda for the PI controller by applying the appropriate meta-rules which trigger a rule in the surface representation knowledge base, after which the consistency enforcer examines the current contents of the blackboard and estimates the effect of the rule and then implements appropriate changes for gains and, when the plant is in a steady state, initiates a closed-loop step response test. Furthermore, by using the closed-loop transient response, the consistency enforcer either verifies or modifies the current hypothesis. Once this is completed, the interpreter executes the PI controller tuning agenda again and the whole process is repeated until the controller is tuned to satisfy the specification.

3. Performance of the controller

The practical performance of the expert tuner was evaluated by tuning a digital PI controller on a laboratory test rig. The diagram of this test rig is shown in Figure 5.10 and it consists of a pump and multiple tank assembly, in which the manipulated input variable is the pump flow rate and the controlled output variable is the level of the second tank. The test rig exhibited significant nonlinearities in both the actuator characteristics and the plant dynamics. In this case, a sampling period of one second was chosen with closed-loop performance limits of 5% maximum overshoot and 5% minimum undershoot. The initial PI controller settings were obtained from an open-loop step response.

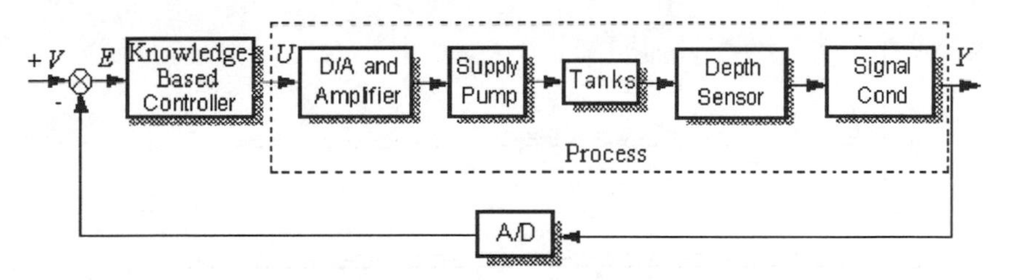

Figure 5.10. Diagram of test rig for expert tuner.

The performance of the initially tuned closed-loop system without any anti-vibration logic is shown in Figure 5.11 from which it is evident that the system has gone into a limit cycle because of the actuator saturation and resulting integrator oscillation (vibration). The performances of initially and finally tuned closed-loop system incorporating the anti-vibration logic are shown in Figure 5.12 and 5.13 respectively. The trajectories of the proportional and integral controller gains are shown in Figure 5.14 [From Poter and Jones, 1987].

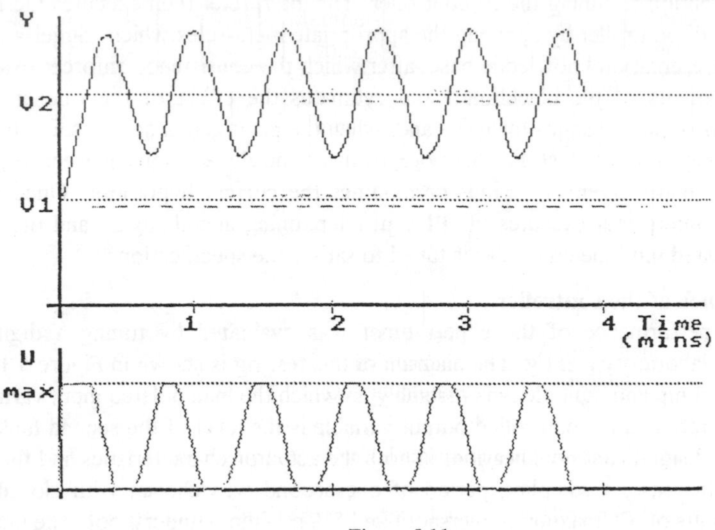

Figure 5.11.

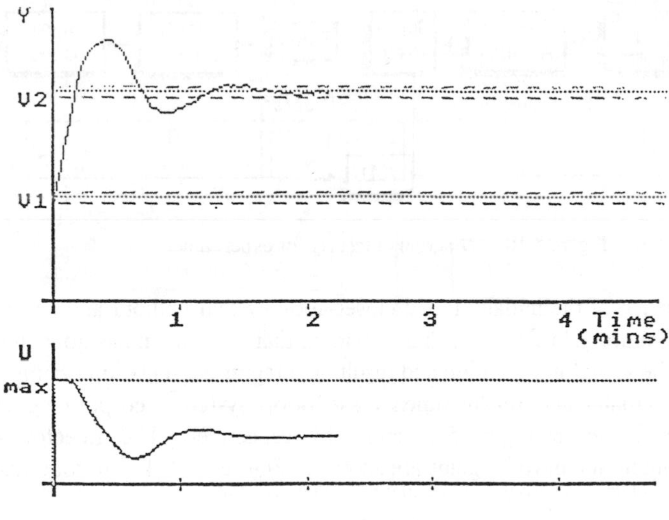

Figure 5.12.

These real-time trials demonstrate the excellent performances of the expert tuner in both setting up the anti-vibration logic on the integral controller and tuning a digital PI controller.

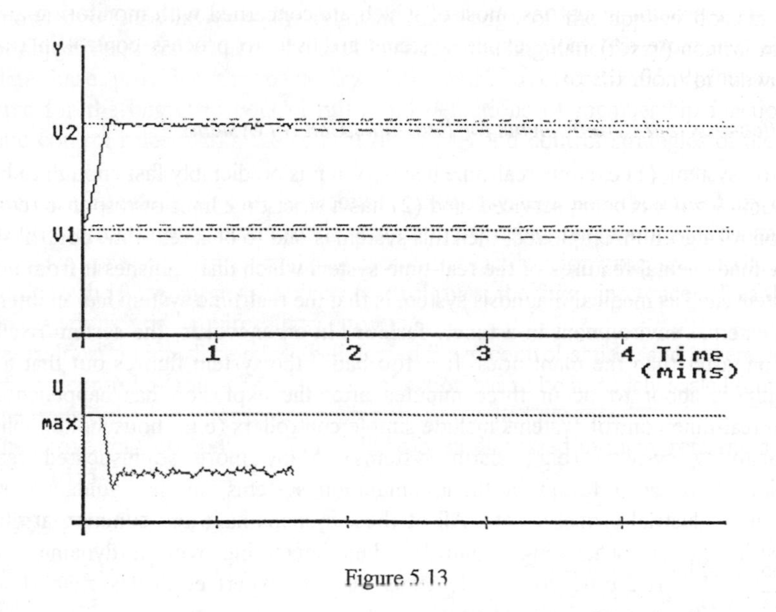

Figure 5.13

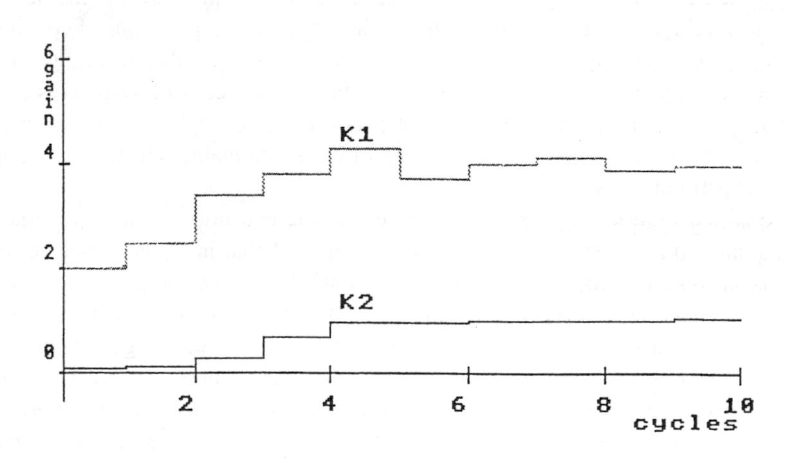

Figure 5.14.

5.3 Real-Time Expert Control Systems

At present, there is a very strong interest in the development and application of expert systems in the process industries, most of which are concerned with monitoring and fault diagnosis, and more and more expert systems are used to process control in real-time control systems[9, 10, 12, 13, 16, 23].

5.3.1 Features and Requirements for Real-time Control Systems

If a control system: (1) exhibits real-time behavior if it is predictably fast enough to be used by the control process being serviced, and (2) has a strict time limit of response regardless of the control algorithm employed, then this system is said to be a real-time control system.

The fundamental features of the real-time system which distinguishes it from nonreal-time system such as medical diagnosis system is that the real-time system has an interaction with an external environment in a timely fashion. In other words, the system reaches its conclusion faster than the plant does. It is too bad if the system figures out that a major catastrophe is about to occur three minutes after the explosion has happened! Some common real-time control systems include simple controllers (e.g., household appliances) and monitoring systems (e.g., alarm systems). Many more sophisticated real-time applications have been found in flight simulation systems, missile guidance, robotic systems and industrial processes etc. All of these systems have the common attribute of time restrictions on processing (control) when interacting with a dynamic external environment. The real-time constraint means that the expert control system should be adapted to suit the process under control.

The integration of expert systems and real-time systems in control is the next logical step in the development of these two technologies. Real-time expert control systems could be used to replace or assist human operators in a wide range of applications. A related reason which supports the development of real-time expert control systems is to reduce cognitive overload on operators to improve productivity. Complex environment that are potentially dangerous or pose threat to human lives are domains which will benefit from real-time expert control systems.

Special techniques are employed to speed up the execution of the real-time expert control system. As we have pointed out in Section 5.2.1 that in order to realize real-time reasoning and decision-making, the size of the knowledge base of an expert control system should not be too large, and the inference mechanism should be as simple as possible. Some of the critical rules may be written in a lower level of the language such as C or Assembler. Demons are used in some software packages. These are special procedures which are executed when specified data values are received or specific inferences are drawn. The knowledge base may be partitioned so that different parts of it can be executed on separate processors, which is known as the blackboard technique introduced previously. Each separate processor is viewed as a stand-alone expert, and they communicate with each other by placing the results of their inference procedures on a blackboard, where another expert system can use these results for their applications [18, 21].

The specific requirements and design features necessary in real-time expert control systems are as follows:

(i) to represent the knowledge related to time exactly.
(ii) to have high-speed context sensitive rule activation.
(iii) to control arbitrary time-varying non-linear process.
(iv) to make temporal reasoning, parallel reasoning and non-monotonic reasoning.
(v) to revise in-line basic control knowledge.
(vi) to be capable of interuptive processing and asynchronous event processing.
(vii) to acquire dynamic and static information of process on time so that real-time in-line diagnosis for the control system can be made.
(viii) efficient recycling of memory elements that are no longer needed and maintenance of sensor histories.
(ix) to accept command sequences from the operators interactively.
(x) to interface with conventional controllers and other application softwares.
(xi) to be able to communicate among multiple expert systems and between expert systems and users.

In this section, we would like to present a new and more general Real-time Expert Intelligent Control System (**REICS**) that we designed and developed recently [3, 21].

5.3.2 Structure of REICS

The Real-time Expert Intelligent Control System, REICS, is a combination of expert system, fuzzy sets and control theory, and a newly developing direction with promising application for intelligent control. This control method is based on the following techniques: utilization of expertise, knowledge model, knowledge base, knowledge inference, control decision and control strategy; combination of knowledge model and traditional mathematical model, and combination of knowledge information processing technique and control technique; simulation and human intelligent behavior, etc. This method can solve the control problems of large-scale and complex systems with time-varying, nonlinearity and multi-disturbance in real-time control processes.

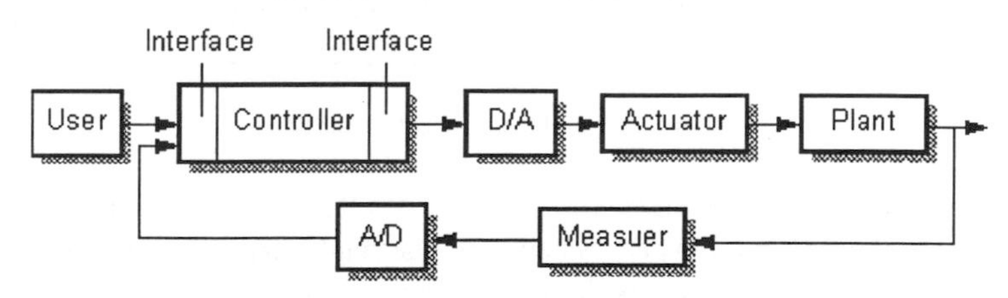

Figure 5.15. Structure of hardware for REICS.

The global structure of the hardware and software of the REICS are shown in Figure 5.15 and 5.16 respectively.

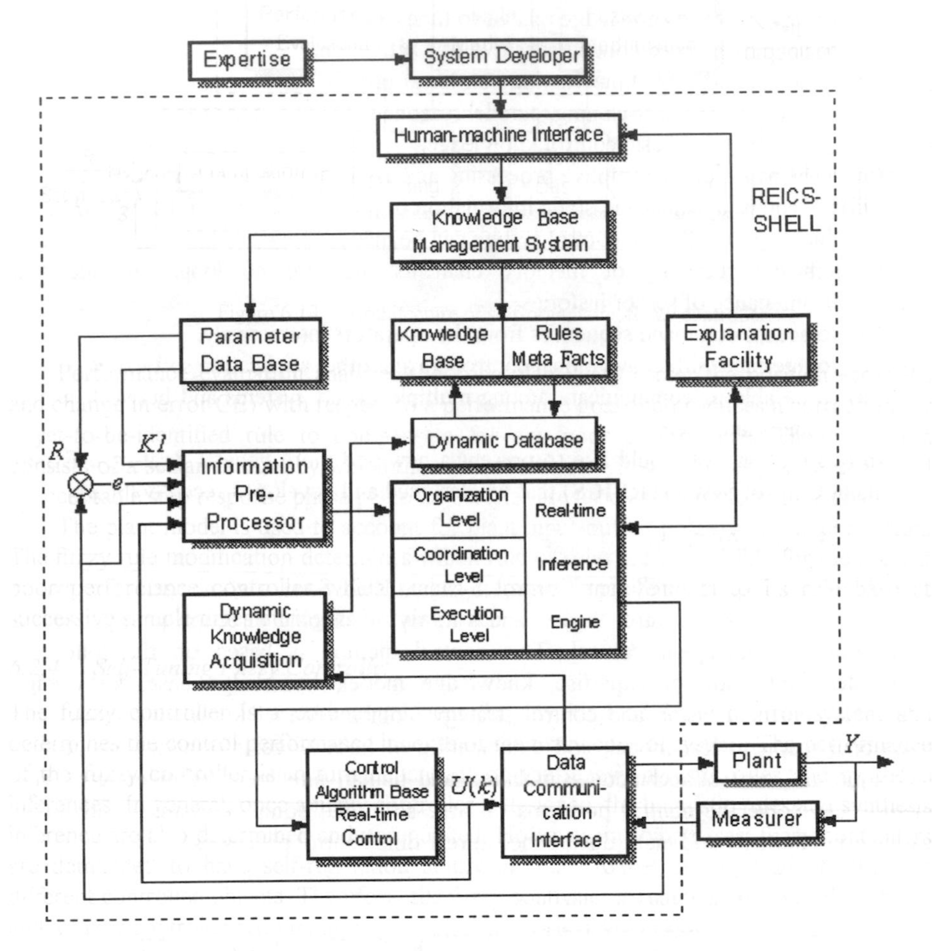

Figure. 5.16. Architecture of software for REICS.

For the system hardware the microprocessor IBM PC/At-386 is used for decision and computation in the high-level expert system, and the single chip microprocessor MCS-51, A/D and D/A converters as well as the interfaces are used in the low level circuits. The software of REICS consists of the modules for the dynamic database, real-time inference engine, information pre-processor, interpreter, set of control algorithms, data communication interface and human-machine interface, etc. The inference engine and knowledge representation are implemented using the programming language Turbo-prolog, and the programming language C is used to compute the set of control algorithms and data communications.

In the global design of REICS, we have proposed a generalized production system based on a frame, an production rules and real-time procedure; this generalized production system is used to represent control knowledge and algorithms for the control system. In the highest control level, the organization level, the reasoning knowledge is represented by the frame, and the imprecise reasoning and the choice of control strategy are carried out by fuzzy approximate reasoning. In the coordination level, the coordination control knowledge is represented by the generalized production rules; the choice of control algorithms, parameter self-tuning rules, and parameter evaluation are performed by backward reasoning and refined uncertainty reasoning. In the real-time control level, the forward real-time reasoning is done according to the reasoning and decision-making results of the above two higher control levels and the sampled real-time data (facts); as a result the real-time control input of the controlled object at any time is given.

5.3.3 Design and Implementation of REICS

In designing and implementing the REICS, the following are the most important.

1. Knowledge representation

The key to expert intelligent control is to make intelligent decisions through problem-solving. Therefore it is important to implement intelligent control to build a good model of knowledge representation for problem-solving. Because the knowledge of object characteristics and parameter rules in the control domain are very complex, the different control objects may use different controls and knowledge. As mentioned above, a generalized production system tool, REICS-SHELL, was developed and used to represent the multi-level knowledge.

Definition 5.1. A generalized production rule has the form:

$$\begin{aligned}&\text{IF }\{\text{frame}((X_1 \text{ is } a_1) \wedge (X_2 \text{ is } a_2) \wedge \ldots \wedge (X_n \text{ is } a_n) \rightarrow (Y_i \text{ is } B_i))\}\\ &\quad\text{THEN }\{\text{IF}\langle \text{rule}(K_1 \wedge K_2 \wedge \ldots \wedge K_n) \rightarrow P_i\, \text{CF}(P_i, K_i)\rangle\\ &\quad\text{THEN}\langle \text{process}((S_1(t_1) \wedge S_2(t_2) \wedge \ldots \wedge S_n(t_n)) \rightarrow U_i(t_i)\rangle\}\end{aligned} \tag{5.9}$$

where: $Y_i \in$ control strategies, and includes ES-PID, ES-Fuzzy, ES-Predict, and ES-Adaptive, which stand for the expert intelligent PID control, expert fuzzy control, expert intelligent predictive control and expert intelligent adaptive control, respectively; $X_i \in$ characteristic knowledge of control object; B_i, a_i are variable values of the fuzzy set; $K_i \in$ precondition of rules including facts, parameters, and real-time sample data, etc.; $P_i \in$ combination of control algorithms, evaluated algorithms, control rules, and parameter regulation rules; $\text{CF}(P_i, K_i) \in$ imprecise reasoning function; $S_i(t_i) \in$ control process at a given time; $U_i(t_i) \in$ output of the control system REICS at a given time.

Definition 5.2. The frame rule of the knowledge representation in the organization level has the form:

$$
\begin{aligned}
(\langle \text{plant}\rangle\ (&(\langle \text{mathematical model } X_1\rangle\ \langle \text{character}\rangle)\\
&(\langle \text{nonlinearity } X_2\rangle\ \langle \text{character}\rangle)\\
&(\langle \text{time-delaying } X_3\rangle\ \langle \text{character}\rangle)\\
&(\langle \text{time-varying } X_4\rangle\ \langle \text{character}\rangle)\\
&(\langle \text{parameter variation } X_5\rangle\ \langle \text{character}\rangle)\\
&(\langle \text{inertia } X_6\rangle\ \langle \text{character}\rangle)\\
&(\langle \text{disturbance } X_7\rangle\ \langle \text{character}\rangle)\\
&(\langle \text{dynamic response } X_8\rangle\ \langle \text{character}\rangle)\\
&\qquad \ldots\ldots\\
&\qquad \ldots\ldots\\
&(\langle \text{control strategies}\rangle\ \ \langle \mathit{ES\text{-}PID}\rangle\\
&\qquad\qquad\qquad\qquad\ \langle \mathit{ES\text{-}Fuzzy}\rangle\\
&\qquad\qquad\qquad\qquad\ \langle \mathit{ES\text{-}Predict}\rangle\\
&\qquad\qquad\qquad\qquad\ \langle \mathit{ES\text{-}Adaptive}\rangle)))
\end{aligned}
\tag{5.10}
$$

where ⟨plant⟩ stands for controlled object. The structure of the knowledge base in the organization level corresponds to the following form:

$$\text{frame}([X_1(A_1), \ldots, X_n(A_n)], Y(B_i));\ \text{predicate} \tag{5.11}$$

where $A_i, B_i \in \{BO, MO, SO, ZO, DO\}$, i.e.,{big, medium, small, zero, undefined} and are variable values of fuzzy statement.

For example, the frame rule corresponding to characteristics of an object is

IF (unknown mathematical model (*ZO*)
  medium nonlinearity (*MO*)
  big time-delaying (*BO*)
  big variation of process parameter (*BO*))

THEN (big probability for using control strategy *ES-PID* (*BO*))

This frame rule can be rewritten as

$$
\begin{aligned}
\text{frame}\ (\ [&X_1(ZO), X_2(MO), X_3(BO), X_4(\_), X_5(BO), \ldots],\\
&Y(ES\text{-}PID), B_i(BO));\ \text{predicate}
\end{aligned}
\tag{5.12}
$$

Definition 5.3. The refined rule

The refined rule for the coordination level has the form:

$$\text{rule}\quad \text{IF}\ (A, C(A))\quad \text{THEN}(B, C(B))\quad \text{CF}(B, A)$$

$$A \in \{K_1 \wedge \ldots \wedge K_n\}$$
$$B \in \{P_1, P_2, \ldots P_m\}$$
$$\mathrm{CF}(B,A) : A \rightarrow B \in [-1, 1]$$
$$C(\_) : U \rightarrow [0,1] \quad (5.13)$$

where A is antecedent of the rule and presents facts, evidences, hypotheses, the real-time data base and objectives; B is consequent of the rule and comprises starting a control or evaluation algorithm or adding a new element to knowledge base; $C(\_)$ represents the uncertain level of evidence; CF(B, A) describes the knowledge uncertainty as the confidence (certainty factor) value of a rule, i.e., the confidence with which the facts in antecedent A support the conclusion in consequent B. CF(B, A) $\in [-1, 1]$, if $CF(B, A) = -1$, then A is false and the operation in B is negative absolutely; if CF(B, A) = 1, then A is true and the operation in B is positive absolutely; if CF(B, A) = 0, then B is independent of A.

In REICS, the knowledge about facts, rules and procedures is stored in the knowledge base in the form of predicate, and is independent of the inference engine.

The inner form of fact is:

$$\mathrm{data}(N, Exp, Value) \quad (5.14)$$

where N stands for the rule number, Exp — parameter expression, $Value$ — real-time measured data of parameter.

The inner form of rule is:

$$\mathrm{rule}(N, Ex, Action, P, [\text{premise } 1, \ldots, \text{premise } n], \mathrm{CF}) \quad (5.15)$$

where Ex — rule type, $Action$ — rule conclusion, premise i — premise condition, P — conclusion type.

The process knowledge in REICS is divided into two kinds: static knowledge and dynamic one. The former mainly is objective knowledge, and the latter is the knowledge for problem-solving and is used to a process such as computational evaluation, calling a computation of control algorithm, and data sampling, etc. The procedure expression is suited for representing the dynamic knowledge, especially for real-time expert control. The procedure sequence is shown in Figure 5.17, where P_0, P_1, ... , P_n are subprocedures of procedure P; S_1, S_2, ... , S_n are sub-sentences of procedure knowledge S.

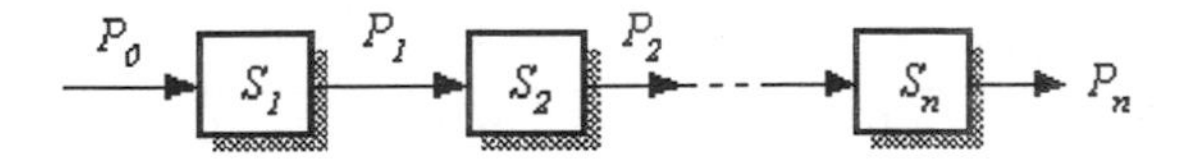

Figure 5.17. Sequence of procedure knowledge.

The logic expression of S is:

$$S = S_1 \cap S_2 \cap \ldots \cap S_n \tag{5.16}$$

The predicate form of this procedure knowledge writing by Turbo-prolog is as:

$$\text{process} \quad :- \quad p(t), p(t+1), \ldots, p(t+n) \tag{5.17}$$

The knowledge base has a hierarchical and modular structure that can decrease the reasoning and searching space, and to make the knowledge base much more flexible.

2. Inference mechanism

The core of REICS is the inference engine that can select related knowledge according to a reasoning strategy and make inference for control algorithms, facts, evidences provided by control experts and data acquired from real-time sampling. As a result the corresponding decision for intelligent control can be given out and used to guide the control action. The inference mechanism includes the inference method and the control strategy; it possesses the capability: (1) to process imprecise knowledge, (2) to do real-time reasoning rapidly, (3) of high reliability of control in the on-line operation, and (4) of wide versatility.

The reasoning consists of three levels; the global flow diagram of the control strategy is shown in Figure 5.18. We will discuss the reasoning mechanism for the three levels in detail.

(i) Reasoning in organization control level

In the highest control level, organization level, the approximate reasoning of fuzzy sets is employed to handle the fuzzy knowledge in frames, and the reduction heuristic matching search technique in AI is used to search for the goal. The algorithm for handling the fuzzy knowledge is explained as follows.

The frame searching rules for matching are:

$$\begin{aligned}
&\text{frame}((X_1 \text{ is } a_{11} \wedge X_2 \text{ is } a_{12} \wedge \ldots \wedge X_n \text{ is } a_{1n}) \rightarrow Y_1 \text{ is } B_1) \\
&\text{frame}((X_1 \text{ is } a_{21} \wedge X_2 \text{ is } a_{22} \wedge \ldots \wedge X_n \text{ is } a_{2n}) \rightarrow Y_1 \text{ is } B_2) \\
&\vdots \\
&\text{frame}((X_1 \text{ is } a_{n1} \wedge X_2 \text{ is } a_{n2} \wedge \ldots \wedge X_n \text{ is } a_{nn}) \rightarrow Y_1 \text{ is } B_n)
\end{aligned} \tag{5.18}$$

where $a_{ij} \in [ZO, SO, MO, BO, DO]$, and the computation of membership function μ_i of fuzzy subsets is listed in Table 5.2.

Table 5.2. Membership function of fuzzy subsets

a_{ij}	0	1	2	3	4	5
ZO	1.0	0.6	0.1	0.0	0.0	0.0
SO	0.6	1.0	0.6	0.1	0.0	0.0
MO	0.1	0.6	1.0	0.8	1.0	0.0
BO	0.0	0.1	0.4	0.8	1.0	0.0
DO	1.0	1.0	1.0	1.0	1.0	1.0

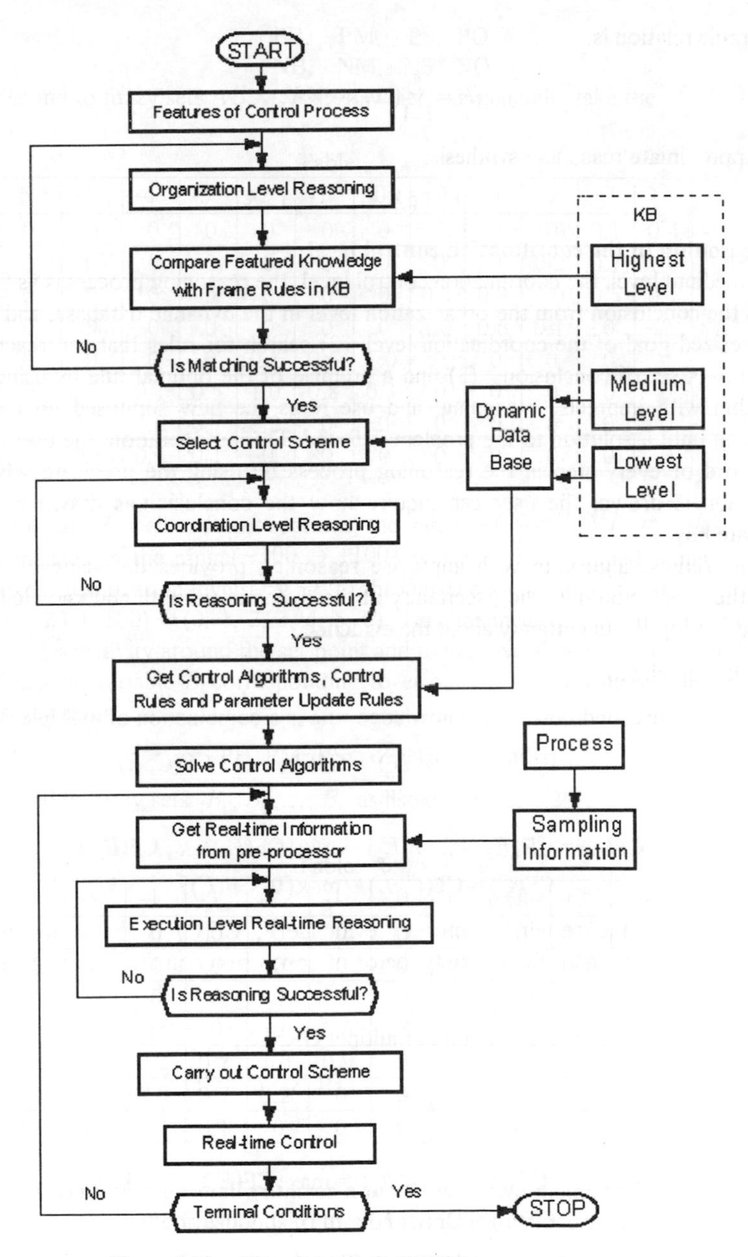

Figure 5.18. Flow diagram for REICS programming.

Frame relation is:

$$R = \overset{n}{\underset{i=1}{\vee}} (X_1 \times \cdots \times X_n \times Y_i). \tag{5.19}$$

Approximate reasoning synthesis:

$$Y_i = (X_1 * \cdots * X_n) \circ R. \tag{5.20}$$

(ii) Reasoning in the coordination control level

In the medium level, the coordination control level, the reasoning process is as follows: (1) access the conclusion from the organization level in the dynamic database, and use it as a hypothesized goal of the coordination level; (2) search for rules that can reach the goal, then draw a set of conclusions; (3) find a premise of the optimal rule by using a refined algorithm with imprecise reasoning, and use it as the new supposed goal for further reasoning until a solution to the problem is found. During operation, the user can exhibit the record of every step in the reasoning process by using the predicate **why**. After a conclusion is drawn, the user can inquire how the conclusion is drawn by means of predicate **how.**

The refined algorithm with imprecise reasoning provides the value of uncertainty about the conclusion, i.e., the uncertainty of the dynamic strength and knowledge , and is derived using the uncertainty about the evidence.

Refined Algorithm

(1) If the precondition of the knowledge rule is a combination of multiple AND, i.e.,

$$\text{IF } (E_1 \wedge \dots \wedge E_n) \quad \text{THEN} \quad C \quad \text{CF}(C, E)$$

then

$$\begin{aligned} \text{CF}(E) = \text{CF}\{E_1 \wedge \dots \wedge E_n\} &= \min\{\text{CF}(E_1), \dots, \text{CF}(E_n)\} \\ \text{CF}(C) = \text{CF}(C, E) * &\max\{0, \text{CF}(E)\} \end{aligned} \tag{5.21}$$

where CF(E) is the certainty factor of evidence e; CF(C) is the certainty factor of conclusion c; CF(C, E) is the certainty factor of conclusion c in the case of given evidence e.

(2) If evidence is a combination of multiple OR, i.e.,

$$\text{IF } (E_1 \vee \dots \vee E_n) \quad \text{THEN} \quad C \quad \text{CF}(C, E)$$

then

$$\begin{aligned} \text{CF}(E) = \text{CF}\{E_1 \vee \dots \vee E_n\} &= \max\{\text{CF}(E_1), \dots, \text{CF}(E_n)\} \\ \text{CF}(C) = \text{CF}(C, E) * &\max\{0, \text{CF}(E)\} \end{aligned} \tag{5.22}$$

(3) If one knowledge rule has multiple conclusions, i.e.,

$$\begin{aligned}&\text{IF }\{\text{IF}\quad E\quad \text{THEN }(C_1, \dots, C_n)\quad \text{CF}(C, E)\}\\&\text{THEN}\quad \text{CF}(C_1) = \dots = \text{CF}(C_n)\\&\qquad = \text{CF}(C, E)*\ \max\{0, \text{CF}(E)\}\end{aligned}\tag{5.23}$$

(4) If multiple knowledge rules have the same conclusion, then two rules with the same conclusion are computed, and a recurrent computation is done afterward, i.e.,

$$\begin{aligned}&\text{IF}\quad E_1\quad \text{THEN}\quad C\quad \text{CF}(C,E_1)\\&\text{IF}\quad E_2\quad \text{THEN}\quad C\quad \text{CF}(C,E_2)\\&\text{THEN}\quad \text{CF}_1(C) = \text{CF}(C,E_1)*\max\{0,CF(E)\}\\&\qquad\quad \text{CF}_2(C) = \text{CF}(C,E_1)*\max\{0,\text{CF}(E)\}\end{aligned}\tag{5.24}$$

$$\text{CF}_{12}(C) = \begin{cases}\text{CF}_1(C)+\text{CF}_2(C)-\text{CF}_1(C)*\text{CF}_2(C),\text{if CF}_1(C)\text{ and CF}_2(C)\geq 0\\ \text{CF}_1(C)+\text{CF}_2(C)+\text{CF}_1(C)*\text{CF}_2(C),\text{if CF}_1(C)\text{ and CF}_2(C)\leq 0\\ |\text{CF}_1(C)+\text{CF}_2(C)|\,/\,\{1-\min(|\text{CF}_1(C)|,|\text{CF}_2(C)|)\},\text{otherwise}\end{cases}$$

where CF(E), CF(C) have the same meanings as above.

(iii) Reasoning in the real-time control level

During real-time operation in REICS, forward reasoning is used, i.e., reasoning from initial data to control goal. First, based on reasoning in the two higher control levels, the current information $E = \{e, \dot{e}, Y, U, R, K_p, \dots\}$ provided by the information pre-processor and the dynamic database is used as the precondition. Then the control rule that matches this precondition in the dynamic database is sought. If the match is successful, then the state goal is found and a series of control actions about the conclusion of the rule is carried out; e.g., computation of current control parameters and control value, D/A and A/D conversion, data sampling, control action transfer, and information reception etc. If the match is not successful, then the search for a matched rule should be continued.

(iv) Design of evaluation interpreter

In order to overcome the difficulty in reasoning for the real-time expert control and to speed up the control response, an evaluation interpreter is designed and applied to evaluate the expression of the computation-type control rules, to judge the logic function and make the matching for the real-time reasoning during the forward and backward reasoning processes. If the promise of the rule is tenable, the value of the conclusion, control volume, can be derived directly. For example, there are some rules:

$$\begin{aligned}&\text{rule 100}\quad \text{IF}((-R < e)\ \wedge (e < R)\\&\qquad\qquad\quad \text{THEN } U_n = K_p * e + K_p \Sigma\, e_i + K_d * \dot{e}\\&\text{rule 200}\quad \text{IF}((a + b * c/d)\ \geq 0.5 * e)\\&\qquad\qquad\quad \text{THEN } K_n(n+1) = 0.98 * K_p\,(n-1)\end{aligned}\tag{5.25}$$

called in major predicate:

$$\text{expr-eval}(\text{"}((a + b * c/d) \leq 0.5 * e)\text{"}, TF)$$

return with

$$TF = 1$$

presenting that the promise of the rule is tenable; with

$$TF = -1$$

representing that the promise of the rule is not tenable. If tenable, then the value can be derived by calling it in major predicate for evaluating arithmetic expression

$$\text{infix - eval}(\text{"}0.98 * K_p\text{"}, A_n)$$

return with A_n = (expression value).

5.3.4 Simulation and Applications of REICS

REICS-SHELL is a new development tool with a multi-level reasoning and a generalized knowledge base for intelligent control systems. In order to examine the range and effect of application after design and implementation, we have done an experimental study of the system on the IBM PC-386 microprocessor by inserting a control interface with multiple functions. This menu and submenu of REICS are shown in Table 5.3 and Table 5.4.

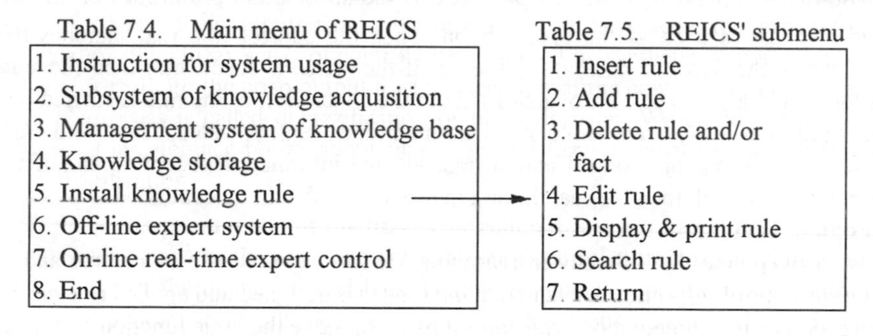

Table 7.4. Main menu of REICS

1. Instruction for system usage
2. Subsystem of knowledge acquisition
3. Management system of knowledge base
4. Knowledge storage
5. Install knowledge rule
6. Off-line expert system
7. On-line real-time expert control
8. End

Table 7.5. REICS' submenu

1. Insert rule
2. Add rule
3. Delete rule and/or fact
4. Edit rule
5. Display & print rule
6. Search rule
7. Return

For a special application, item 2 in the main menu has to be input in order to build its knowledge base, which includes the input control rule, set of system parameters and initial value. Then items 5, 6 and 7 should be input into the computer respectively through a keyboard for installing knowledge base, self-optimizing parameters, defining control algorithm, computing control value in real-time and outputting on-line intelligent control.

Four kinds of intelligent control algorithms have been included in the REICS: they are ES-PID, ES-Fuzzy, ES-Predict, and ES-Adaptive, all installed in control algorithm base. Meanwhile some control rules should be input such as:

rule 1 IF($e > \alpha R$) THEN($U_n = U_m$) CF=0.9

rule 2 IF($e<-\alpha R$) THEN($U_n = -U_m$) CF=0.9

rule 3 IF($|e| < a_1 \wedge |\dot{e}| < \delta_1$) THEN($U_n = U_{n-1} + K_p\dot{e} + K_i e$) CF=0.8

rule 4 IF($|e| < \delta_1 \wedge |\dot{e}| < \delta_2$) THEN($U_n = U_{n-1}$) CF=0.95

rule 5 IF($e * \dot{e} \leq 0$) THEN(IF($|\dot{e}| \leq a_2 \wedge |e| \leq b_2$)

THEN PID(K_p, K_i, K_d)) CF=0.94

⋮

rule 50 IF($e * \dot{e} \geq 0$) THEN(IF($|e| > a_2 \wedge |\dot{e}| > b_2$)

THEN FC(K_e, K_c, K_i)) CF=0.95

Parameter rules:

rule 51 IF(big time-delay) THEN(K_e is small, K_c is small) CF=0.89

rule 52 IF(divergent oscillation response)

THEN(K_e, K_c, K_i decrease, a_i, b_i increase) CF = 0.78

⋮

rule 110 IF($c_1 = 0 \wedge c_2 = 0 \wedge c_3 = 0$)

$$\text{THEN} \quad \left. \begin{aligned} K_p(n) &= 0.98 * K_p(n-1) \\ T_i(n) &= 0.618 * T_i(n-1) \\ T_d(n) &= 0.25 * T_d(n-1) \end{aligned} \right\} \tag{5.26}$$

where e, $\dot{e}$ - error and its rate for a controlled object; δ_1, δ_2, a_i, b_i are parameters given by experts; c_1, c_2, c_3 - the first, second and third peak of the dynamic response curve measured in real-time.

Three different controls, i.e., (a) conventional PID control, (b) Fuzzy control and (c) REICS control have been compared through simulation for which a simplified diagram is shown in Figure 5.19.

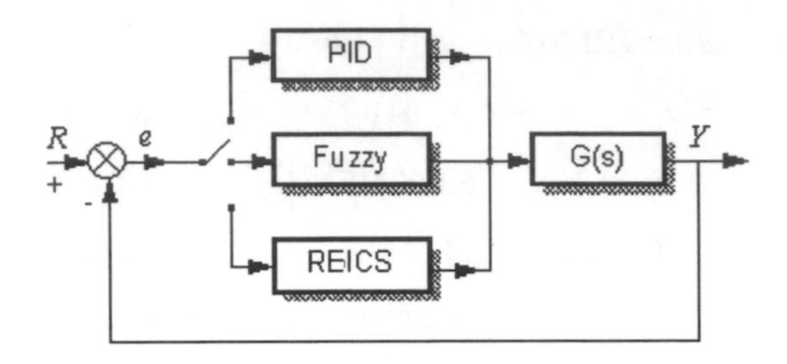

Figure 5.19. Simulation diagram for comparison.

1. Examples of Simulation

Example 1.

The controlled plant is a nonlinear system with a stochastic disturbance acting on it. The mathematical model of a plant is in the form:

$$y(t)=\frac{y(t-1)e^{-y(t-1)}+u(t-1)}{1+u(t-1)e^{-y(t-1)}}+\omega(t)$$

where $\omega(t)$ is a white noise with 0.15 standard variance.

The step response of the plant is illustrated in Figure 5.20.

Example 2.

The plant is a nonlinear system with large time delay, i.e.,

$$y(t)=\frac{y(t-1)y(t-2)y(t-3)y(t-4)u(t-4)-y(t-1)y(t-2)y(t-3)y(t-4)+u(t-5)}{1+y^2(t-2)+y^2(t-4)}$$

where time delay d=4. Figure 5.21 illustrates the step response curves of the plant.

In Figure 5.20 and Figure 5.21, three different controls, were compared through simulation, i.e., (1) conventional PID control , (2) Fuzzy control and (3) REICS control.

A comparison of robustness among PID, Fuzzy and REICS was simulated for two cases: (a) when the object's parameter was changed, (b) when exotic disturbance was inducted. The simulation results show that the REICS has a better adaptivity to parameter's change and a better capability of anti-disturbance than that of PID and fuzzy control. From the experiments it can be drawn that REICS possesses a stronger robustness than PID and fuzzy control systems do.

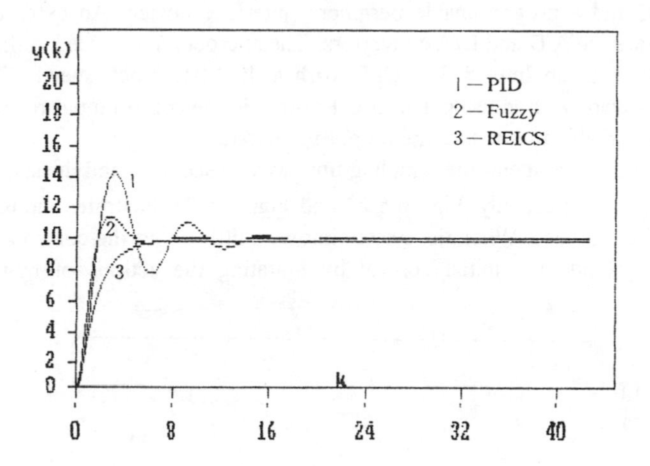

Figure 5.20. Output of the plant for example 1.

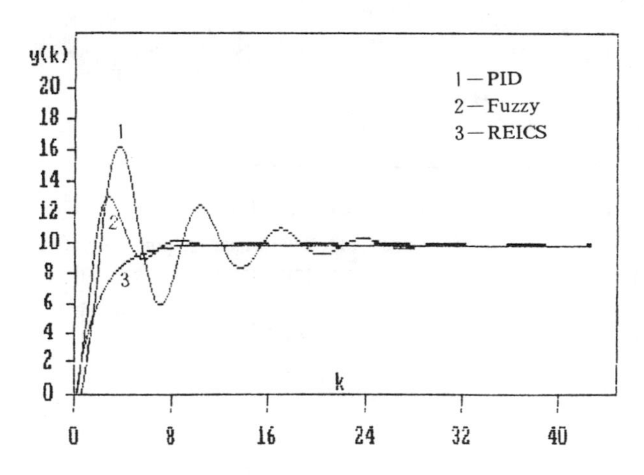

Figure 5.21. Output of the plant for example 2.

2. Application for temperature control of a rotary kiln furnace

For an application of REICS, the proposed control scheme was applied to the temperature control of an industrial rotary kiln. The temperature control system can be divided into five main components: the rotary kiln furnace, the temperature sensor module, the programmable input-output interface board, the microcomputer, and the actuator. The interface circuit board consists of an analogue-digital (A/D) converter, a digital-analogue (D/A) converter, and a programmable peripheral interface device. An external clock is designed to operate the A/D and D/A converters. The microcomputer used in the system is the IBM PC-386 with an Intel 80386 CPU with a 40 MHz clock speed. The REICS control programs were written using Turbo-C to provide the control input to the actuator through the D/A and also to measure the output temperature.

In the industrial experiment, the sampling time was 30 seconds, and the setpoints were $500^{\circ}C$ and $1000^{\circ}C$ respectively. Figure 5.22 and Figure 5.23 illustrates the temperature response of the kiln furnace. When the system knows a little about the controlled process, the REICS can provide the initial control by imitating the actions of human expert operators.

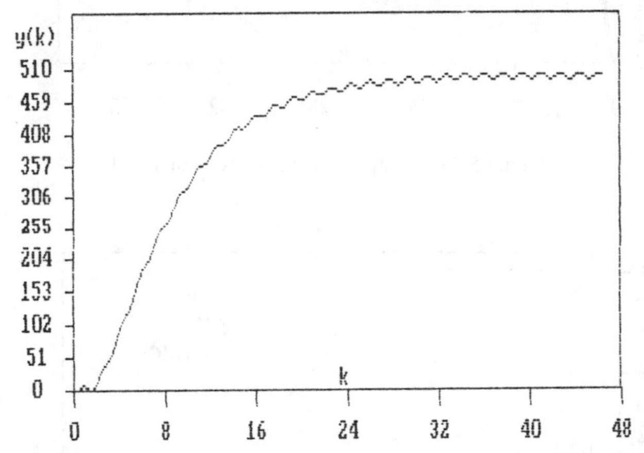

Figure 5.22. Temperature response of a kiln furnace (setpoint=$500^{\circ}C$).

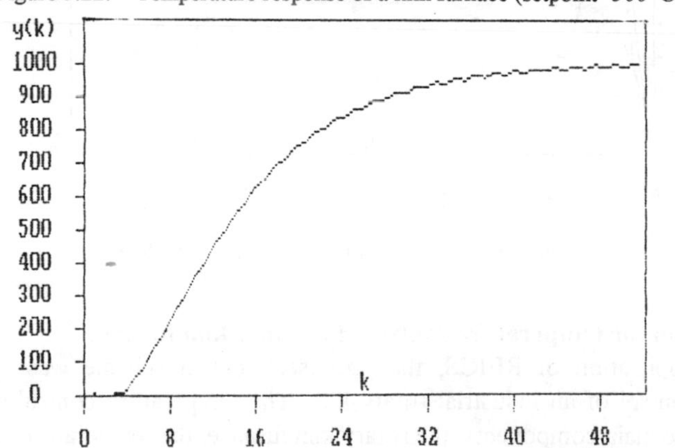

Figure 5.23. Temperature response of a kiln furnace (setpoint=$1000^{\circ}C$).

The REICS controlled industrial rotary kiln furnace has been operated successfully in Zhengzhou Aluminum Plant, China since August 1994. If the initial weight values and the expert controlled parameters, such as control error and threshold of error ratio, are selected properly, then the control system can be implemented with the following performances: (1) very small or almost no overshoot; (2) less regulating time; (3) without oscillation; (4) high control accuracy of temperature (within $\pm 2^{\circ}C$). Industrial running has shown that the REICS is effective for real-time control for systems with large time delay and strong nonlinearity [20].

REICS-SHELL is a development tool for real-time control systems implemented in an IBM PC-386 microprocessor system. The results of research both in simulation and applications have shown that REICS possesses many advantages such as flexible control strategy, easy knowledge acquisition, the ability to process control knowledge with uncertainty, good real-time control capability, adaptivity, and versatility. When using this REICS-SHELL tool, one simply adds control rules and revises rules, facts and control algorithm parameters in the knowledge base. Therefore this tool is practical for industrial process control systems with uncertainty and imprecise mathematical models.

5.4 Summary

Expert control systems have been the most actively researched and widely applied area of intelligent control in recent years. According to the complexity of the system, basically, there are two types of expert controllers (control systems): expert controller and expert control system. The former has been used even much more widely, especially in industrial process control. Based on the control mechanism of the system, the expert control system can be sorted into two kinds: the direct expert control system and the indirect expert control system. In the former, the controller provides a control signal to the system directly. In the latter, the controller effects the controlled process indirectly.

According to the different application fields and specific control requirements, engineers may design and employ different types of expert controllers.

For simpler control process/plant with uncertainty, non-linearity, and/or imprecise and incomplete information, etc., the general industrial expert controller can be used. For more complex control process/plant, the expert control system such as blackboard expert control system should be applied.

The integration of expert systems and real-time systems in control is a next step logically in the development of expert control systems: real-time expert control systems. Special techniques are used to speed up the execution of these control systems.

Most expert controllers/control systems are with hierarchical structure. In designing an expert controller, the knowledge base and inference engine or blackboard are specially emphasized. In studying these contents, readers had better go over Section 5.1 (Expert Systems) and Blackboard Approach.

Some design principles have been presented in Section 5.2.1. These principles should be followed in designing the expert controllers.

We have demonstrated two application examples/paradigms in this chapter. They involve (a) an expert tuner for PI controller on a laboratory test rig to control the liquid level of a multiple tank assembly; (b) a real-time expert intelligent control system for temperature of an industrial kiln furnace. From these examples the reader would have a better understanding for the structures, design methodologies and implementation of the expert controllers. The results of simulations and applications have shown that the expert control systems/controllers possess excellent performances and can be applied to a wide range of areas.

References

1. K. J. Åström, J. J. Anton, and K. E. Arzen, "Expert Control," *Automatica*, 22 (1986) 277.
2. Z.-X. Cai, *Intelligent Control* (Electronic Industry Press, Beijing, 1990).
3. Z.-X. Cai, Y.-N. Wang, and J.-F. Cai, "A real-time expert control system," *AI in Engineering, Special Issue: AI in Engineering in China*, Elsevier, England, 1996.
4. R. Engelmore, and T. Morgan, eds. *Blackboard System* (Addison-Wesley, Reading, MA, 1988).
5. E. A. Feigenbaum, and P. McCorduch, *The fifth generation, Artificial Intelligence and Japan's Computer Challenge to the World* (Addison-Wesley, Reading, MA, 1983).
6. F. Hayes-Roth, D. A. Waterman, and D. B. Lenat, eds. *Building Expert Systems* (Addison-Wesley, Reading, Mass., 1983) 16-18.
7. F. Hayes-Roth, "A blackboard architecture for control," *Artificial Intelligence*, 1985.
8. R. F. Hodson, and A. Kandel, *Real-Time Expert Computer Architecture* (CRC, New York, 1991).
9. J. Jiang, and R. Dorainswemi, "Information acquisition in expert control system design using adaptive filters," *Proc. IEEE ISIC*, 1987, 165-170.
10. J. Jiang, and R. Dorainswemi, "Design, implementation and performance evaluation of a real-time knowledge-based controller," *Proc. IEEE ISIC*, 1988, 233-238.
11. A. H. Jones, and B. Porter, "Expert tuners for PID controllers," *Proc. IASTED Int. Conf. Computer-Aided Design and Application*, Paris, 1985.
12. J. McGhee, M. J. Grimble, and P. Mowforth, eds. *Knowledge-Based Systems for Industrial Control* (Peter Pergrinus, UK, 1990).
13. R. L. Moore, L. B. Hawkinson, C. G. Knickerbocker, and L. M. Churchman, "A real-time expert system for process control," *Proc.IEEE First Conference on AI*

Applications, 1984, 569-576.

14. L. A. Nguyen, "Real-time simulation of the space station freedom mobile servicing center," *Proc. IEEE Int. Conf.Robotics and Automation*, Nice, France, 1992, 872-877.
15. B.-M. Pfeiffer, D. H. Owens, and A. Farouq, "CODES: a knowledge base for controller design with an expert system." *Proc. IEEE Int.Conf. Robotics and Automation*, 1992, 537-541.
16. B. Porter, A. H. Jones, and C. B. McKeown, "Real-time expert controllers for plants with non-linerities," *Proc.IEEE ISIC*, 1987, 171-177.
17. P. Smith, *Expert System Development in PROLOG and TURBO-PROLOG* (Sigma, Wilmslow, Cheshire, UK, 1988)1-26.
18. S. G. Tzafestas, ed. *Engineering Systems with Intelligence: Concepts, Tools, and Applications* (Kluwer Academic Publishers, The Netherlands, 1991).
19. T. C. Walker, and R. K. Miller, *Expert Systems Handbook: An Assessment of Technology Applications* (Fairmont, Lilburn, GA, USA, 1990).
20. Y.-N. Wang, *Intelligent Control Theory and Its Application Research on Computer Integrated Control Systems*, Ph.D. Dissertation (in Chinese), Hunan University, Changsha, China, October, 1994.
21. Y.-N. Wang, T.-S. Tong, and Z.-X. Cai, "A real-time expert intelligent control system REICS," *Algorithms and Architectures of IFAC*, Vol. 51, No. 2, Pergaman Press, 1992 , 307-312.
22. S. M. Weiss, and C. A. Kulikowski, *A Practical Guide to Designing Expert Systems* (Rowman & Allanheld, NJ, 1984) 1-4.
23. M. Wu, W.-H. Cui, Y.-F. Xie, and Z.-X. Cai, "Expert Control of the Hydrometallurgical Process of Zinc," *Preprints of IFAC YAC*, Beijing, 1995, 785-791.

CHAPTER 6

FUZZY CONTROL SYSTEMS

Fuzzy control has been an active research and application area of intelligent controls for the past twenty years. As we have mentioned in Section 1.2.2, fuzzy sets were first proposed by Zadeh [49] as a method of handling real world classes of objects. Since then, a great amount of research efforts has been carried out both in the theoretical investigations and practical applications of fuzzy sets and fuzzy control.

Fuzzy control is a kind of control approach that uses the fuzzy set theory. The usefulness of fuzzy control can be considered in two respects. On one hand, fuzzy control offers a novel mechanism to implement such control laws that are often knowledge-based (rule-based) or even in linguistic description. On the other hand, fuzzy control provides an alternative methodology to facilitate the design of nonlinear controllers for such plants being controlled, that are usually uncertain and very difficult to cope with by using conventional nonlinear control theory.

Expert Control Systems and Fuzzy Logic Control (FLC) systems have certainly one thing in common: both aim to model human experience, human decision-making behavior. There are, however, clear differences between expert control systems and fuzzy logic control systems: (1) The existing FLC systems originated in control engineering rather than in Artificial Intelligence; (2) FLC models are mostly rule-based systems; (3) The application domains of FLC are narrower than those of expert control systems; (4) The rules of FLC systems are generally not extracted from the human expert through the system but formulated explicitly by the FLC designer. For these reasons, it is necessary to discuss the fuzzy control system separately from the expert control systems in this chapter.

In this section, the foundation of fuzzy set and fuzzy logic for control is briefly reviewed at first, then the types, structure, design and properties of fuzzy logic controller are presented, and an application example of FLC is demonstrated at last.

6.1 Mathematical Foundation for Fuzzy Control

6.1.1 Fuzzy Sets and Their Operations

1. Fuzzy sets

Definition 6.1. Let X be an interval containing the possible values of a variable, with an individual value in X denoted by x. Thus X is the set of all possible values of x, i.e., $X = \{x\}$, or $x \in X$. A fuzzy set F defined in interval X is characterized by a membership function $\mu_F(x)$ which associates each value in X to a real number in the interval [0,1]. The particular value of $\mu_F(x)$ at x is known as the "grade of membership" of x in F.

If F were a conventional set, the membership function could only take on the values 0 and 1, with a value of zero indicating that an element does not belong to the set and value one indicating that it does. Thus, fuzzy sets can be regarded as extensions to conventional set theory, and which obey many of the conventional set identities.

Example 6.1. Class of tall men

A fuzzy set representing "tall men" would assign a grade of membership of zero to a man four feet tall, and an intermediate grade of membership, perhaps 0.8, to a man who was five feet ten inches in height. The choice of membership function for a particular fuzzy set is completely subjective, and most certainly it is not made statistically. Although the membership function of a fuzzy set appears superficially resembling a cumulative probability distribution, there are important differences between the two concepts. Considering the "tall men" example, if we gather together a group of men belonging to the set "tall men", it would be incorrect to say that the probability of a given man being less than five feet ten inches in height is 0.8, since the probability of membership is, in fact, one.

Figure 6.1 graphically illustrates three common types of membership function. Figure 6.1(a) shows a quantized membership function, where the range of a variable value is divided into a number of sub-ranges and the membership function of the fuzzy set consists of a set of grades of membership for each sub-range. Quantized membership functions are usually represented in tabular form, and may, sometimes, offer computational advantages. Figure 6.1(b) shows a bell-shaped, continuous, membership function. Although there must be some difficulty in deciding an appropriate shape and spread for the function, they do have the capability of including a wide range of values at low grades of membership. A triangular membership function is shown in Figure 6.1(c). This is a semi-continuous function which has the major advantage of being completely specified by just three values.

2. Fuzzy sets operations

Definition 6.2. A fuzzy set can be said to be empty if, and only if, its membership function is zero for all values of x.

Definition 6.3. The complement of a fuzzy set F is denoted by F' and defined by

$$\mu_{F'}(x) = 1 - \mu_F(x) \tag{6.1}$$

Linguistically, the complement can be represented as the operator **NOT**, e.g., the complement of the set of tall men is the set of men who are not tall.

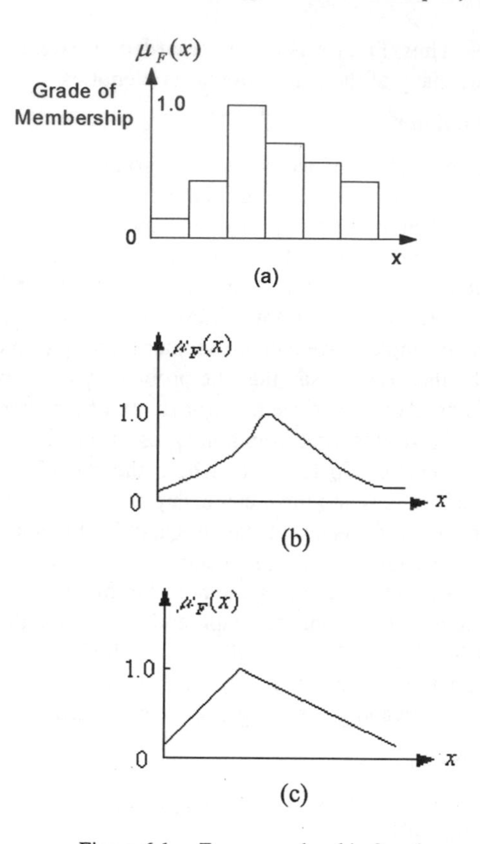

Figure 6.1. Fuzzy membership functions.

Definition 6.4. A fuzzy set F is a subset of fuzzy set G if, and only if,

$$\mu_F(x) \le \mu_G(x)$$

for all x, or

$$F \subset G \quad \Leftrightarrow \quad \mu_F(x) \le \mu_G(x). \tag{6.2}$$

Definition 6.5. Two fuzzy sets, F and G, are equal if and only if

$$\mu_F(x) = \mu_G(x) \quad \text{for all } x \text{, or}$$
$$F = G \Leftrightarrow \mu_F(x) = \mu_G(x). \tag{6.3}$$

Definition 6.6. The union of two fuzzy sets F and G is the third fuzzy set H, written as $H = F \cup G$, with the membership function

$$\mu_H(x) = \max[\mu_F(x), \mu_G(x)], \; x \in X \tag{6.4}$$

or in abbreviated form written by disjunction :

$$\mu_H(x) = \mu_F(x) \vee \mu_G(x) \; . \tag{6.5}$$

Linguistically, the union of fuzzy sets can be interpreted as the operator **OR**, e.g., if F is the set of high temperatures and G is the set of medium temperatures, then the union of the sets is the set of temperatures which are high or medium. This is illustrated graphically in Figure 6.2.

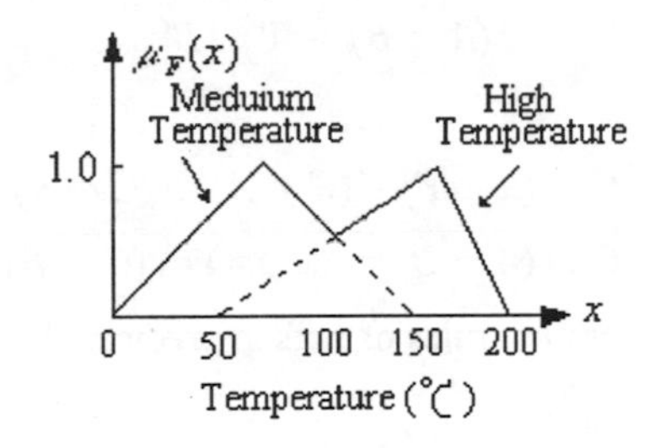

Figure 6.2. Fuzzy union.

Definition 6.7. The intersection of two fuzzy sets F and G is the third fuzzy set H, written as $H = F \cap G$, with the membership function given by

$$\mu_H(x) = \min[\mu_F(x), \mu_G(x)], \; x \in X \tag{6.6}$$

or in abbreviated form written by conjunction:

$$\mu_H(x) = \mu_F(x) \wedge \mu_G(x) \, . \tag{6.7}$$

Linguistically, the intersection of fuzzy sets can be interpreted as the logical operator **AND**, e.g., if F is the set of high temperatures and G is the set of medium temperatures, then the intersection of F and G is the set of temperatures which are both medium and high. The fuzzy intersection operation is illustrated graphically in Figure 6.3.

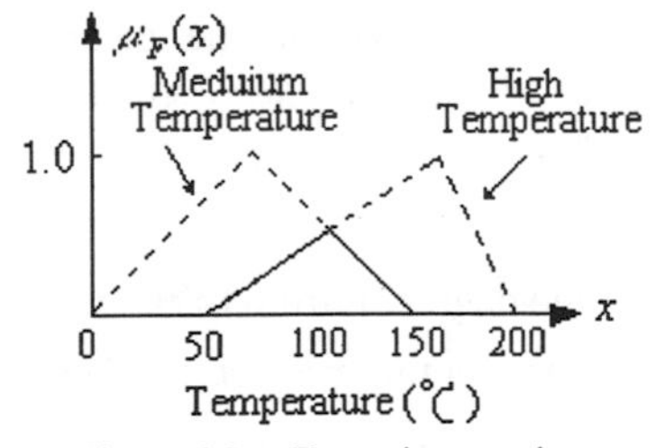

Figure 6.3. Fuzzy intersection.

Definition 6.8. Many of the basic ordinary-set identities hold for fuzzy sets when the complementation, union and intersection operations are defined as above. For example,

Associative Laws:

$$A \cup (B \cup C) = (A \cup B) \cup C \tag{6.8}$$

$$A \cap (B \cap C) = (A \cap B) \cap C \tag{6.9}$$

DeMorgan's Laws:

$$(A \cup B)' = A' \cap B' \tag{6.10}$$

$$(A \cap B)' = A' \cup B' \tag{6.11}$$

Distributive Laws:

$$C \cap (A \cup B) = (C \cap A) \cup (C \cap B) \tag{6.12}$$

$$C \cup (A \cap B) = (C \cup A) \cap (C \cup B) \tag{6.13}$$

These are similar to the operation rules of wffs in predicate logic as shown in Table 2.2.

6.1.2 Fuzzy Logic Operations

Fuzzy logic is an extension of fuzzy set theory. Conventional set theory is based on Boolean logic where set membership is either true or false, but in fuzzy set theory membership is defined over a continuous range of grades of membership, which is equivalent to grades of truth in a logical sense. Conventional logic would allow the statement:

"The temperature is high"

to be either true or false, depending on whether or not the current numerical value of the temperature belongs to the set "high temperature". This is true regardless of whether "the temperature" is high, or is not, in fuzzy set. Fuzzy logic, however, allows intermedium grades of truth to exist between true (logic 1) and false (logic 0), and thus it makes sense to say that the statement is true with a grade of truth equal to, say, 0.7.

1. Fuzzy logic operations based on fuzzy set theory

We can use fuzzy logical operator (NOT, AND, OR) based on the definitions produced from fuzzy set theory. The definitions for the compound operators are intuitively defensible, but can not be said to be theoretically correct since many compound conditions will be defined in different spaces.

(i) Logic NOT

Definition 6.9. Let γ_A be the degree of truth of the proposition for "x is A", $\gamma_{NOT\,A}$ be the degree of truth of the proposition for "x is not A", then

$$\gamma_{NOT\,A} = 1 - \gamma_A$$

This definition comes directly from fuzzy set theory since γ_A represents the membership of the value of the variable in A, and $\gamma_{NOT\,A}$ represents the membership of the value of the variable in the set NOT A, i.e., if the variables value is x, then

$$\left.\begin{aligned}\gamma_A &= \mu_A(x)\\ \gamma_{NOT\,A} &= \mu_{NOT\,A}(x) = 1-\mu_A(x) = 1-\gamma_A\end{aligned}\right\} \tag{6.14}$$

(ii) Logical AND

Definition 6.10. Let γ_A be the degree of truth of the proposition for "x is A", γ_B be the degree of truth of the proposition for "x is B", $\gamma_{A\,AND\,B}$ be the degree of truth of the proposition for "x is A AND y is B", then

$$\gamma_{A\,AND\,B} = \min(\gamma_A, \gamma_B). \tag{6.15}$$

Note that this definition does not come directly from fuzzy intersection since the two conditions may be defined in different spaces, e.g., high temperature and low pressure. For example, let γ_T be the degree of truth that the temperature is high and equal to 0.6, γ_P be the degree of truth that the pressure is low and equal to 0.4, then the degree of truth of the compound condition "the temperature is high and the pressure is low" is as:

$$\gamma_{T\,AND\,P} = \min(0.6, 0.4) = 0.4$$

(iii) Logical OR

Definition 6.11. Let γ_A be the degree of truth of the proposition for "x is A", γ_B be the degree of truth of the proposition for "y is B", $\gamma_{A\,OR\,B}$ be the degree of truth of the proposition for "x is A OR y is B", then

$$\gamma_{A\,OR\,B} = \max(\gamma_A, \gamma_B) \tag{6.16}$$

For example, using the degrees of truth from the previous example, the truth of condition

"the temperature is high OR the pressure is low"

is $\gamma_{T\,OR\,P} = \max(0.6, 0.4) = 0.6$.

2. Fuzzy logic operations based on probabilistic expressions

Several investigations have discovered that systems using the fuzzy set based on logical operators exhibit discontinuities, an undesirable characteristic for a controller. Alternative means of implementing the logical operators, based on probabilistic expressions, have therefore been produced, which are used to improve the characteristics of fuzzy logic systems.

(i) Logical NOT

The expressions are the same as Expression (6.14).

(ii) Logical AND
By using the nomenclature in **1.**(ii), we have

$$\gamma_{A\ AND\ B} = \gamma_A \cdot \gamma_B \tag{6.17}$$

and using the same example as in **1.**(ii), we have

$$\gamma_{T\ AND\ P} = 0.6 \times 0.4 = 0.24$$

(iii) Logical OR
By using the nomenclature of **1.**(iii), we get

$$\gamma_{A\ OR\ B} = \gamma_A + \gamma_B - \gamma_A \cdot \gamma_B \tag{6.18}$$

and using **1.**(iii) example to get

$$\gamma_{T\ OR\ P} = 0.6 + 0.4 - 0.24 = 0.76$$

3. Fuzzy state descriptions
Terms of linguistic variables (see Definition 6.2 to Definition 6.7) are used to describe the states of the process. In practical terms, the error value (E) and the change of error values (C) calculated are quantized into a number of points corresponding to the elements of a universe of discourse, and the values are then assigned grades of membership in seven or eight fuzzy subsets as follows:

- PB is 'positive big'
- PM is 'positive medium'
- PS is 'positive small'
- PO is 'positive nil'
- NO is 'negative nil'
- NS is 'negative small'
- NM is 'negative medium'
- NB is 'negative big'

In some literature of fuzzy sets, the PO and NO are merged into one subset ZE that means 'zero error'.

The relationship between measured error or change in error value and grade of membership are defined by look-up tables. We will discuss this issue in some detail within application examples later.

6.1.3 *Fuzzy Sets and Fuzzy Logic for Control*

1. Reference sets
The parameters and solutions for control problems are usually expressed in crisp, measurable values such as 20°C, 10^5Nm^{-2}, etc. but systems which operate using fuzzy logic make use of qualitative concepts like low temperature or high pressure. A means of associating crisp values with qualitative terms is therefore required. This is often achieved by defining several reference fuzzy sets for each of the variables of the interest. These

reference sets provide a quantitative description of the qualitative terms which are used in the fuzzy system. A particularly crisp numerical value of a variable can, and often does, belong to several of the fuzzy reference sets defined for the variable, and the assignment of set grades of membership to a particular value is known as fuzzification. An example of fuzzification is given in Figure 6.4. In this example, three fuzzy sets have been defined to represent qualitative values of temperature. The crisp value (70°C) belongs to the fuzzy set 'low temperature' (is low) with a grade of membership of 0.3, and belongs to 'medium' and 'high' temperatures with memberships of 0.9 and 0.2 respectively.

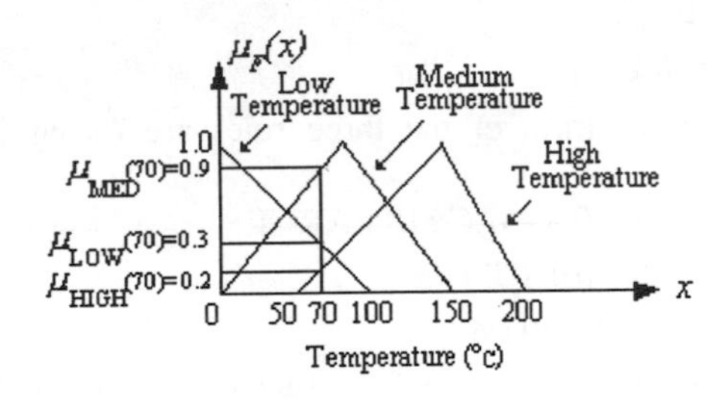

Figure 6.4. An example of fuzzification.

The definition of appropriate reference sets always has to be the first step in constructing any type of fuzzy control system. The choice of reference sets is an important one, since a poor choice of reference sets will result in a system with a poor performance.

2. Rule-based systems

The rule-based system is the most common application of fuzzy logic and fuzzy sets to control problems. As its name implies, the rule-based system is constructed according to a set of rules which describe, in qualitative terms, how an output behaves when subjected to various inputs. In the case of a rule-based controller, the inputs may be the change and rate of change in error measurement and the output may be the change in control action. A rule-based process model would use process inputs as input values and return an estimated process output.

The rules which make up a rule-based system are usually expressed in the form of "**IF... THEN...**" statements. The conditional part of the statement can make full use of the fuzzy logical operators to produce quite complicated conditions. The consequent part of a particular rule always assigns, to some degree, a particular qualitative value to the output. The qualitative value assigned is an output reference set and the degree of assignment is the overall degree of truth of the conditional expression. Each rule will thus produce its own fuzzy output set and, since all of these sets are defined in the same space, the overall output from all these rules can be arrived at by carrying out a fuzzy union.

Example 6.2

Consider a rule-based controller described by three rules:

(a) IF ε IS LARGE AND $\Delta\varepsilon$ IS LARGE, THEN ΔCa IS LARGE.

(b) IF ε IS MEDIUM, THEN ΔCa IS MEDIUM.

(c) IF $\Delta\varepsilon$ IS MEDIUM, THEN ΔCa IS MEDIUM.

If we assume that fuzzification gives the following truth values for the inputs

γ_ε IS LARGE = 0.3

γ_ε IS MEDIUM = 0.5

$\gamma_{\Delta\varepsilon}$ IS LARGE = 0.6

$\gamma_{\Delta\varepsilon}$ IS MEDIUM = 0.2

then the overall degrees of truth of the three rules are (using the logical operators described previously):

(a) $\gamma_a = \min(0.3, 0.6) = 0.3 \rightarrow \Delta Ca$ IS LARGE

(b) $\gamma_b = 0.5 \rightarrow \Delta Ca$ IS MEDIUM

(c) $\gamma_c = 0.2 \rightarrow \Delta Ca$ IS MEDIUM

Figure 6.5 shows the resulting output sets from each of the rules and the resultant overall output from the rule base, derived by carrying out a fuzzy union. The solid line in Figure 6.5 indicates overall output set.

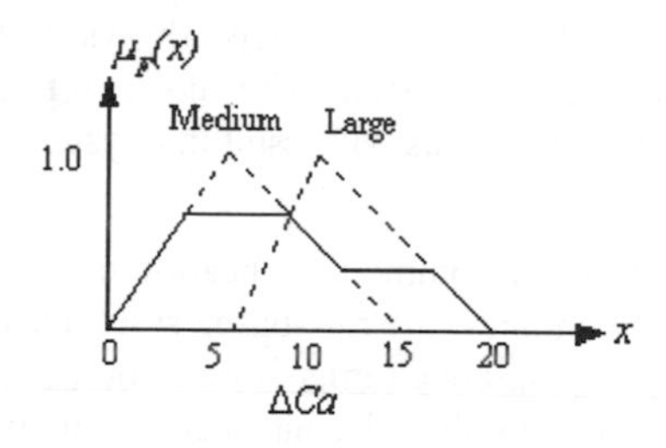

Figure 6.5. Output from a fuzzy rule base.

3. Relational systems

Another way that fuzzy logic can be applied to control problems is through fuzzy relational equations. Basically, when expressed in terms of reference sets, a fuzzy relational equation is a way of representing relationships which may exist between qualitative input and qualitative output states. The core of a relational model is the relational array, which is a matrix made up of values which represent the degrees of truth of all the possible cause-and-effect relationships between the input and output reference sets. The principal advantage of relational systems over rule-based systems is that several techniques exist which allow relational array entirety, and hence the relational model, to be identified directly from input-output data.

Algorithm 6.1.

A fuzzy relational equation is represented by

$$y = x \circ R \tag{6.19}$$

where y is the array of degrees of truth of output sets;
x is the array of degrees of truth of input sets;
R is the fuzzy relational array.

Given an input, the fuzzy relational equation is resolved to produce an output by using

$$\gamma_y(i) = \max_{j \in x}[\min(\gamma_R(i,j), \gamma_x(j))] \tag{6.20}$$

where $\gamma_y(i)$ is the degree of truth that the ith output set applies, $\gamma_x(i)$ is the degree of truth that the input belongs to the jth input set, $\gamma_R(i,j)$ is the relational array entry representing the degree of truth of a cause-and-effect relationship between the jth input set and the ith output set.

Example 6.3.

A system has been modeled by using a relational equation with the output and input described by two fuzzy reference sets. The resultant relational array produced by identification is:

$$R = \begin{bmatrix} 0.2 & 0.7 \\ 0 & 0.1 \end{bmatrix}$$

Given that an input to the system, described in terms of its reference sets, is x = [0.8, 0.2] , what is the output?

$$y = \begin{bmatrix} \gamma_y(1) \\ \gamma_y(2) \end{bmatrix} = \begin{bmatrix} \max(\min(0.2, 0.8), \min(0.2, 0.7)) \\ \max(\min(0, \ 0.8), \min(0.1, 0.2)) \end{bmatrix} = \begin{bmatrix} 0.2 \\ 0.1 \end{bmatrix}$$

A discussion of the various identification algorithms for relational systems is beyond the scope of this book and the interested reader may refer to the literature [35, 44, 45].

4. Defuzzification

The output from both rule-based and relational fuzzy systems is a fuzzy set, which usually must be converted back into a single crisp value before it is used to a control system. The procedure for doing this is called defuzzification, and there are three popular types of defuzzification, i.e., mean of maxima, center of area, and fuzzy mean. Figure 6.6 illustrates an output defuzzified by each of the three methods.

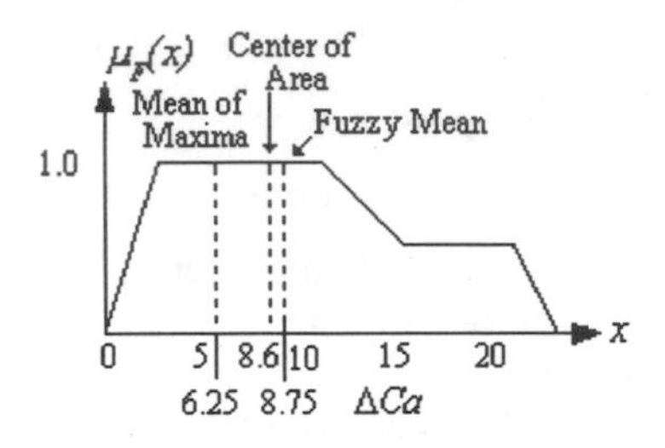

Figure 6.6. An output defuzzification.

Method 1 Mean of maxima. If there is a single maximum in the output membership function, then the value of the output at this point is taken as the defuzzified output. If there are several points at which the membership function is at its maximum, then the defuzzified value is taken to be the mean of all the output values where the membership function is at its maximum.
Method 2 Center of area. This method of defuzzification involves finding the value of output which bisects the area under the membership function curve.
Method 3 Fuzzy mean. Each output reference set is assigned a characteristic output value. With triangular reference sets the center point is taken as the characteristic output, and with other types of reference sets the value where the membership function is the largest is normally taken as the characteristic output. To generate a defuzzified output a weighted mean of the characteristic values is needed, where the weighting factors are the degrees of truth for the relevant reference sets, i.e.,

$$y = \sum y_i \cdot \gamma_i / \sum \gamma_i \qquad (6.21)$$

where y_i is characteristic value of set i, γ_i is degree of truth that set i applies, y is defuzzified output.

6.2 Architectures of Fuzzy Controllers

The development of fuzzy logic controller (FLC) is similar to that of a knowledge-based system application [29]. After defining design requirements and performing system identification we need to develop a knowledge base (KB) which consists of rule base contents and structure, termset definitions, and scaling factors. The validated knowledge base is compiled to minimize memory requirements and run-time search, and it is deployed on the target microprocessor [2]. The KB can be derived from knowledge engineering sessions with process operators including analysis of observed operator responses [33], from published rule bases for standard control policies such as PI and PD [41], and from linguistic models of the open and closed loop systems [3].

We have mentioned in the beginning of this section that the expert control system and the fuzzy logic control system have at least one thing in common, i.e., both want to model human experiences and human decision-making behaviors. In addition, both the expert control system and the FLC contain the KB and inference engine, and most of them have been the rule-based systems until now. Therefore, the FLC is usually said to be **Fuzzy Expert Controller (FEC)** or **Fuzzy Expert Control System**. Sometimes, the fuzzy expert systems are called the second generation of expert systems because they provide two essential and unique advantages in the design, development and implementation of the expert system, i.e., (1) fuzzy knowledge representation, (2) fuzzy inference methods.

6.2.1 General Structures of FLC

In theory, a fuzzy controller is represented by a N-dimensional relation R. This relation R can be seen as a function restricted between 0 and 1 of N variables. R is a combination of several N-dimensional relations R_i, each representing a rule (r_i : IF... THEH...). The inputs x of the controller are fuzzified into a relation X, in the case of multiple input and single output (MISO) control (N-1) dimensional. By applying compositional rule of inference [49, 50], the fuzzy output Y is calculated. A crisp, numerical output y is obtained by defuzzification of the fuzzy output.

Figure 6.7 shows the schematic diagram of the theoretical approach of a fuzzy controller with input x and output y. Because in this theoretical approach multi-dimensional functions are used to describe the relations R, X and Y, a great amount of memory is needed when implementing a discrete approximation.

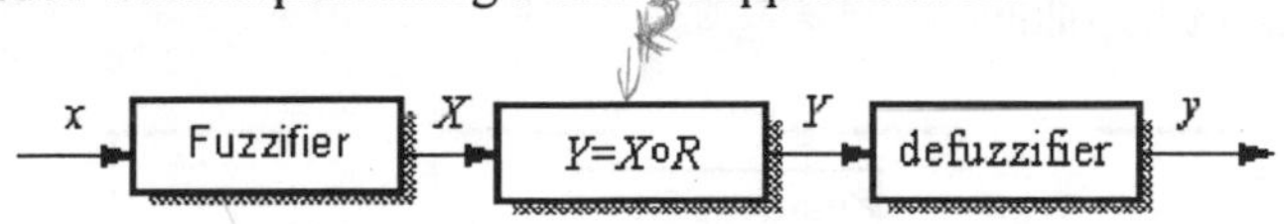

Figure 6.7. Schematic of theoretical fuzzy controller.

Figure 6.8 shows a general architecture of the fuzzy logic controller, which consists of input and output scaling, fuzzification, fuzzy decision process, and defuzzification. The scaling factors map the controller inputs and outputs to and from normalized intervals on which the fuzzy reasoning takes place. Fuzzification (or quantization) makes the measured controller inputs dimensionally compatible with the Left-Hand Side (LHS) of the rules. No information loss, however, occurs at this step. The fuzzy decision process is performed by an inference engine that matches the LHS of all the rules with the input; determines the partial degree of matching of each rule; and aggregates the weighted output of the rules, generating a probability distribution of values on the output space. The defuzzification summarizes this distribution into a point that is used by the actuator after scaling .

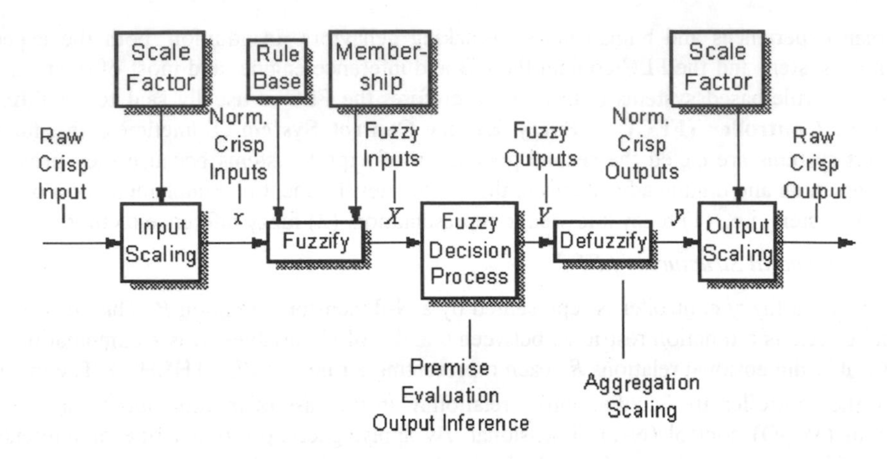

Figure 6.8. General architecture of FLC.

The essential structure for the fuzzy control system is shown in Figure 6.9. The fuzzy controller consists of four essential components, i.e., fuzzification interface, knowledge base, inference engine (fuzzy inference and decision unit), and defuzzification interface. Their functions are as follows.

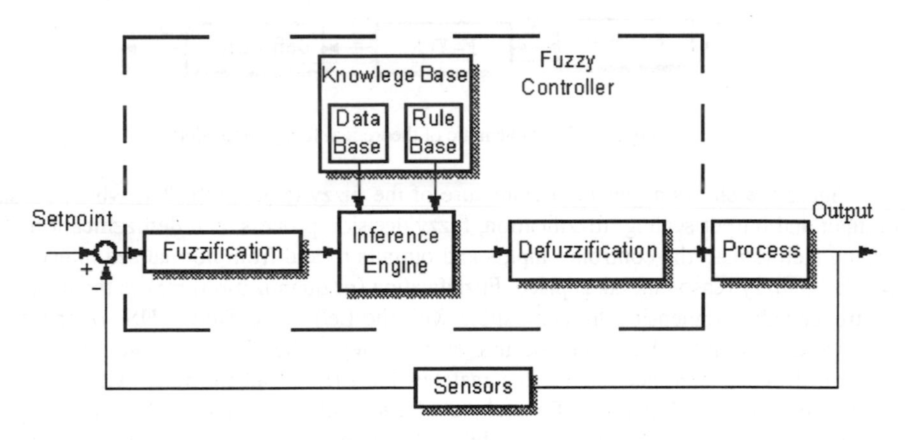

Figure 6.9. Essential structure of fuzzy control system.

(i) Fuzzification interface measures the input variable (setpoint) and output variable(s) of the controlled system, and maps them into a suitable range that corresponds to the universe of discourse, then the crisp input data is transformed into suitable linguistic values or labels of fuzzy sets. This unit can be seen as a mark of the fuzzy sets.

(ii) Knowledge Base involves the relevant knowledge of the applied domain and the control objectives, it consists of the data base and the linguistic(fuzzy) control rule base. The data base provides the necessary definitions on discretization of universes of discourse for the linguistic control rules and definitions of membership functions. The linguistic control rules marks the control objectives and control strategies of the domain experts.
(iii) Inference Engine is the core of the fuzzy control system. Based on the fuzzy concept, the fuzzy control information can be acquired by using fuzzy implication and fuzzy inference rules of fuzzy logic, and the anthropomorphic decision processes can be realized. According to the fuzzy inputs and fuzzy control rules, the fuzzy inference solves the fuzzy relation equations, and acquires fuzzy outputs.
(iv) Defuzzification interface takes the inferred fuzzy control action, and generates a crisp or nonfuzzy control action. The crisp control action has to be inversely scaled (out scaling) by range transformation prior to controlling the controlled process.

The above four components (units) have been discussed in many references in more detail [14, 16, 54].

6.2.2 PID Fuzzy Controller

In general for fuzzy control systems, the structure of the two-dimensional fuzzy control is usually used. This structure can ensure the simplicity and rapidity of the system. The input linguistic variables of this kind of controllers are the error E and error variation EC of the system. Therefore, this controller has a function similar to the traditional PD controller, and possesses good dynamic properties. However, the static error of this controller can not be eliminated. In order to improve the static performance, a fuzzy integration is added to the controller to form a PID fuzzy controller.

There are several methods for introducing the fuzzy integration factor to the fuzzy controller.
(i) Braae-Rutherford Method
This method was developed by Braae and Rutherford in 1978, and its structure is shown in Figure 6.10 [3]. The integration units are placed before the fuzzification unit and after the defuzzification unit. Although this fuzzy controller can decrease the error of the system to a certain extent, a vibration phenomenon of the limit cycle can not be guaranteed to be eliminated.

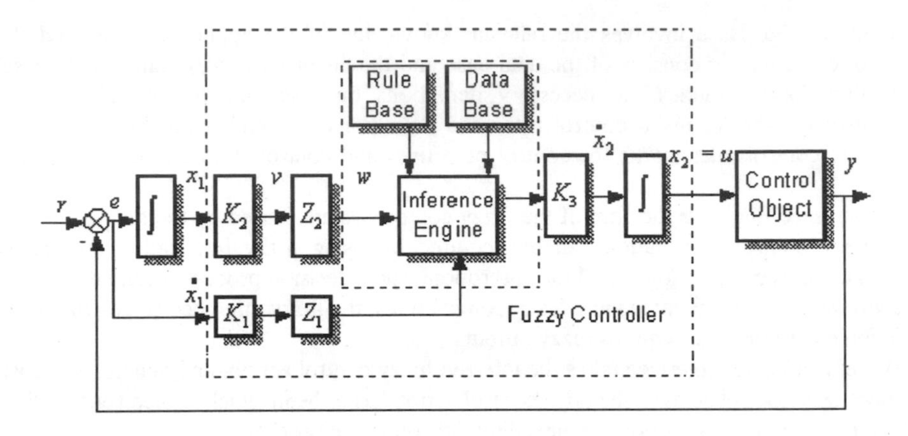

Figure 6.10. Braae-Rutherford fuzzy controller.

(ii) Bialkowski Method

This fuzzy controller was proposed by Bialkowski in 1983, and is shown in Figure 6.11 [54]. The controller consists of PI controller and 2D fuzzy controller connected parallelly. The output of the hybrid fuzzy/PID controller is equal to the output u_i of the conventional PI controller added to the output u_f of the 2D fuzzy controller, i.e.,

$$u_n = u_i + u_f \tag{6.22}$$

where $u_i = k_I \sum_i e_i$, $u_f = f(e, \dot{e})$ a fuzzy relationship function. In the hybrid controller, both e(t) and $k_I \sum_i e_i$ are continuous variables. This controller can eliminate both the limit cycle and the system error, and leads the system to a non-error fuzzy control system, i.e., e→0.

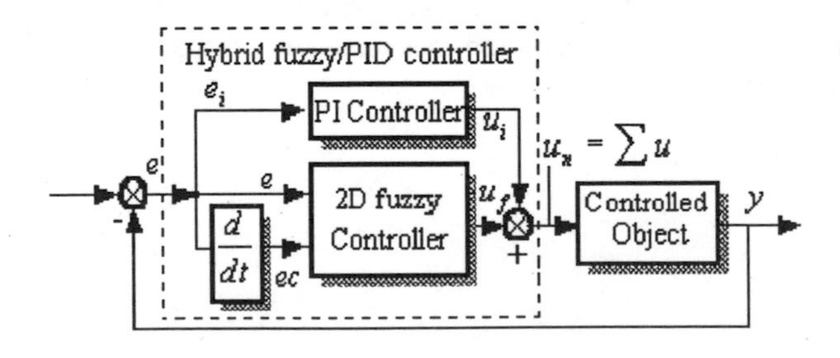

Figure 6.11. Hybrid fuzzy/PID controller.

(iii) Basseville Method

M. Basseville proposed a PID fuzzy controller in 1988, as shown in Figure 6.12. This controller integrates the fuzzification variable of error e, and it can also eliminate the system error but is difficult to eliminate the limit cycle near zero.

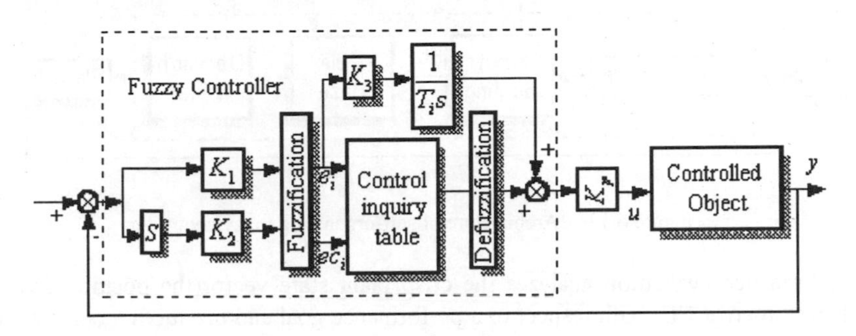

Figure 6.12. Basseville's PID fuzzy controller.

6.2.3 *Self-Organizing Fuzzy Controller*

The parameters of the direct adaptive controller are directly synthesized from the error between the desired and actual response of the plant. At present no analytical methods exist for direct adaptive control of nonlinear systems, however the self-organizing fuzzy controllers attempt to address the direct adaptive nonlinear control problem in which the reference model of ideal closed loop behavior is encapsulated in the performance index that directs the control rule updating process. A self-organizing fuzzy controller is a fuzzy controller whose control policy adapts in accord with changes in the process or its environment, and is an experientially determined processor which has to perform two tasks simultaneously: identification and control. These controllers can cope with multivariable as well as single input-output systems, nonlinearities, temporal variations in parameters, and random disturbances [16].

The theory of the self-organizing fuzzy controllers was originally proposed by Mamdani and his students at Queen Mary College, London [36, 39, 48], and has been recently extend by others [2, 9, 10, 15, 32, 37, 40].

Figure 6.13 shows the general architecture of the self-organizing fuzzy controller which is consists of two levels, the basic level and the self-organizing level. The former is a conventional fuzzy linguistic control rule base, and the latter makes evaluation to each sampled input/output response and generates a modification to the controller. The self-organizing FLC is an automatic approach to obtaining the rule base for a fuzzy controller. Rules are generated and modified, as new conditions arise, to control the plant to some desired response with the FLC.

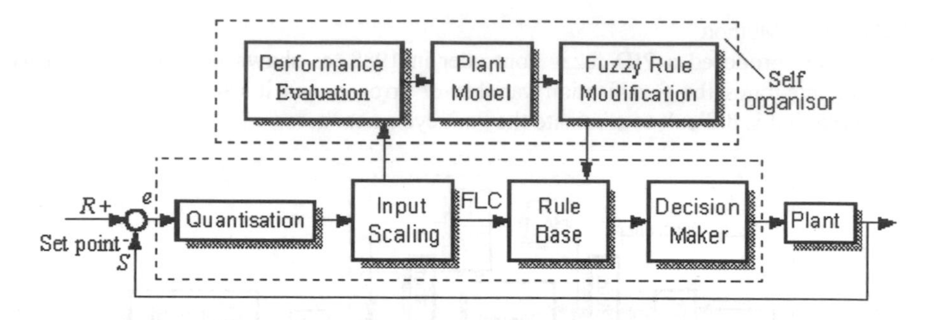

Figure 6.13. Architecture of self-organizing fuzzy controller.

Performance evaluation analyzes the crisp plant state vector(the position error PE and change in error CE) with respect to a performance goal and produces a correction for a yet-to-be-identified rule to compensate for any poor performance. The correction consists of a scalar amount by which the rule conclusion will be adjusted. Acceptable and unacceptable step response phase plane trajectories are used as the performance goals.

The plant model is used to account for plant input-output polarity in rule correction. The fuzzy rule modification determines which rule or rules are responsible for the present poor performance controller which makes iterative improvements to its rule base at successive sample times during its learning trails.

6.2.4 Self-Tuning Fuzzy Controller

The fuzzy controller is a core component for any kind of fuzzy control system and determines the control performance indexes of the fuzzy control system. The performance of the fuzzy controller is in turn dependent on the fuzzy linguistic rules and synthesis inferences. In general, once a fuzzy controller is designed, the linguistic rules and synthesis inference are also determined and unregulated. However, in some cases, fuzzy controllers are demanded to have self-regulation ability in order to get stronger adaptability for different controlled objects. Therefore, the fuzzy controller is required to have adaptability, and a fuzzy controller with multiple regulated factors is developed.

The fuzzy controller with multiple regulated factors can find optimum parameters and get satisfactory control results by the use of self-optimizing methods. However, the self-optimizing process is more complex, with large amounts of computation, and is inconvenient for on-line regulation. Self-tuning fuzzy controller is another method with regulated factors and self-optimizing abilities.

The structure of self-tuning fuzzy controller is shown in Figure 6.14, where the input/output relation of the controlled object is

$$Y(kT) = f_1[u(kT)]. \tag{6.23}$$

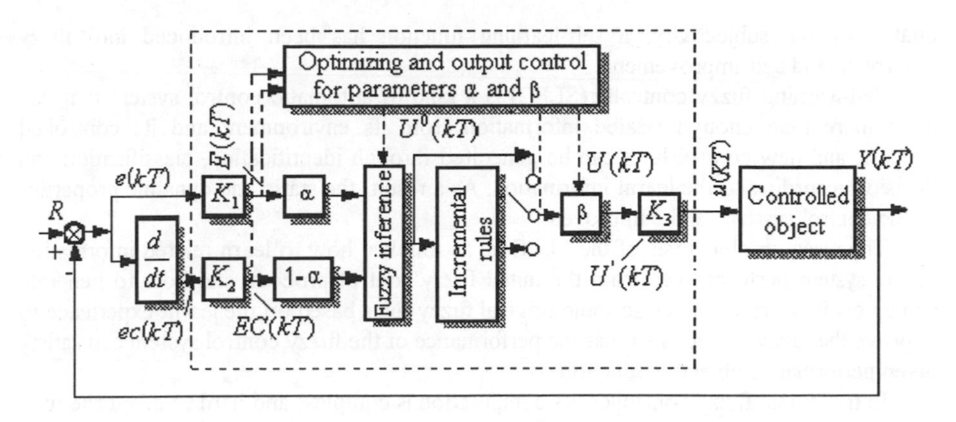

Figure 6.14. Structure of self-tuning fuzzy controller.

The input error signal and error ratio of the fuzzy controller are as respectively follows:

$$\alpha E(kT) = \text{int}[\alpha K_1 e(kT) + \varepsilon] \tag{6.24}$$

$$(1-\alpha)EC(kT) = \text{int}[(1-\alpha)K_2 ec(kT) + \varepsilon] \tag{6.25}$$

The output of the fuzzy controller is

$$u(kT) = K_3 U(kT) \tag{6.26}$$

The output of the control system is

$$Y(kT) = f_1\{K_3 \beta\, [\alpha E(kT) + (1-\alpha)EC(kT)]\} \tag{6.27}$$

The error of the system is

$$e(kT) = R - \dot{Y}(kT) \tag{6.28}$$

where $\alpha \in (0,1)$, $\beta \in (0,1)$ are regulated weighted factors; ε is integral variable. We chose ε=0.5 or ε=0 depending upon the situation of the controlled object; $ec(kT)=e(kT)-e(kT-T)$; K_1 *and* K_2 are fuzzification factors of the input error and the error ratio of the fuzzy controller respectively; K_3 is fuzzy decision factor of the output of the fuzzy controller.

6.2.5 Self-Learning Fuzzy Controller

The quality of a fuzzy control system depends basically on the establishment of fuzzy control rules and the reality of fuzzy relations. However, the establishment of fuzzy rules usually are with some degree of subjectivity, and a fine dynamic property of the system cannot be guaranteed. In order to overcome the negative influence on the system control quality by the subjectivity, a self-learning function has been introduced into fuzzy controller, and self-improvement.

Self-learning fuzzy controller(SLFC) is a kind of automatic control system that can learn more than enough related information from its environment and its controlled process, and new control laws can be generated through identification, classification and decision according to the learnt information. As a result, the static and dynamic properties of the control system can be improved.

The research objectives of the SLFC lie as follows: how to learn related information of the system performances from the initial fuzzy control process, and how to generate some new fuzzy rules or revise some original fuzzy rules based on the learnt experience to improve the fuzzy relations so that the performance of the fuzzy control system can satisfy given performance objective function.

In traditional fuzzy controller, its computation is complex, and hard to adapt the real-time requirement for the self-learning fuzzy control. The computation is mainly expended in the processing of the fuzzy relation R. Therefore, in order to establish an effective algorithm for the self-learning fuzzy controller, it is necessary to improve the processing of R. Some improved fuzzy relations R and inference algorithms for self-learning fuzzy control have been developed and used. A structure of the self-learning fuzzy control system is shown in Figure 6.15. From Figure 6.15 we can see that the expected output $Y^*(kT)$ is the output of the reference model, the real output of the control system is $Y(kT)$, the error $E(kT)= Y^*(kT)-Y(kT)$, and the error change $EC(kT)=E(kT)-E(kT-1)$.

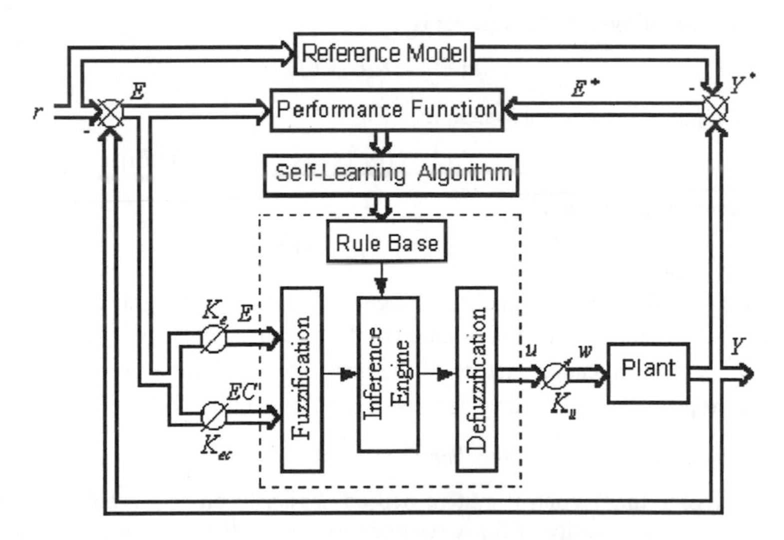

Figure 6.15. A Structure of self-learning fuzzy controller.

6.2.6 *Expert Fuzzy Controller*

Fuzzy control possesses the advantages of small overshoot, strong robustness and good adaptability to the nonlinearity of system. Therefore, fuzzy control is an effective control strategy. However, the fuzzy control is also with its own shortages. The first shortage is lower precision of system owing to simple fuzzy processing of information. In order to increase the precision, the quantized degree has to be increased; so the searching scope of the system would increase. The second one is the singleness, both for the structure of fuzzy controller and for the form of knowledge representation; so it is difficult to treat heuristic knowledge needed to control complex systems, and its application fields are limited. The third one, as the rank of system nonlinearity is higher, is that the established fuzzy control rules would become incomplete or indefinite; so the control result would be worse. These shortages could be overcome by integrating the techniques of fuzzy control and expert systems, and constructing a new control system-expert fuzzy control system. Figure 6.16 depicts a configuration of the expert fuzzy control system based on fuzzy controller, where a module of expert controller is integrated to form the expert fuzzy control system. During the operating process of the control system, the dynamic output performance of the controlled object (process) is supervised continuously through the module of performance identification, and the processed parameters are sent to the expert controller. The expert controller makes the inference and decision to modify the coefficients K_1, K_2, K_3 and the parameters of the control table of the fuzzy controller according to the current known knowledge of the system dynamic properties in the knowledge base until a satisfactory dynamic control property is acquired.

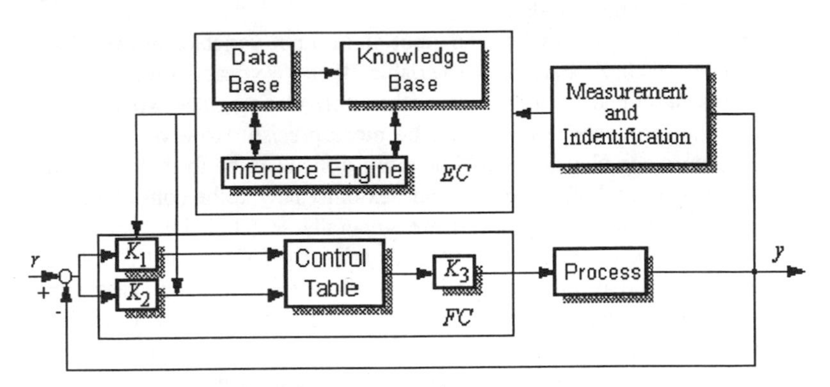

Figure 6.16. Configuration of expert fuzzy control system.

Besides the above discussed fuzzy-based integrated controllers which involve PID fuzzy controller, self-organizing fuzzy controller, self-tuning fuzzy controller, self-learning fuzzy controller and expert fuzzy controller, some other integrated fuzzy controllers such as adaptive fuzzy controller, optimal fuzzy controller [8, 11, 18, 34] and fuzzy-neurocontroller have been proposed and developed in recent years. We will introduce the fuzzy-neurocontroller in Chapter 7.

6.3 Design of Fuzzy Controllers

We are going to focus on the fundamental aspects of the design of fuzzy controllers. Before discussing issues for designing the fuzzy control systems, the design requirements of a fuzzy controller will be introduced first.

6.3.1 Design Requirements for Fuzzy Controllers

In designing a fuzzy controller, the following issues are required to be involved:

1. Select a reasonable structure for the fuzzy controller

The first step for designing a fuzzy controller is to select and determine a reasonable structure for the fuzzy controller. Selecting a structure for the fuzzy controller means that the input and output variables of the fuzzy controller are required to be determined. In general, the error signal E(or e) and error rate EC (or $\Delta E, \Delta e$) are selected as input variables of the fuzzy controller, and the change of controlled variable as output variable. The structure of a fuzzy controller has a significant influence on the performance of the controlled system and hence it is required to be reasonable selective reasonably according to the specific situation of the controlled object.

2. Select and extract fuzzy control rules

The fuzzy control rules are the core of the fuzzy controller. In designing the rules, the following issues should be considered:

(i) Select vocabularies that describe the input and output variables of the controller. We call the vocabulary a fuzzy state of the variable. If more vocabularies are selected, i.e., every variable is described by more states, then formulating rules would be with more flexibility, and the formulated rules would be more precise. However, the control rules would become more complex and would be more difficult to be formulated. Therefore, in selecting the fuzzy states, both simplicity and flexibility have to be considered. In practical applications, seven to nine fuzzy states are generally selected, i.e., positive big (PB), positive medium (PM), positive small (PS), negative big (NB), negative medium (NM), negative small (NS), and average zero (AZ) or positive zero (PO) and negative zero (NO).

(ii) Define fuzzy sets which represent fuzzy states. In defining a fuzzy set, we consider the queer shape of membership function of the fuzzy set first. As an input error varies in fuzzy subsets with higher resolution, the output change caused by the input error would be more acute. Inversely, as an input error varies in fuzzy subsets with lower resolution, the change would be more smooth. Therefore, a fuzzy set with low resolution would be used in the case of a large scope of error, and a fuzzy set with high resolution would be used when the error is close to zero.

As described above, the linguistic variable (fuzzy state) A corresponding to error E is selected as follows:

PB, PM, PS, PO
NB, NM, NS, NO

They correspond to fuzzy sets $A_1, A_2, \cdots, A_8$ which can generally take the values as listed in Table 6.1

Table 6.1.

		-6	-5	-4	-3	-2	-1	-0	+0	+1	+2	+3	+4	+5	+6
A_1	PB	0	0	0	0	0	0	0	0	0	0	0.1	0.4	0.8	1.0
A_2	PM	0	0	0	0	0	0	0	0	0	0.2	0.7	1.0	0.7	0.2
A_3	PS	0	0	0	0	0	0	0	0.3	0.8	1.0	0.5	0.1	0	0
A_4	PO	0	0	0	0	0	0	0	1.0	0.6	0.1	0	0	0	0
A_5	NO	0	0	0	0	0.1	0.6	1.0	0	0	0	0	0	0	0
A_6	NS	0	0	0.1	0.5	1.0	0.8	0.3	0	0	0	0	0	0	0
A_7	NM	0.2	0.7	1.0	0.7	0.2	0	0	0	0	0	0	0	0	0
A_8	NB	1.0	0.8	0.4	0.1	0	0	0	0	0	0	0	0	0	0

In this section as well as latter sections, the variables($E, \Delta E, \Delta^2 E$ *or* e, de, d^2e) are initially normalized in the range($-100 \rightarrow +100$) which is then mapped nonlinearly into thirteen integer quantization levels($-6 \rightarrow 6$) through the logarithmic transformation(0, 1, 2.36, 5, 10.8, 23.4, 100)→(0, 1, 2, 3, 4, 5, 6). This nonlinear quantization is to provide greater control sensitivity around the set point and to improve signal to noise ratio.

The linguistic variable B corresponding to error rate EC is usually selected as the following seven levels:

PB, PM, PS, AZ, NB, NM, NS

that correspond to fuzzy sets $B_1, B_2, \ldots, B_8$ as listed in Table 6.2.

Table 6.2.

		-6	-5	-4	-3	-2	-1	0	+1	+2	+3	+4	+5	+6
B_1	PB	0	0	0	0	0	0	0	0	0	0.1	0.4	0.8	1.0
B_2	PM	0	0	0	0	0	0	0	0	0.2	0.7	1.0	0.7	0.2
B_3	PS	0	0	0	0	0	0	0	0.9	1.0	0.7	0.2	0	0
B_4	AZ	0	0	0	0	0	0.5	1.0	0.5	0	0	0	0	0
B_5	NS	0	0	0.2	0.7	1.0	0.9	0	0	0	0	0	0	0
B_6	NM	0.2	0.7	1.0	0.7	0.2	0	0	0	0	0	0	0	0
B_7	NB	1.0	0.8	0.4	0.1	0	0	0	0	0	0	0	0	0

The fuzzy controller generally utilizes a model with two inputs and single output as shown in Figure 6.17 that corresponds to the following linguistic formula:

$$\text{IF } A \text{ AND } B \text{ THEN } C \tag{6.29}$$

and the structure shown in Figure 6.18.

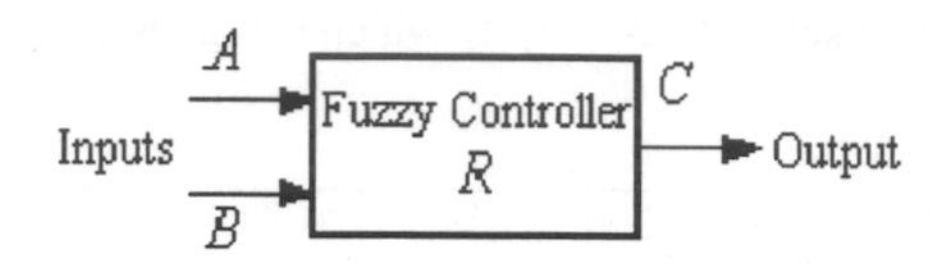

Figure 6.17. FC with two inputs and single output.

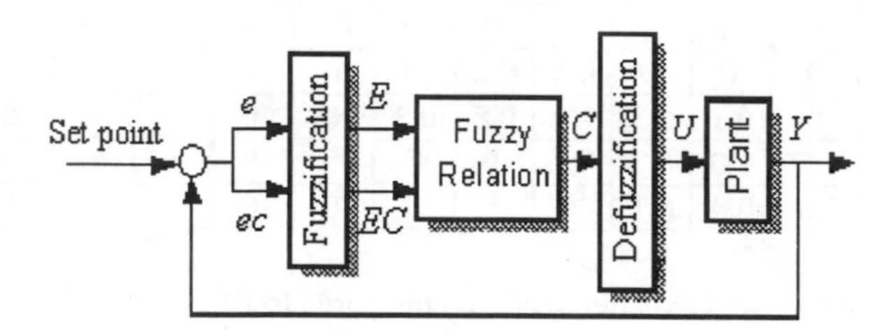

Figure 6.18. Structure of fuzzy control system with two inputs and single output.

The linguistic variable C corresponding to control decision U usually also has seven levels in which fuzzy sets $C_1, C_2, \ldots, C_7$ have the values listed in Table 6.3.

Table 6.3.

		-7	-6	-5	-4	-3	-2	-1	0	+1	+2	+3	+4	+5	+6	+7
C_1	PB	0	0	0	0	0	0	0	0	0	0	0	0.1	0.4	0.8	1.0
C_2	PM	0	0	0	0	0	0	0	0	0	0.2	0.7	1.0	0.7	0.2	0
C_3	PS	0	0	0	0	0	0	0	0.4	1.0	0.8	0.4	0.1	0	0	0
C_4	AZ	0	0	0	0	0	0	0.5	1.0	0.5	0	0	0	0	0	0
C_5	NS	0	0	0	0.1	0.4	0.8	1.0	0.4	0	0	0	0	0	0	0
C_6	NM	0	0.2	0.7	1.0	0.7	0.2	0	0	0	0	0	0	0	0	0
C_7	NB	1.0	0.8	0.4	0.1	0	0	0	0	0	0	0	0	0	0	0

(iii) Determine table of fuzzy control states

We usually write the inference linguistic rules, i.e., a fuzzy condition statement, into a table of fuzzy control states according to the human's practical experience in the control process. Table 6.4 lists the fuzzy control states.

Generally speaking, the design principles of the control rules for the fuzzy controller are as follows: when the error is bigger, the control rate should decrease the error as soon as possible; when the error is smaller, besides eliminating the error, the system stability has also to be considered in order to prevent unnecessary overshoot and oscillation.

The determination principles for Table 6.4 are as follows.

Table 6.4.

U \ ΔE / E	NB NM NS	AZ PS	PM PB
NB NM	PB	PM	AZ
NS	PM	PM AZ	NS
AZ	PM PS	AZ NS	NS
PS	PS AZ	NM	NM
PM PB	AZ	NM	NB

For the negative error, if the error is NB and the error rate is also negative (NB, NM or NS), then PB is selected as the control rate in order to eliminate the error as soon as possible and to increase the control response fast enough; if the error rate is positive, it means that the error tends to decrease, then a smaller control rate should be selected, i.e., if error rate is PS, then the control rate is AZ; if error is NM, the control principle is the same as above in order to eliminate error as soon as possible, so the control rate is selected in the same control level as shown in Table 6.4.

For smaller errors (NS, AZ or PS), the major problem is to bring the control system to its steady state as soon as possible and to prevent the overshooting phenomenon. The control rate is determined according to the error rate. For example, if the error is NS, the error rate is PB, then NS is selected as the control rate. According to the operating features of the control system, when the error and error rate change their signs simultaneously, then the control rate also has to change its sign. Therefore, for the positive error, the corresponding control rate can be determined symmetrically as listed in Table 6.4. Fuzzy condition statements can be written by means of Table 6.4 as of now.

3. Determine strategies for fuzzification and defuzzification and define control table

After deriving the fuzzy sets of error and error rate, the fuzzy set U of the control rate can be acquired by synthetic algorithm of fuzzy inference

$$U = (E \times EC) \circ R \tag{6.30}$$

where R is a matrix of fuzzy relations; U is the fuzzy set of control rates which will be transformed into crisp value U as shown in Table 6.5.

Table 6.5.

EC / U / E	-6	-5	-4	-3	-2	-1	0	+1	+2	+3	+4	+5	+6
-6	7	6	7	6	7	7	7	4	4	2	0	0	0
-5	6	6	6	6	6	6	6	4	4	2	0	0	0
-4	7	6	7	6	7	7	7	4	4	2	0	0	0
-3	6	6	6	6	6	6	6	3	2	0	-1	-1	-1
-2	4	4	4	5	4	4	4	1	0	0	-1	-1	-1
-1	4	4	4	5	4	4	1	0	0	0	-3	-2	-1
-0	4	4	4	5	1	1	0	-1	-1	-1	-4	-4	-4
+0	4	4	4	5	1	1	0	-1	-1	-1	-4	-4	-4
+1	2	2	2	2	0	0	-1	-4	-4	-3	-4	-4	-4
+2	1	1	1	-2	0	-3	-4	-4	-4	-3	-4	-4	-4
+3	0	0	0	0	-3	-3	-6	-6	-6	-6	-6	-6	-6
+4	0	0	0	-2	-4	-4	-7	-7	-7	-6	-7	-6	-7
+5	0	0	0	-2	-4	-4	-6	-6	-6	-6	-6	-6	-6
+6	0	0	0	-2	-4	-4	-7	-7	-7	-6	-7	-6	-7

In building a fuzzy control system, at first, all errors and error rates are needed to be transformed into fuzzy inputs from crisp inputs. As we have seen, this process is called fuzzification. The fuzzy set U of control rates acquired by synthetic computation will then be transformed into the crisp output for the execution of the control. We also have known that this process is called defuzzification or fuzzy decision. We could use any methods introduced in reference books about fuzzy sets and fuzzy control to realize fuzzification and defuzzification.

4. Determine parameters of the fuzzy controller

The practical scope of the error and error rate in a control system is called basic universe of discourse of the quantized variables. During designing a specific fuzzy controller the universe of discourse for all inputs and output(s) has to be determined. For instance, in designing a fuzzy controller for level surface, the control scope of the liquid level, the maximum and minimum capacities of the valves are needed to be determined in the first place. These control requirements will define the scope of voltage or current of the A/D and D/A converters in the fuzzy controller.

The quantized factors and scale factors of the fuzzy controller influence the static and dynamic properties of the fuzzy controller evidently, thereby, all these factors should be selected reasonably. Suppose the basic universe of discourse of the error is $[-x, +x]$, and the universe of discourse of the fuzzy set of the errors is $\{-n, -n+1, \cdots, 0, \cdots, n-1, n\}$, then the quantized factor k_1 can be determined by the following equation:

$$k_1 = \frac{n}{x} \tag{6.31}$$

The quantized factor k_2 of the error rate and scale factor k_3 of the output control rate can be determined in the same way as k_1.

In this subsection we briefly introduce some major issues for designing the fuzzy controller. For a better and deeper insight into the following issues: (1) the structure selection of the single input-single output fuzzy controller and multiple inputs-multiple outputs fuzzy controller; (2) the fuzzy rule selection including the determination of fuzzy linguistic variables and membership function of the linguistic value, and to establish fuzzy control rules by various inference modes; (3) defuzzification methods such as the maximum membership grade (Mamdani inference), middle placed (Lason inference) and weighted average (Tsukamoto inference) and others; (4) determination for the universe of discourse and the scale factors of the fuzzy controller, interested readers could look-up some monographs and books related to fuzzy control [16, 29, 35, 54].

6.3.2 Design Approaches to FLC and EFC

The rules and fuzzy membership functions employed in fuzzy controllers are usually determined heuristically, i.e., they are manually coded on the basis of an intuitive understanding of the functioning of the underlying process to be controlled. A systematic design of the rules and their accompanying fuzzy membership functions has been attempted in the past. For example, a genetic algorithm has been successfully employed to optimize the behavior of a fuzzy controller used in an autonomous spacecraft rendezvous maneuver [12, 19]. More recently, a neural network of the associative memory type was employed to initially train (off-line) and then adapt (on-line) the parameters of a fuzzy controller for an inverted pendulum [24]. Another methodology employed in the design of fuzzy controllers is centered around fuzzy inductive reasoning, a technique geared at the qualitative simulation of dynamical continuous-time processes [6, 7, 30].

Several main design approaches can be overviewed as follows.

1. Linguistic phase plane approach

Let us use fuzzy PID controller as an example to explain this design approach. A fuzzy PID controller owns a large number of parameters. So, the design of such a controller involves many factors that are to be determined. However, in the most practical situations, the number of fuzzy values, the number of classification levels, and the membership distributions of fuzzy variables are fixed intentionally. For example, eight or seven fuzzy values, fourteen or fifteen classification levels and normal distributions are used. Thus, based on the configuration of fuzzy PID controller, the design work can be stated as

- adjust the universes of the fuzzy variables
- modify the elements of the rule matrices

In the same manner as in conventional cases, the fuzzy PD control is used to improve the dynamic behavior and the fuzzy I control is used to eliminate the steady state error.

The latter has less importance in improving the dynamic performance of the closed-loop system than the former. Naturally, too high fuzzy I effect will cause instability.

In order to reveal how the adjustments of the elements and the universes of the fuzzy PD control rule matrix may exert the anticipated desirable influences on the performance of the closed-loop system, the linguistic phase plane analysis [4] can be employed as an effective approach. In the phase plane shown in Figure 6.19, a linguistic trajectory is plotted by connecting those cells on the fuzzy PD control rule matrix where the continuous or discrete trajectory passes, and the meanings of PB, PM, PS, AZ, NS, NM and NB were defined previously in Section 6.1.2. It is well known that the continuous trajectory is directly resulted from the coordinates of the real-time errors (E) and error changes (EC).

The control output values on a linguistic trajectory indicate the real-time fuzzy PD control outputs at the same time as the fuzzy error and error changes. The convergent process along the trajectory towards the origin point represents the response behavior of the closed-loop system. By varying the control output values on the trajectory, the convergent process, i.e., the shape of the trajectory, can be modified. The increase of the control output values will tighten up the trajectory; the decrease of these values will stretch out the trajectory. It is obvious that abrupt variations in the values of adjacent cells of the decision matrix will result in abrupt changes in the control effects.

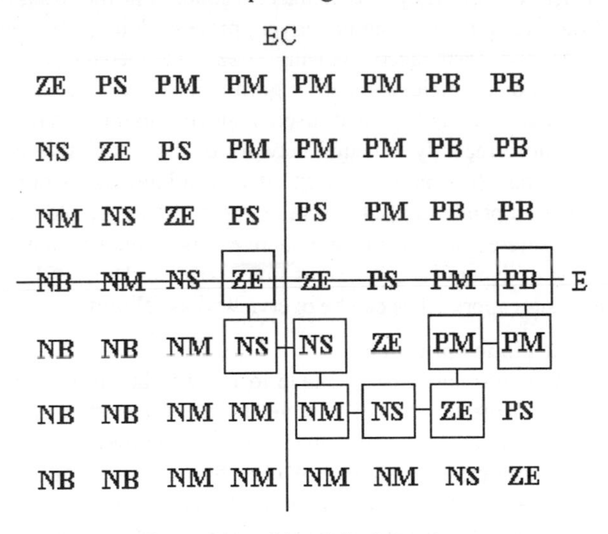

Figure 6.20. Linguistic phase plane.

The above stated synthetic experiences can be used for the modification of the elements of the fuzzy PD control rule matrix. As for the adjustment of the universes, normally the universes of the error change are selected according to their real-time maximum values which might happen in the closed-loop system. But the universe of the control output directly leads to the variation of the controller gains. Therefore, it is concluded that the control gains can be adjusted by altering the universe of the control output, and the control nonlinearity can be adjusted by altering the elements of the rule matrix.

Subject to reference inputs, the whole fuzzy PID controller can be designed in the following way. First, select suitable elements and universes of the rule matrices by experience or by trial-and-error in order that the closed-loop system is at least stable and shows no considerable steady-state errors. Secondly, use the linguistic phase plane approach to modify the fuzzy PD rule matrix, if this cannot give rise to obvious improvement in the transient, adjust the fuzzy I control rule matrix, and then alter the fuzzy PD control rule matrix once again. Repeat these procedures until the closed-loop performances are satisfactory. In addition, since the fuzzy PID controller can be treated as a nonlinear PID controller, the effects of fuzzy PID control can be discussed using the classical PID principle in the fuzzy set sense. The fuzzy P effect is necessary, the fuzzy I effect is to eliminate the steady-state error and the fuzzy D effect is to improve the dynamic characteristic. The design techniques aim at changing the P-I-D effects of the fuzzy PID controller.

2. Expert system approach

As presented above, design expertise on the fuzzy PID controller is usually expressed in the form of linguistic rules rather than mathematical formulae. In such cases, expert system methods are useful to organize human experiences so that the design of the fuzzy PID controller can be performed automatically by computers rather than manually by control engineers. The construction of an expert system for the design of the fuzzy PID controller includes three stages:

- knowledge acquisition
- knowledge representation
- knowledge manipulation

The knowledge acquisition decides the correctness of the expert system. It is of absolute necessity though it involves tedious collective work. The design knowledge for the fuzzy PID controller can be acquired from linguistic phase plane analyses, classical control rules and common engineering senses. Furthermore, this kind of knowledge can be modified and enriched by learning throughout the operations of the expert system.

The knowledge representation determines the efficiency of the expert system. Its importance lies in that it forms a framework for the succeeding knowledge manipulation. With reference to a step input response, the triplet (RT, OS, ST) is adopted as an index to evaluate the performance of the controller in the closed-loop system, where variables RT, OS and ST denote rise time, overshoot and settling time respectively. In the linguistic phase plane, the initial trajectory is separated into three segments which are described by the triplet (IS, MS, FS), where IS, MS and FS denote initial, middle and final segments respectively. A linguistic quartet (O, S, M, B) is defined to indicate the values of (RT, OS, ST) and the modifications of (IS, MS, FS), where O, S, M and B denote OK, small, medium and big respectively. Thus, the design knowledge on the fuzzy PID controller can be formulated using a series of production rules relating (RT, OS, ST) to (U_d, U_i) through (IS, MS, FS), where U_d and U_i denote the fuzzy PD control output and the fuzzy I control output change respectively. These rules are used iteratively to modify the initial phase plane trajectory towards obtaining a better close-loop performance.

The knowledge manipulation accounts for the effectiveness of the expert system. It has the responsibility for producing desirable conclusions. Since the design knowledge is represented by production rules, two inference methods in the expert system theory may be exploited, i.e., forward chaining and backward chaining. Because in this context, conclusions are needed to be inferred from given facts, the forward chaining method should be suitable for the expert system to design the fuzzy PID controller. As the linguistic terms are described by fuzzy sets, the required expert system is considered to be a standard knowledge-based system with production rules based on fuzzy logic. Once a formal step is taken into the expert system theory, many expert system tools or languages can be borrowed. Therefore, the desirable expert system can be readily realized by means of the existing AI technology.

3. CAD environment for fuzzy PID controller

To develop a systematic design methodology for the fuzzy PID controller, a computer aid design (CAD) environment in a HP 9000/370 computer has been set up as shown in Figure 6.20 [27, 28, 47]. This CAD environment consists of an expert system and a simulation package.

The expert system is divided into two levels, i.e., the upper level and the lower one. The upper level uses qualitative reasoning to complete the symbolic manipulation of the design rules. The lower level uses quantitative reasoning to accomplish the numerical manipulation of the design procedures. The major parts and their roles in the expert system are:

(1) Knowledge base, which stores the design heuristics in the form of production rules;
(2) Data base, which stores the original and/or intermediate data in the form of matrices;
(3) Inference engine, which employs the forward chaining method to trigger and match the dcsign rules;
(4) Man-machine interface, which facilitates the interaction between the user and the expert system;
(5) Symbolic-numerical interface, which coordinates the symbolic and numerical manipulations;
(6) Numerical processes, which are used to quantize the real values of the measurements, to expose the difference between the real values and the desired values, to evaluate the performance of the closed-loop system, to calculate the decision matrices, and to incorporate the designed fuzzy PID control algorithms into the simulation package.

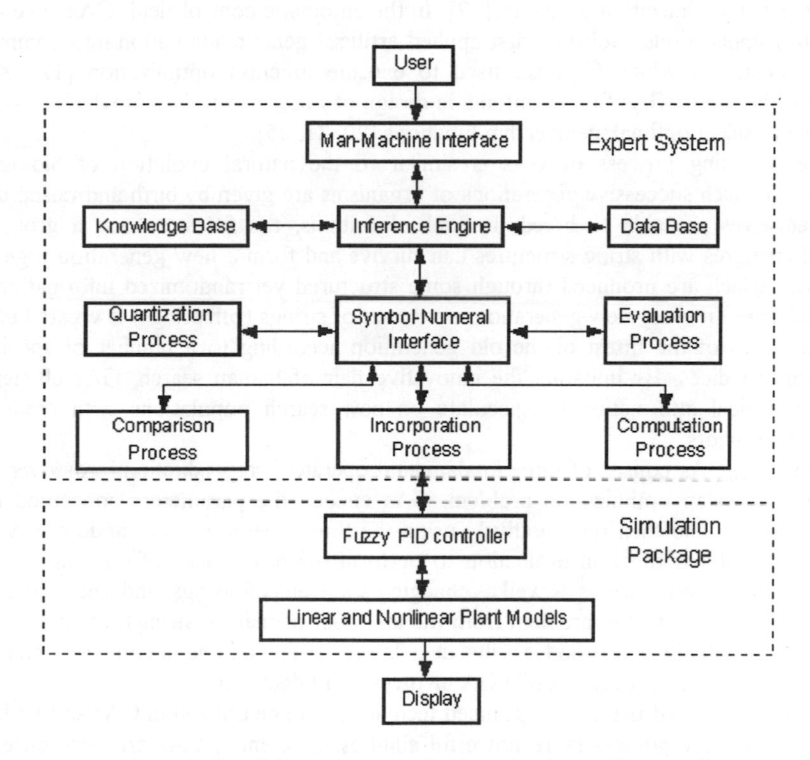

Figure 6.20. CAD environment for designing fuzzy PID controller.

The simulation package is used to emulate the closed-loop system which integrates linear and some nonlinear plants with a designed fuzzy PID controller. The whole dynamic system is simulated by using the fourth-order Runge-Kutta numerical method. The simulation results are displayed by using the HP Starbase graphic facility. The symbolic and numerical manipulations are programmed using the software languages LISP and C respectively.

The numerical processes are expressed in terms of C functions. The connection between LISP sentences and C language is realized by using the DOS command background in LISP language. The data communication is enhanced by using individual data files.

4. Genetic Algorithm-based optimization

Genetic Algorithms (GAs) are search algorithms which use operations similar to that found in natural genetics to guide their trek through a search space. Genetic Algorithms are becoming more and more popular as important mathematical means for nonlinear multi-objective optimization problems [12]. In the automatic control field, GAs have also found their applications. Hollstien first applied artificial genetic adaptation into computer control systems in which GAs are used to execute function optimization [12]. Karr suggested the use of GAs for the automatic design of fuzzy controllers in which GAs are employed to select optimal membership functions [20, 21, 25].

The searching process of GAs is similar to the natural evolution of biological creatures in which successive generations of organisms are given by birth and raised until they themselves are able to breed. In such algorithms, the fittest among a group of artificial creatures with string structures can survive and form a new generation together with those which are produced through some structured yet randomized information or gene exchange. In every new generation, a new set of strings (offsprings) is created using bits and pieces of the fittest of the old generation according to a number of specified performance indices. By imitating the innovative flair of human search, GAs efficiently exploit historical information to speculate on new search populations with gradually improved behaviors.

Generally, GAs consist of three fundamental operators: reproduction, crossover and mutation. Given an optimization problem, GAs encode the parameters concerned into finite bit strings, and then run iteratively using the three operators in a random way but based on the fitness function evaluation to perform the basic tasks of copying strings, exchanging portions of strings as well as changing some bits of strings, and finally find and decode the solutions to the problem from the last pool of mature strings. Even a simple GA can undoubtedly give good results in a large variety of engineering optimization problems. The overall processing of a GA involves more details [6].

It should be noted that more advanced techniques can be utilized in GAs so that such upgraded GAs may possess more powerful abilities. Inherently, GAs are very different from conventional optimization techniques. GAs search for a population of points, not a single point, so that they can arrive at the globally optimal point rapidly and meanwhile avoid locking at local optima. They work with a coding of the parameter sets, not the parameter themselves, so that they can get rid of the analytical limitation of search spaces. They only require objective information, not derivatives, so that they can utilize various kinds of objective functions even multiple, nonlinear or knowledge-based. They exploit probabilistic transition rules, not deterministic ones, so that they can efficiently walk to the neighborhood of the optimal solution. In one word, the robust feature and simple structure of GAs will make them very suitable for a lot of complicated control optimization problems.

6.4 Properties of Fuzzy Controllers

Fuzzy controller is called a linguistic controller whose core component is a set of fuzzy rules. Through a series of computation to fuzzy inference and synthesis of fuzzy relations, a determined control value is acquired finally. Because of the specificity of the controller structure, the fuzzy controller possesses some special properties that are different from the ones of a traditional controller and are difficult to analyze by the general analysis method.

In this section the static and dynamic properties of the fuzzy controller are introduced. These properties include two aspects. For static properties, the completeness, consistency and interaction of the control rules, and the robustness of the fuzzy controller are analyzed. For dynamic properties, the stability, controllability, sensitivity, convergency, reproducibility, accuracy and mapping properties of the fuzzy controller are investigated.

6.4.1 Static Properties of Fuzzy Controller

There are four important properties that are required to be considered in designing a fuzzy controller: (1) completeness of the control rules, (2) interaction of the control rules, consistency of the control rules, and (4) robustness of the fuzzy controller. They are analyzed as follows.

1. Completeness of control rules

From the input/output point of view, the completeness of the control rules means that the fuzzy controller can generate a corresponding control u for any input fuzzy state x, an empty fuzzy set of controls corresponding to a specified non-empty fuzzy set of input can not be accepted. A more formal definition for the completeness is given as follows.

Definition 6.11. A given set of control rules: if x_i, then $u_i (i = 1, 2, \cdots, n)$ is complete in this condition:

$$\underset{x \in X}{\forall} \ \underset{1 \le i \le n}{\exists} X_i(x) > \varepsilon \tag{6.32}$$

where u_i is fuzzy set of control, $X = E \times EC$, $x \in X$, $\varepsilon \in (0,1)$. Then we call this property completeness of fuzzy control rules.

All the fuzzy controllers satisfy the condition of completeness. In other words, Inequality(6.32) means that the union of fuzzy relations X_i of the control rules is greater than zero for all $x \in X$, i.e.

$$\underset{x \in X}{\forall} (\bigcup_{i=1}^{n} X_i(x)) > \varepsilon. \tag{6.33}$$

The requirement seems natural, since we determine fuzzy subsets that are usually overlapped, and every linguistic value of the linguistic variables for the fuzzy controller is in a fuzzy category of the linguistic variables. Violation of Inequality(6.32) occurs only if some categories are missed, it happens if one forgets to describe the relevant condition-action pair.

It is noted that if the fuzzy control is generated by on-line inference in the form of a fuzzy rule base, then the match might not be completed in the case of an incomplete fuzzy rule base.

2. Interaction of control rules

Definition 6.12. If interaction among fuzzy control rules happens, then the following inequality holds:

$$\underset{1\le i\le n}{\exists} X_i \circ R \neq U_i \tag{6.34}$$

i.e.

$$\underset{1\le i\le n}{\exists}\ \underset{u\in U}{\exists} (X_i \circ R)(u) \neq U_i(u) \tag{6.35}$$

and this implies that the control generated by synthetic computation of $X_i \circ R$ is no longer equal to U. We call this property interaction of fuzzy control rules.

Theorem 6.1. If the fuzzy relation R of the fuzzy controller is a Cartesian union of X_i and U_i, and all X_i in control rules are regular fuzzy sets, then the fuzzy control subsets U_i generated by the fuzzy controller satisfy the following relation:

$$\underset{1\le i\le n}{\forall} \cup_i \subseteq X_i \circ R \tag{6.36}$$

that implies:

$$\underset{1\le i\le n}{\forall} X_i,\ \exists x \in X,\ X_i(x) = 1\ . \tag{6.37}$$

Theorem 6.2. If the fuzzy relations $X_i(i = 1,2,\cdots,n)$ and $X_j(j = 1,2,\cdots,n)$ are not intersected each other, i.e.

$$X_i \cap X_j = \phi \quad (i \neq j; i,j = 1,2,\cdots,n) \tag{6.38}$$

and X_j are regular fuzzy sets, then the following relation is satisfied:

$$\underset{1\le i\le n}{\forall} X_i \circ R = U_i \tag{6.39}$$

that means no interaction will happen among the fuzzy control rules.

In determining membership function distribution of fuzzy linguistic values of input variables, the overlapping between two fuzzy subsets should not be too much in order to avoid too strong interaction that may lead to a poor control result.

3. Consistency of control rules

The consistency of fuzzy control rules is another fundamental problem. Implemental and contradictory information in the fuzzy controller may lead to unexpected and unsatisfactory results. The fuzzy control rule base is generated by means of a human operator's experience, however, its implementation may be subject to different performance criteria. This may lead to contradiction of control rules, and the consistency of the control rules must be evaluated prior to implementation.

For a given set of control rules, if the resulting fuzzy set of controls is multimodal, then inconsistency occurs when there are two or more control rules with almost the same state conditions but with different control actions. High accuracy and low consumption are often contradictory. We must study the control rules to get a deeper insight into their interrelationships, and then try to eliminate or replace the main contradictory control rules.

What is inconsistency for a set of control rules? If two input fuzzy sets X_k and X_i are slightly different, and the *k*th and *i*th fuzzy rules corresponding to the linguistic qualifiers S_k and S_i produce a slight deference between control actions U_k and U_i, the fuzzy rules *k* and *i* are consistent, otherwise, they are inconsistent.

Definition 6.13. Consider the following general criterion for inconsistency between *i*th and *k*th fuzzy rules, C_{ik} :

$$C_{ik} = \left| \prod (X_i | X_k) - \prod (U_i | U_k) \right|$$

$$= \left| \sup_{x \in X} [S_i(x) \wedge S_k(x)] - \sup_{u \in U} [U_i(u) \wedge U_k(u)] \right| \tag{6.40}$$

where *sup* represents superior limit. If $S_k = S_i$, and $[U_i(u) \wedge U_k(u)] = 0$, then C_{ik} achieves its maximal value at $C_{ik}=1$. If $S_k = S_i$, and $U_i(u) = U_k(u)$, $C_{ik}=0$, then fuzzy rules *i* and *k* is with its minimal inconsistency, i.e., the *i*th and *k*th fuzzy rules are consistent completely. If we sum up C_{ik} for the second subscript *k*, then we have:

$$C_i = \sum_{k=1}^{n} C_{ik} \tag{6.41}$$

that represents the inconsistency between the ith rule and other rules.

The study of consistency of fuzzy rules provides an alternative standard for establishing fuzzy rules, and can be used to eliminate mutual conflict among fuzzy control rules.

6.4.2 Dynamic Properties of Fuzzy Controller

In order to evaluate the operating characteristics and correctness of fuzzy control systems, the following properties are studied: stability, sensitivity, convergence dynamic, good mapping property, reproducibility and accuracy of the fuzzy controller [5, 16, 22, 35].

1. Stability

Theorem 6.3. Consider the following fuzzy sets:

$$\mathcal{X} = \{(x, \mu(x))\}; \quad \mu: x \to [0, 1]; \quad \text{card} \quad x=n \tag{6.42}$$

$$\mathcal{Y} = \{(y, \mu(y))\}; \quad \mu: y \to [0, 1]; \quad \text{card} \quad y=m \tag{6.43}$$

$$R: \quad \mathcal{X} \times \mathcal{Y} \to [0, 1] \tag{6.44}$$

where $\mathcal{X}$ is a family of all fuzzy sets defined on X, $\mathcal{Y}$ is a family of all fuzzy sets defined on Y, R is a fuzzy relation which can be interpreted as a mapping with its domain in $\mathcal{X}$ and a space of values contained in $\mathcal{Y}$. This mapping can be denoted in terms of the compositional rule of inference given by

$$V = U \circ R \tag{6.45}$$

where $U \in \mathcal{X}$, $V \in \mathcal{Y}$ are fuzzy sets defined on X and Y respectively. The fuzzy sets (6.44) has the α-stability property with respect to some family $P \subset \mathcal{X}$ and some feasible subset $F(Y) \subset Y$ if for every fuzzy set $U \in P$, $U \circ R = V$ is such that $\mu_V(y) \le \alpha$ for every $y \in Y - F(Y)$.

Let the system $R = A \to B = A \times B$ be given, then

$$\bigvee_{i=1}^{n} \mu_A(x_i) \le \alpha, \quad or \quad \mu_B(y_j) \le \alpha$$

for all $y_j \in Y - F(Y)$ if system (6.45) has the α-stability property with respect to a family $\mathcal{X}$ of all fuzzy sets defined on X. Let the fuzzy relation R be given as: $R = A_1 \to B_1$, or ..., or $A_s \to B_s$, where $A_1, \ldots, A_s$ and $B_1, \ldots, B_s$ are fuzzy sets defined on X and Y respectively; $A_1, \ldots, A_s \in \mathcal{X}$; $B_1, \ldots, B_s \in \mathcal{Y}$; $R = (A_1 \times B_1) \cup \ldots \cup (A_s \times B_s)$.

Owing to the specificity of the structure of fuzzy controller, up to now, a satisfactory analysis approach for dynamic stability of fuzzy control system has not been proposed. The following approaches are used to analyze the dynamic stability of fuzzy controller [1, 23, 31, 54].

- describing function analysis;
- phase plane analysis;
- linearlity approximation analysis;
- analysis based on Lyapunov's Direct Method.

All of these analysis methods have their limitations, and better methods are expected to come out.

There are some useful methods for improving dynamic properties of fuzzy controllers. They are regulation of fuzzy rules, regulation of scale factors, introduction of integration function, and self-tuning of fuzzy rules. There exist three essential methods to deal with fuzzy logic PID controller, nested or windowing FLC, PID and performance index based FLC [16, 54].

2. Sensitivity

Theorem 6.4. Let $C \in \mathcal{X}$, $B \in \mathcal{Y}$, and R be a fuzzy relation defined on $X \times Y$, and $B =$

$R \circ C$. If a family of fuzzy sets

$$\mathcal{A}_0 = \{A \in \mathcal{X}: A \neq C, B = R \circ A\}$$

is non-empty, then the fuzzy system is said to be non-sensitive with regard to the family $\mathcal{A}_0$ for output fuzzy set B.

Let a, $b \in [0,1]$, then

$$a \propto b = \max\{x \in [0,1]: a \wedge b \leq b\}.$$

Let us consider $A \in \mathcal{X}$ and R to be a fuzzy relation defined on X and Y, then $A \propto R = B$, $B \in \mathcal{Y}$, and

$$\mu_B(y) = \underset{x \in X}{\wedge} [\mu_A(x) \propto \mu_B(x,y)]. \tag{6.46}$$

Let R be a fuzzy relation on X and Y, then a fuzzy relation R^{-1}, the inverse of R, is defined on $Y \times X$, and

$$\mu_{R(-1)}(y,x) = \mu_R(x,y)$$

for all $(y, x) \in Y \times X$, where $R(-1)$ represents R^{-1}. In the case of a fuzzy set $B \in \mathcal{Y}$, $B^{-1} = B$, and

$$\mu_{B(-1)}(y) = \mu_B(y).$$

Let R be a fuzzy relation on $X \times Y$, $B \in \mathcal{Y}$, and $\mathcal{A} = \{A: R \circ A = B\}$, then $\mathcal{A} \neq \varnothing$ if $(R \propto B^{-1})^{-1} \in \mathcal{A}$ and if $\mathcal{A} \neq \varnothing$, $(R \propto B^{-1})^{-1}$ is the greatest element. If $R \circ A = B$, then for every $D \in \mathcal{X}$, we have $R \circ D = B$ provided that $A \subseteq D \subseteq (R \propto B^{-1})^{-1}$.

If $R \circ A = B$ and $(R \propto B^{-1})^{-1} = A$, then each increase of any component of $[\mu_A(x_1), \ldots, \mu_A(x_n)]$ implies $R \circ A \neq B$.

If $(R \propto B^{-1})^{-1} \supset A$, then $\mu_A(x)$ may be increased up to $\mu_{(R \propto B^{-1})^{-1}}(x)$ without any change in the output fuzzy set B.

In this way, a sufficient condition for the upper evaluation on the input fuzzy sets can be obtained, under which the fuzzy system is non-sensitive for output fuzzy set B.

Let a, $b \in [0,1]$, then

$$a \delta b = \begin{cases} 1, a = b \\ 0, a \neq b \end{cases} \tag{6.47}$$

Let $A \in \mathcal{X}$, R be a fuzzy relation defined on $X \times Y$, $B \in \mathcal{Y}$ and $R \circ A = B$. Then let A^* be following fuzzy set

$$A^* \in \mathcal{X}$$

$$\mu_{A^*}(x) = \underset{y \in Y}{\vee} \{[(\mu_A(x) \wedge \mu_R(x,y))\delta\mu_B(y)] \wedge [\mu_A(x) \wedge \mu_R(x,y)]\} \quad (6.48)$$

for all $x \in X$.

Let C, A, $A^* \in \mathcal{X}$, $B \in \mathcal{Y}$, R be a fuzzy relation on $X \times Y$ and $R \circ A = B$. If $C \subset A^*$, $R \circ C \neq B$, and let A, $A^* \in \mathcal{X}$, $B \in \mathcal{Y}$, R be a fuzzy relation on $X \times Y$ and $R \circ A = B$, then any fuzzy set $D \in \mathcal{X} : A \subseteq D \subseteq (R \propto B^{-1})^{-1}$ satisfies the equality $R \circ D = B$.

In this way, the sufficient conditions for the lower and upper evaluation of the input fuzzy sets family can be obtained, under which the fuzzy system in non-sensitive for the output fuzzy set B.

3. Good mapping property

Theorem 6.5. Let $R = (A_1 \times B_1) \cup \ldots \cup (A_s \times B_s)$ be given, $A_i \in \mathcal{X}, B_i \in \mathcal{Y}$, i=1, 2,..., s. We say that R has the good mapping property if for every i=1, 2,...,s, $R \circ A_i = B_i$.

Let $A_1, \ldots, A_s$ be fuzzy sets defined on X. $\sum_{i=1}^{s} A_i(x) = 1$ for any $x \in X$, and let $B_1, \ldots, B_s$ be fuzzy sets defined on Y. Let $R = (A_1 \times B_1) \cup \ldots \cup (A_s \times B_s)$. Then $\mu_{R \circ A}(y_j) = \mu_{B(k)}(y_j)$, if for some $k \in \{1, 2, \ldots, s\}$ and some $j \in \{1, 2, \ldots, m\}$, we have

$$\mu_{B(k)}(y_j) \leq \max \mu_{A(k)}(x) \text{ and } \mu_{B(k)}(y_j) \geq \mu_{B(l)}(y_j), l \neq k \quad (6.49)$$

For a given fuzzy set A defined on elements of some space X, $0 \leq \xi(A) \leq 0.5$ is the grade of fuzziness of A if $\xi(A) = \max_{x \in X}\{\mu_A(x) \wedge [1 - \mu_A(x)]\}$. Let $A_1, \ldots, A_s$ be fuzzy sets defined on X, $\sum_{i=1}^{s} A_i(x) = 1$ for any $x \in X$, and let $B_1, \ldots, B_s$ be fuzzy sets defined on Y. Let $R = (A_1 \times B_1) \cup \ldots \cup (A_s \times B_s)$, then $\mu_{R \circ A(i)} = \mu_{B(i)}$ if:

(1) $\mu_{B(i)}(y) \leq \max_{x \in X} \mu_{A(i)}(x)$ for i=1, 2,...,s and $y \in Y$

(2) $\mu_{B(i)}(y) \leq \mu_{B(j)}(y)$ for some $y \in Y$ and some $j \neq i$ implies

$$\xi(A_i) \leq \xi(B_i).$$

A fuzzy relation matrix R has the good mapping property if the grade of fuzziness B_i is not less than that of A_i except for the case $\mu_{B(i)} \geq \mu_{B(j)}$ for $j \neq i$. In this case, no restriction on B_i and A_i is necessary in order to maintain the good mapping property of R.

4. Convergence property

Theorem 6.6. Let $X = \{x_i\}$, i=1, 2,...,m be a family of fuzzy sets defined on a multidimensional finite discrete space $\mathcal{X} = \{x_j\}$, j=1, 2,...,n. A fuzzy metric space is the pair of elements (x, ρ) such that $\rho: X \times X \to R^*$ is a mapping which satisfies the following conditions:

$$\left.\begin{aligned} &\rho(x_1,x_2)=0 \leftrightarrow x_1=x_2 \text{(in a fuzzy sense)} \\ &\rho(x_1,x_2)=\rho(x_2,x_1); x_1,x_2 \in X \\ &\rho(x_1,x_2) \le \rho(x_1,x_3)+\rho(x_3,x_2); x_1,x_2,x_3 \in X \end{aligned}\right\} \tag{6.50}$$

Let z be a fuzzy neighborhood of fuzzy set $x_i \in X$ if $x_i \subset z!(\mu_{x(i)} \le \mu_z)$. The sequence of fuzzy sets $\{x_n\}$ converges to the fuzzy set (point) x_o if for each neighborhood z almost all terms of $\{x_n\}$ satisfy $x_n \subset z$.

Fuzzy sequence $\{x_n\}$ of terms of fuzzy metric space X is said to be a Cauchy fuzzy sequence iff $\forall\varepsilon > 0$, $\exists k \in N$, $\forall n > k$, $\forall m > k$, $\rho(x_n,x_m) < \varepsilon$ or $\lim_{n,m\to\infty} \rho(x_n,x_m)=0$. Every fuzzy convergent sequence is a Cauchy sequence. If in the fuzzy metric space (x, ρ) every Cauchy fuzzy sequence converges to an element of this space, then it is called a fuzzy complete metric space.

Let (x, ρ) be a fuzzy complete space, and R: $X \times U \times X \to [0,1]$ be a fuzzy relation such that a fuzzy dynamic system is given by

$$x_n = x_{n-1}U_{n-1} \circ R, \quad n=1,2,3,\ldots \tag{6.51}$$

where U_{n-1} = constant for all n=1, 2, 3,..., and

$$\begin{aligned} \rho(x_1U_{n-1} \circ R,\, x_2U_{n-1} \circ R) \le \lambda_\rho(x_1,x_2) \\ \forall x_1,x_2 \in X,\ \forall\lambda \in [0,1]. \end{aligned} \tag{6.52}$$

Then there exists one and only one fuzzy point (set) $X^* \in X$ such that

$$X^* = X^*U_{n-1} \circ R \tag{6.53}$$

Hence the fuzzy dynamic system has the convergence property.

5. Reproducibility property

Definition 6.13. Informally speaking, reproducibility property occurs whenever fuzzy sets representing output of a system share properties of fuzzy sets representing input. $\mathfrak{I}_X$ will denote the family of all fuzzy sets defined on a set X. All fuzzy sets which we consider later will be defined on a finite set. A fuzzy set X is called normal if $\mu_X(x)=1$ for some $x \in X$. The family of all normal fuzzy sets on the fuzzy set $\mathcal{X}$ (Cartesian product) will be denoted $\mathfrak{I}^N_{\mathcal{X}}$.

Theorem 6.7. Let R be a maxmin fuzzy relation represented by a matrix $\{\gamma_{xy}\}$,

$$\left.\begin{array}{l} R\colon \Im_{\mathcal{X}} \to \Im_{\mathcal{Y}}, X \to Y = X \circ R \\ \mu_Y(y) = \max\limits_{X \in \mathcal{X}} \min\{\mu_X(x), \gamma_{xy}\} \end{array}\right\} \tag{6.54}$$

Let $\Im \subset \Im_{\mathcal{X}}$ and $\vartheta \subset \Im_{\mathcal{Y}}$ be some families of fuzzy sets. A fuzzy relation R has the reproducibility property with respect to $(\Im, \vartheta)$, abbreviated as $(\Im, \vartheta_N)$- RP, if $X \circ R \subset \vartheta$, for all $X \in \Im$. A fuzzy relation R has $(\Im_{\mathcal{X}}^N, \Im_{\mathcal{Y}}^N)$ - RP if

$$\max_{y \in Y} \gamma_{xy} = 1 \quad \text{for all } X \in \mathcal{X}. \tag{6.55}$$

A fuzzy set $X \in \Im_{\mathcal{X}}$ is called a fuzzy interval if $\mu_X(x)$ is a non-decreasing function of x for $x \le x_o$ and a non-increasing function of x for $x_o \le x$, where x_o is some point in $\mathcal{X}$.

A fuzzy set $X \in \Im_{\mathcal{X}}$ is called a fuzzy n-dimensional interval (or simply a fuzzy interval) if for all $x_0, x_1 \in \mathcal{X}$ and all $x \in \overline{x_0, x_1}$ we have $\mu_X(x) \ge \min(\mu_X(x_0), \mu_X(x_1))$. The families of all fuzzy n-dimensional intervals and all n-dimensional intervals will be denoted by $\Im_{\mathcal{X}}^I$ and $\Im_{\mathcal{X}}^i$ respectively. A fuzzy relation R has the $(\Im_{\mathcal{X}}, \Im_{\mathcal{Y}}^I)$ - RP iff it has $(\Im_{\mathcal{X}}^i, \Im_{\mathcal{Y}})$ - RP.

Let $\mathcal{X} = \mathcal{X}^1 \times \mathcal{X}^2 \times \cdots \times \mathcal{X}^n$, and $\mathcal{X}_0 \subset \Im$. Points $x_0, x_1 \in \mathcal{X}_0$ are called $\mathcal{X}_0$-close (which will be denoted as $x_0 \sim x_1 (\text{mod } \mathcal{X}_0)$ if $x_0, x_1 \cap \mathcal{X}_0 = \{x_0, x_1\}$). A fuzzy relation R has the $(\Im_{\mathcal{X}}^I, \Im_{\mathcal{Y}}^I)$ - RP iff

(1) for all $x \in \mathcal{X}$ the fuzzy set $Y \in \Im_{\mathcal{Y}}$ defined by $\mu_Y(y) = \gamma_{xy}$ is a fuzzy interval, and

(2) for all s=1, 2,...,k and all $x_0, x_1 \in \mathcal{X}(s)$ which are $\mathcal{X}(s)$-close the fuzzy set $Y \in \Im_{\mathcal{Y}}$ defined by $\mu_Y(y) = \max(\gamma_{x(0)}, \gamma_{x(1)})$ is a fuzzy interval.

If a fuzzy relation R has the $(\Im_{\mathcal{X}}^N, \Im_{\mathcal{X}}^N)$ -RP then it has $(\Im_{\mathcal{X}}^I, \Im_{\mathcal{Y}}^I)$ -RP iff

(1) for all $x \in \mathcal{X}$ the fuzzy set $Y \in \Im_{\mathcal{Y}}$ defined by $\mu_Y(y) = \gamma_{xy}$ is a fuzzy interval, and

(2) the fuzzy set Y defined by $\mu_Y(y) = \max(\gamma_{x(0)y}, \gamma_{x(1)y})$ is a fuzzy interval for all pairs x_0, x_1 of adjacent points.

6. Accuracy of fuzzy model

The following factors have a significant influence on the accuracy of the fuzzy model:

(1) The form of the mathematical definition of fuzzy implication

IF ... AND ... THEN ...;

(2) The form of the mathematical definition of the sentence connective **AND** ;

(3) The form of the mathematical definition of the sentence connective **ALSO** ;

(4) The form of the mathematical definition of relation composition.

Definition 6.14. The root-mean-square error is assumed to be the criterion to the estimation of the accuracy of a fuzzy model

$$\Delta^2 e = \sum_{i=1}^{v} (y_r - y_m) / \sum_{i=1}^{v} y_r \tag{6.56}$$

where y_r is the output of the real system, and y_m is the output of the fuzzy model.

In order to ensure the highest accuracy of the model with the least computational operations, the problem of defining fuzzy implications was being investigated. The following fuzzy implications fulfill the above mentioned conditions:

$$\mu_{R2*}(i_t, n_j) = \overset{6}{\underset{s=1}{\Lambda}} \begin{Bmatrix} 1, \mu_t(i_t) \le \mu_N(n_j) \\ 1, \text{ otherwise} \end{Bmatrix} \tag{6.57}$$

$$\mu_{R3*}(i_t, n_j) = \overset{6}{\underset{s=1}{\Lambda}} \begin{Bmatrix} 1, \mu_I(i_t) \le \mu_N(n_j) \\ \mu_N(n_j), \text{ otherwise} \end{Bmatrix} \tag{6.58}$$

$$\mu_{R4*}(i_t, n_j) = \overset{6}{\underset{s=1}{\Lambda}} \left\{ 1 \, \Lambda \, \frac{\mu_N(n_j)}{\mu_I(i_t)} \right\} \tag{6.59}$$

$$\mu_{R5*}(i_t, n_j) = \overset{6}{\underset{s=1}{\Lambda}} \left\{ 1 \, \Lambda \, \{ 1 - \mu_I(i_t) + \mu_N(n_j) \} \right\} \tag{6.60}$$

$$\mu_{R6*}(i_t, n_j) = \overset{6}{\underset{s=1}{\Lambda}} \left\{ 1 \, \Lambda \, \frac{\mu_N(n_j) * (1 - \mu_I(i_t)}{\mu_I(i_t) * (1 - \mu_N(n_j))} \right\} \tag{6.61}$$

It is suggested that in constructing the fuzzy models of real systems, fuzzy implications *R2** and *R3** should be used [23].

6.4.3 Controllability of Fuzzy Control System

Controllability or reachability is a significant concept in state space theory of linear time invariant control systems, this concept implies whether a control can be synthesized to make a process from one state to another in finite time.

Definition 6.15. In a fuzzy control system the process is defined by the first order difference equation

$$V_{k+1} = U_k \circ V_k \circ R \tag{6.62}$$

and the specified goal set $V \equiv G$, where k denotes sample period, if a finite control sequence$\{U_0, U_1, \cdots, U_n, n<\infty\}$ exists such that $V_0 \to G = V_n$, then the system is reachable.

In practice, if the model R is imperfect, $V_n \neq G$, then some measure $\|V_n - G\| \leq \beta$ of how well the goal is achieved under closed loop is to be ascertained for a given(β, G). A family of fuzzy sets ($G \in (G_{\min}, G_{\max})$) are generally defined to ensure that a non-empty set of solutions to (6.62) exist to satisfy $G = V_n$ for some n.

For optimal control, some dynamic conflicting constraints such as minimum peak overshoot, fast rise time and minimum energy usage are needed to incorporate in a practical fuzzy logic control system. The fuzzy constraint conditions can be represented as $C{:}U \to [0, 1]$ and incorporated in goal condition through

$$\max_{U \to [0,1]} \{\|U - C\| \text{ and } \|V_n - G\|\} \tag{6.63}$$

This dual control condition can be satisfied by defining a fuzzy relation Z as the Cartesian product of the goal G_i and a constant constraint condition C, such that $Z = G \times C$ with membership function $\mu_z(u, v) = \mu_c(v) \wedge \mu_c(u)$. Therefore, from Equation (6.63) for some k in which $V_k \to G$,

$$Z = (U_k \circ V_k \circ R) \times C \tag{6.64}$$

or

$$\begin{aligned} \mu_z(u, v) &= \max_u [\mu_{v_k}(u) \wedge \mu_{v_k}(v) \wedge \mu_R(u, v)] \wedge \mu_c(u) \\ &= \max_u [\mu_{v_k}(u) \wedge \mu_{v_k}(v) \wedge \mu_R(u, v) \wedge \mu_c(u)] \\ & u \in U, v \in V. \end{aligned} \tag{6.65}$$

From the above equation and related property of the composition operator $\varnothing$, the greatest fuzzy set of control U_k is given as:

$$\mu_{U_k}(u) = \min_v \{[\mu_c(u) \wedge \mu_{V_k}(v) \wedge \mu_R(u,v)] \varnothing \mu_z(u,v)\}. \tag{6.66}$$

An alternative technique for goal and constraint satisfaction prescribes a level $Y_{ij} \in [0,1]$ to each linguistic pair of controls $U(i)$ and states $V(j)$ that in some manner incorporates both the goal and the constraint conditions. An unsymmetrical matrix of $\{Y_{ij}\}$ can be assigned for the complete set of linguistic pairs $\{U^{(i)}, V^{(j)}\}$, i, j=1, 2,$\cdots$, n. For given $V_k^{(i)}$, consider the finite discrete one order fuzzy model

$$V_{k+1}^{(ij)} = U_k^{(j)} \circ V_k^{(i)} \circ R. \tag{6.67}$$

The resultant next state $V_{k+1}^{(ij)}$ is matched against the desired value $D_{k+1}^{(i)}$, with consequent distance measure $\lambda_k^{(i)}$ that is a normalized finite convex fuzzy set

$\lambda = \mu_{\lambda} \in [0,1]$ defined on the difference $(V_{k+1}^{(ij)} - D_{k+1}^{(i)})$. The distance matching is carried out for all states (i=1, 2,$\cdots$,n) to produce the distance set $\{\lambda_k^{(i)}\}$. The overall suitability of the control $U_k^{(i)}$ is determined by the weighted sum:

$$Y_j = \frac{\sum_{i=1}^{n} \lambda_k^{(i)} Y_{ji}}{\sum_{i=1}^{n} \lambda_k^{(i)}}. \tag{6.68}$$

The fuzzy control rule "IF $V_k^{(i)}$ THEN $U_k^{(i)}$ with max Y_j" gives the optimal control for Equation(6.67). This approach considers both the reachability and suitability of the controller and its construction through a prescribed state-control table indicating the quality of control achieved. The distance measures λ_k between actual system response and the fuzzy model prediction by Equation(6.67) can also be used to ascertain model quality or fitness during system identification [46].

6.4.4 Robustness of Fuzzy Control Systems

We are interested in the reaction of the controller to input disturbances, system parameters and structural change. If the probabilistic characteristics of the inputs, and the operator involved in the determination of the fuzzy control are known, one can obtain the relevant probabilistic characteristics of the control [35]. Assume that the input variable X is random within a given distribution function F_x. This input implies that the level of activation of the ith control rule is also random. It is denoted by Λ_i and its distribution function F_{Λ_i} is derived from a basic formula of probability calculus:

$$F_{\Lambda_i}(w) = P\{w | \Lambda_i(w) < w\} = \int_{\{x = x_i(x) < w\}} dF_x(x)$$
$$w \in [0,1]. \tag{6.69}$$

Considering Λ_i is random in its character, the control U is also a random variable, so we have:

$$U(u,w) = \bigvee_{i=1}^{n} [\Lambda_i(w) \wedge U_i(u)], \quad u \in U. \tag{6.70}$$

For calculating the distribution function of U, we assume that $u \in U$ is fixed. According to the computation formula of random variable distribution, the minimum of two random variables, say A and B, has a distribution function:

$$F_{\min(A,B)}(w) = F_A(w) + F_B(w) - F_{AB}(w,w). \tag{6.71}$$

If A and B are independent, then it holds:

$$F_{\min(A,B)}(w) = F_A(w) + F_B(w) - F_A(w)F_B(w). \tag{6.72}$$

Applying Formula (6.72) directly to the problem above, since U_i is not a random variable (only the activation of the ith control rule is random), it is:

$$F_{U_i}(u) = \begin{cases} 0, & \text{if } w < U_i(u) \\ 1, & \text{otherwise} \end{cases}. \tag{6.73}$$

According to Formula (6.72), we have:

$$F_{\min(\Lambda_i,U_i)}(w) = F_{\Lambda_i}(w) + F_{U_i}(u) - F_{\Lambda_i}(w)F_{U_i}(u). \tag{6.74}$$

If $w < U_i(u)$, then

$$F_{\min(\Lambda_i,U_i)}(w) = F_{\Lambda_i}(w)$$

otherwise,

$$F_{\min(\Lambda_i,U_i)}(w) = 1.$$

Next, we consider the distribution function of the union of $\Lambda_i \cap U_i$, and assume that the random variables are independent, then it is:

$$F_U(w) = \prod_{i=1}^{n} F_{\min(\Lambda_i,U_i)}(w) \tag{6.75}$$

by denoting $w_{\min} = \min\ U_i(u)$, then it is satisfied:

$$F_U(w) = \begin{cases} \prod_{i=1}^{n} F_{\Lambda_i}(w), & \text{if } w < w_{\min} \\ 1, & \text{otherwise} \end{cases}. \tag{6.76}$$

Executing this procedure for every element of U_i, then the required distribution function of the control U can be derived.

The obtained formula enables us to study the robustness of the fuzzy controller, and find how the disturbances influence the fuzzy controller. The ratio of the mean value and the standard deviation can be taken as a measure of the randomness of the variable. ξ_u and m_u denote the standard deviation and the mean value of u respectively, i.e.,

$$\begin{aligned} m_u &= \int_0^1 w\,dF_U(w) \\ \xi_u &= \int_0^1 (w - m_u)^2 dF_U(w) \end{aligned} \tag{6.77}$$

then a plot of this ratio versus the same ratio of the input variable

$$\frac{\xi_u}{m_u} = g(\frac{\xi_x}{m_x}) \tag{6.78}$$

may serve as an indicator of the robustness of the fuzzy controller, where $g(\_)$ is a corresponding relation and determines the robustness of the fuzzy controller. If m_u / ξ_u is fairly constant for increasing m_x / ξ_x, the fuzzy controller is robust for the specified range of variance of input. The robustness of the fuzzy controller can be analyzed by changing the shape of the membership function of input variable x_i that influences the distribution function F_u.

6.4.5 Controllability and Robustness of A Fuzzy Control System under Directional Disturbance [42]

In this subsection, first a T-fuzzy system based on related fuzzy control theories is proposed, then an L-controllable condition of the T-System is proven, in the end, an L-controllable condition of T-System under the L-condition and disturbance of $\xi \neq 0, |\xi| < d$ is derived. During the discussion, the capital letters *X*, *Y*, *U* are used to represent spaces, the small letters *a*, *b*, *c* to represent elements of the spaces, *x*, *y*, *z* — fuzzy sets, *t*, *o*, δ, ε — real, *I*, *k*, *n*, *m* — integers.

1. Related theories of fuzzy systems and T-System

Assume that *X* and *U* are limited subspaces of the Euclidean spaces R^n and R^m; x_k and u_k are fuzzy sets in *X*, *U*; x_k— system state; u_k— control variable. Then a fuzzy control system can have the form:

$$\begin{aligned} &x_{k+1} = x_k \circ u_k \circ R \\ &R\text{: } X \times U \times X \to [0,1]. \end{aligned} \tag{6.79}$$

Gupta proposed a method for describing system controllability of fuzzy relation in the energetic controllable point of view [13].

Definition 6.16. If x_0 and x_e are normal fuzzy sets, and minimum fuzzy number $c(x_0, x_e) > x_0 \cup x_e$, then the Liapunov energy is:

$$L(x_0, x_e) = \int_X \mu_c(x_0, x_e)(a)\, da\,. \tag{6.80}$$

Definition 6.17. For a given positive number ε, if $x_1 \in F(x)$ is normal, and there exist normal fuzzy sets

$$\begin{aligned} &u_1 \cdots u_{n-1} \in F(U) \\ &x_{i+1} = x_i \circ u_i \circ R \end{aligned}$$

then

$$L(x_n, x_e) < L(x_e, x_e) + \varepsilon$$

as a result x_1 is L-controllable.

Definition 6.18. For all normal fuzzy numbers x_k, u_k, if fuzzy sets($x_k \circ u_k \circ R$) are normal fuzzy numbers, then R is called (*N*, *I*)-RP fuzzy relation [5].

Definition 6.19. A fuzzy system is called T-System if it satisfies the following conditions:

(i) X and U are limited subspaces of R^n and R^m.

(ii) R is fuzzy relation of (*N*, *I*)-RP;

(iii) V_1 and V_2 are sets consisting of some normal fuzzy numbers in fuzzy sets *F*(*X*) and *F*(*U*), and

(1) $\forall x_1, x_2 \in V_1, \quad x_1 \cap x_2$ is not – normal;

(2) $\bigcup_{x \in V_1} \{a \in X | \mu_x(a) = 1\} = X$

$\bigcup_{u \in V_2} \{b \in U | \mu_u(b) = 1\} = U$

(3) $\forall x \in V_1, \{a | \mu_x(a) > 0\} \subseteq S_r(c(x)), r > 0$

where $c(x)$ represents the center of X, $S_r(d)$— a neighborhood with its center in d and radius of r in X.

(4) $Y_k = x_k \circ u_k \circ R, \quad x_{k+1} = Y_k \circ V_1$

where $Y_k \circ V_1 = \bigcup \{x \in V_1 | x \cap Y_k \text{ normal}\}$, and $Y_k \circ V_1 \in V_1$.

Definition 6.20. If x is a normal fuzzy number, and $c(x)$ is the center of x, $r(x)$=Max{r}, r>0, such that $\lambda S_r(c(x)) \subset x \circ \lambda_U \circ R$, where λ represents the characteristic function.

Definition 6.21. If $\forall x \in V_1$, then Energy is defined as:

$$D(x) = \|c(x) - c(x_e)\|$$

where x_e is fixed point.

Property 6.1. For T-System with smaller δ_1, if $x \in V_1$ is controllable for energy D, i.e., $u_1, \cdots u_n \in V_2$ and $D(x_{n+1}) < \varepsilon$ hold, then x is also controllable for energy L.

2. Analysis for L-controllability of T-System

For T-System the following conclusion can be drawn.

Lemma 6.1. Assume that $u = \lambda_U, x \in V_1$, and $x \circ u \circ R \supset \lambda_{Sr}(c(x))$, there exist $y \in V_1, u_2 \in V_2$ for $\forall a \in Sr(c(x))$, then

$$\|c(y) - a\| < \delta_1$$

$$y = (x \circ u_2 \circ R) \circ V_1$$

where U is a point in the fuzzy system, and λ_U is the characteristic function of U.

[Proof] $\because \mu_{x \circ u \circ R}(a) = 1$

$\therefore$ exist $b \in U, c \in x$, such that

$R(c,b,a)=1$, and $\mu_x(c)=1$

According to Definition 6.19(iii)(2),

$\Rightarrow u_2 \in V_2, u_2 \supset \lambda_b$

$\Rightarrow (x \circ u_2 \circ R) \supset x \circ \lambda_b \circ R$

$\Rightarrow (x \circ u_2 \circ R) \circ V_1 \supset (x \circ \lambda_b \circ R) \circ V_1$

$\because (x \circ u_2 \circ R) \circ V_1 \in V_1$

$\Rightarrow (x \circ u_2 \circ R) \circ V_1 = (x \circ \lambda_b \circ R) \circ V_1$

$\therefore$ the intersection is normal. Since $(x \circ u_2 \circ R) \circ V_1$ is a fuzzy number with radius less than δ_1 .

$\Rightarrow \|c((x \circ u_2 \circ R) \circ V_1) - a\| < \delta_1$

Let $y = (x \circ u_2 \circ R) \circ V_1$, then $\|c(y) - a\| < \delta_1$.

Theorem 6.8. In the T-System, if $\forall x \in V_1, r(x) > 2 \cdot \delta_1$, then $x \in V_1$ is L- controllable.

[Proof] If $\forall x \in V_1$ is not L-controllable, then there exists $y \in V_1$,

$\forall u \in V_2, \quad D((y \circ u \circ R) \circ V_1) \geq D(y)$

$\because \quad r(y) > 2 \cdot \delta_1$, let

$$f' = c(y) - 2 \cdot \delta_1 \frac{1}{\|c(y)c(x_e)\|} \overrightarrow{c(y)c(x_e)}$$

From Lemma 6.1, we can get $u_2 \in V_2$, so

$$\|c((y \circ u \circ R) \circ V_1) - f\| < \delta_1$$

$$\Rightarrow \|c((y \circ u \circ R) \circ V_1) - c(x_e)\|$$

$$\leq \|f - c(x_e)\| + \|c((y \circ u \circ R) \circ V_1) - f\|$$

$$\leq \|c(y) - c(x_e)\| - 2 \cdot \delta_1 + \delta_1$$

$$\leq D(y) - \delta_1$$

this results in a contradiction. Therefore, this theorem holds.

3. Analysis for L-controllability of T-System under directional disturbance ξ

Let us consider a disturbance ξ: if ξ=0, T-System is represented by R, and if $\xi \neq 0$, T-System is represented by R_ξ, in which the variation of ξ is $\|\xi_n - \xi_{n-1}\| < \delta_2$, where δ_2 is a small enough positive number.

Definition 6.22. Let $\vec{g}$ is a unit vector in x, and

$$R_\xi(x,u,y) = R(x, u, (y - |\xi| \cdot \vec{g}))$$

since $(x \circ u \circ R)$ is simply a displacement of $(x \circ u \circ R)$. Many properties of T-System are invariant under disturbance ξ, $|\xi| < d$.

Lemma 6.2. If $x \in V_1$, and $r(x) > 2\cdot(\delta_1+\delta_2)+d$, $|\xi|<d$, then

$$\forall a \in S_{2(\delta_1+\delta_2)}(c(x))$$

exists $u \in V_2$, such that $\left\| c((x \circ u \circ R_\xi)\circ V_1) - a \right\| < \delta_1 + \delta_2$.

[Prove] $\because \left\| a - c(x) \right\| < 2\cdot(\delta_1+\delta_2)$

$$\Rightarrow \left\| a - |\xi| \vec{g} - c(x) \right\| < 2\cdot(\delta_1+\delta_2)+d$$

From Lemma 6.1, there exists $u \in V_2$, so

$$u_{x\circ u\circ R}(a - |\xi| \cdot \vec{g}) = 1$$

From the definition of R_ξ, $u_{x\circ u\circ R_\xi}(a) = 1$

$$\Rightarrow \left\| c((x \circ u \circ R_\xi)\circ V_1) - a \right\| < \delta_1 + \delta_2$$

During the proving process, ξ is viewed as invariant in a period of time.

Inference 6.1. If $\forall x \in V_1$, and $r(x) > 2(\delta_1+\delta_2)+d$, then *L*-controllable of the T-System holds when $\left\| \xi_n - \xi_{n-1} \right\| < \xi_2$.

For more general case encountered, $\forall x \in V_1$,

$$r(x) > 2(\delta_1 + \delta_2) + d$$

is not always held. We want to realize that the T-System is controllable under disturbance $\xi(|\xi| < d)$ and for weaker conditions $r(x) > 2(\xi_1 + \xi_2)$ what are the fuzzy numbers of V_1. We would like to analyze the fuzzy number in V_1 geometrically as follows.

Definition 6.23. Assume *h*(*t*), *t*∈[0, 1], is a continuous curve in *X*, and satisfies the following two conditions: (1) if $t_1 = t_2$, then $h(t_1) \neq h(t_2)$, (2) $\forall x \in V_1$, there exist *t*∈[0, 1], then $\delta_1+\delta_2 > \left\| c(x) - h(t) \right\|$, so we get

$$r(x) > 2\cdot(\delta_1+\delta_2)+d.$$

Definition 6.24. If *H*={*a*∈*X*} there exists cure *h* satisfied condition of Definition 6.23 and *t*∈[0, 1], so that *a*=*h*(*t*), $c(x_e)$=*h*(0), $L(h, t_1, t_2)$ expresses path length of curve *h* from h(t_1) to h(t_2), $x_1, y \in V_1$, and}

(i) if there exists $u_1, \ldots, u_n \in V_2$, then

$$x_{i+1} = (x_i \circ v_i \circ R_{\xi i}) \circ V_1, \quad \left\| c(x_{n+1} - c(y) \right\| < \delta_1 + \delta_2$$

noted as $x_1 \sim y$.

(ii) if *h* and t_1, t_2 exist, then

$$\left\| c(x_1) - h(t_1) \right\| < \delta_1 + \delta_2, \quad \left\| c(y) - h(t_2) \right\| < \delta_1 + \delta_2$$

noted as $x_1 \overset{h}{\sim} y$.

Lemma 6.3. $x_1 \overset{h}{\sim} x_0, x_1, x_0 \in V_1$, then $x_1 \sim x_0$

[Proof] From condition

$$x_1 \overset{h}{\sim} x_0, \Rightarrow t_0, t_1 \in [0,1], \quad t_0 < t_1, \quad \|c(x_i - h(t_i))\| > \delta_1 + \delta_2, i = 0, 1$$

We prove this lemma in three cases as follows:

(i) $\|c(x_1) - h(t_0)\| < \delta_1 + \delta_2$

$$\because \|c(x_0) - h(t_0)\| < \delta_1 + \delta_2$$

$$\Rightarrow \|c(x_1) - h(x_0)\| < 2 \cdot (\delta_1 + \delta_2)$$

From Lemma 6.2, $u \in V_2$, then

$$\|c((x_1 \circ u \circ R_\xi) \circ V_1) - c(x_0)\| < \delta_1 + \delta_2$$

(ii) $\|c(x_1) - h(t_0)\| \in [(\delta_1 + \delta_2), 2(\delta_1 + \delta_2)]$.

From Lemma 6.2, have $u_1 \in V_2$, such that

$$\|c((x_1 \circ u_1 \circ R_\xi) \circ V_1) - h(t_0)\| < \delta_1 + \delta_2.$$

Let $x_2 = (x_1 \circ u_1 \circ R_\xi) \circ V_1$, then x_2 is in accord with case (i)

(iii) $\|c(x_1) - h(t_0)\| > 2(\delta_1 + \delta_2)$.

According to the above conditions, have

$$\begin{aligned} L(h, t_0, t_1) &\geq \|c(x_1) - h(t_0)\| - \|c(x_1) - h(t_1)\| \\ &\geq 2(\delta_1 + \delta_2) - (\delta_1 + \delta_2) \\ &= \delta_1 + \delta_2. \end{aligned}$$

Therefore, if we can prove that. there exists $u_1, \cdots, u_n \in V_2$, then $\|c(x_{n+1}) - h(t_{n+1})\| < \delta_1 + \delta_2$, and the out of following two inequalities:

$$L(h, t_0, t_{n+1}) < \delta_1 + \delta_2$$

$$\|c(x_{n+1} - h(0))\| < 2(\delta_1 + \delta_2)$$

at least one hold(s). Let us disprove it.

[Disproof] If the lemma is not held, then there exist

$$x \in V_1, \quad \|c(x) - h(t_x)\| < \delta_1 + \delta_2$$

$$\|c(x) - h(0)\| \geq 2(\delta_1 + \delta_2)$$

and $\forall u \in V_2$, let $y = (x \circ u \circ R_\xi) \circ V_1$, there exists $t_2 \in (t_1, t_x)$, so

$$\|c(y) - h(t_2)\| < \delta_1 + \delta_2$$

$$L(h, t_1, t_2) < L(h, t_1, t_x) - (\delta_1 + \delta_2)$$

are held. Make t_2 satisfy:

$$\|c(x)-h(t_2)\| = 2(\delta_1+\delta_2)$$

$\because h$ is continuous $\Rightarrow$ there exist $t_2 \in [t_0, t_x]$, so

$$\|c((x\circ u\circ R_\xi)\circ V_1)-h(t_2)\| < \delta_1+\delta_2$$

$$\Rightarrow L(h,t_2,t_x) \geq \|c(x)-h(t_2)\| - \|c(x)-h(t_x)\| \geq \delta_1+\delta_2$$

$$\Rightarrow L(h,t_1,t_2) < L(h,t_1,t_x)-(\delta_1+\delta_2)$$

$\Rightarrow$ contradiction

Hence, the lemma holds.

Theorem 6.9. Let $SH=\{a\in X$, there exists $b\in H$ and $\|a-b\|<(\delta_1+\delta_2)\}$. If G is a convex set of X, and the boundary of G, i.e.,$\partial G\subset SH$. For the T-System, $r(x)>2(\delta_1+\delta_2)$, $\forall x\in V_1$, and $|\xi|<3(\delta_1+\delta_2)$, then

$$\forall x\in V_1, c(x)\in G$$

the T-System with disturbance ξ has: $x\sim x_0$

[Proof] if $x_1\in V_1, c(x_1)\in G$,and

(i) $c(x_1)\in SH$, then from Lemma 6.2 $\Rightarrow x\sim x_e$.

(ii) $c(x_1)\notin SH$, then there exist $P\geq\delta_1+\delta_2$, a∈H, so $c(x_1)+P\cdot\vec{g}=a$. Let $q_\xi=|\xi|-\min(P,2(\delta_1+\delta_2))$, then $|q_\xi|<2\cdot(\delta_1+\delta_2)$. From Lemma 6.1, there exists u∈V_2, so

$$u_{x\circ u\circ R}(c(x_1)-q_\xi\cdot\vec{g})=1$$

$$\Rightarrow u_{x_1\circ u\circ R_\xi}(c(x_1)-q_\xi\cdot\vec{g}+|\xi|\cdot\vec{g})=1$$

$$\Rightarrow u_{x_1\circ u\circ R_\xi}(c(x_1)+\min(P,2\cdot(\delta_1+\delta_2))=1$$

$\because G$ is a convex set

$$\Rightarrow c(x\circ u\circ R_\xi)(DV_1)\in G\cup SH$$

Let $x_2=(x_1\circ u\circ R_\xi)\circ V_1$, if $c(x_2)\in SH$, then $x_1\sim x_n$.

$\because$ x is finited, $\therefore$ h exists.

6.5 Application Example of Fuzzy Controller

As an example, let us introduce a fuzzy expert force control for a biped robot [38].

6.5.1 Biped Force Control in the Double-support Phase

The double-support and single-support phases in biped locomotion alternate. A biped usually starts and stops motion at the double-support configuration. Most studies in biped control, however, have concentrated on the single-support phase because it is the predominant portion of the locomotion period. Nevertheless, the double-support phase plays an important role in achieving smooth locomotion, especially at low speeds when the center of gravity is moved from one leg to the other. In the double-support phase, both legs dynamically interact with the ground. In the single-support phase, one leg lifts while the other leg touches the ground. Position control is sufficient for lifting the leg. When two legs come into contact with the ground, however, position control may fail due to incomplete information about the environment. Force control is required, especially when the center of gravity is introduced to provide the biped with balance.

Modern control theory has been applied to biped control, its effectiveness is diminished when complete knowledge of the environment is unavailable; e.g., when the structure of the biped system is not completely known, when parameter uncertainty is excessive, or multiple objectives are specified. Besides, goals and constraints are not always quantified by numerical values. For example, the control object may distribute the weight to the biped equally to both feet. Fuzzy sets theory can be an effective means of dealing with linguistically specified objects. By incorporating linguistic control rules and fuzzy set theory, we can accommodate uncertainty and achieve control objectives.

In this example, the primary objective is to develop a variable gain fuzzy logic control method for a biped robot in the double-support phase and show that the proposed system can be used to statically balance the robot; furthermore, in the example the characteristics of the biped in the double-support phase are formulated; the proposed variable gain fuzzy control system is derived; and hardware implementation and experimental results are described.

An m-link biped model is shown in Figure 6.21. From force and torque balance, we obtain

$$\sum_{i=0}^{m-1} \mathbf{r}_i \times m_i \mathbf{q} = \mathbf{p}_1 \times \mathbf{f}_1 + \mathbf{p}_{m-1} \times \mathbf{f}_2 \tag{6.81}$$

$$\sum_{i=0}^{m-1} m_i \mathbf{q} = \mathbf{f}_1 + \mathbf{f}_2 \tag{6.82}$$

where m_i and $\mathbf{r}_i$ are the mass and the position of the center of gravity of link i; $\mathbf{p}_i$ is the position at the junction of link (i - 1) and link i; and $\mathbf{f}_1$ and $\mathbf{f}_2$ are the reaction forces acting through the ankles $\mathbf{p}_1$ and $\mathbf{p}_{m-1}$. We assume that the feet do not experience point contact; however the total reaction force of each foot acts through the ankle. Let the reference coordinate be assigned to foot 1. In the double-support phase, the position of both feet are stationary on the ground, and we obtain the following positional and rotational constraints

$$\mathbf{p}_0 = 0, \quad \mathbf{a}_0 = 0, \quad \text{and} \quad \mathbf{p}_m = \mathbf{p}, \ \mathbf{a}_m = \alpha \tag{6.83}$$

where $\mathbf{p}$ and α are the constant positional vector and the Euler rotational angle vector of foot 2 respectively. The above constraints simply can be written as

$$\mathbf{C}(\mathbf{q}) = \mathbf{C}_0 \tag{6.85}$$

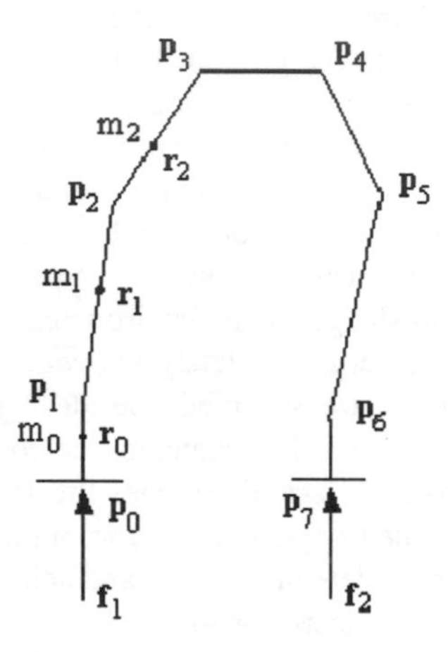

Figure 6.21. A m-link biped walking robot.

where $\mathbf{q}$ is a joint $2n$-vector and n is the number of degrees of freedom of each leg, $\mathbf{c}(\mathbf{q})$ is a 6-vector, and $\mathbf{c}_0$ is the vector of constraints. From Equations (6.81), (6.82) and (6.84), we have

$$\mathbf{J}(\mathbf{q})\dot{\mathbf{q}} = \dot{\mathbf{y}} \tag{6.86}$$

where

$$\mathbf{J}(\mathbf{q}) = [\mathbf{J}_r(\mathbf{q}), \mathbf{J}_p(\mathbf{q})]^T$$

$$\mathbf{J}_r(\mathbf{q}) = \partial / \partial \mathbf{q}(\sum \mathbf{r}_i \times \mathbf{m}_i \mathbf{g}))$$

$$\mathbf{J}_p(\mathbf{q}) = \partial / \partial \mathbf{q}(\mathbf{c}(\mathbf{q}))$$

and

$$\dot{\mathbf{y}} = [(\mathbf{p}_{m-1} - \mathbf{p}_1) \times \dot{\mathbf{f}}_2, 0]^T .$$

Solving Equation (6.86), we can express the joint trajectories $\dot{\mathbf{q}}$ in terms of the desired reaction force trajectories $\dot{\mathbf{f}}_1$ and $\dot{\mathbf{f}}_2$. It is assumed that the number of degrees of freedom is $n \geq 5$. Equation (6.85) can be used for force control in the double-support phase. In this case, we write

$$\mathbf{J}(\mathbf{q})\delta\,\mathbf{q} = \delta\,\mathbf{y} \tag{6.86}$$

where

$$\delta\mathbf{y} = [(\mathbf{p}_{m-1} - \mathbf{p}_1) \times \delta\mathbf{f}_2,\;\; \mathbf{c}_0 - \mathbf{c}(\mathbf{q})]^T$$

is the force and positional constraint error. Using this error the configuration of the biped can be controlled.

6.5.2 Fuzzy Variable Gain Force Control

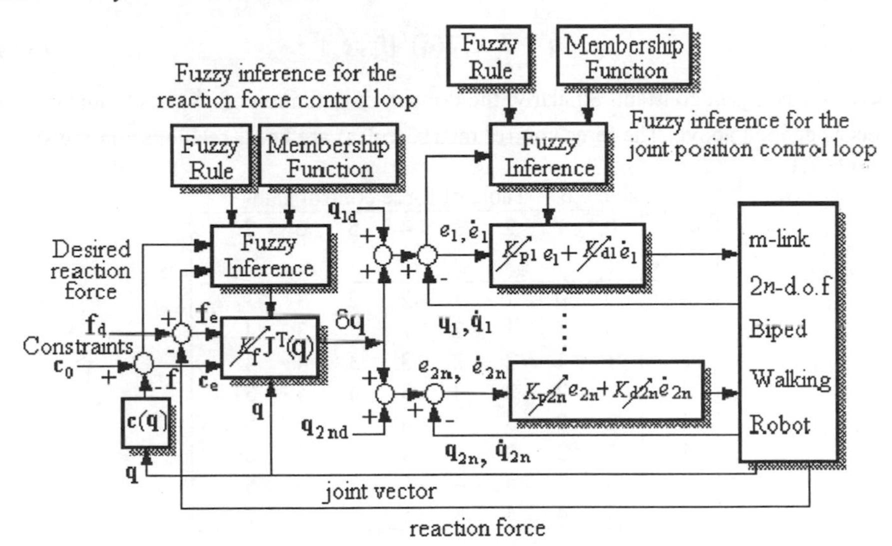

Figure 6.22. Structure of fuzzy variable gain force control system. (with Shih *et al.*, 1991)

Figure 6.22 shows the structure of the proposed variable gain force control. The inputs to the system are desired reaction forces $\mathbf{f}_d = [\mathbf{f}_{1d}{}^T, \mathbf{f}_{2d}{}^T]^T$, the constraint vector is $\mathbf{c}_0$ and the desired joint trajectories $\mathbf{q}_d$, which are derived from Equation (6.85). There are loops for force control and position control.

In the force control loop, the objective is to minimize the tracking force error $\mathbf{f}_e$. The control input is a function of the error. The constraint error $\mathbf{c}_e$ is also important in determining the control input because it delineates the desired and actual biped configurations. Therefore, the root-mean-square (RMS) values of $\mathbf{f}_e$ and $\mathbf{c}_e$ are selected as linguistic variables to be used in the force control rules and these quantities are formally defined as

$$\mathbf{f_e} = G_f \, \mathrm{RMS}(\mathbf{f_d} - \mathbf{f})$$

$$\mathbf{c_e} = G_c \, \mathrm{RMS}(\mathbf{c}_0 - \mathbf{c}(\mathbf{q}))$$

where G_f and G_c are gain constants. The domains of definition of $\mathbf{f_e}$ and $\mathbf{c_e}$ have 8 linear quantization levels that ensure small errors between the actual and desired force trajectories. The linguistic values are 0 (zero), 1 (very small), 2 (small), 3 (medium), 4 (big), 5 (large), 6 (very large), and 7 (huge).

The consequence of each force control rule is defined in terms of the force control gain K_f, and the motion control input $\delta\mathbf{q}$ is given by the expression

$$\delta\mathbf{q} = G_q K_f \mathbf{J}(\mathbf{q})^T [\mathbf{f}_e, \mathbf{c}_e]^T \tag{6.87}$$

where G_q is a gain constant. Similarly, the domain of definition of K_f has 8 quantization levels as defined above. The force control matrix (rules) are linear relations and are shown in Table 7.1.

Table 6.6. Table of force control gains

$\mathbf{c}_e$ \ $\mathbf{f}_e$	0	1	2	3	4	5	6	7
0	0	0	1	1	2	2	3	3
1	0	1	1	2	2	3	3	4
2	1	1	2	2	3	3	4	4
3	1	2	2	3	3	4	4	5
4	2	2	3	3	4	4	5	5
5	2	3	3	4	4	5	5	6
6	3	3	4	4	5	5	6	6
7	3	4	4	5	5	6	6	7

In the position control loop, the objective is to minimize the tracking joint position error $\mathbf{e}$. The control input is a function of this error. The change of this tracking error $\mathbf{d_e}$ is also important in determining the control input, because it delineates the desired and actual tracking velocities. Therefore, $\mathbf{e}$ and $\mathbf{d_e}$ are selected as linguistic variables to be used in the control rules and these quantities are formally defined as

$$\mathbf{e} = \mathrm{G_e}(\text{desired point} - \text{current point})$$

$$\mathbf{d}_e = \mathrm{G_d}(\text{current error} - \text{previous error})$$

where G_e and G_d are gain constants. The domains of definition of $\mathbf{e}$ and $\mathbf{d}_e$ have 15 linear quantization levels. The linguistic values are + (positive), - (negative), 0 (zero), 1 (very small), 2 (small), 3 (medium), 4 (big), 5 (large), 6 (very large), 7 (huge).

The consequence of each position control rule, on the other hand, is defined in terms of the PD gain changes δK_e and δK_d. The actuator control input u is given by the expression

$$u = G_u(K_e e + K_d e) \tag{6.88}$$

where

$$K_e = k_e + k_1 * \delta K_e$$

$$K_d = k_d + k_2 * \delta K_d$$

where k_e and k_d are the bias gains, G_u is the control constant gain that scales the control action, and k_1 and k_2 are the gain constants that determine the range of variation of each term. The range of variation should be matched with a stability interval in order to guarantee stability. The fuzzy variable gain controller is a nonlinear PD feedback law. A conventional PD controller is a special case of the fuzzy controller with constant gains. Similarly, the domains of the fuzzy value of δK_e and δK_d also involve 15 quantization level defined as above.

The fuzzy control rules used in the position control loop are listed in Table 6.7 and are based on the following rules: (1) if the output has the desired value and the error derivative is zero, maintain constant output of the controller; and (2) if the output diverges from the desired value, the action depends on the sign and the value of the error, and its derivative. If the conditions are such that the error can be corrected quickly, maintain constant or nearly constant controller output. Otherwise, change the controller output to achieve satisfactory results.

Table 6.7. Table of position control PD gains

d_c \ e	-7	-6	-5	-4	-3	-2	-1	0	1	2	3	4	5	6	7
-7	-7	-6	-6	-5	-5	-4	-4	-3	-3	-2	-2	-1	-1	0	0
-6	-6	-6	-5	-5	-4	-4	-3	-3	-2	-2	-1	-1	0	0	0
-5	-6	-5	-5	-4	-4	-3	-3	-2	-2	-1	-1	0	0	0	1
-4	-5	-5	-4	-4	-3	-3	-2	-2	-1	-1	0	0	0	1	1
-3	-5	-4	-4	-3	-3	-2	-2	-1	-1	0	0	0	1	1	2
-2	-4	-4	-3	-3	-2	-2	-1	-1	0	0	0	1	1	2	2
-1	-4	-3	-3	-2	-2	-1	-1	0	0	0	1	1	2	2	3
0	-3	-3	-2	-2	-1	-1	0	0	0	1	1	2	2	3	3
1	-3	-2	-2	-1	-1	0	0	0	1	1	2	2	3	3	4
2	-2	-2	-1	-1	0	0	0	1	1	2	2	3	3	4	4
3	-2	-1	-1	0	0	0	1	1	2	2	3	3	4	4	5
4	-1	-1	0	0	0	1	1	2	2	3	3	4	4	5	5
5	-1	0	0	0	1	1	2	2	3	3	4	4	5	5	6
6	0	0	0	1	1	2	2	3	3	4	4	5	5	6	6
7	0	0	1	1	2	2	3	3	4	4	5	5	6	6	7

6.5.3 Implementation and Results of the Biped Robot Control

The total height of the biped machine is 92 cm and the total weight is about 45 kg. Each leg has three segments: upper leg, lower leg, and foot. Twelve actuators provide the capability to change the position and orientation of the legs. There are eight loadcell sensors located under each foot to detect the vertical ground reaction force. Twelve DC servo motors are employed as actuators. The relative angles between two adjacent joints are monitored by optical encoders and their speeds are measured by tachometers. The motors are driven by servo-amplifiers for velocity control. The resolution of each encoder is 500 pulses per turn. A harmonic drive between each motor and its joint provides 100 : 1 reduction. Two micro-switches detect joint limits.

The hardware system is implemented with an IBM PC/AT and thirteen Intel 8097 microcontrollers. The PC is used as a host computer for trajectory planning, joint coordination, and data collection as well as for fuzzy control table development. Twelve microcontrollers are used in the joint controllers which interface the DC servo motors, optical position encoders and joint sensors. The other controller is employed to collect the loadcell sensor data. During operation the host computer performs trajectory planning and force control. The joint controller is based on a modular design that, because of the I/O features of the 8097, was easily implemented. The joint controller communicates with the host computer by both serial and parallel ports and carries out the low level position control. Except for the communication driver, other high level software for the PC was programmed in C. The joint controller software was developed in ASM-96.

In the following primary biped experiment, both feet are on the ground and the body sways to the right, to the left, and then returns to its original position. Let the joint vector $\mathbf{q}$ be represented by

$$\begin{aligned}\mathbf{q} &= [q_1 \cdots q_{12}]^T \\ &= [\theta_6^{(1)} \cdots \theta_1^{(1)}\ \theta_1^{(2)} \cdots \theta_6^{(2)}]^T.\end{aligned}$$

The desired $\mathbf{q}_1$ joint trajectory is shown in Figure 6.23(a), and other desired joint trajectories are

$$\mathbf{q}_6 = -\mathbf{q}_1,\ \mathbf{q}_7 = -\mathbf{q}_1,\ \mathbf{q}_{12} = \mathbf{q}_1, \text{ and } \mathbf{q}_i = 0 \text{ for } i = 2, 3, 4, 5, 8, 9, 10, 11.$$

The desired vertical reaction forces $\mathbf{f}_{1z}$ and $\mathbf{f}_{2z}$ are shown in Figure 6.23(b) respectively. Figure 6.23(c) shows the resulting vertical reaction forces $\mathbf{f}_{1z}$ and $\mathbf{f}_{2z}$ respectively with only position control . Without force control, the reaction force may not return to their initial values; besides, the force trajectories experience sudden changes if the position error changes during the motion. Figure 6.23(d) shows the resulting vertical reaction forces $\mathbf{f}_{1z}$ and $\mathbf{f}_{2z}$ with position and force control respectively. After including the force control loop, the reaction force trajectories are much closer to the desired force trajectories.

Fuzzy variable gain control has been applied to control biped robot weight force distribution. Experimental results show that the proposed technique is easy to be implemented and the system performance improved [38, 53].

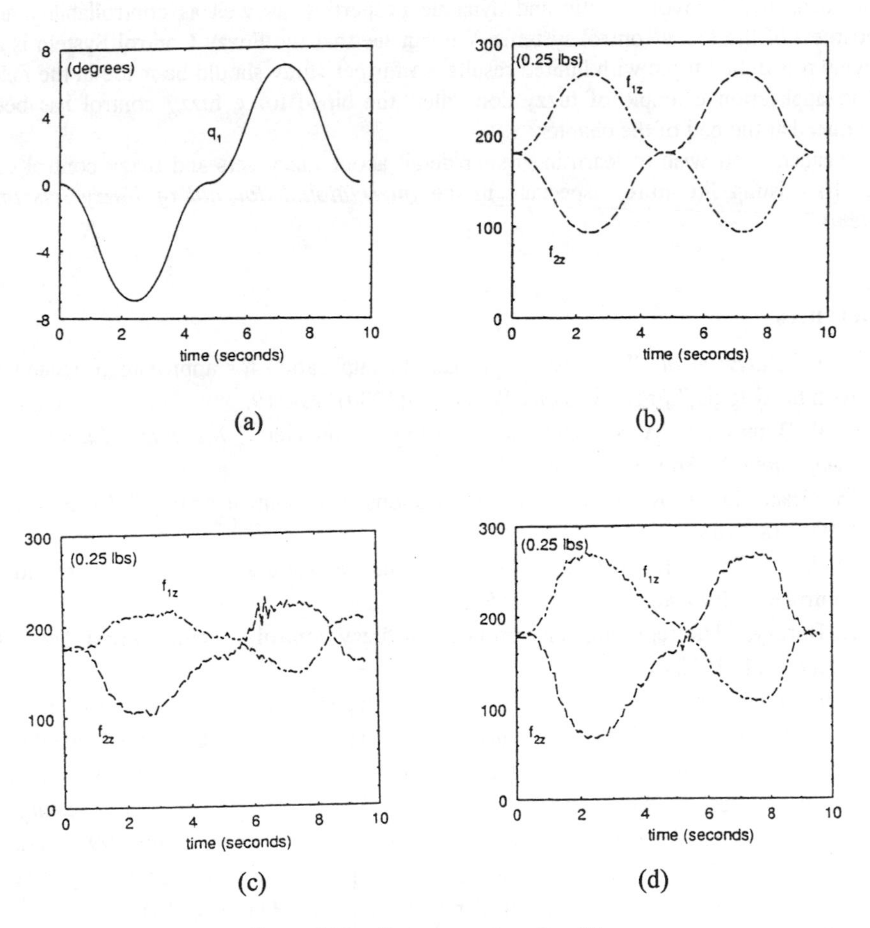

Figure 6.23. Experimental results of fuzzy force control for a biped robot. (with Shih et al, 1991)

6.6 Summary

This chapter has discussed the main ideas of fuzzy sets, fuzzy logic and their representation for control. The reader can get an overall view of the structures of fuzzy controllers including various structural schemes such as PID fuzzy controller, self-organizing fuzzy controller, and expert fuzzy controller. The integration of different control ideas to build a new structure of controller is very important and effective. Significant emphasis has been put on the design approaches and properties of fuzzy controllers, which involve static and dynamic properties, as well as controllability and robustness of the fuzzy control systems. We can see that the Fuzzy Control System is an actively researched topic with limited results, so further study should be made in the field. As an application example of fuzzy controller, the biped force fuzzy control has been introduced at the end of the chapter.

Readers who wish to learn in greater detail about fuzzy sets and fuzzy control can refer to existing literature, especially in the *International Journal of Fuzzy sets and Systems*.

References

1. J. F. Baldwin *et al.*, "Axiomatic approach to implication for approximate reasoning with fuzzy logic," *Fuzzy Sets and Systems*, **3**(1980)193-219.
2. P. P. Bonissone, "A compiler for fuzzy logic controllers," *Int. Fuzzy Engineering Symposium*, Yokohams, Japan, 1991.
3. M. Braae, D. A. Rutherford, "Fuzzy relations in a control setting," *Kybernets*, 7 (1978) 185-188.
4. M. Braae, D. A. Rutherford, "Theoretical and linguistic aspects of the fuzzy logic controller," *Automatica*, **15** (1979) 553-577.
5. K. Burdzy, "The reproducibility property of fuzzy control systems," *Fuzzy Sets and Systems*, **11** (1983) 161-177.
6. F. E. Cellier, *Continuous System Modeling* (Springer-Verlay, New York, 1991).
7. F. E. Cellier, A. Nebot, F. Mugica and A. de Albornoz, "Combined qualitaive/ quantitative simulation models of continuous-time processes using fuzzy inductive reasoning techniques," *Proc. SICICA'92 IFAC Symposium on Intelligent Components and Instruments for Control Applications* (Malaga, Spain, 1992) 22-24.
8. C.-L. Chen, P.-C. Chen and C.-K. Chen, "A pneumatic model-following control system using a fuzzy adaptive controller," *Automatica*, **29** (1993) 1101.
9. S. Daley, and K. F. Gill, "Comparison of a fuzzy logic controller with PID controller law," *Trans. ASME*, **111** (1989) 128-137.
10. H. N. Fairbrother, B. A. Stacey, and R. Sutton, "Fuzzy self-organising controller of a

ROV," *Proc. IEE Control 91'*, Conf. Pub. no.332 (1991) 499-504.

11. J. Fei and C. Isik, "Adaptive fuzzy control via modification of linguistic variables," *Proc. IEEE Fuzzy'92*, San Diego, CA, USA, 1992, 399-406.
12. D. E. Goldberg, *Genetic Algorithms in Search, Optimization and Machine Learning* (Addison-Wesley, Reading Mass., 1989).
13. M. M. Gupta, "Controllability of fuzzy control systems," *IEEE Trans. SMC*, **16** (1986).
14. M. M. Gupta *et. al.*, "Multivariable structure of fuzzy control systems," *IEEE Trans. SMC*, **16** (1986) 638-656.
15. C. J. Harris and A. B. Read, "Knowledge based motion control of autonomous vehicles," *First IFAC Workshop on AI in Real-Time Control*, Swansea, 1988, 149-154.
16. C. J. Harris, C. G. Moore and M. Brown, *Intelligent Control: Aspects of Fuzzy Logic and Neural Nets* (World Scientific, Singapore, 1993)113-130.
17. J. H. Holland, *Adaptation in Natural and Artificial Systems* (The University Michigan Press, 1975; MIT Press, 1992).
18. S. Isaka and A. V. Sebald, "An optimization for fuzzy controller design," IEEE Trans. SMC, **22** (1992) 1469.
19. C. L. Karr, L. M. Freeman, D. L. Meredith, "Improved fuzzy process control of spacecraft autonomous rendezvous using a genetic algorithm," *Proc. SPIE Intelligent Control and Adaptive Systems Conference* (Philadelphia, Penn., 1989) 274-288.
20. C. L. Karr, and D. A. Stanley, "Fuzzy logic and genetic algorithms in time-varying control problems," *Proc. North American Fuzzy Information Processing Society*, Columbia, MO, 1991, 285-290.
21. C. L. Karr, "Design of an adaptive fuzzy logic controller using a genetic algorithm," *Proc. of 4th Int. Conf. Genetic Algorithms*, San Diego, CA, 1991, 450-457.
22. J. B. Kiszka, M. Kohanska and D. Sliwinska, "The influence of some fuzzy implication operators on the accuracy of a fuzzy model-II," *Fuzzy Sets and Systems*, **15** (1985) 223-240.
23. J. B. Kiszka, M. M. Gupta, and P. N. Nikiforuk, "Energistic stability of fuzzy systems," *IEEE Trans. SMC*, **15** (1985) 783-792.
24. B. Kosko, *Neural Networks and Fuzzy Systems — A Dynamical Systems Approach to Machine Intelligence* (Prentice-Hall, Englewood Cliffs, NJ, 1992).
25. K. Kristinsson and G. A. Dumont, "System identification and control using genetic algorithsm," *IEEE Trans. SMC*, 22(1992)1033-1046.
26. D. P. Kwok, T. P. Leung, and S. Feng, "Genetic algorithms for the optimal dynamic control of robot arms," *Proc. IECON* (Maui, USA, 1993) 381-385.
27. D. P. Kwok, P. K. S. Tam, C. K. Li, and P. Wang, "Synthesis of fuzzy PI controllers

for robot arms," *Proc. of 13th IASTED Int. Symp. on Robotics and Manufacturing* (California, USA, 1990)114-117.

28. D. P. Kwok, P. K. S. Tam, C. K. Li and P. Wang, "Analysis and design of fuzzy PID control systems," *Proc. IEE 3rd Int. Conf. on Control*, Vol. 2 (Heriot-Watt. Univ. Edingburgh, 1991)955-960.
29. C. C. Lee, "Fuzzy logic in control systems: fuzzy logic controller, part I & II," *IEEE Trans. SMC*, **20** (1990) 405-435.
30. D. Li and F. E. Cellier, "Fuzzy measures in inductive reasoning," *Proc. WSC'90 Winter Simulation Conference* (New Orleans, Louisiana, 1990) 527-538.
31. R.-H. Li and S.-Y. Qin, *Theories and methods for intelligent control* (in Chinese) (Xi'an Jiaotung University Press, Xi'an , China, 1994).
32. D. A. Linkers, M. F. Abbod, "Self-organising fuzzy logic controllers for real-time processes," *Proc. IEE Control'91*, (1991) 971-976.
33. S. Masui, T. Terano, and Y. Sugaya, "Indentificaiton of fuzzy rules in a manual control system," *Proc. 3rd Int. Fuzzy Systems Association*, 1987, pp.143-146.
34. C. G. Moore and C. J. Harris, "Indirect adaptive fuzzy control," *Int. J. of Control*, **56** (1992) 441.
35. W. Pedrycz, *Fuzzy Control and Fuzzy Systems* (Research Studies Press, New York, 1989) 111-140.
36. T. J. Procyk and E. H. Mamdani, "A linguistic self-organizing process controller," *Automatica*, **15** (1979)15-30.
37. S. Shao, "Fuzzy self-organising control and its application for dynamical systems," *Fuzzy Sets and Systems*, **26** (1988)151-164.
38. C. L. Shih, W. A. Gruver and Y. Zhu, "Fuzzy logic force control for a biped robot," in *Proc. IEEE Int. Symp. Intelligent Control* (Arlington, Virginia, USA, 1991), 269-274.
39. K. Sugiyama, "Rule-based self-organising control," in *Fuzzy Computing, eds.* M.M.Gupta, and Yamakawa (Elsevier Science, North Holland, Amsterdam, 1988).
40. R. Sutton and I. M. Jess, "A design study of a self-organising fuzzy autopilot for ship control," *Proc. Inst. Mech. Engrs.*, **205** (1991) 35-47.
41. K. L. Tang and R. J. Mulholland, "Comparing fuzzy logic with classical controller designs," *IEEE Trans. SMC*, **17** (1987) 1085-1087.
42. S.-S. Tang, and Z.-X. Cai, "Robustness of a fuzzy control system under directional disturbance (in Chinese)," in *Intelligent Control and Intelligent Automation* (Science Press, Beijing, 1993)918-923.
43. R. M. Tong, "Analysis of fuzzy control algorithms using the relation matirx," *Int. J. Man-Machine Studies*, **8** (1976) 679-686.

44. R. M. Tong, "A control engineering review of fuzzy systems," *Automatica*, **13** (1977) 559-569.
45. R. M. Tong, "A retrospective view of fuzzy control systems," *Fuzzy Sets and Systems*, **14** (1984) 199-210.
46. F. Van der Rhee, H. R. Van Nauta Lemke, and J. G. Dijkman, "Knowledge-based fuzzy modeling of systems," *Proc. 11th IFAC World Congress*, 7 (1990)186-191.
47. P. Wang, and D. P. Kwok, "Analysis and synthesis of an intelligent control system based on fuzzy logic and PID principle," *Intelligent Systems Engineering*, **2** (1992) 157-171.
48. Y. Yamazaki, *An improved algorithm for self-organising control*, Ph.D.Dissertation, Queen Mary College, London, 1982.
49. L. A. Zadeh, "Fuzzy sets," *Information and Control*, **8** (1965) 338-353.
50. L. A. Zadeh, "Fuzzy algorithm," *Information and Control*, **12** (1965) 94-102.
51. L. A. Zadeh, "A rationale for fuzzy control," *Trans. ASME. J. Dynamic Systems, Measurement and Control*, **94** (1972) 3-4.
52. L. A. Zadeh, "Outline of a new approach to the analysis of complex systems and decision processes," *IEEE Trans. SMC*, **3** (1973) 28-44.
53. Y. F. Zhang and J. Shen, "Gait synthesis for the SD-2 biped robot to climb sloping surface," *IEEE Trans. Robotics and Automation*, **6** (1990) 86-96.
54. J. Zhu, *Fuzzy Control Principles and Applications* (in Chinese) (Mechanical Industry Press, Beijing, 1995).

CHAPTER 7
NEUROCONTROL SYSTEMS

Over the past ten years, there have been many attempts and considerable progress in Artificial Neural Networks (ANN, or NN) as a new and interested tool for identification, modeling and control of the dynamic systems [19, 53, 59, 69, 70, 75, 76, 78]. The number of papers in journals and conferences with regard to ANN has increased sharply; monographs, textbooks, proceedings and special issues on the Artificial Neural Networks and ANN-based control have been published one after another. The special issues on the subject are regarded to play an important role in promoting this intellectual trend of thought [33, 35-38, 40].

In this chapter, firstly, the origins, characteristics, properties, structures, models and algorithms of ANN will be introduced; then various structures of neurocontrol will be analyzed; finally, a paradigm of NN-based control system will be presented, and some open research and application topics will be discussed.

7.1 Introduction to Artificial Neural Networks

In this section, the origins of ANN research and its characteristics and properties will be briefly introduced, as well as the relationship between ANN and control will be discussed.

7.1.1 Origins of ANN Research

The pioneers in the field of ANN, McCulloch and Pitts, proposed an idea that a "mindlike machine" could be manufactured by interconnecting models based on the behavior of biological neurons, this is the concept of neurological networks [62]. They made a neuron model representing a basic component of the brain and it shows its versatility as a logical operation system. With the progress of research on brain and computers, the research objective was changed from the "mindlike machine" to the "learning machine", for which a learning model was proposed by Hebb [26] who was concerned with the adaptation laws involved in neural systems. Rosenblatt coined the name "Perceptron" and devised an architecture to which subsequently much attention has been paid [77]. By early 1960's, specific design guidelines for learning systems were given by Widrow and Hoff's Adaline (adaptive linear element) and Steinbuch's Learning Matrix [81, 89]. The perceptron received considerable excitement when it was first introduced because of its conceptual simplicity. However, Minsky and Papert had proven mathematically that the Perceptron cannot be used for complex logic function [67].

Grossberg and Kohonen made significant contributions to NN research in 1970's. Based on biological and psychological evidence, Grossberg proposed several architectures of nonlinear dynamic systems with novel characteristics [23]. The dynamics of the network were modeled by first order differentiable equations, and the architectures of the network are self-organizing neural implementations of pattern clustering algorithms. Kohonen developed his work on self-organizing maps, based on the idea that neurons organize themselves to tune various and specific patterns [51]. Werbos developed a back-propagation algorithm in the 1970's [87]. Hopfield introduced a recurrent-type neural network based on the interaction of neurons [31]. This model is known as Hopfield net. As a learning algorithm of the feedforward neural network, the back-propagation algorithm was also rediscovered by Parker [74] and Rumelhart *et. al.* [78] in the middle 1980's. Kosko developed his adaptive Bidirectional Associative Memory [52]. Recently, a wide range of applications from household apparatus to industries have been found.

7.1.2 ANN for Control

The following characteristics and properties of NN are important for control:

(1) **Parallel distributed processing**. NN's have a highly parallel structure and a capability of parallel implementation that can lead to a better fault tolerance and a faster overall processing. This is suitable for real-time control and dynamic control.

(2) **Nonlinear mapping**. NN's possess inherent nonlinear property that stems from their theoretical ability to approximate arbitrary nonlinear mapping. It brings a new promise to nonlinear control problems.

(3) **Learning by training**. NN's are trained by using past data records from the system under study. A suitably trained NN has the ability to generalize all data. Thus, NN's can solve such control processes that are hard to deal with by mathematical models or descriptive rules.

(4) **Adaptation and integration**. NN's can be adapted on-line and operated simultaneously on both quantitative and qualitative data. The strong abilities of adaptation and information fusion make the networks process massive and different control inputs simultaneously, solve the problems such as complementarity and redundancy among the input information, and implement information integration and fusion processing. These are specially suited for the control of complex, large-scale and multivariable systems.

(5) **Hardware implementation**. NN's can be implemented in parallel not only by software but also by hardware. Recently, some VLSI hardware implementations have been introduced and are available for commercial purposes. This brings NN's with higher speed and of a larger scale for the implemented networks.

It is clear that the NN's have potentialities for intelligent control systems by their learning and adaptation, self-organization, function approximation, and massively paralleled processing capabilities.

Applications of NN's to pattern recognition, signal processing, system identification and optimization have been widely studied. In the control field, many attempts have been made to apply the NN to control systems where the NN's are used to deal with the nonlinearities and the uncertainty of control systems and to approximate functions such as system identification. [58, 59, 66, 69].

Research on applications of NN-based control can be classified into several principal methods depending on structures of the control systems such as supervised control, direct inverse control, neural adaptive control, and predictive control that will be discussed in some more detail later.

7.2 Structures of Artificial Neural Networks

The NN structure is defined by the basic processing elements and the way in which they are interconnected.

7.2.1 *Neuron and Its Properties*

The basic processing element of the connectionist architecture is often called a neuron by analogy with neurophysiology. Each neuron model building up a network simulates a biological neuron as shown in Figure 7.1. The neuron unit consists of multiple inputs x_i, i=1,2,...,n, and one output y. The internal state is represented by the weighted sum of the input signals, and the output is as follows:

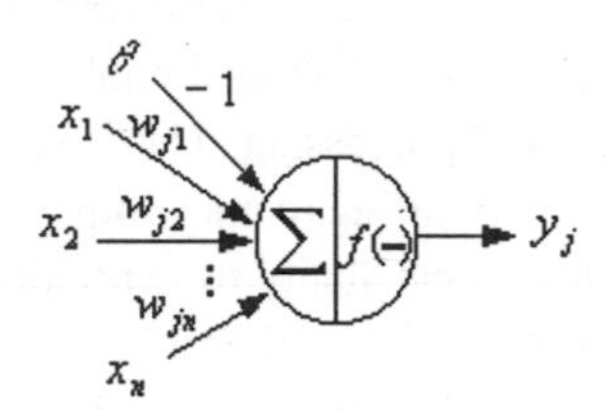

Figure 7.1 Neuron model

$$y_j(t) = f(\sum_{i=1}^{n} w_{ji}x_i - \theta_j) \tag{7.1}$$

where θ_j is a bias (threshold) of the neuron unit, w_{ji} is a weight coefficient of connection, n is the number of input, y_j is the output of the neuron, t is time, f(-) is an output transformation function that often uses the two-valued function of 1 and 0 using threshold fend-limiter or the sigmoid function as shown in Figure 7.2, all of these functions are continuous and nonlinear. A two-value function is represented by

$$f(x) = \begin{cases} 1, & x \geq x_0 \\ 0, & x < x_0 \end{cases} \tag{7.2}$$

as seen in Figure 7.2(a). A conventional sigmoid function is as follows:

$$f(x) = \frac{1}{1+e^{-ax}}, \quad 0 < f(x) < 1. \tag{7.3}$$

A hyperbolic tangent function is often used to replace the conventional sigmoid function since the output value of the sigmoid function is positive (see Figure 7.2(b)), whereas that of the hyperbolic tangent function is both positive and negative (see Figure 7.2(c)). The hyperbolic tangent function is represented by the following expressions:

$$f(x)=\frac{1-e^{-ax}}{1+e^{-ax}},\quad -1<f(x)<1 \tag{7.4}$$

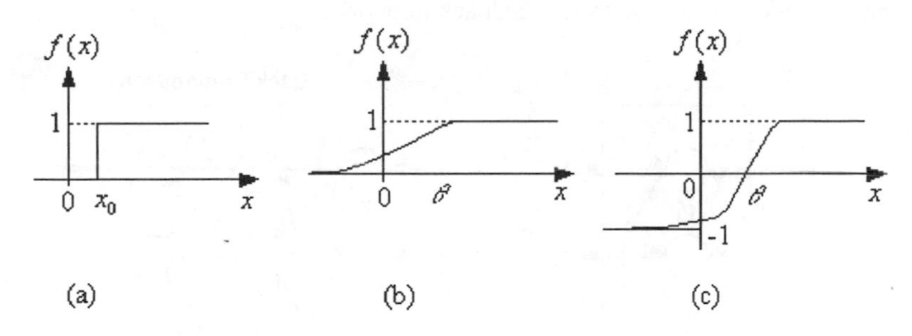

Figure 7.2. Some transformation functions in neurons.

The w_{ji} takes a positive value for excitation and a negative one for inhibition.

7.2.2 Basic Types of ANN

1. Basic properties and structures of ANN

In the human brain, a tremendous number of neurons are interconnected to form the neural networks and perform advanced intelligent activities of problem solving.

The ANN is built by neuron models and is a parallelly distributed structure of information processing network that consists of many neurons. Every neuron has a single output, that can be connected to another neuron, and there are many (multiple) input connection ways, each connection way corresponds to a weight coefficient of connection. In the strict sense, ANN is a directed graph with the following properties:

(1) There exists a state variable x_i for every node i;

(2) There exists a weight coefficient of connection w_{ij} from node j to node i;

(3) There is a Threshold value θ_i for every node i;

(4) For every node i, a transformation function $f_i(x_i, w_{ji}, \theta_i), i \neq j$ is defined, in the most general case, the function takes the form: $f_i(\sum_j w_{ij}x_j - \theta_i)$.

The structures of the ANN is basically classified into two types: recurrent networks and feedforward networks.

(1) Recurrent (Feedback) networks

In the recurrent networks, multiple neurons are interconnected to organize an interconnecting neural network as shown in Figure 7.3. The output of some neurons are feedback to the same neurons or to neurons in preceding layers. Therefore, signals can flow in both forward and backward directions. The Hopfield network [31], the Elman network [12] and the Jordan network [44, 45] are examples of recurrent networks. The recurrent networks can also be called feedback networks.

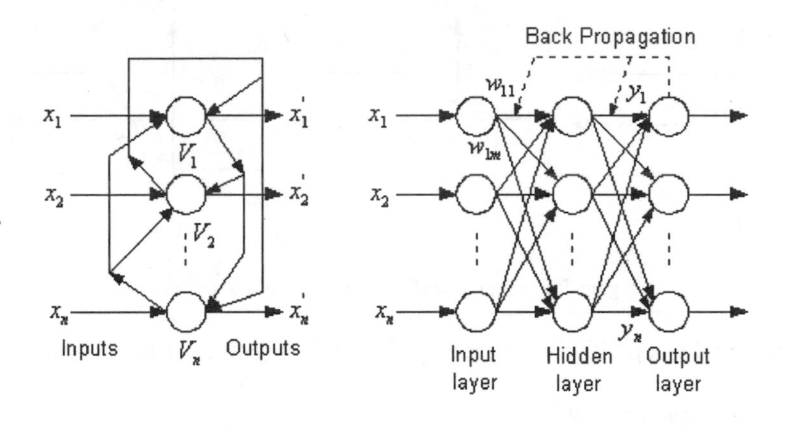

Figure 7.3. Recurrent (feedback) NN.

Figure 7.4. Feedforward (multilayer) NN.

In Figure 7.3, v_i indicates the state of the node, x_i the input (initial) value of the node, and x_i' the output value after convergence.

(2) Feedforward networks

The feedforward networks have a hierarchical structure that consists of some layers without interconnections among neurons in the same layer. The signals flow from the input layer to the output layer via unidirectional connections, the neurons being connected from one layer to the next, but not within the same layer, as shown in Figure 7.4. The feedforward networks also have another name called multilayer networks. In Figure 7.4, the solid line indicates physical signal flow, whereas the dotted line indicates back propagation. Examples of feedfoward networks include the multi-layer perceptron (MLP) [78], the learning vector quantization (LVQ) network [50], the cerebellar model articulation control (CMAC) network [1, 2], and the group-method of data handling

(GMDH) network [27].

2. Basic learning algorithms

Neural networks are trained by two main types of learning algorithms: supervised and unsupervised learning algorithms. In addition, there exists a third type, reinforcement learning, which can be regarded as a special form of supervised learning.

(1) Supervised learning

A supervised learning algorithm adjusts the strengths or weights of the inter-neuron connections according to the difference between the desired and actual network outputs corresponding to a given input. Thus, supervised learning requires a teacher or supervisor to provide desired or target output signals. Examples of supervised learning algorithms include the delta rule [89], the generalized delta rule or backpropagation algorithm [78] and the LVQ algorithm [50].

(2) Unsupervised learning

Unsupervised learning algorithms do not require the desired outputs to be known. During training, only input patterns are presented to the neural network which automatically adapts the weights of its connections to cluster the input patterns into groups with similar features. Examples of unsupervised learning algorithms include the Kohonen [50] and Carpenter-Grossberg Adaptive Resonancc Theory (ART) [13] competitive learning algorithms.

(3) Reinforcement learning

As mentioned before, reinforcement learning is a special case of supervised learning. Instead of using a teacher to give target outputs, a reinforcement learning algorithm employs a critic only to evaluate the goodness of the neural network output corresponding to a given input. An example of a reinforcement learning algorithm is the genetic algorithm (GA) [21, 30].

7.2.3 Typical Models of ANN

Up to now, there exist over thirty models of ANN in development and application. The followings are representative ones:

(1) Adaptive Resonance Theory (ART) proposed by Grossberg is a network of rough classification for input data according to selective parameters [23]. ART-I is used for two-value input, whereas ART-II is for continuous inputs. The disadvantage of this network is its excessive sensitivity.

(2) Bidirectional Associative Memory (BAM) developed by Kosko is a single-state interconnective associative network that possesses learning ability [52]. The disadvantage of this network is that the density of storage is too small, and can cause vibration easily.

(3) Boltzmann Machine (BM) developed by Hinton *et. al.* is based on the Hopfield net

and with learning capability that searches for solutions by a simulated annealing process as a stochastic technique [29]. Its training time is longer than that of the BP network.
(4) Back-Propagation (BP) training algorithm developed originally by Werbos is an iterative gradient algorithm designed to minimize the mean square error between the actual output of a feed-forward net and the desired output [87]. The BP net is a multilayer mapping network. If the parameter is suitable, this network can converge the output to a smaller value of mean square error, and it has become the most widely used network in recent years. The longer training time also is the disadvantage of the BP net.
(5) Counter Propagation Network (CPN) proposed by Hecht-Nielson is a connective net, and consists of five layers in general. CPN can be used for associative memory, but needs many processing units [28].
(6) Hopfield network developed by Hopfield is a kind of single-layer self-associative net without learning capability. The model consists of a set of first-order (nonlinear) differentiable equations that minimize a certain energy function. The drawbacks of Hopfield net are that the cost of computer memory is high and the connection has to be symmetrical.
(7) Madaline algorithm is a developed form of Adaline that in turn is a weighed sum of the inputs, together with a LMS algorithm to adjust the weights to minimize the difference between the desired signal and the output. Both the Adaline and Madaline algorithms are powerful tools for adaptive signal processing and adaptive control. Madaline algorithm possesses stronger learning capability, but the linear function between input and output has to be satisfied.
(8) Neocognitron proposed by Fukushima [20] was the most complex multilayer net in structure until now. Without a teacher during learning, the Neocognitron had capability of selection, and has no sensitivity to movement and rotation of sample parts. However, this net exhausts too many nodes and interconnections; its parameters are in quantity and hard to select.
(9) Perceptron developed by Rosenblatt in 1958 is a set of trainable linear classifiers [77]. The perceptron network is the oldest net and seldom used today.
(10) Self-Organizing Maps (SOM) network proposed by Kohonen is based on the idea that neurons organize themselves to tune various and specific patterns. The SOM can form continuous mapping among families of nodes which play the role of vectorial quantizer.

According to the reference materials by Illingworth [39], the most typical models (algorithms) of ANN, their learning rules and application fields are listed in Table 7.1.

Table 7.1. Typical models of ANN

Name of Model	With/without teacher	Learning rule	Forward/back propagation	Application field
AG	without	Hebb Law	Back	Data classification
SG	without	Hebb Law	Back	Information processing
ART-I	without	Competition Law	Back	Pattern classification
DH	without	Hebb Law	Back	Voice processing
CH	without	Hebb/competition	Back	Combinatory optimization
BAM	without	Hebb/competition	Back	Image processing
AM	without	Hebb Law	Back	Pattern storage
ABAM	without	Hebb Law	Back	Signal processing
CABAM	without	Hebb Law	Back	Combinatory optimization
FCM	without	Hebb Law	Back	Combinatory optimization
LM	with	Hebb Law	Forward	Process supervision
DR	with	Hebb Law	Forward	Process prediction/control
LAM	with	Hebb Law	Forward	System control
OLAM	with	Hebb Law	Forward	Signal processing
FAM	with	Hebb Law	Forward	Knowledge processing
BSB	with	Error Revision	Forward	real-time classification
Perceptron	with	Error Revision	Forward	Linear classif./prediction
Adaline/ Madaline	with	Error Revision	Back	Classif., noise inhibition
BP	with	Error Revision	Back	Classification
AVQ	with	Error Revision	Back	Data self-organization
CPN	with	Hebb Law	Back	Self-organizing mapping
BM	with	Hebb/simul. Annealing	Back	Combinatory optimization
CM	with	Hebb/simul. Annealing	Back	Combinatory optimization
AHC	with	Error Revision	Back	Control
ARP	with	Stochastic Increase	Back	Pattern matching/control
SNMF	with	Hebb Law	Back	Voice/Image processing

7.3 Examples of ANN

In the following, the representative examples of artificial neural networks and associated learning algorithms are briefly described [73].

7.3.1 *Multilayer Perceptron*

Multilayer perceptrons (MLPs) are perhaps the best known type of feedforward networks. Figure 7.4 is an MLP with three layers: an input layer, an output layer and an intermediate or hidden layer. Neurons in the input layer only act as buffers for distributing the input signals x_i to neurons in the hidden layer. Each neuron *j* (see Figure 7.1) in the hidden layer sums up its input signals x_i after weighting them with the strengths of the respective connections w_{ij} from the input layer and computes its output y_j as a function *f* of the sum, viz.

$$y_j = f(\sum w_{ij} x_i) \tag{7.5}$$

f can be a simple threshold function or a sigmoidal, hyperbolic tangent or radial basis function as listed in Table 7.2.

Table 7.2. Activation functions

Type of Functions	Functions
Linear	$f(s) = s$
Threshold	$f(s) = \begin{cases} +, & \text{if } s > s_t \\ -, & \text{otherwise} \end{cases}$
Sigmoid	$f(s) = 1/(1+\exp(-s))$
Hyperbolic tangent	$f(s) = (1-\exp(-2s))/(1+\exp(2s))$
Radial basis function	$f(s) = \exp(-s^2/\beta^2)$

The output of neurons in the output layer is computed similarly.

The back propagation (BP) algorithm, a gradient descent algorithm, is the most commonly adopted MLP training algorithm. It gives the change Δw_{ji} in the weight of a connection between neurons *i* and *j* as follows:

$$\Delta w_{ji} = \eta \delta_j x_i \tag{7.6}$$

where η is a parameter called the learning rate and δ_j is a factor depending on whether the neuron *j* is an output neuron or a hidden neuron. For output neurons,

$$\delta_j = (\frac{\partial f}{\partial net_j})(y_j^{(t)} - y_j) \tag{7.7}$$

and for hidden neurons,

$$\delta_j = (\frac{\partial f}{\partial net_j})\sum_q w_{qj}\delta_q \tag{7.8}$$

In Equation (7.7) and Equation (7.8), net_j is the total weighted sum of input signals to the neuron j and $y_j^{(t)}$ is the target output for neuron j.

For all but the most trivial problems, several epochs are required for the MLP to be properly trained. A commonly adopted method to speed up the training is to add a "momentum" term to Equation (7.6) which effectively lets the previous weight change influence the new weight change, viz.:

$$\Delta w_{ji}(k+1) = \eta\delta_j x_i + \mu\Delta w_{ji}(k) \tag{7.9}$$

where $\Delta w_{ji}(k+1)$ and $\Delta w_{ji}(k)$ are weight changes in epochs (k+1) and (k) respectively and μ is the "momentum" coefficient.

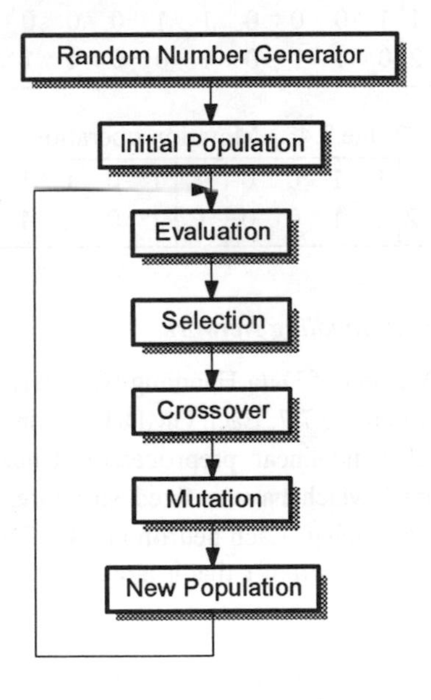

Figure 7.5. A simple genetic algorithm.

Another learning algorithm suitable for training MLPs is the GA (see Figure 7.5). This is an optimization algorithm based on evolution principles. The weights of the connections are considered genes in a chromosome. The goodness or fitness of the chromosome is directly related to how well trained the MLP is. The algorithm starts with a randomly generated population of chromosomes and applies genetic operators to create new and fitter populations. The most common genetic operators are the selection, crossover and mutation operators. The selection operator chooses chromosomes from the current population for reproduction. Usually, a biased selection procedure is adopted which favours the fitter chromosomes. The crossover operator (see Table 7.3) creates two new chromosomes from two existing chromosomes by cutting them at a random position and exchanging the parts following the cut. The mutation operator (see Table 7.4) produces a new chromosome by randomly changing the genes of an existing chromosome. Together, these operators simulate a guided random search method which can eventually yield the target outputs of the neural network.

Table 7.3. Crossover operation

Parent 1	1	0	0	0	1	0	0	1	1	1	1	0
Parent 2	0	1	1	0	1	1	0	0	0	1	1	0
New string 1	1	0	0	0	1	1	0	0	0	1	1	0
New string 2	0	1	1	0	1	0	0	1	1	1	1	0

Table 7.4. Mutation operation

Old string 1	1	1	0	0	0	1	0	1	1	1	0	1
New string 2	1	1	0	0	1	1	0	1	1	1	0	1

7.3.2 Group Method of Data Handling Network

Figure 7.6 shows a Group Method of Data Handling (GMDH) network, and the details of one of its neurons is shown in Figure 7.7. Each GMDH neuron is an N-Adaline which is an adaptive linear element with a nonlinear preprocessor. Unlike the feedforward neural networks previously described which have a fixed structure, a GMDH network has a structure which grows during training. Each neuron in a GMDH network usually has two inputs x_1 and x_2 and produces an output y that is a quadratic combination of these inputs, viz.

$$y = w_0 + w_1 x_1 + w_2 x_1^2 + w_3 x_1 x_2 + w_4 x_2^2 + w_5 x_2 \tag{7.10}$$

as seen in Figure 7.7.

Training a GMDH network consists of configuring the network starting with the input layer, adjusting the weights of each neuron, and increasing the number of layers until the accuracy of the mapping achieved with the network deteriorates.

The number of neurons in the first layer depends on the number of external inputs available. For each pair of external inputs, one neuron is used.

Training proceeds with presenting an input pattern to the input layer and adapting the weights of each neuron according to a suitable learning algorithm.

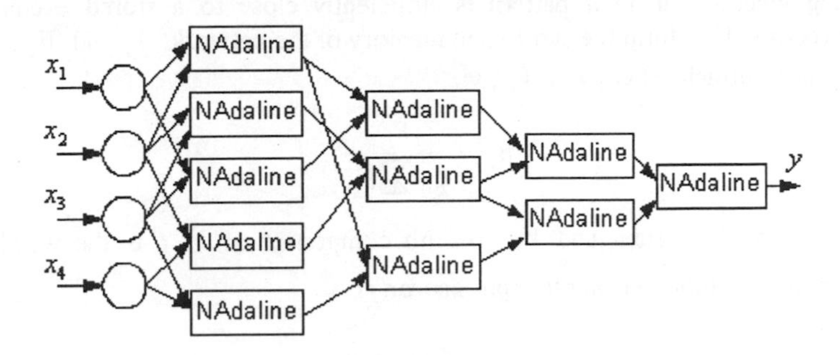

Figure 7.6. A trained GMDH network.

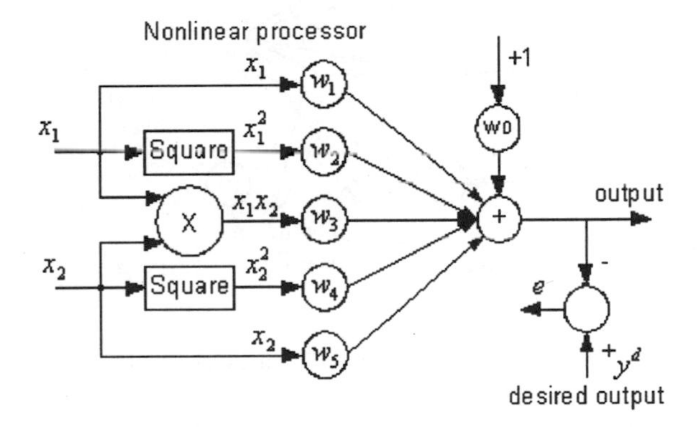

Figure 7.7. Details of a GMDH neuron.

7.3.3 *Adaptive Resonance Theory Network*

There are different versions of the Adaptive Resonance Theory (ART) network. Figure 7.8 shows the ART-1 version for dealing with binary inputs. Later versions, such as ART-2, can also handle continuous-valued inputs.

As illustrated in Figure 7.8, an ART-1 network has two layers, an input layer and an output layer. The two layers are fully interconnected, the connections are in both the forward (or bottom-up) direction and the feedback (or top-down) direction. The vector W_i of weights of the bottom-up connections to an output neuron i forms an exemplar of the class it represents. All the W_i vectors constitute the long-term memory of the network. They are employed to select the winning neuron, the latter again being the neuron whose

W_i vector is most similar to the current input pattern. The vector V_i of the weights of the top-down connections from an output neuron *i* is used for vigilance testing, that is, determining whether an input pattern is sufficiently close to a stored exemplar. The vigilance vectors V_i form the short-term memory of the network. V_i and W_i are related in that W_i is a normalized copy of V_i , viz.

$$W_i = \frac{V_i}{\varepsilon + \sum V_{ji}} \tag{7.11}$$

where ε is a small constant and V_{ji}, the *j*th component of V_i (i.e. the weight of the connection from output neuron *i* to input neuron *j*).

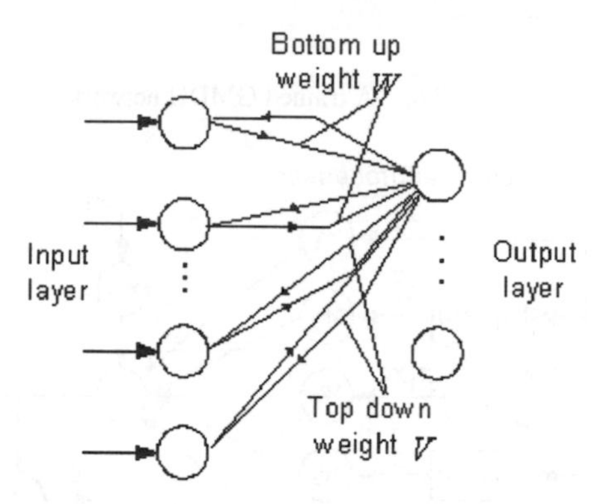

Figure 7.8. An ART-1 network.

Training an ART-I network occurs continuously when the network is in use and involves the following steps:

(1) initialize the exemplar and vigilance vectors W_i and V_i for all output neurons, setting all the components of each V_i to 1 and computing W_i according to Equation (7.11). An output neuron with all its vigilance weights set to 1 is known as an uncommitted neuron in the sense that it is not assigned to represent any pattern classes;

(2) present a new input pattern x ;

(3) enable all output neurons so that they can participate in the competition for activation;

(4) find the winning output neuron among the competing neurons, i.e. the neuron for which $x \cdot w_i$ is largest ; a winning neuron can be an uncommitted neuron as is the case at the beginning of training or if there are no better output neurons;

(5) test whether the input pattern x is sufficiently similar to the vigilance vector V_i of the winning neuron. Similarity is measured by the fraction r of bits in x that are also in W_i , viz.

$$r = \frac{x \cdot V_i}{\sum x_i} \tag{7.12}$$

is deemed to be sufficiently similar to V_i if r is at least equal to *vigilance threshold* $\rho(0< \rho <1)$;
(6) go to step (7) if $r \geq \rho$ (i.e. there is *resonance*); else disable the winning neuron temporarily from further competition and to step (4) repeating this procedure until there are no further enabled neurons;
(7) adjust the vigilance vector V_i of the most recent winning neuron by logically ANDing it with x, thus deleting bits in V_i that are not also in x; compute the bottom-up exemplar vector W_i using the new V_i according to Equation (7.11); activate the winning output neuron;
(8) go to step (2).

The above training procedure ensures that if the same sequence of training patterns is repeatedly presented to the network, its long-term and short-term memories are unchanged (i.e. the network is *stable*). Also, provided there are sufficient output neurons to represent all the different classes, new patterns can always be learnt, as a new pattern can be assigned to an uncommitted output neuron if it does not match previously stored exemplars well (i.e. the network is *plastic*).

7.3.4 Learning Vector Quantization Network

Figure 7.9 shows an Learning Vector Quantization (LVQ) network which comprises three layers of neurons: an input buffer layer, a hidden layer and an output layer. The network is fully connected between the input and hidden layers and partially connected between the hidden and output layers, with each output neuron linked to a different cluster of hidden neurons. The weights of the connections between the hidden and output neurons are fixed to 1. The weights of the input-hidden neuron connections form the components of *reference* vectors (one reference vector is assigned to each hidden neuron). They are modified during the training of the network. Both the hidden neurons (also known as Kohonen neurons) and the output neurons have binary outputs. When an input pattern is supplied to the network, the hidden neuron whose reference vector is closest to the input pattern is said to win the competition for being activated and thus allowed to produce a "1". All other hidden neurons are forced to produce a "0". The output neuron connected to the cluster of hidden neurons that contains the winning neuron also emits a "1" and all other output neurons a "0". The output neuron that produces a "1" gives the class of the input pattern, each output neuron being dedicated to a different class. The simplest LVQ training procedure is as follows:

(1) initialize the weights of the reference vectors;

(2) present a training input pattern to the network;

(3) calculate the (Euclidean) distance between the input pattern and each reference vector;

(4) update the weights of the reference vector that is closest to the input pattern, that is, the reference vector of the winning hidden neuron. If the latter belongs to the cluster connected to the output neuron in the class that the input pattern is known to belong to , the reference vector is brought closer to the input pattern. Otherwise, the reference vector is moved away from the input pattern;

(5) return to (2) with a new training input pattern and repeat the procedure until all training patterns are correctly classified (or a stopping criterion is met).

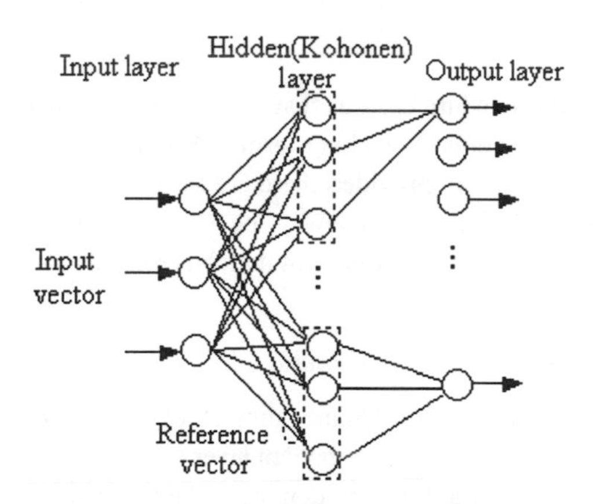

Figure 7.9. Learning Vector Quantization network.

7.3.5 *Kohonen Network*

A Kohonen network or a self-organizing feature map has two layers, an input buffer layer to receive the input pattern and an output layer as shown in Figure 7.10. Neurons in the output layer are usually arranged into a regular two-dimensional array. Each output neuron is connected to all input neurons. The weights of the connections form the components of the reference vector associated with the given output neuron.

Training a Kohonen network involves the following steps:

(1) initialize the reference vectors of all output neurons to small random values;

(2) present a training input pattern;

(3) determine the winning output neuron, i.e. the neuron whose reference vector is closest to the input pattern. The Euclidean distance between a reference vector and the input vector is usually adopted as the distance measure;

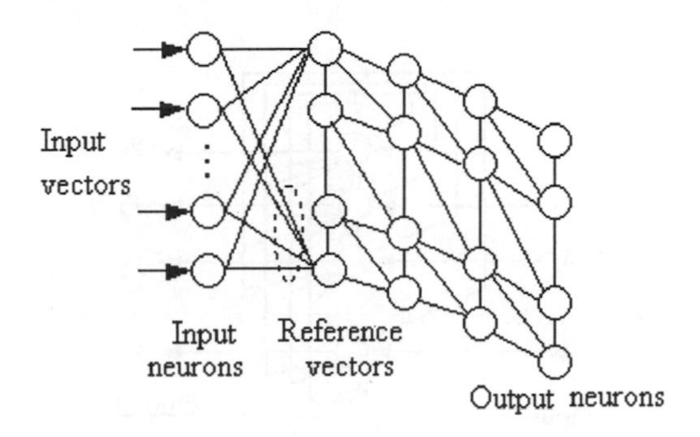

Figure 7.10. A Kohonen network.

(4) update the reference vector of the winning neuron and those of its neighbours. These reference vectors are brought closer to the input vector. The adjustment is greatest for the reference vector of the winning neuron and decreased for reference vectors of neurons further away. The size of the neighbourhood of a neuron is reduced as training proceeds until, towards the end of training, only the reference vector of a winning neuron is adjusted.

In a well-trained Kohonen network, output neurons that are close to one another have similar reference vectors. After training, a labeling procedure is adopted where input patterns of known classes are fed to the network and class labels are assigned to output neurons that are activated by those input patterns. As with the LVQ network, an output neuron is activated by an input pattern if it wins the competition against other output neurons, that is , if its reference vector is closest to the input pattern.

7.3.6 Hopfield Network

Hopfield network is a typical recurrent network (see Figure 7.3). Figure 7.11 shows one version of a Hopfield network. This network normally accepts binary and bipolar inputs (+1 or -1). It has a single "layer" of neurons, each connected to all the others, giving it a recurrent structure, as mentioned earlier. The training of a Hopfield network takes only one step, the weights w_{ij} of the network being assigned directly as follows:

$$w_{ij} = \begin{cases} \frac{1}{N}\sum_{c=1}^{p} x_i^c x_j^c, & i \neq j \\ 0 & , \ i = j \end{cases} \tag{7.13}$$

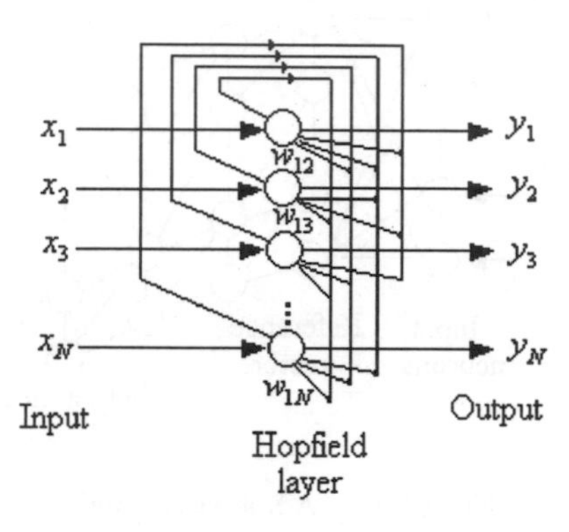

Figure 7.11. A Hopfield network.

where w_{ij} is the connection weight from neuron i to neuron j, and x_i^c (which is either +1 or -1) is the ith component of the training input pattern for class c, p the number of classes and N the number of neurons (or the number of components in the input pattern). Note from Equation 7.13 that $w_{ij} = w_{ji}$ and $w_{ii} = 0$, a set of conditions that guarantee the stability of the network. When an unknown pattern is input to the network, its outputs are initially set equal to the components of the unknown pattern, viz.

$$y_i(0) = x_i \qquad 1 \le i \le N \tag{7.14}$$

Starting with these initial values, the network iterates according to the following equation until it reaches a minimum energy state, i.e. its outputs stabilize to constant values.

$$y_i(k+1) = f\left[\sum_{j=1}^{N} w_{ij} y_i(k)\right] \qquad 1 < i \le N \tag{7.15}$$

where f is a hard limiting function defined as

$$f(x) = \begin{cases} -1, & \text{for } x < 0 \\ 1\,, & \text{for } x > 0 \end{cases} \tag{7.16}$$

7.3.7 *Elman and Jordan Nets*

Figures 7.12 and Figure 7.13 show an Elman net and a Jordan net, respectively. These networks have a multi-layered structure similar to the structure of MLPs. In both nets in addition to an ordinary hidden layer, there is another special hidden layer sometimes called the context or state layer. This layer receives feedback signals from the ordinary hidden layer (in the case of an Elman net) or from the output layer (in the case of a Jordan net). The Jordan net also has connections from each neuron in the context layer back to itself. With both nets, the outputs of neurons in the context layer, are fed forward to the hidden layer. If only the forward connections are to be adapted and the feedback connections are preset to constant values, these networks can be considered ordinary feedforward networks and the BP algorithm is used to train them. Otherwise, a GA could be employed [46, 72].

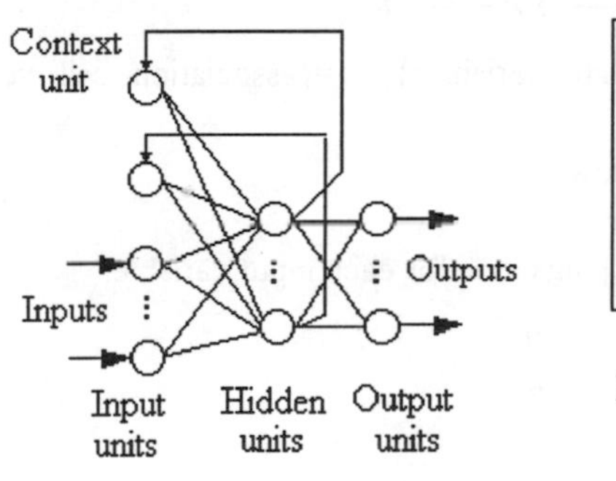

Figure 7.12. An Elman network.

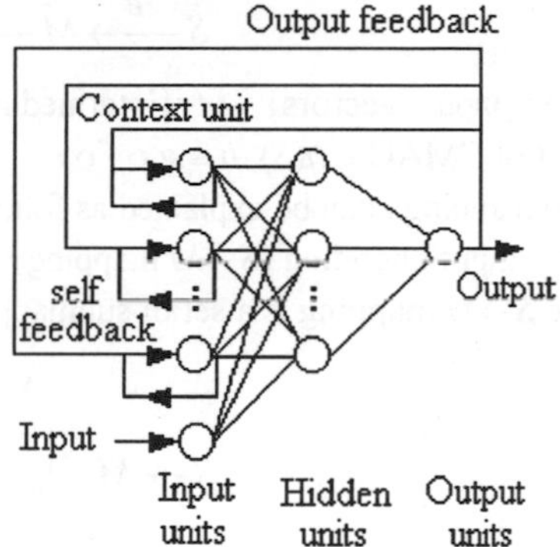

Figure 7.13. A Jordan network.

7.3.8 *Cerebellar Model Articulation Control Network*

Cerebellar Model Articulation Control (CMAC) [1-5] can be considered a supervised feedforward neural network with the characteristics of a fuzzy associative memory. A basic CMAC module is shown in Figure 7.14.

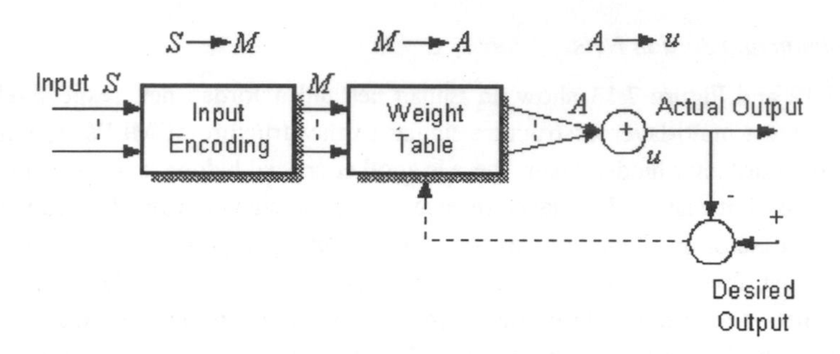

Figure 7.14. A basic CMAC module.

CMAC consists of a series of mappings:

$$S \xrightarrow{e} M \xrightarrow{f} A \xrightarrow{g} u \tag{7.16}$$

where S={input vectors}, M={intermediate variables}, A={association cell vectors}, u=output of CMAC $\equiv h(S)$, $h \equiv g \circ f \circ e$.

The mappings can be explained as follows.

(1) Input encoding ($S \to M$ mapping)

The $S \to M$ mapping is a set of submappings, one for each input variable:

$$S \to M = \begin{bmatrix} s_1 \to m_1 \\ s_2 \to m_2 \\ \vdots \\ s_n \to m_n \end{bmatrix} \tag{7.17}$$

The range of s_i is coarsely discretised using the quantising functions $q_1, q_2, \ldots, q_k$. Each function divides the range into k intervals. The intervals produced by function q_{j+1} are offset by one kth of the range compared to their counterparts produced by function q_j, m_i which has a set of k intervals generated by q_1 to q_k respectively.

An example is given in Figure 7.15 to illustrate the internal mappings within a CMAC module. The $S \to M$ mapping is shown in the left most part of the figure.

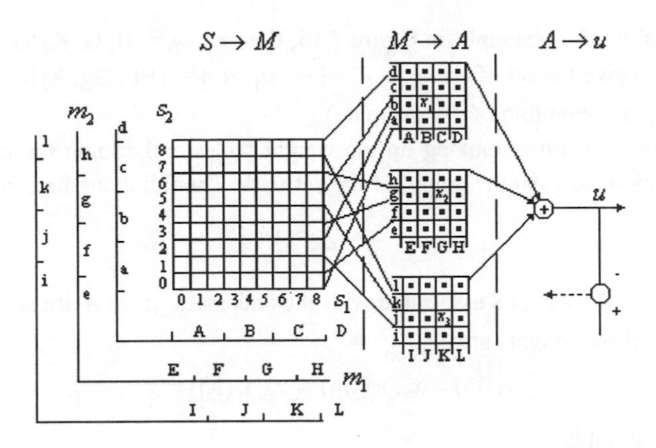

Figure 7.15. Internal mappings within a CMAC module.
(From Pham and Liu, 1995)

In Figure 7.15 two input variables s_1 and s_2 are represented with unity resolution in the range of 0 to 8. The range of each input variable is described using three quantizing functions. For example, the range of s_1 is described by functions q_1, q_2 and q_3. q_1 divides the range into intervals A, B, C and D. q_2 gives intervals E, F, G, and H and q_3 provides intervals I, J, K and L. That is,

$$q_1 = \{A, B, C, D\}$$

$$q_2 = \{E, F, G, H\}$$

$$q_3 = \{I, J, K, L\}.$$

For every value of s_1, there exists a set of elements, m_1, which are the intersection of the functions q_1 to q_3, such that the value of s_1 uniquely defines set m_1 and vice versa. For example, value $s_1 = 5$ maps to set $m_1 = \{B, G, K\}$ and vice versa. Similarly, value $s_2 = 4$ maps to set $m_2 = \{b, g, j\}$ and vice versa.

The *S*→*M* mapping gives CMAC two advantages: the first is that a single precise variable s_i can be transmitted over several imprecise information channels. Each channel carries only a small part of the information of s_i. This increases the reliability of the information transmission. The other advantage is that small changes in the value of s_i have no influence on most of the elements in m_i. This leads to the property of input generalization which is important in an environment where random noise exists.

(2) Address computing (*M*→*A* mapping)

A is a set of address vectors associated with weight tables. *A* is obtained by combining the elements of m_i. For example, in Figure 7.15, the sets m_1={B, G, K} and m_2={b, g, j} are combined to give the set of elements $A = \{a_1, a_2, a_3\}$ = {Bb, Gg, Kj}.

(3) Output computing ($A \rightarrow u$ mapping)

This mapping involves looking up the weight tables and adding the contents of the addressed locations to yield the output of the network. The following formula is employed:

$$u = \sum_i w_i(a_i) \tag{7.18}$$

That is, only the weights associated with the addresses a_i in A are summed. For this given example, these weights are:

$$w(\text{Bb}) = x_1,\ w(\text{Gg}) = x_2,\ w(\text{Kj}) = x_3$$

Thus the output is:

$$u = x_1 + x_2 + x_3 \tag{7.19}$$

Training a CMAC module consists of adjusting the stored weights. Assuming that f is the function that CMAC has to learn, the following training steps could be adopted:

(1) Select a point S in the input space and obtain the current output u corresponding to S;

(2) Let $\bar{u}$ be the desired output of CMAC, that is, $\bar{u} = f(S)$;

(3) If $|\bar{u} - u| \leq \xi$, where ξ is an acceptable error, then do nothing; the desired value is already stored in CMAC. However, if $|\bar{u} - u| > \xi$, then add to every weight which contributed to u the quantity

$$\Delta = \alpha \frac{\bar{u} - u}{|A|} \tag{7.20}$$

where $|A|$ = the number of weights which contributed to u and α is the learning rate.

7.4 Structural Schemes of Neurocontrol

Owing to different methods of classification, the structures of controller may naturally be different.

In this section, we briefly introduce some typical schemes of control structure that involve NN-based learning control, NN-based direct inverse control, NN-based adaptive control, NN-based internal model control, NN-based predictive control, NN-based optimal decision control, NN-based re-excitative control, CMAC based control, hierarchical NN-based control, and multilayer NN based control, etc.[33, 73].

7.4.1 NN-based Learning Control

Since the dynamic property of a controlled system is unknown or only partly known, it is necessary to search some laws to govern the system action and behavior, so that the system can be controlled effectively. In some situations it may be desirable to design an automatic controller which mimics the action of the human. The rule-base expert control and fuzzy control are two approaches to implementing this control, whereas NN is another one; we call this control NN-based learning control, supervised neural control, or NN-based supervised control. Figure 7.16 shows a structure of NN-based learning control, where a supervisor and a trainable NN controller (NNC) are involved. The input of the controller corresponds to the sensory input information received by the human, the outputs used for training correspond to the human control input to the system.

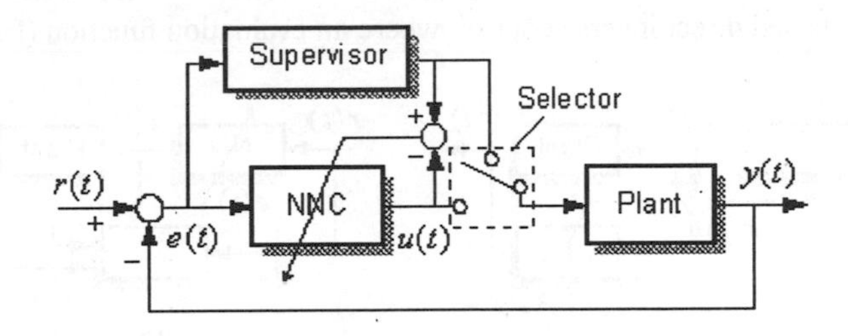

Figure 7.16. NN-based supervised control.

For implementing NN-based supervised control, the following steps are involved:

(1) Require necessary and useful control information through sensors and sensory information processing.

(2) Construct neural networks to select NN type, structural parameters and learning algorithm etc.

(3) Train NN controller to implement a mapping between inputs and output so as to make a correct control. In training process, the linear law, feedback linearization, or nonlinear feedback of decoupling transformation can be used as a supervisor to train the NN controller.

The NN-based supervised control has been used in the standard pole-cart control system [22].

7.4.2 NN-based Direct Inverse Control

As the term suggests, NN-based direct inverse control utilizes an inverse model of the controlled system, which is simply cascaded with the controlled system in order that the system results in an identity mapping between desired response (the network inputs) and the output of the controlled system. Thus, the network acts directly as a feedforward controller, and the output of the controlled system is equal to the desired output. This control scheme has been used in robotics, e.g., in Miller's CMAC network in which the direct inverse control was used to increase the degree of tracking accuracy of a PUMA robotic manipulator to 10^{-2} [66]. This approach relies heavily on the usability of the degree of accuracy of the inverse model which is used as the controller. Because of the absence of feedback, this approach lacks robustness. The parameters of the inverse model can be adjusted by using on-line learning so that the robustness of the controlled system can be increased to some extent.

Figure 7.17 gives two structural schemes of NN-based direct inverse control. In Figure 7.17(a), the networks NN1 and NN2 have the same network structure of the inverse model, and use the same learning algorithm. Figure 7.17(b) is another structural scheme of NN-based direct inverse control, where an evaluation function (EF) is used.

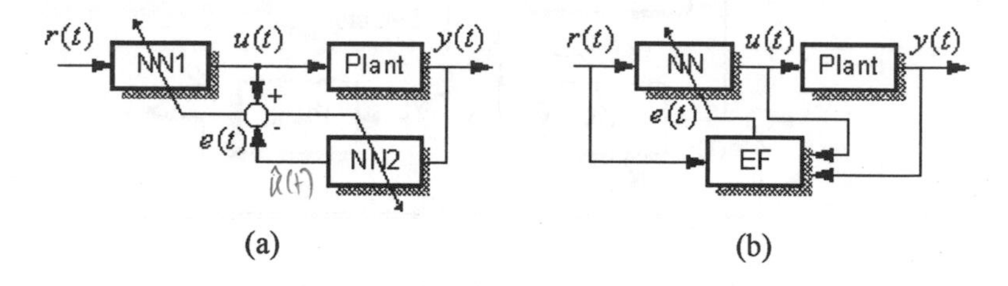

Figure 7.17. NN-based direct inverse control.

7.4.3 *NN-based Adaptive Control*

In the same way as the traditional adaptive control, the NN-based adaptive control is classified into two types: the self-tuning control (STC) and the model reference adaptive control (MRAC). The difference between the STC and MRAC is that: the STC regulates inner parameters of the controller directly according to the results of forward and/or inverse model identification of the controlled system so that a given performance index of the system can be satisfied; in the MRAC, the desired performance of the closed-loop control system is described through a stable reference model which is defined by its input-output pair $\{r(t), y^r(t)\}$. The objective of the control system is to make the output of plant y(t) match the output of reference model asymptotically, i.e.,

$$\lim_{t\to\infty}\|y^r(t)-y(t)\| \le \varepsilon \tag{7.21}$$

where ε is a specified constant and $\varepsilon \ge 0$.

1. NN-based self-tuning control

NN-based STC has two types, the direct one and the indirect one.

(i) NN-Based Direct STC

It consists of a traditional controller and a NN identifier with off-line identification in which a very high modeling accuracy is regained. The structure of STC basically is the same as direct inverse control.

(ii) NN-Based Indirect STC

It consists of a NN controller and a NN identifier that can be revised on-line. Figure 7.18 shows the structure of NN-based indirect STC.

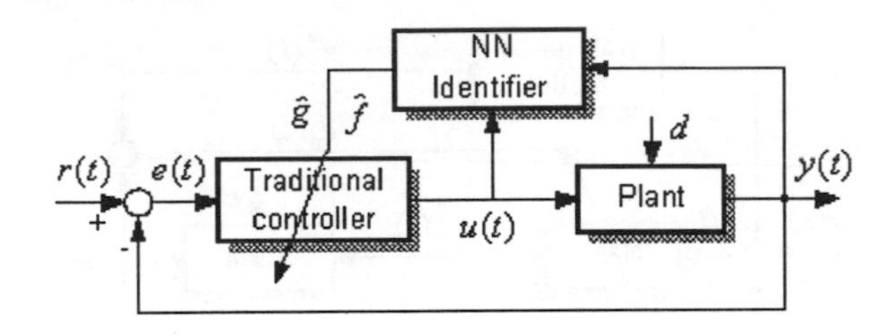

Figure 7.18. NN-based indirect STC.

In general, we assume that the controlled objective (plant) is a single variable nonlinear system as follows:

$$y_{k+1} = f(y_k) + g(y_k)u_k \tag{7.22}$$

where $f(y_k)$ and $g(y_k)$ are nonlinear functions. Let $\hat{f}(y_k)$ and $\hat{g}(y_k)$ stand for evaluation values of $f(y_k)$ and $g(y_k)$ respectively. If $f(y_k)$ and $g(y_k)$ are identified by neural network off-line, then $\hat{f}(y_k)$ and $\hat{g}(y_k)$ with enough approximated accuracy can be obtained, and the conventional control law can be given directly as:

$$u_k = \left[y_{d,k+1} - \hat{f}(y_k)\right] / \hat{g}(y_k) \tag{7.23}$$

where $y_{d,k+1}$ is the desired output at time (k+1).

Similarly, we can use the neural network to evaluate performance parameters of the output response such as rising time, overshoot, or the natural vibration frequency and damping coefficient of a second-order system and so on, the conventional method of pole placement can be used to regulate the parameters of the controller.

2. NN-Based model reference adaptive control

NN-based MRAC also has two types, the NN-based direct MRAC and the NN-based indirect MRAC.

(i) NN-Based Direct MRAC

From the structure of Figure 7.19 we learn that NN-based controller in the direct MRAC attempts to keep the difference between the output of the controlled object and the output of the reference model $e_c(t) = y(t) - y^m(t) \to \infty$. Because the back-propagation needs to know the mathematical model of the controlled object, learning and revision of NN controllers have encountered many problems.

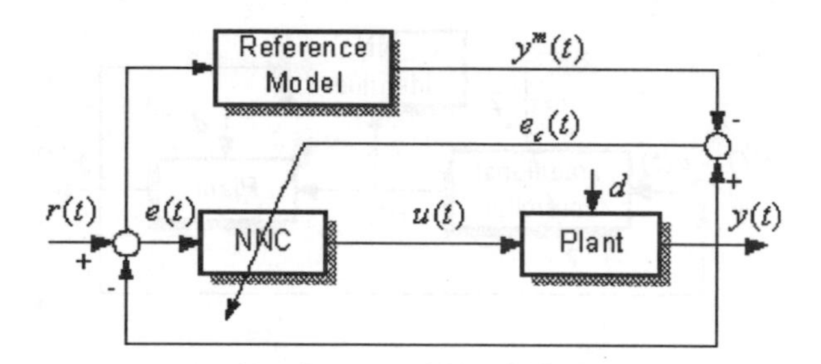

Figure 7.19. NN-based direct MRAC.

(ii) NN-Based Indirect MRAC

Its structure can be seen in Figure 7.20 where the NN identifier (NNI) identifies the feed-forward model of the controlled object first off-line, then on-line learning and revision are done by $e_i(t)$. Clearly, NNI can provide the error $e_c(t)$ or back-propagation path of its gradient.

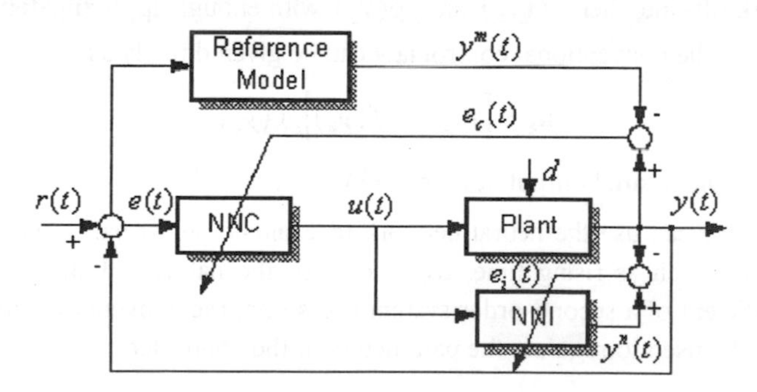

Figure 7.20. NN-based indirect MRAC.

7.4.4 *NN-Based Internal Model Control*

In conventional internal model control (IMC), the forward and inverse models of a controlled system are used as elements in the feedback loop. IMC has been thoroughly examined and has shown that it can be employed for transparent robustness and stability analysis [68], and it is a new and important approach to nonlinear system control [16].

The structure of NN-based IMC is illustrated in Figure 7.21 in which a system model is placed in parallel with the real system. The feedback signal is obtained from the deference between the system output and model output, and is then processed by the NN1, a NN-based controller with inverse model in the forward control path; the NN1 controller should be related to the inverse of the system.

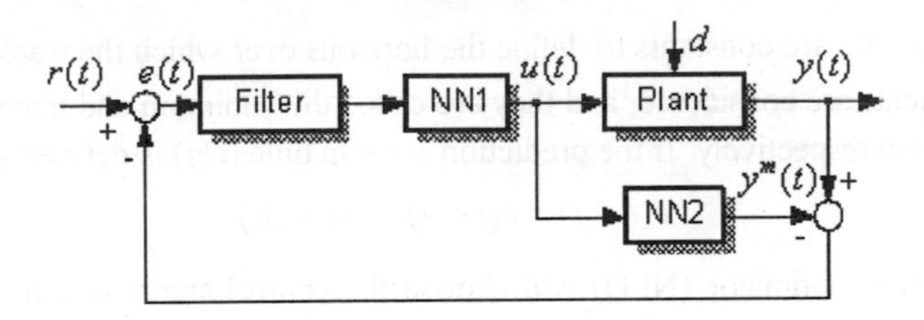

Figurc 7.21. NN-based internal model control.

In Figure 7.21, NN2 is also NN-based but with a forward model of the system. The filter is usually a linear one and can be designed to introduce desirable robustness and tracking response to the closed-loop system [32].

7.4.5 *NN-Based Predictive Control*

Predictive control, i.e., model-based control, is a new control algorithm developed in the late 1970's, and has the particulars of predictive model, rolling optimization and feedback correction. It has been proven that the approach has desirable stability properties for nonlinear systems [49, 61].

A structural scheme of the NN-based predictive control is shown in Figure 7.22 in which a neural network model, neural network predictor (NNP) provides prediction of the future plant response over the specified horizon:

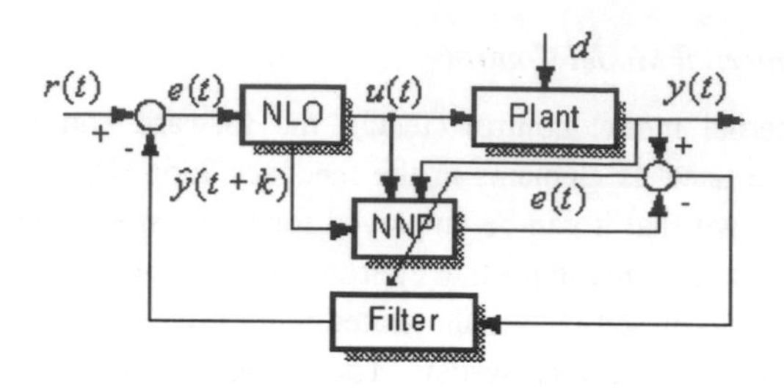

Figure 7.22. NN-based predictive control.

$$y(t+j|t),\ j = N_1, N_1+1, \ldots, N_2$$

where N_1 and N_2 are constants to define the horizons over which the tracking error and control increments are considered, and they are called the minimum and maximum level of output prediction respectively. If the prediction error in time $(t+j)$ is defined as:

$$e(t+j) = r(t+j) - y(t+j|t) \tag{7.24}$$

then the nonlinear optimizor (NLO) will choose the control signal $u(t)$ to minimize the quadratic performance criterion J

$$J = \sum_{j=N_1}^{N_2} e^2(t+j) + \sum_{j=1}^{N_2} \lambda_j \Delta^2 u(t+j-1) \tag{7.25}$$

where $\Delta u(t+j-1) = u(t+j-1) - u(t+j-2)$, and λ is the value of the control weights.

Algorithm 7.1.

Algorithm of NN-based predictive control:

Step 1 Compute the future expected output sequence

$r(t+j),\ j = N_1, N_1+1, \ldots, N_2$;

Step 2 Using NN prediction model to generate prediction output

$y(t+j|t),\ j = N_1, N_1+1, \ldots, N_2$;

Step 3 Compute the predictive error

$e(t+j) = r(t+j) - y(t+j|t),\ j = N_1, N_1+1, \ldots, N_2$;

Step 4 Minimize the performance criterion J, and get an optimal control sequence

$u(t+j), j=0, 1, 2, \ldots, N$;

Step 5 Use u(t) as the first control, then return to step 1.

It is noted that the NLO is practically an optimal algorithm, therefore, a dynamic feedback network can be used to replace the NLO implemented by this algorithm and the NNP consisting of feed-forward neural network.

7.4.6 NN-Based Adaptive Judgment Control

Whatever the NN control structure may be, all of the control approaches have an essential identity, i.e., the desired or expected input of the controlled object has to be provided. In case the system model is unknown or partly unknown, the desired input is hard to be provided.

NN-based adaptive judgment control or reinforcement control proposed by Barto et al. and developed by Anderson uses the mechanism of reinforcement learning [6, 8].

The NN-based adaptive reinforcement control usually consists of two networks, the adaptive judgment network (AJN) and control selection network (CSN), as shown in Figure 7.23.

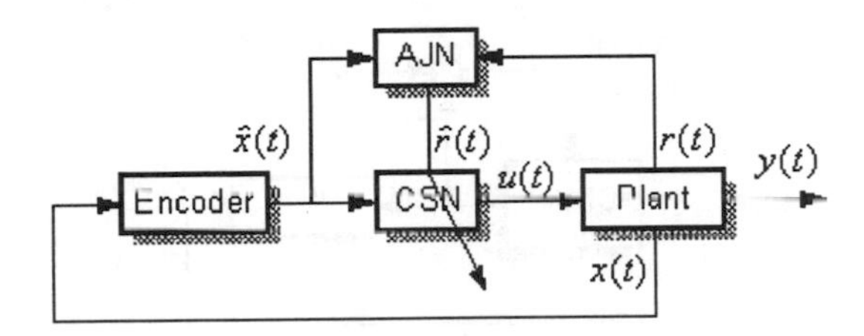

Figure 7.23. NN-based adaptive reinforcement control.

In the control system, the AJN corresponds to a "teacher" needing reinforcement learning, and plays two roles: (1) through continuous reinforcement learning with rewards and punishments, the AJN will be gradually trained to be a "skilled" teacher; (2) after learning, the AJN generates a reinforcement signal according to current situation of the controlled system and feedback signal $r(t)$ of the outside reinforcement, then provides the internal reinforcement signal $\hat{r}(t)$, so that the effect of the current control action can be judged. The CSN corresponds to a multilayered feed-forward neural network controller learning under the guidance of internal reinforcement signal. Through learning the CSN selects the next control action according to the system state after encoding.

7.4.7 CMAC Based Control

CMAC was developed by Albus. It is one of the main NN controllers found applications in recent years [5, 55, 65, 66, 84, 86].

There are two schemes to use CMAC for control. The first scheme has the structure as depicted in Figure 7.24 [1, 2]. In the control system, it shows that both the command and feedback signals are used as inputs to the CMAC controller. The output of the controller is fed directly to the plant. The desired output of the neural network controller has to be supplied. The training of the controller is based on the error between the desired and actual controller output. Two stages are needed to make the system work. The first stage is training the controller. When CMAC receives the command and feedback signals, it produces an output. This output is compared with the desired output. If there are differences between them, then the weights are adjusted to eliminate the differences. On completion of this stage, CMAC has learnt how to produce a suitable output to control the plant according to the given command and the measured feedback signals. The second stage is control. CMAC can work well when the required control is close to that with which it has been trained. Both stages are completed without the need to analyze the dynamics of the plant and to solve complex equations. However, in the training stage, this scheme requires the desired plant input to be known.

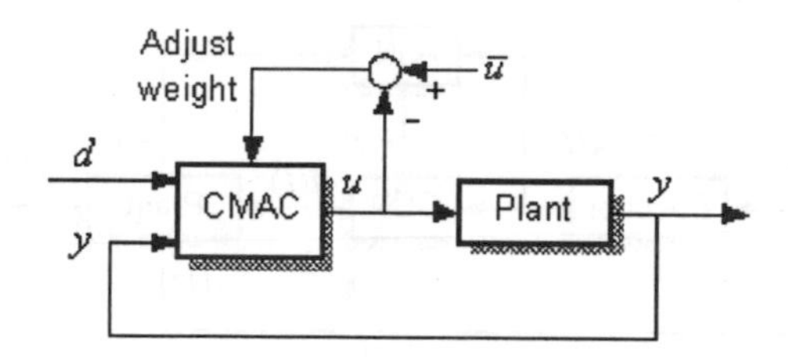

Figure 7.24. CMAC based Control(I).

The second control scheme is illustrated in Figure 7.25 [65, 66]. In this scheme, the reference output block produces a desired output at each control cycle. The desired output is sent to the CMAC module which provides a signal to supplement the control signal from a fixed gain conventional error feedback controller. At the end of each control cycle, a training step is executed. The observed plant output during the previous control cycle is used as input to the CMAC module. The difference between the computed plant input u^* and the actual input u is used to compute the weight adjustment. As CMAC is trained continually following successive control cycles, the CMAC function forms an approximation of the plant inverse transfer function over particular regions of the input space. If the future desired outputs are in regions similar to previous observed outputs, the CMAC output will be similar to the actual plant input required. As a result, the output errors will be small and CMAC will take over from the fixed-gain conventional controller.

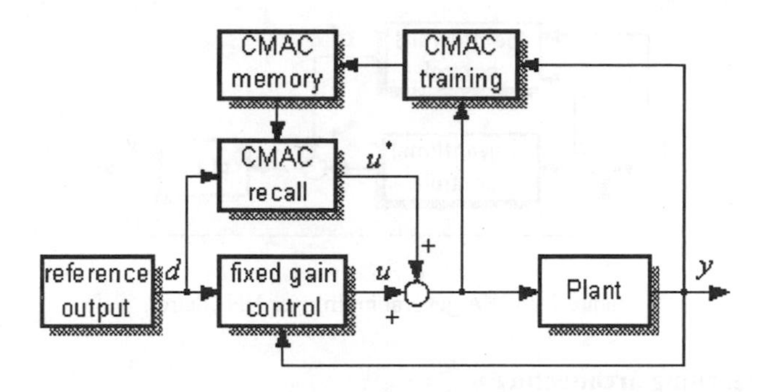

Figure 7.25. CMAC based control (II).

From the above description, scheme 1 is a closed-loop control system, because besides the command variables, the feedback variables are used as inputs to the CAMC module to be encoded so that any variations in the plant output can cause variations in the input it receives. In scheme1, the adjustment of weights is based on the error between the desired controller output and the actual controller output, rather than the error between desired plant output and the actual plant output. As already mentioned, this requires the designer to assign the desired controller output and will cause problems because usually only the desired plant output is known to the designer. The training in scheme 1 can be considered to be the identification of a proper feedback controller. In scheme 2, the CMAC module is used for learning an inverse transfer function with the assistance of a conventional fixed-gain feedback controller. After training, CMAC will be the principal controller. In this scheme, control and learning proceed at the same time. The disadvantage of this scheme is that it requires a fixed-gain controller to be designed for the plant.

7.4.8 Multilayered NN Control

The multilayered neural network controller is essentially a feedforward controller [76].

Consider a general control system shown in Figure 7.26. In this system, there are two kinds of control action: feedforward and conventional feedback control. The feedforward control part is implemented by a neural network. The aim of training the feedforward part is to minimize the error between the desired and actual plant output. This error is the input to the feedback controller, the feedback and feedforward actions are considered separately, concentrating on the training of the feedforward controller by ignoring the existence of the feedback control. Three learning architectures were proposed: indirect, general, and specialized.

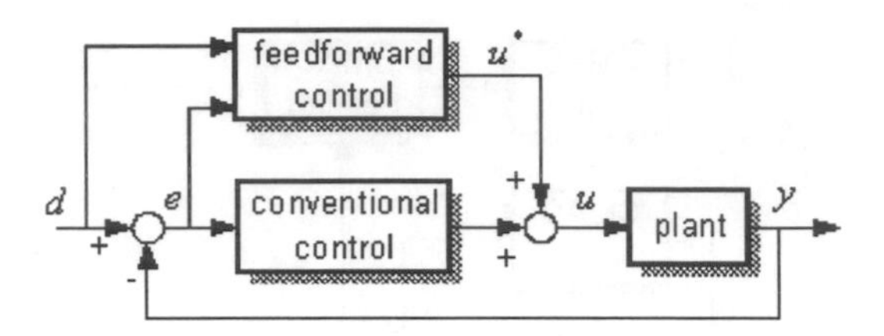

Figure 7.26. A general multilayer NN control.

1. Indirect learning architecture

The indirect learning architecture shown in Figure 7.27 has two identical neural networks for training. In this architecture, each network acts as an inverse dynamics identifier. The goal of training is to find an appropriate plant control u from the desired response d. The weights are adjusted based on the difference between the outputs of networks I and II to minimize error e, because if network I can be trained so that $y = d$, then $u = u^*$. However, this cannot guarantee that the error between the desired output d and the actual output y is minimized [75, 76].

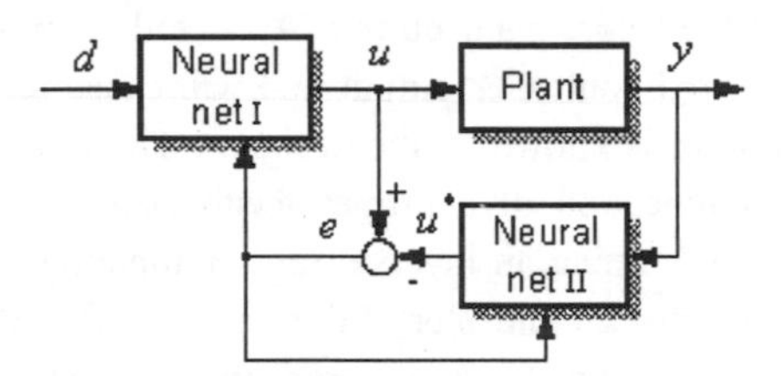

Figure 7.27. Indirect learning architecture.

2. General learning architecture

The general learning architecture is depicted in figure 7.28 and does minimize $e = d\text{-}y$ in Figure 7.27. The network is trained to minimize the error between the plant input u and the network output u^*. In training, u should be in such a range that y covers the desired output d. After training, the network is able to provide an appropriate u to the plant if a desired output d is sent to it. The limitation of this architecture is that generally which u corresponds to the desired output d is not known so that the network has to be trained over a large range of u to enable the plant output y to include the desired value d during learning.

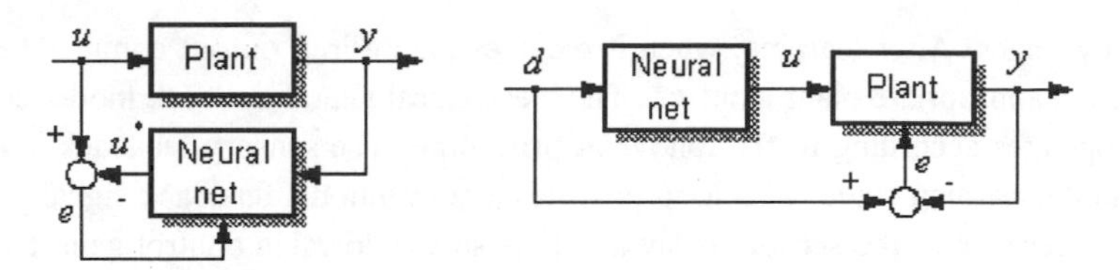

Figure 7.28. General learning architecture.

Figure 7.29. Specialized learning architecture.

3. Specialized learning architecture

The specialized learning architecture is shown in Figure 7.29. The desired output *d* is the input to the network when it is being trained. By applying the error back propagation method, the difference *e* between the desired output *d* and the actual output *y* of the plant is minimized through training. Thus, not only can a good plant output be expected but also the training can be executed in the region of the desired output, without having to know the proper range of plant input. However, in this architecture the plant is treated as a layer of the network. To be able to train the network, either the plant dynamics model has to be known or some approximation has to be made. Learning in the multilayered neural network controller is accomplished by the error back propagation training algorithm [66]. The error can be the difference between the desired and actual plant output, or the difference between the correct plant input and the input calculated by the neural networks.

7.4.9 Hierarchical NN control

The hierarchical NN control model is depicted in Figure 7.30 [47, 48]. In the figure, *d* is the desired plant output, *u* is the control input of the plant, *y* is the actual plant output, u^* and y^* are the computed plant input and output as given by the neural networks. The system can be considered to consist of three parts. The first part is a conventional feedback loop known as external feedback. The feedback control is based on the error *e* between the desired plant output *d* and the actual plant output *y* measured by sensors, i.e. $e = (d\text{-}y)$. Usually the external conventional feedback controller is a proportional-derivative controller. The second part is the path connected with neural network I which is an internal model of the plant dynamics. This neural network monitors the plant input *u* and output *y* and learns the plant dynamics. After learning, it can provide an approximate plant output y^* when it receives the plant input *u*. In this sense, this part acts as a system dynamics identifier. Based on the error $d\text{-}y^*$, this part provides an internal feedback loop which is much faster than the external feedback loop as the latter usually has sensory delays in the feedback path. The third part of the system is the neural network II which monitors the desired output *d* and the plant input *u*. This neural network learns to model the plant inverse dynamics. After learning, when it receives the desired output command d, it can produce the appropriate plant input u^*. The hierarchical neural network model controlled system operates according to the following procedure. The sensory feedback is effective mainly in the learning stage. This loop provides a conventional feedback signal to control the plant. Because of the sensory delay and thus small allowable control gain, the system response is slow, which limits the speed of the learning stage. During the learning stage, neural network I learns the system dynamics, while neural network II learns the inverse dynamics. As learning proceeds, the internal feedback gradually takes over the role of the external feedback as the main controller. Then, as learning proceeds further, the inverse dynamics part will replace the internal feedback control. The final result is that the plant is controlled mainly by a feedforward controller since the plant output error is nearly absent with the internal feedback providing fast control to deal with random disturbances. In the above procedure, control and learning are executed simultaneously. The neural networks function as identifiers: one for the identification of plant dynamics, and the other for the identification of inverse dynamics.

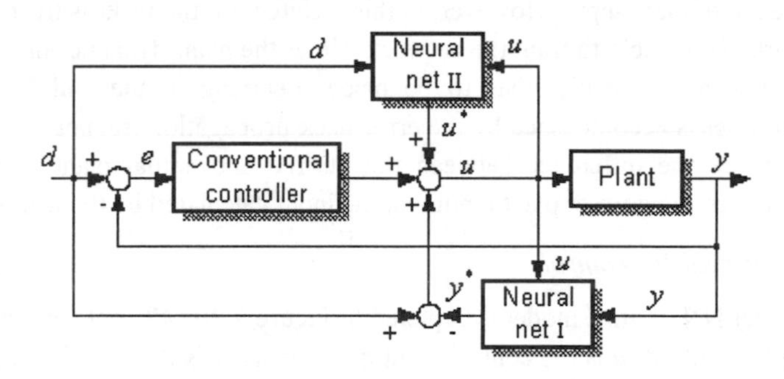

Figure 7.30. Hierarchical NN controller.

Importantly, a hierarchical neural network model based control system can be separated into two subsystems, the (forward) dynamics identifier based system (see Figure 7.31) and the inverse dynamics identifier based system (see Figure 7.32), which can be applied individually.

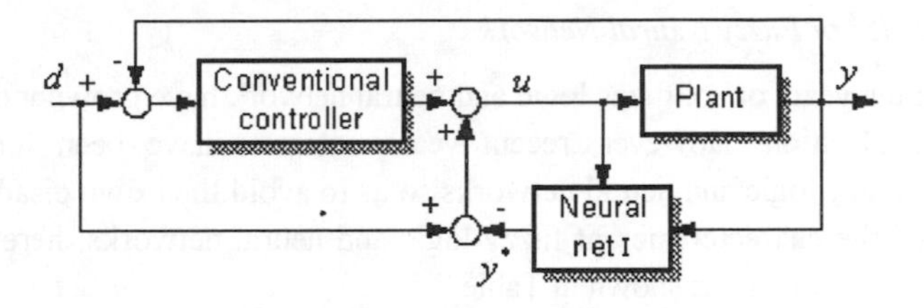

Figure 7.31. Forward dynamics identifier based control system.

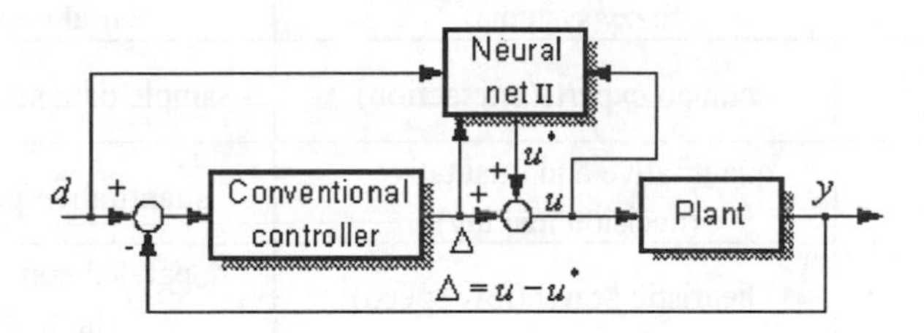

Figure 7.32. Inverse dynamics identifier based control system.

In summary, a hierarchical neural network model based system has the following characteristics:

(1) The system has two identifiers, one for the identification of plant dynamics, the other for inverse plant dynamics;

(2) There is a main feedback loop, which is important in the training of the neural networks;

(3) As training proceeds, the inverse dynamics part becomes the main controller;

(4) The final effects of hierarchical neural network model based control are similar to feedforward control.

7.5 Integration of Fuzzy Logic, Expert System and NN for Control

Similar to fuzzy control, the neural control can be combined with other control principles to form new integration of control such as PDP- neuro hybrid control, NN-based learning control, NN-based adaptive control, NN-based self-tuning control, NN-based self-organizing control, NN-based optimal control, and NN-based expert control, etc. Some of them have been discussed in the former section, Section 7.4.

We would like to introduce the integration of neurocontrol with fuzzy logic and expert systems in the following subsection.

7.5.1 Principles of Fuzzy Neural Network

Over the past ten years or so, fuzzy logic and neural network have grown independently in theory and application. However, recent years, interests have been focused on the integration of fuzzy logic and neural networks so as to avoid their own disadvantages. We have discussed the characteristics of fuzzy logic and neural networks, here we present a comparison between them as shown in Table 7.5.

Table 7.5. Comparison of fuzzy systems and neural networks

skills	fuzzy systems	neural networks
knowledge acquisition	human experts(interaction)	sample data sets(algorithms)
uncertainty	quantitative and qualitative (decision making)	quantitative(perception)
reasoning	heuristic search(low speed)	parallel computations (high speed)
adaptation	low	very high (adjusting link weights)

To enable a system to deal with cognitive uncertainties in a manner more like humans, one may integrate fuzzy logic and neural network to form a new research area — fuzzy neural network(FNN). Basically, the two approaches performing this combination can be distinguished. The first aims at seeking a functional mapping between fuzzy reasoning algorithms and neural network paradigms [70], while the second approach is intended to find a structural mapping from a fuzzy reasoning system to a kind of neural network [84, 86, 95]. In the following, we discuss the concept, algorithm and application schemes of fuzzy neural networks in detail.

1. Concepts and architectures

Buckely proposed the definition of fuzzy neural net in his paper [11, 12]. For simplicity, consider the three layered feedforward neural networks shown in Fig. 7.33.

The different types of FNNs is defined as follows:

Definition 7.1. A regular fuzzy neural net (RFNN) is a neural network with fuzzy signals and/or fuzzy weights, i.e., (1) FFN1 has real number input signals but fuzzy weights; (2) FNN2 has fuzzy set input signals and real number weights; (3) FNN3 has both fuzzy set input signals and fuzzy weights.

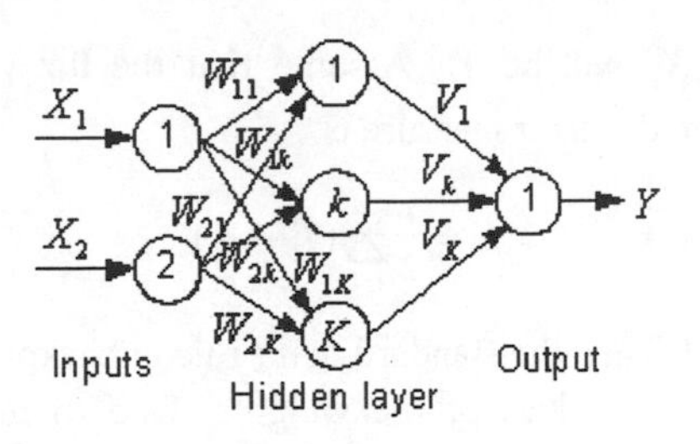

Figure 7.33. Neural network FNN3.

Definition 7.2. A hybrid fuzzy neural net (HFNN) is another kind of FNN which combine the fuzzy signals and weights using other operations besides addition and multiplication to obtain neural net input.

In the following, we describe the internal computations of FNN3 in more detail. Suppose FNN3 is the architecture which is the same as in Fig. 7.33. The input to the input neuron 1(neuron 2) is fuzzy signals $X_1(X_2)$. So the input to the hidden neuron k is

$$I_k = X_1 W_{1k} + X_2 W_{2k}\ , \quad k = 1,2,\ldots,K \tag{7.26}$$

The output from the kth hidden neuron will be

$$Z_k = f(I_k)\ , \quad k = 1,2,\ldots,K \tag{7.27}$$

If f is a sigmoidal function, it follows that the input to the output neuron is

$$I_0 = Z_1 V_1 + Z_2 V_2 + \ldots + Z_k V_k \tag{7.28}$$

and the final output will be

$$Y = f(I_0) \tag{7.29}$$

where regular fuzzy arithmetic are used.

2. Learning algorithms

For a regular neural network, learning algorithms can be classified into two main groups [33]: supervised learning, which needs external teacher signals (data sample pairs), and unsupervised learning, which relies only upon internal signals of neural network. These algorithms may be directly generalized into FNN. The main learning algorithms for FNNI, I = 1,2,3, in recent research works are summarized as follows:

(i) Fuzzy backpropagation algorithm

Fuzzy backprogagation algorithm which is based on FNN3 was developed by Barkley. Let the training set be (X_l, T_l), $X_l = (X_{l1}, X_{l2})$ for inputs and T_l desired output, $1 \le l \le L$.

The actual output for X_l will be Y_l. Assume that the fuzzy signals and weights are triangular fuzzy sets, then the error measure is:

$$E = \frac{1}{2}\sum_{l=1}^{L}(T_l - Y_l)^2 \tag{7.30}$$

which is to be minimized. Then the standard delta rule in backpropagation can be fuzzfied and adopted to update the values of the weights. Due to fuzzy arithmetic, a special stopping rule for iterations was also proposed [12]. However, convergence of this algorithm is still an open topic.

(ii) α-cut based backpropagation algorithms

Some efforts have been made to improve the properties of fuzzy backpropagation algorithms. In Ref. 9, a backpropagation algorithm for the individual α-cuts of the weights in a FNN3 was discussed. Usually, the α-cuts of a fuzzy set A is defined as:

$$A[\alpha] = \{x | \mu_A(x) \geq \alpha\} \text{ for } 0 < \alpha \leq 1. \tag{7.31}$$

Also, there are other α-cut based backpropagation proposed in Ref. 11, *etc*. However, the most eminent drawback of these methods is that these algorithms have limitations in the types of input fuzzy signals and fuzzy weights. Usually, these fuzzy membership functions have triangular shapes.

(iii) Genetic algorithms

In order to improve the performance of fuzzy control systems, the applications of genetic algorithms in fuzzy systems have been widely developed [71]. The genetic algorithm enables us to generate an optimal set of parameters for the fuzzy reasoning model based either on their initial subjective selection or on a random selection. A genetic algorithm into the training of fuzzy neural networks was introduced [12]. The type of genetic algorithm used will depend on the kinds of fuzzy sets used for input and weights and the error measure that needs to be minimized.

(iv) Other learning algorithms

Fuzzy chaos and algorithms based on other fuzzy neurons would be an interesting topic for future research.

3. Approximation ability of fuzzy neural networks

It has been proved that a regular feedforward multilayered neural network has the property to approximate arbitrarily nonlinear function with well accuracy which is an attractive ability for its application in nonlinear, ill-defined control system. Next appears the result that fuzzy systems can also be universal approximators. Now some interests have been focused on the fuzzy neural network's approximator ability. Conclusions have been drawn up that regular fuzzy neural nets, based on standard fuzzy arithmetic and the extension principle, cannot be universal approximators, whilst the hybrid fuzzy neural nets, not necessarily based on standard fuzzy arithmetic, can be universal approximators. These conclusions may be useful for building FNN controllers.

7.5.2 Fuzzy Neural Control Schemes

Many schemes of integrating fuzzy logic description and neural network implementation have been proposed which are mostly inspired by the investigation of adaptive fuzzy control systems. These schemes can be mainly divided into two aspects: structural equivalent to fuzzy system and functional equivalent to fuzzy system.

The followings are some typical schemes for integration of fuzzy logic and neural network that are used for control, modeling and identification.

1. Mixed FNN

In this scheme, the neural network and the fuzzy set are used independently in a system, or either one serves as a preprocessor for the other. For instance, the fuzzy set is used as a supervisor or a critic for the neural network in order to improve the convergence of learning as shown in Figure 7.34 [18].

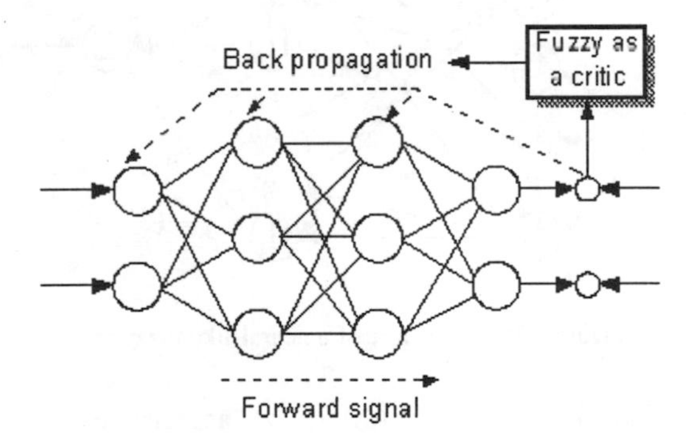

Figure 7.34. Mixed FNN.

2. Neural-like fuzzy set (FAM)

In this case, the FNN uses the fuzzy neurons described by fuzzy sets, instead of the nonfuzzy neuron. In knowledge-based systems, a set of conditional statements, IF-THEN rules, are often used to represent human knowledge extracted form human experts. This knowledge is usually associated with uncertain and fuzzy terms, such as big, small, high, low, many, often and sometimes, and so forth. Thus, the antecedents and consequence in the "IF-THEN" rules are treated as fuzzy sets.

Figure 7.35 presents a structure of FNN in which FNN clusters input parameters into fuzzy subspaces and identifies input-output relationships by using a weighted network [34], that is based on the simplified fuzzy inference [10]. In Figure 7.35, A_{ip} is the membership function for the ith input variable x_i , (i=1,...,n), in the pth rule; w_p is the consequence of the pth rule. The results of individual rules are as follows:

$$\mu_p = \prod A_{ip}(x_i) = A_{1p}(x_1) \times A_{2p}(x_2) \times \ldots \times A_{np}(x_n) \tag{7.32}$$

Therefore, the output of the FNN is:

$$y = \sum \mu_p \cdot w_p \,. \tag{7.33}$$

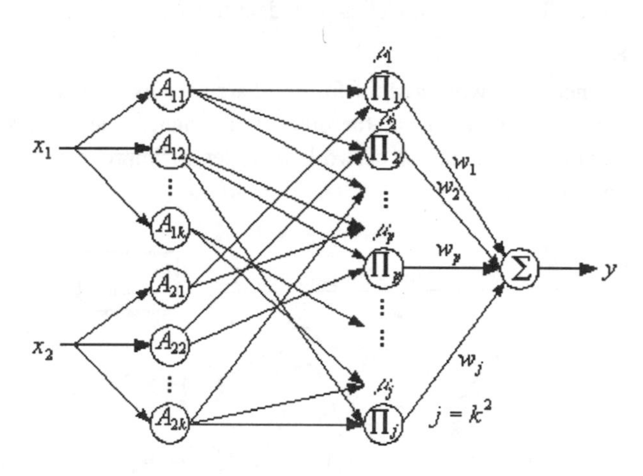

Figure 7.35. Structure of a neural-like fuzzy set.

Another neural-like fuzzy system FAM, fuzzy associative memory systems were proposed by Kosko [54]. FAMs are fuzzy systems $S: I^n \rightarrow I^p$ constructed in a simple neural-like manner, and a two-layered feedforward heteroassociative and fuzzy classifier is used to store an arbitrary fuzzy spatial pattern pair through fuzzy Hebbian learning. In a self-organizing fuzzy controller, two dynamical FAM rule bases are used: control rule base and adjusting rule base. This approach has an improved control performance but the drawbacks are the use of fuzzy arithmetic and the absence of any learning algorithm. Future research should be focused on the more effective and appropriate learning algorithm for FAM.

3. Adaptive FNN

In the development of FNN, an adaptive fuzzy logic controller implemented with neural networks was proposed [84-86]. The neural networks can be considered as a structural mapping from a fuzzy system to neural networks. The procedure of decision-making of a fuzzy logic controller leads to a neuro-fuzzy network consisting of three types of subnets for pattern recognition, fuzzy reasoning, and control synthesis respectively, as shown in Figure 7.36. The unique knowledge structure embedded in this structured network enables it to carry out adaptive changes of fuzzy reasoning methods and membership functions for both input signal patterns and output control actions. Wang also presented off-line training rules and on-line learning algorithms based on gradient methods for optimization of networks [86].

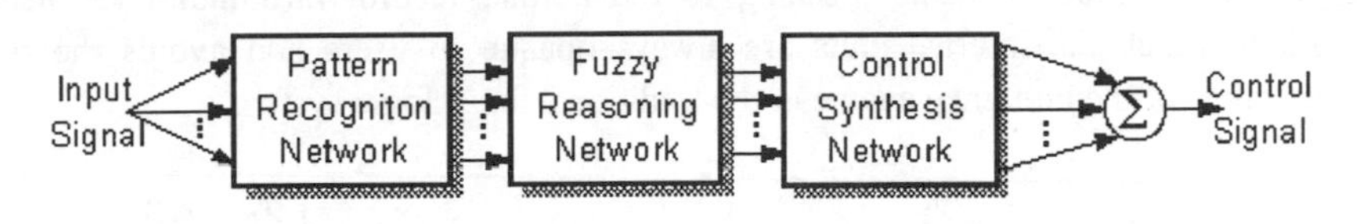

Figure 7.36. Three subsets of FNN.

Another adaptive FNN is adaptive-network-based fuzzy inference system ANFIS that can construct an input - output mapping based on both human knowledge (in the form of fuzzy IF-THEN rules) and stipulated input-output data pairs [41]. In the simulation, the ANFIS architecture is employed to on-line model nonlinear functions and identify nonlinear components in a control system, and predict a chaotic time series.

4. Multilayer FNN

A special multilayer perceptron architecture known as FuNe was used to generate fuzzy systems for a number of real world applications [25]. The FuNe trained with supervised learning can be employed to extract fuzzy rules from a given representative input/output data set. Moreover, the optimization of the knowledge base is possible through the tuning of membership functions. The FuNe is also a one-to-one structural mapping between the fuzzy logic controllers and the neural networks that is composed by three subnets, i.e. fuzzification net, rule generation net, and defuzzification net in which every net may also be a multilayered neural network. Figure 7.37 gives a five-layer fuzzy neural network for control and decision-making system in which Layer 1 consists of input linguistic nodes, Layer 2 — input term nodes, Layer 3 — rule nodes, Layer 4 — output term nodes, and Layer 5 — output linguistic nodes [60]. As shown in Figure 7.37, two linguistic nodes for each output variable are included. One in the left is for training data (desired output) to be fed into this net, and the other in the right is for decision signal (actual output) to be pumped out of this net. Layers 2 and 4 act as membership functions to represent the terms of the respective linguistic variables. Besides, a node in Layer 2 can be either a single node to perform a simple membership function or composed of multilayer nodes (a subneural net) to perform a complex membership function. Therefore, the total number of layers in the model can be more than five. The proposed connectionist model can be contrasted with the traditional fuzzy logic control and decision system in their network structure and learning ability. Such fuzzy networks can be constructed from training examples by machine learning techniques, and the connetionist structure can be trained to develop fuzzy logic rules and find optimal input/output membership functions. The learning speed can be converged much faster than the original backpropagation learning algorithm by combining both unsupervised (self-organized) and supervised learning schemes. This model also provides human-understandable meaning to the normal feedforward multilayer neural network in which the internal units are always opaque to users and avoids the rule-matching time of the inference engine in the traditional fuzzy logic system.

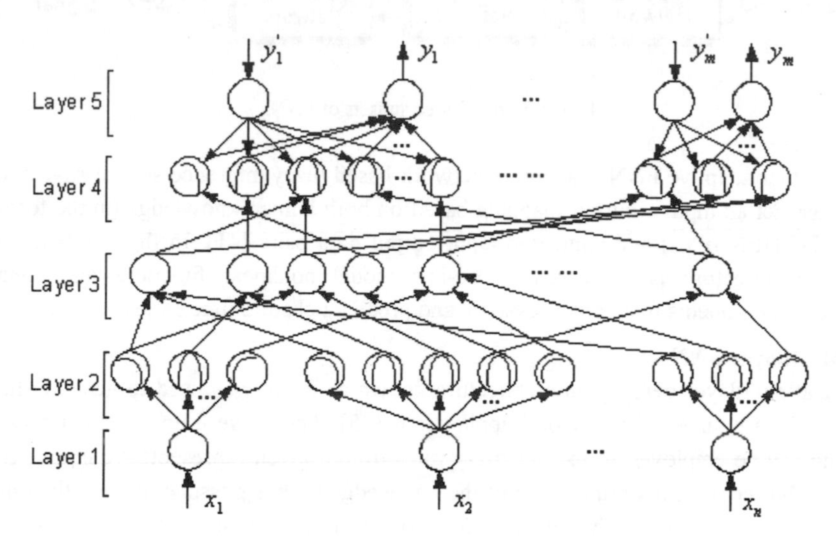

Figure 7.37. Example of multilayer FNN.

7.6 Paradigms of NN-Based Control Systems

Neural networks have been widely used in industry, business and science in recent years, especially in the areas of pattern recognition, image processing and signal identification [80, 88]. Although successful applications in control area are relatively limited until now, reports on representative examples of NN-based control are growing. In this section, two samples for NN-based control will be presented; one is the double-neuron synchro control system for hydraulic turbine generator [15], another is the direct fuzzy neurocontrol system for train traveling process [83].

7.6.1 Double-Neuron Synchro Control System for Hydraulic Turbine Generator

1. Synchro control of hydraulic turbine generator

The ideal conditions for connecting a hydraulic turbine generator with an electric power net in parallel are as follows:

$$\Delta u=0, \quad \Delta f=0, \quad \Delta\varphi=0 \tag{7.34}$$

where Δu, Δf and $\Delta\varphi$ are the voltage difference, frequency difference and phase difference between the generator and the net respectively. Moreover, the period of time for connecting should be as short as possible. Under these conditions the synchro control system can realize a rapid and shock-free connection. In order to satisfy these control conditions, a composite control scheme with frequence tracking and phase tracking is applied. The structural diagram of the composite control system is shown in Figure 7.38, where f_N and φ_N are the frequency and phase of the power net, f_G and φ_G are the frequency and phase of the generator set.

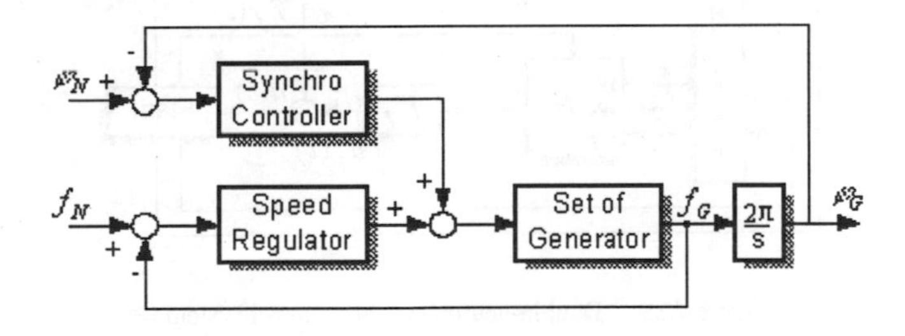

Figure 7.38. Structure of composite control system.

The load dynamic characteristics of the generator is represented as follows:

$$T_a \frac{dx}{dt} + e_g x = m_t - m_{go} \tag{7.35}$$

where x presents load current of the generator; T_a is the sum of time constant of the generator set and that of the load; e_g is self-regulation coefficient of the generator's load; m_t is the hydraulic driven torque; and m_{go} is the change of the load torque caused by turning on or off.

The transfer function for the control system of hydraulic turbine generator is:

$$G(s)=\frac{1}{T_y s+1}\cdot\frac{e_y-(e_{qy}e_h-e_{qh}e_y)T_w s}{e_{qh}T_w s+1}\cdot\frac{1}{T_a s+e_g-e_x} \tag{7.36}$$

where T_y is the time constant of the electro-hydraulic driven system; T_w is the time constant of the water current of pressured diversion system; the others, e_y, e_{qy}, e_h, e_{qh} and e_x are parameters (coefficients) related to the water torque, water quantity, water height, rotation speed of generator, and opening of leaf [14].

2. Double-neuron synchro control

The architecture of the double-neuron synchro control system for hydraulic tubine generator is presented in Figure 7.39, where N_f and N_φ are frequence neuron and phase neuron, $K_f>0$ and $K_\varphi>0$ are scale factors of the neurons, $Z_f(t)$ and $Z_\varphi(t)$ are performance indices, x_i is input of neuron, w_i is the weight factor of x_i, i =1,..., 5.

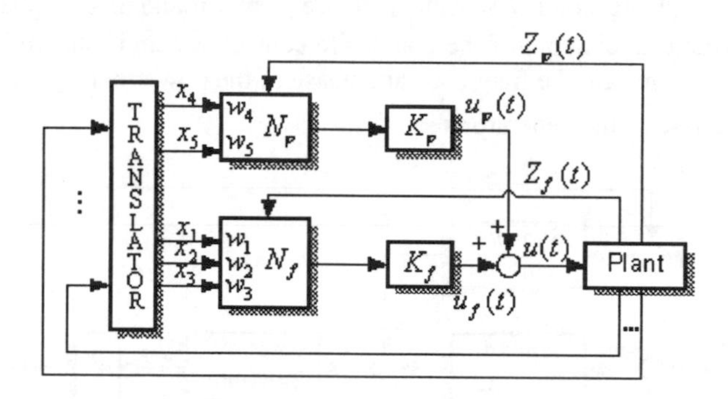

Figure 7.39. Double-neuron synchro control system.

The neural model-free learning algorithm is as follows [7, 14, 15]:

$$u(t)=K\sum_{i=1}^{n}w_i'(t)x_i(t) \tag{7.37}$$

$$w_i'(t)=w_i(t)\Big/\sum_{i=1}^{n}|w_i(t)| \tag{7.38}$$

$$w_i(t+1)=w_i(t)+d[r(t)-y(t)]u(t)x_i(t),\quad i=1,2,\ldots,n \tag{7.39}$$

where K and d are constants to be determined, and the input of neuron $x_i(t)$ is selected with i =3, i.e.,

$$
\begin{aligned}
x_1(t) &= r(t) \\
x_2(t) &= r(t) - y(t) \\
x_3(t) &= x_2(t) - x_2(t-1).
\end{aligned}
\tag{7.40}
$$

According to the Equations (7.37) to (7.40) and structure shown in Figure 7.39, the following neuron synchro-control algorithms are derived.

For Neuron N_f:

$$
\left.
\begin{aligned}
& u_f(t) = \frac{K_f \sum_{i=1}^{3}\left[w_i(t)x_i(t)\right]}{\sum_{i=1}^{3}\left|w_i(t)\right|} \\
& w_i(t+1) = w_i(t) + d_f\left[f_N(t) - f_G(t)\right]x_i(t) \\
& \qquad i = 1, 2, 3 \\
& x_1(t) = f_N(t) \\
& x_2(t) = f_N(t) - f_G(t) \\
& x_3(t) = x_2(t) - x_2(t-1)
\end{aligned}
\right\}
\tag{7.41}
$$

For Neuron N_φ:

$$
\left.
\begin{aligned}
& u_\varphi(t) = \frac{K_\varphi \sum_{i=4}^{5}\left[w_i(t)x_i(t)\right]}{\sum_{i=4}^{5}\left|w_i(t)\right|} \\
& w_i(t+1) = w_i(t) + d_\varphi\left[\varphi_N(t) - \varphi_G(t)\right]x_i(t) \\
& \qquad i = 4, 5 \\
& x_4(t) = \varphi_N(t) - \varphi_G(t) \\
& x_5(t) = x_5(t-1) + x_4(t)
\end{aligned}
\right\}
\tag{7.42}
$$

The total output of the neuron controller is

$$
u(t) = \begin{cases} u_f(t) + u_\varphi(t), & for \ |x_2(t)| \le 1\ Hz \\ u_f(t), & for \ |x_2(t)| > 1\ Hz \end{cases}
\tag{7.43}
$$

where d_f and d_φ are learning rates of neuron N_f and neuron N_φ respectively.

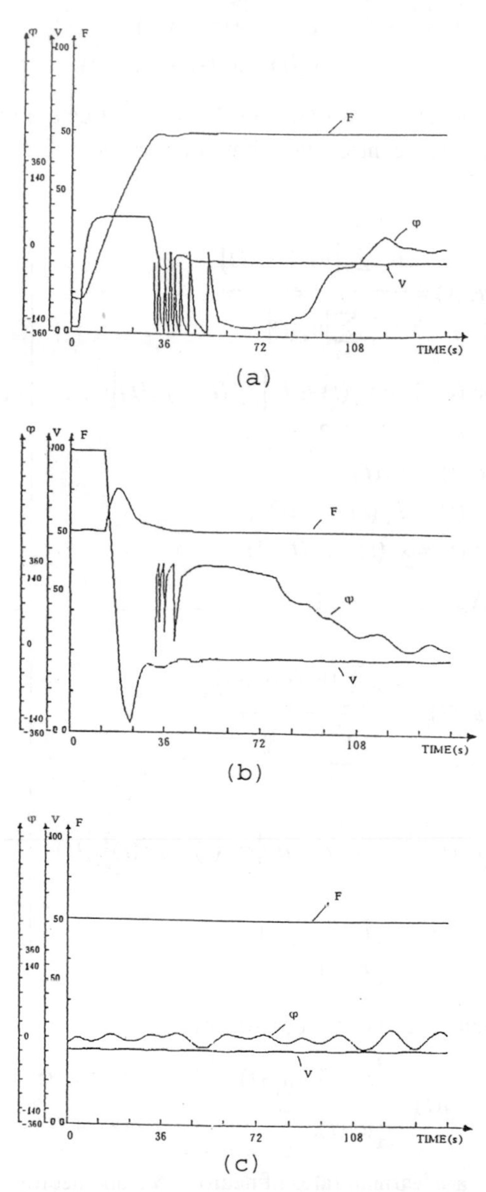

Figure 7.40. Real-time dynamic simulation curves of neuron sychro-control.

3. Real-Time Dynamic Simulation

A hydraulic turbine generator set in a hydraulic power station is used as an object of real-time simulation. The model parameters of the set are as follows: $e_{qy} = 1, e_{qh} = 0.5, e_h = 1.5,\quad e_g - e_x = 0.25, T_y = 0.8s, T_w = 1.5s$ and $T_a = 8s$. Substitute these parameters into Equation (7.36), then the transfer function of the object is:

$$G(s) = \frac{1}{0.8s+1} \cdot \frac{1-1.5s}{0.75s+1} \cdot \frac{1}{8s+0.25} \tag{7.44}$$

The simulation on auto-starting, unloading of 100% load, and empty load state is carried out by using the neuron synchro-control algorithms of Equations (7.41)-(7.43) and by selecting the parameters of K_f=5, K_φ=1/65536, d_f=4, d_φ=2, sample period τ=0.04s. The phase controller is put into operation when frequency difference $|\Delta f| < 1Hz$. Figure 7.40 (a)-(c) show the simulation results, where F is the frequency of the generator set, V — opening of leaf, φ — phase difference.

From Figure 7.40 we can see that: (a) after auto-starting for 70 seconds, an evident effect of phase control is reached, then the phase is approaching its synchronous point, and after 120 seconds, $\Delta\varphi \approx 0$ (b) after unloading 100% load for 60 seconds, the phase control acts obviously, and $\Delta\varphi \to 0$ rapidly (c) in empty load state, the phase difference slowly swings near by $\Delta\varphi = 0$, and an opportunity of connection to power net about 8 times per minute is provided.

The simulation has shown that the double-neuron synchno control of the dydralic turbine generator possesses good control results, and could be directly applied to the control of hydraulic turbine generators in the water power stations.

7.6.2 Direct Fuzzy Neurocontrol for Train Traveling Process

With the development of micro-computer technology in recent years, the automatic train operation (ATO) system has been one of the research focuses in the field of railway automation in the world. However, these ATO systems are usually inferior to skilled human operator due to the complexity of the controlled process. It is well known that the train traveling process is affected by many uncertain factors and belongs to a complex dynamic process which is hard to model using conventional identification method. Under different working conditions, the control objectives and control strategies are so different with the varying process characteristics [42] that conventional control theory based ATO systems are hard to meet with the requirements of the process. Therefore, experienced driver's knowledge based intelligent control systems for train traveling process have been proposed during the past ten years, including Yasumobu's fuzzy ATO [91, 92] and Jia's FMOC (Fuzzy Multiobjective Optimal Control) ATO. These intelligent control ATO aims to form a controller using the control rules extracted from skilled driver's experiences based on the fuzzy sets theory. Although these systems have approached encouraging results with the computer simulations and practical applications, including field-test, there exist the following two common drawbacks: (1) The partition of fuzzy linguistic variables and the shapes of membership functions which are excessively dependent on the expert's experience and are usually hard to adjust on-line; (2) Fuzzy inference methods are not adaptive enough for control. In one word, it is difficult to further improve performance of the ATO system simply by using fuzzy control methods.

In this paradigm, we incorporate the learning ability of neural network into the fuzzy system to form a direct fuzzy-neurocontrol for train traveling processes. This integrated controller comprises the complematory characteristics of fuzzy system and neural network, and improves the adaptive ability of the conventional fuzzy system for the system parameters variation [82, 83].

1. Fuzzy-neurocontroller

In recent years, the research on fusion of fuzzy control and neural network has been an active area because they are complementary and ideal tools to achieve linguistic knowledge representation and the adaptive knowledge evolution, the two essential features in human controls [24, 25]. Currently, the integrating methods can be mainly divided into two aspects: one is to realize the fuzzy control system directly by being structurally equivalent to the fuzzifier, fuzzy rule base and defuzzifier by using corresponding neural network respectively [57, 60]. In this method, the knowledge structure of the original fuzzy system is constrained with the burden of slowly learning speed. Another aspect is determining the functional equivalent to the fuzzy system by neural network [43, 82]. In our fuzzy-neurocontroller, the second integrating method is considered.

(i) Architecture

The similarity of fuzzy system and neural networks in nonlinear mapping and approximation ability provide a chance for integrating the learning ability of neural network into the fuzzy system. Therefore, we adopt a multilayered forward neural network to implement the mapping in the fuzzy system. The architecture diagram of the fuzzy neurocontroller is shown in Figure 7.41. Without losing generality, we consider the fuzzy system which has two input variables and one output variable.

As shown in Figure 7.41, the fuzzy neuro system has a total of five layers. Nodes at layer one are input nodes (linguistic nodes) which represent input linguistic variables of the train traveling process and no weights relating to layer two. Layer five is the output layer which performs the defuzzification process. In this situation, the COA (center of area) method [56] is used and the defuzzification output of the system is:

$$y = \frac{\sum_{i=-6}^{6} \mu_Y(i)*i}{\sum_{i=-6}^{6} \mu_Y(i)}. \tag{7.45}$$

The layer two consists of membership function nodes which represent all the fuzzy sets of input linguistic variables and complete the mapping from the crisp input values to fuzzy values. Layer three is the middle layer whose nodes have no clear meaning. Nodes at layer four represent the points in the discrete universe of discourse of output variables which range from -6 to 6. It is obvious that the links $[w_{ij}]$ and $[w_{jk}]$ are trained to represent the fuzzy control rules. The active functions of nodes at layer three and layer four are sigmoid functions as follow:

$$f(x) = 1/(1+e^{-x})$$

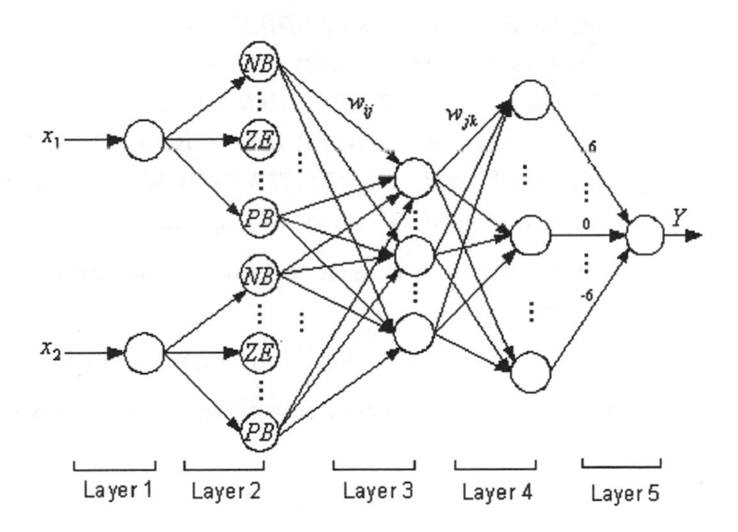

Figure 7.41. The architecture of fuzzy-neurocontroller. for train traveling process.

With this five-layered structure of the proposed connectionist model, the whole process of fuzzy systems from fuzzification, fuzzy inference to defuzzification can be performed through the forward calculation of the neural network.

(ii) Building fuzzy relations

The fuzzy relations of the fuzzy system, i.e., the fuzzy rule base can be stored in parallel in the weights of the neural network by the learning procedure. For convenience of discussion, supposed we have a fuzzy controller with two inputs (A and B) and one output C. The fuzzy sets of A, B and C are defined as $\{NB, NM, NS, ZE, PS, PM, PB\}$ whose membership functions are triangular shapes as shown in Figure 7.42.

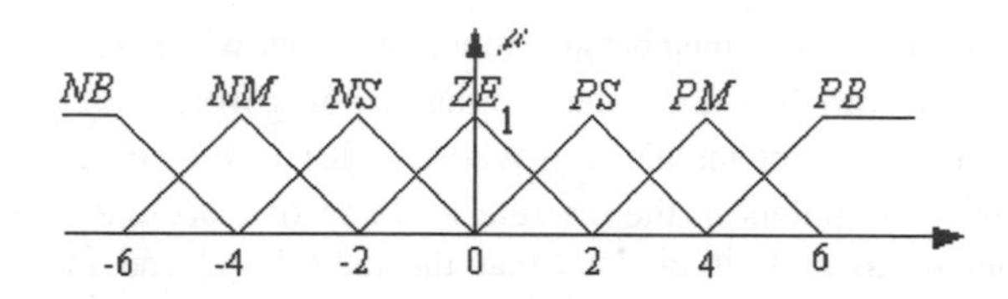

Figure 7.42. Membership function of the fuzzy-neuro system.

The fuzzy rule base comprises the following 6 rules:

$$\begin{aligned}
&R_1: \; IF(A \text{ is } PB \quad and \quad B \text{ is } PB) \; THEN \; C \text{ is } PB.\\
&R_2: \; IF(A \text{ is } PM \quad and \quad B \text{ is } PB) \; THEN \; C \text{ is } PM.\\
&R_3: \; IF(A \text{ is } PS \quad and \quad B \text{ is } PS) \; THEN \; C \text{ is } ZE.\\
&R_4: \; IF(A \text{ is } PM \quad and \quad B \text{ is } NB) \; THEN \; C \text{ is } NM.\\
&R_5: \; IF(A \text{ is } PS \quad and \quad B \text{ is } NM) \; THEN \; C \text{ is } NS.\\
&R_6: \; IF(A \text{ is } PS \quad and \quad B \text{ is } NS) \; THEN \; C \text{ is } ZE.
\end{aligned} \tag{7.46}$$

Thus, the outputs of layer two corresponding to the degrees of membership of input fuzzy sets can be represented as:

$$\left[\mu_{NB}(a), \mu_{NM}(a), \ldots, \mu_{PB}(a), \mu_{NB}(b), \ldots, \mu_{PM}(b), \mu_{PB}(b)\right] \tag{7.47}$$

The outputs of layer four are the degrees of membership of output fuzzy sets which can be represented as

$$\left[\mu_C(-6), \mu_C(-5), \ldots, \mu_C(-1), \mu_C(0), \mu_C(1), \ldots, \mu_C(5), \mu_C(6)\right] \tag{7.48}$$

The corresponding training samples can be expressed: e.g., for rule R_1, there are

$$\begin{aligned}
&\text{input sample: } [0,0,0,0,0,0,1;\, 0,0,0,0,0,0,1]\\
&\text{output sample: } [0,0,0,0,0,0,0,0,0,0.5,1,0.5,0]
\end{aligned} \tag{7.49}$$

For other rules, the training samples are similar to (7.49)

Based on the learning samples, the error backpropagation learning algorithm is adopted to train the neural network. After learning, the whole fuzzy rules can be kept in the weights of network. The increasing and/or updating of the rules can be completed by increasing and/or updating the train data sets. Moreover, the calculation burden is relatively small.

(iii) Fuzzy inference

The fuzzy inference of the original fuzzy system can be implemented by the parallel calculation of the fuzzy neural network based on the following two principles:

(1) When the input fuzzy sets A and B are similar to A_k and B_k, the fuzzy implication ($A_k, B_K \rightarrow C_K$) is activated, then the output fuzzy set C is similar to C_K.

(2) When the input fuzzy sets are different from the sample fuzzy sets, a series of fuzzy implication will be activated with different degrees, then the output is the nonlinear interpolation of the corresponding activated rules. The emphasis degree of each rules can be described by:

$$d(X, X_k) = (\sum_{i=1}^{n} (x_i, x_{ik})^2)^{\frac{1}{2}} \tag{7.50}$$

where X_k is training input signal while X is the current input signal.

2. Mathematical description of train traveling process

In the following, an approximate mathematical description of train traveling process based on the idea of process partition of complex dynamic process is provided for the convenience of simulation.

The train traveling process is very complex and affected by many uncertainty factors, such as railway conditions (curve and gradient), a traveling speed, environment (weather) and working conditions. It is hard to give an accurate mathematical model of the train traveling process. Therefore, from the engineering practice point of view, the following model is used:

$$\frac{dv}{dt} = \xi * f(n,v) = \xi * \frac{F(n,v)}{G+P} \tag{7.51}$$

where ξ : acceleration coefficient (km/h), usually equal to 120 for electric train, 250-300 for high speed train.

$f(n,v)$: unitary joint effort (knf)

n: control notch

v: traveling speed (km/h)

P: weight of locomotive (knf)

G: total weight of wagons (knf)

$F(n,v)$: joint effort which can be calculated by the following equation:

$$F(n,v) = F_q(n,v) - B_d(n,v) - B_p(r,v) - (P+G)[W_0(v) + W_1(v)] \tag{7.52}$$

where $F_q(n,v)$: the tractive force of locomotive (knf)

$B_d(n,v)$: the power-braking force of locomotive

$B_p(n,v)$: the pneumatic braking force

r: the pressure decrement on pneumatic pipe

$W_0(v)$: the basic resistance of the train (nf / kn)

$W_1(v)$: the additional resistance due to the curve, gradient and tunnels (nf / kn) *etc*.

According to the features of the train traveling process under different working conditions, it is partitioned into five characteristically distinguishable subprocesses with different control objectives which are SUPI (speed-up-from-still subprocess), SUP (speed-up subprocess), CSP (constant speeds subprocess), SAP (speed-adjusting subprocess) and TSP (train stopping process). In different subprocess, the force $F(n,v)$ is different. Therefore, we have five different models of process corresponding to different subprocesses and need five fuzzy-neurocontrollers which comprise five groups of different control rules.

3. Simulations results

The close-loop control system of train traveling process based on the proposed fuzzy-neuro controller is shown in Figure. 7.43.

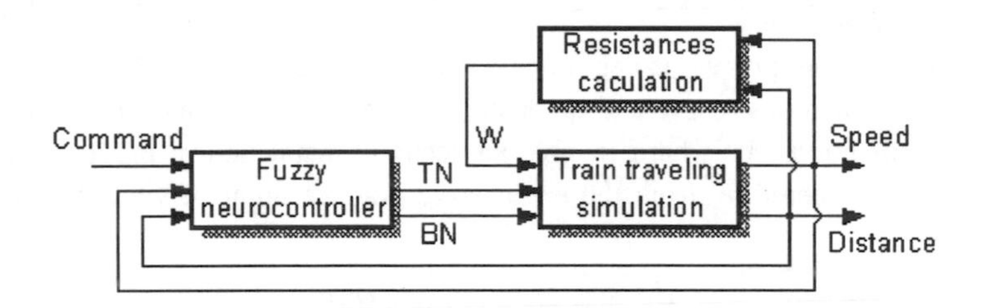

Figure 7.43. The diagram of close-loop control system for train traveling process.

The "8k" electrical locomotive is chosen as a typical simulation model that drives a train with 1000 tons traveling on a representative line including several sections with different environmental conditions. There are five fuzzy-neurocontrollers corresponding to different subprocesses. In general, the architectures of all fuzzy-neurocontrollers are similar to Figure 7.41, except for the adjustments of the input variable and the universe of discourse of the output variable. For instance, for SUP subprocess, $V_p = V_0 - V$ (the difference of the given speed and the traveling speed) and $V_s = V_0 - V_d$ (the difference of the given speed and control degree designing speed) are adopted as the input variable of the network, while the change of traction notch DPN is taken as the output variable. The training rules are shown in the formula (7.46). Layer three has 8 nodes and the initialized weights of the neural network are random value [-0.5, 0.5]. BP learning algorithm is used to train the controllers and the learning rate is 0.15 while the error tolerance is 0.01.

A group of simulation results are shown in Figure 7.44 (a)-7.44(d) for five subprocesses compared with a human driver's control results. Figure 7.44(a) represents SUP subprocess, where train travels on a two-kilometer railway which comprises a gradient and a curve. Figure 7.44(b) represents train speed up from 10km to 80km. Figure 7.44(c) shows that the train kept the speed of 70km to travel without the control notch being exerted. Figure 7.44(d) represents the speed adjusting process. From the curves, it can be found that the control results of the proposed fuzzy-neurocontroller is satisfied with the decrease of the change times of the notch compared with a skilled driver's control, thus the riding comfort, energy saving and traceability performance indices can be met with simultaneously.

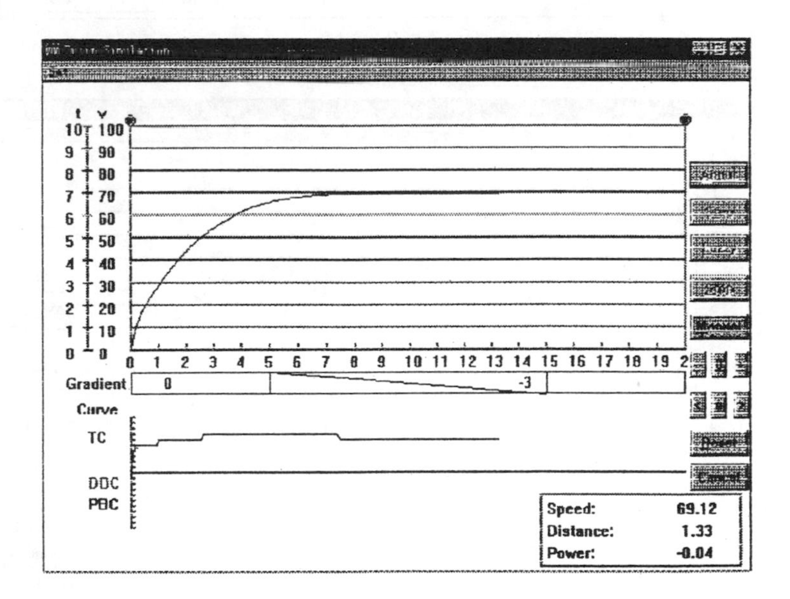

(a)

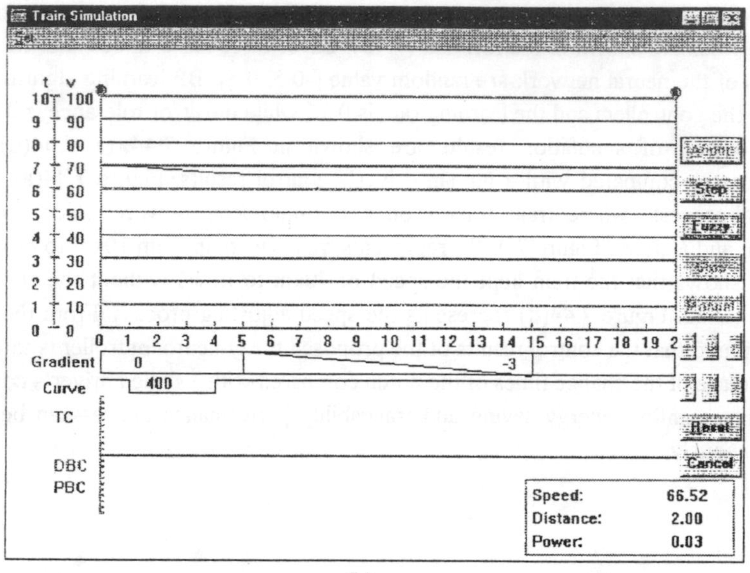
Train Simulation
Set
t v
Gradient 0 -3
Curve 400
TC
DBC
PBC
About
Step
Fuzzy
Manual
Reset
Cancel
Speed: 66.52
Distance: 2.00
Power: 0.03

(b)

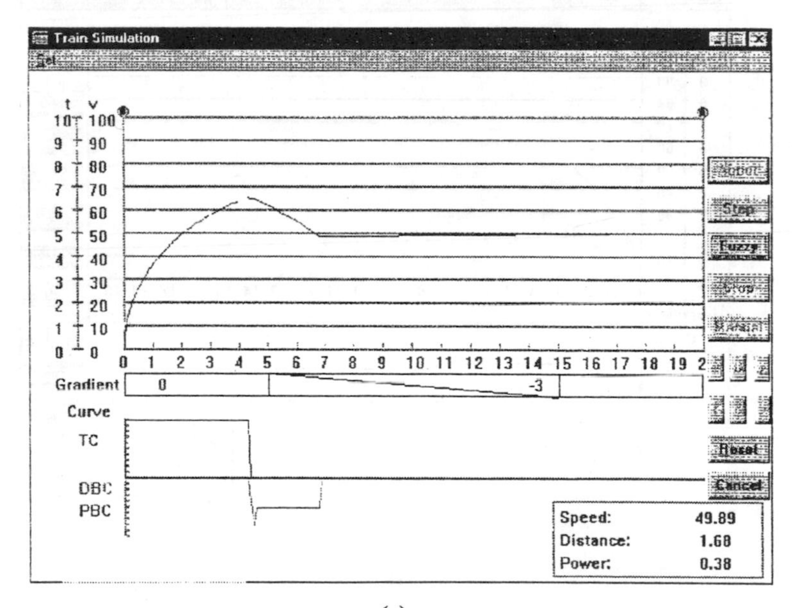
Train Simulation
Set
t v
Gradient 0 -3
Curve
TC
DBC
PBC
Step
Fuzzy
Reset
Cancel
Speed: 49.89
Distance: 1.68
Power: 0.38

(c)

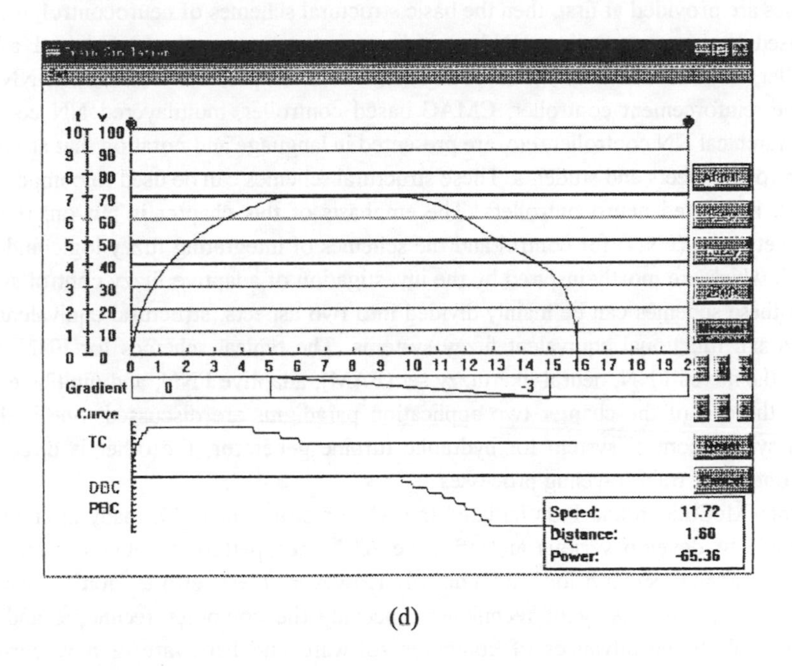

(d)

Figure 7.44. Simulation results of fuzzy neurocontoller.

4. Further research

A novel scheme for implementing automatic train traveling operations based on a fuzzy neural controller was proposed and the simulation results was satisfying. The proposed approach provided a meaningful attempt to achieve the adaptive fuzzy system by incorporating neural network into the fuzzy system. Further research will focus on the following aspects:

(1) Modeling of complex dynamic systems based on fuzzy neural networks.

(2) Research on more efficient learning algorithm superiority over the traditional backpropagation learning algorithm adopted in this paradigm.

(3) Conversion from the control rules extracted from expert experiences to training data sets, for example, one fuzzy control rule may correspond to a group of training samples.

7.7 Summary

In this chapter a brief introduction of artificial neural networks, their structures and examples are provided at first, then the basic structural schemes of neurocontrol, including NN-based learning controller, NN-based direct inverse controller, NN-based adaptive controller, NN-based internal model controller, NN-based predictive controller, NN-based adaptive reinforcement controller, CMAC based controller, multilayered NN controller, and hierarchical NN controller, etc. are presented in language and notation that are familiar for control engineers and students. These structural schemes can be used to compose more complex integrated neurocontrollers. The emphasis of this chapter is put on, the fuzzy neural networks (FNN) for control and the schemes of integrating fuzzy logic and neural network which are mostly inspired by the investigation of adaptive fuzzy control systems. In sum these schemes can be mainly divided into two aspects: structural equivalent fuzzy systems and functional equivalent fuzzy systems. The typical schemes for FNN control involve the mixed FNN, neural-like fuzzy set (FAM), adaptive FNN, and multilayer FNN, etc. At the end of the chapter two application paradigms are discussed, one is double-neuron sychro control system for hydraulic turbine generator, the other is direct fuzzy neurocontrol for train traveling processes.

Since McCulloch and Pitts initiated the ANN research in 1943, many attempts have been made to develop various and effective ANNs for pattern recognition, image and signal processing, supervision and control. However, these efforts don't always get rewards due to the reality of techniques, specially the computer technique and VLSI technique. With the advances of computer software and hardware, a new current of development of ANN has come out since late 1980s. More than ten years has gone by, however, many attempts for applications in control field are still without any results. The main difficulty is in the design and fabrication of the ANN in VLSI meaning. To solve this problem, researchers may continue to go a long way. The integration of ANN with fuzzy logic, expert systems, adaptive control and even PID control would create a good product for intelligent control.

References

1. J. S. Albus, "A new approach to manipulator control: Cerebellar model articulation control (CMAC)," *Trans. ASME, J. of Dynamics Syst., Meas. and Contr.*, **97** (1975) 220-227.
2. J. S. Albus, "Data storage in the cerebellar model articulation controller(CMAC)," *Trans. ASME, J. of Dynamics Syst., Meas. and Contr.*, **97** (1975) 228-233.
3. J. S. Albus, "A model of the brain for robot control," *Byte*, (1979) 54-95.
4. J. S. Albus, "Mechanisms of planning and problem solving in the brain," *Math. Biosci.*, **45** (1979) 247-293.

5. P. E. An, M. Brown, C. J. Harris, A. J. Lawrence, C. J. Moore, "Associative memory neural networks: Adaptive modelling theory, software implementations and graphical user," *Engng. Appli. Artif. Intell.*, **7(1)**, 1994, 1-21.
6. C. W. Anderson, "Learning to control an inverted pendulum using neural networks," *IEEE Control System Magazine*, **9** (1989) 31-37.
7. P. Baldi, "Gradient descent learning algorithm overview: A general dynamic system perspective," *IEEE Trans. Neural Networks*, **6** (1995)182.
8. A. G. Barto, R. S. Sutton, and C. W. Anderson, "Neuron-like adaptive elements that can solve difficult learning control problems," *IEEE Trans. SMC*, **13**(1983) 834-846.
9. C. Batur, "Adaptive expert control," *Int. J. Control*, **54** (1991) 867-881.
10. M. Braae, and D. A. Rutherford, "Fuzzy relations in a control setting," *Kybernetes*, 7 (1978) 185-188.
11. J. J. Buckley, "Can fuzzy neural nets approximate continuous fuzzy functions?" *Fuzzy Sets and Systems*, **61** (1993) 43-52.
12. J. J. Buckley, and Y. Hayashi, "Fuzzy neural network: A survey," *Fuzzy Sets and Systems*, **66** (1994) 1-13.
13. G. A. Carpenter, and S. Grossberg, "The ART of adaptive pattern recognition by a self-organising neural network," *Computer*, March 1988, 77-88.
14. J. Chen, *A Study of Neurocontrol Theory and Its Industrial Applications* (in Chinese), Ph. D. Dissertation, Zhejiang University, Hangzhou, China, 1996.
15. J. Chen, S. Q. Wang, N. Wang, and J. C. Wang, "Application of neuron intelligent control in synchronization of hydraulic turbine generators," *Proc. IEEE Int. Conf. NNSP*, Nanjing, China, 1995, 546-549.
16. C. G. Economou, M. Morari, and B. O. Palsson, "Internal model control, 5. Extension to nonlinear systems," *Ind. Eng. Chem. Process Des.* Dev. **25** (1986)403-411.
17. J. L. Elman, "Finding structure in time," *Cognitive Science*, **14** (1990) 179-211.
18. T. Fukuda, T. Shibata, M. Tokita, and T. Mitsuoka, "Neural network application for robotic motion control: Adaptation and learning," in: Proc. *Int. Joint Conf. Neural Networks*, San Diego, CA, 1990, 447-451.
19. T. Fukuda, and T. Shibata, "Research trends in neuromorphic control," *J. Robotics Mechatron*, **2(4)**, 1991, 4-18.
20. K. Fukushima, S. Miyake, and T. Ito, "Neocognitron: A neural network model for a mechanism of visual pattern recognition," *IEEE Trans. SMC*, **13**(1983) 826-834.
21. D. Goldberg, *Genetic Algorithms in Search, Optimization and Machine Learning*: (Addison-Wesley, Reading, MA, 1989).
22. E. Grant, B. Zhang, "A neural net approach to supervised learning of pole balancing," in: *Proc. IEEE Int. Symp. on Intelligent Control*, 1989, 123-129.
23. S. Grossburg, "Adaptive pattern Classification and universal recording, II: Feedback,

expectation, olfaction, and illusions," *Biol. Cybern.* **23**(1976)187-202.
24. M. M. Gupta, and D. H. Rao, "On the principles of fuzzy neural networks," *Fuzzy Sets and Systems*, **61**(1991) 1-18.
25. S. K. Halgamuge, and M. Glesner, "Neural networks in designing fuzzy systems for real world applications," *Fuzzy Sets and Systems*, **65**(1994)1-12.
26. D. O. Hebb, *The Organization of Behavior* (Wiley, New York, 1949) 60-78.
27. R. Hecht-Nielsen, *Neurocomputing*, (Addison-Wesley, Reading. MA, 1990).
28. R. Hecht-Nielson, "Neurocomputer applications," in: *Neural Computers*, R. Eckmiller and Ch. V. D. Malsburg (Springer-Verlay, Berlin, 1988) 445-453.
29. G. E. Hinton, T. J. Sejnowski, and D. H. Ackley, "Boltzmann machine: Constaint satisfaction networks that learn," *Tech. Rep.* CMU-CS-84-119, 1984.
30. J. H. Holland, *Adaptation in Natural and Artificial Systems* (University of Michigan Press, Ann Arbor, MI, 1975).
31. J. J. Hopfield, "Neural networks and physical systems with emergent collective computational abilities," *Proc. of the National Academy of Sciences*, **79** (1982) 2554-2558.
32. K. J. Hunt, and D. Sharbaro, "Neural networks for nonlinear internal model control," *Proc. IEE Pt. D.* **138** (1991) 431-438.
33. K. J. Hunt, D. Sbarbaro, R. Zbikowshi, and P. J. Gawthrop, "Neural networks for control systems — a survey," *Automatica*, **28(6)**, 1992:1083-1112.
34. H. Ichihashi, "Learning in hierarchical fuzzy models by conjugate gradient method using BP errors," in: *Proc. Symp. Intelligent Systems*, 1991, 235-240.
35. *IEEE Control Systems Magazine*, Special Issue on Neural Networks, **8**(1988).
36. *IEEE Control Systems Magazine*, Special Issue on Neural Netowrks, **9**(1989).
37. *IEEE Control Systems Magazine*, Special Issue on Neural Networks, **10**(1990).
38. *IEEE Trans. Industrial Electronics*, Special Issue on Neural Networks Applications, **39(6)**, 1992.
39. W. T. Illingworth, "Beginnes guide to neural networks," *IEEE ASE Magazine*, **9** (1989).
40. *Int. J. Control*, Special Issue on NN-Based Control, **56(2)**, 1992.
41. J-S. Roger Jang, "ANFIS: Adaptive-network-based fuzzy inference system," *IEEE Trans. SMC*-**23** (1993)665-685.
42. L. M. Jia, and X. D. Zhang, "On fuzzy multiobjective predictive control and its adaptive alternative," *Proc. IFAC 12th World Congress*, Sydney, 1993, 895-899.
43. L. M. Jia, and X. D. Zhang, "On fuzzy multiobjective optimal control," *Engineering Applications of AI*, **6(2)**, 1993, 153-164.
44. M. I. Jordan, "Attractor dynamics and parallelism in a connectionist sequential machines," *Proc. 8th Annual Conf. Cognitive Science Society*, 1986:531-546.

45. M. I. Jordan, "Supervised learning and system with excess degree of freedom," *COINS Tech. Rep.*, 1988, 27-88.
46. D. Karaboga, *Design of Fuzzy Logic Controllers Using Genetic Algorithms*, Ph.D. thesis, University of Wales, Cardiff, UK, 1994.
47. M. Kawato, K. Furukawa, and R. Suzuki, "A hierarchical neural network model for control and learning of voluntary movement," *Biol. Cybern.*, **57**(1987)169-185.
48. M. Kawato, Y. Uno, M. Isobe, and R. Suzuki, "Hierarchical neural network model for voluntary movements with application to robotics," *IEEE Contr. Mag.*, 1988, 8-15.
49. S. S. Keerthi, and E. G. Gilbert, "Moving-horizon approximations for a general class of optimal nonlinear infinate-horizon discrete-time Systems," *Proc. 20th Annual Conference on Information Science and Systems*, 1986, 301-306.
50. T. Kohonen, *Self-Organising and Associative Memory* (3rd ed.) (Springer-Verlag, Berlin, 1989).
51. T. Kohonen, "Self-organized formation of topologically correct feature maps," *Biol. Cybernetics*, **43** (1982) 59-69.
52. B. Kosko, "Adaptive bidirectional associative memories," *Appl. Optics.*, **26** (1987) 4947-4960.
53. B. Kosko, *Neural Networks and Fuzzy Systems* (Prentice-Hall, Englewood Cliffs, NJ, 1992)
54. B. Kosko, "Fuzzy associative memory systems," in: *Fuzzy Expert Systems*, A. Kandel, (Addison-Wesley, Reading, MA, 1986).
55. L. G. Kraft and D. P. Campagna, "A comparison between CMAC neural network control and two traditional adaptive control systems," *IEEE Control System Magazine*, **10**(1990)36-43.
56. C. C. Lee, "Fuzzy logic in control system: Fuzzy logic controller, Part I and II," *IEEE Trans. on SMC*, **20(2)**, 1990, 404-435.
57. T. P. Leung, and Q. J. Zhou, *et. al*, "An Optimization design method of fuzzy logic controller," *Control Theory and Applications*, **12(1)**, 1995, 492-497.
58. A. U. Levin, and K. S. Narendra, "Recursive identification using feed-forward neural networks," *Int. J. of Control*, **61**(1995)533-547.
59. G. Lightbody, and G. W. Irwin, "Direct neural model reference adaptive control," *IEE Proc. Control Theory Applications*, **142(1)**, 1995:31-43.
60. C.-T. Lin, and C. S. G. Lee, "Neural network based fuzzy logic control and decision system," *IEEE Trans. Computer*, **40**(1991) 1320-1336.
61. D. Q. Mayne, and H. Michalska, "Receding horizon control of nonlinear systems," *Trans. IEEE AC*-**35** (1990) 814-824.
62. W. S. McCulloch, and W. H. Pitts, "A logical calculus of the ideas immanent in nervous activity," *Bull. Math. Biophysics*, **5**(1943)115-123.

63. W. T. Miller III, R. S. Sutton, and P. J. Werbos, *Neural Network for Control* (MIT Press, Cambridge, MA, 1990).
64. W. T. Miller, F. H. Glanz, and L. G. Kraft III, "Application of a general learning algorithm to the control of robotic manipulators," *Int. J. of Robotics Research*, **6(2)**, 1987, 84-98.
65. W. T. Miller, "Real time application of neural networks for sensor based control of robots with vision," *IEEE Trans. Syst. Man, and Cybern.*, **19(4)**, 1989, 825-831.
66. W. T. Miller III, R. P. Henes, F. H. Glanz, and L. G. Kraft III, "Real time dynamic control of an industrial manipulator using a neural network based learning controller," *IEEE Trans. on Robotics and Automation*, **6(1)**, 1990, 1-9.
67. M. L. Minsky, and S. A. Papert, *Perceptrons* (expended edition) (MIT Press, Cambridge, MA, 1969)1-20.
68. M. Morari and E. Zafirou, *Robust Process Control* (Prentice-Hall, Englewood, NJ, 1989).
69. K. S. Narenda, and K. Parthasaratny, "Identification and control of dynamical system using neural networks," *IEEE Trans. Neural Networks*, **1**(1990) 4-27.
70. J. Nie, and D. A. Linkens, "A hybrid NN-based self-organizing controller," *Int. J. of Control*, **60(2)**, 1994, 197-222.
71. D. Park, A. Kandel, and G. Langholz, "Genetic-based new fuzzy reasoning modes with application to fuzzy control," *IEEE Trans. SMC.* **24**(1994)39-47.
72. D. T. Pham, and D. Karaboga, "Dynamic system identification using recurrent neural networks and genetic algorithms," *Proc. 9th Int. Conf. on Mathematical and Computer Modelling*, San Francisco, July 1993.
73. D. T. Pham, and X. Liu, *Neural Networks for Identification, Prediction and Control* (Springer-Verlag, London,1995).
74. D. Praker, *Learning-logic*, MIT Tech. Rep. TR-47, Cambridge, MA, 1985.
75. D. Psaltis, A. Sideris, and A. A. Yamamura, "Neural controller," in: *Proc. IEEE Int. Conf. Neural Networks* (San Diego, CA, 1987) 551-558.
76. D. Psaltis, A. Sideris, and A. A. Yamamura, "A multi-layered neural network controller," *IEEE Control Systems Magazine*. 1988, 17-20.
77. F. Rosenblatt, "The perceptron: A probabilistic model for information storage and organization in the brain," *Psychlogical Review*, **65**(1958) 368-408.
78. D. E. Rumelhart, and J. L. McClelland, *Parallel Distributed Processing: Explorations in the Microstructures of Cognition, Vol. 1: Foundations* (MIT Press, Cambridge, MA, 1986).
79. A. C. Sanderson, and R. J. Peterka, *Neural modeling and model identification, CRC Critical Reviews in Biomedical Engineering*, **12**(1985) 237-309.
80. Q. Shen, D.-W. Hu, and C. Shi, *Neural Networks: Application Techniques* (National

Defence Univ. of S&T Press, Changsha, China, 1994).

81. K. Steinbuck, and V. A. W. Piske, "Learning matrix and their applications," *IEEE Trans. Electronic Computers, EC*-**12**(1963) 846-862.
82. J. Wang, and Z.-X. Cai, "Research on advanced train traveling control system based on neural network," *Proc. Chinese Intelligent Robotics*, Changsha, 1995, 380-385.
83. J. Wang, and Z-X. Cai, "Direct fuzzy neurocontrol for train traveling process," *Trans. of Chinese Non-Ferrous Metals*, **3**(1996) 36.
84. F.-Y. Wang, and D. D. Chen, "Rule generation and modification for intelligent controls using fuzzy logic and neural networks," *Intelligent Automation and Soft computing*, 1993.
85. F-Y Wang, "Design of adaptive fuzzy control systems using neural networks," *Proc. Chinese World Congress Intell. Control Intell. Automation*, (Science Press, Beijing, 1993)516-522.
86. F-Y. Wang, and H-M Kim, "Implementing adaptive fuzzy logic controllers with neural networks: A design Paradigm." *J. of Intelligent and Fuzzy Systems*, **1**(1995) 110-128.
87. P. Werbos, *Beyond Regression: New Tools for Prediction and Analysis in the Behavioural Sciences*, Ph.D. Dissertation, Harvard University, Cambridge, MA, 1974.
88. B. Widrow, D. E. Rumelbart, and M. A. Lehr, "Neural networks: applications in industry, business and science," *communication of the ACM*, **37**(1994)93.
89. B. Widrow, and M. E. Hoff, Jr., "Adaptive Switching circuits," *Proc. IRE WESCON Convention Record*, Part 4, (IRE, New York, 1960) 96-104.
90. T. Yabuta, and T. Yamada, "Neural network controller characteristics with regard to adaptive control," *IEEE Trans. SMC*, **22** (1992) 170-176.
91. S. Yasmdiu, and S. Miymoto, "Fuzzy control for automatic train operation system," In: *Industrial Application of fuzzy control*, (North-Holland,1985,) 1-18.
92. S. Yasumobu, S. Miyamoto and H. Ihara, "Fuzzy control for automatic train operation system," *IFAC Control in Transportation Systems*, 1983, 33-39.

CHAPTER 8

LEARNING CONTROL SYSTEM

Learning control is one of the earliest research areas of intelligent control. The concepts of learning and learning control were introduced about a quarter of century ago [17, 18, 64, 68]. During the past ten years, learning control with applications to dynamic systems, such as robot manipulation and flight guidance, has become an increasingly important research topic. A set of learning control schemes and better control processes have also been investigated and proposed [4, 5, 9, 24, 25, 36, 65].

In this chapter, some basic concepts including definition, reasons and history of learning control followed by machine learning and its application in control will be first introduced. Then the schemes of learning control, which involve pattern recognition-based learning control, iterative learning control, repetitive learning control, NN-based learning control and rule-based learning control, etc., will be discussed. After that, some issues for learning control, which deal with design principles, control rules, stability analysis and convergence analysis of learning control, will be analyzed. Finally, an example of learning control systems, the self-learning fuzzy neurocontrol for arc welding process, will be presented.

8.1 Introduction to Learning Control

8.1.1 What is Learning Control ?

Learning is a very general term that denotes the way in which people and computers enrich their knowledge and improve their skills. Until now, no unified definition has been established for learning control, because people with different backgrounds have different opinions on 'learning'.

Learning is one of the main intelligence areas of human beings. Human activities (including taking control for an automated device) also need learning. During the evolution process of human beings, the learning function has played an important role, and learning control is really an attempt to model fine regulation-control mechanisms of the human being. In the following, various definitions on learning and learning control will be presented.

N. Wiener gave a general definition for learning in 1965.

Definition 8.1. An animal that possesses living capability is one that can be renovated by the environment in which it undergoes throughout its individual life. An animal that can reproduce successors is one that can at least produce animals similar to itself, although this similarity may change with time. If the change is self-geneticable, then there exists a material that can be influenced by the natural selection. If the change comes out in a type of behavior and provided this behavior is harmless, then this change would be continued generation after generation. This change of behavioral type from one generation to its next generation is called racial learning or system growth learning, and the behavior change or behavior learning that happens in a specific individual is called individual growth learning.

C. Shannon defined learning with more limitation in 1953.

Definition 8.2 (Shannon). Suppose: (1) an organism or a machine is in a kind of environment; (2) There exists a "successful" measure or "adaptive" measure to the environment; (3) This measure is a relative local measure with respect to time, i.e., people can test this successful measure in a time shorter than the life time of the organism. If this local successful measure is improved with the time for the environment considered, then it is said that, for the selected successful measure, the organism or machine is in learning for adapting this kind of environment.

Osgood presented learning from the point of view of psychology in 1953.

Definition 8.3. In repetitive circumstances with same features, the organism individual continuously changes its own behaviors and enhances its selection in competitive reaction depending upon its own adaptability. This kind of selective variation formed by the individual experience is the so-called learning.

Tsypkin gave learning and self-learning a more general definition as follows [68].

Definition 8.4. Learning can be understood as a process that repeats various input signals, corrects system states outside, and makes the system have specific responses to specific inputs. Self-learning is a learning without outside correction and without any additional information about the correctness of the system responses.

Simon made a more precise definition on learning [64].

Definition 8.5. Learning denotes changes in the system that are adaptive in the sense that they enable the system to do the same task or tasks drawn from the same population more efficiently and more effectively the next time.

Minsky replaced the improvement criterion of learning with a more general one requiring that the changes are merely useful [54]:

Definition 8.6. Learning is making useful changes in the workings of our minds.

In the following, we overview the definitions for learning system, learning control and learning control system.

Definition 8.7. Learning system is a system that can learn unknown information of the process and use the information as experience for further decision or control so that the system performance can be gradually improved [17].

Definition 8.8. If a system can learn the unknown intrinsic and featured information of a process or its environment, and applies the learnt experience for further evaluation, classification, decision or control to improve the quality of the system, then the system is called a learning system [61].

Definition 8.9. Learning control can evaluate unknown information during the operation process of the system and determine an optimal control based on the evaluated information so that the performance of the system can be improved step by step [61].

Definition 8.10. Learning control is a method of control in which the actual experience plays a part in the nature of control parameters and algorithms [71].

Definition 8.11. If a learning system controls a process with unknown features by use of learnt information, this system is called the learning control system [17].

The definition of learning control can be described mathematically as follows:

Definition 8.12. In a limited time domain[0, T], give the expected response $y_d(t)$ of a controlled object, and search for a given input $u_k(t)$, so that the response $y_k(t)$ of the $u_k(t)$ is improved than $y_0(t)$ in a sense, where k is the number of search, $t \in [0, T]$. The searching process is called learning control process. As $k \to \infty$, $y_k(t) \to y_d(t)$, the learning control process is convergent.

According to the above definitions, the mechanism of learning control can be summarized as follows [11, 25].

(1) Search for and find a relatively simple relation between input and output of dynamic control systems.

(2) Execute each control process updated by 'learning' results of the previous control process.

(3) Improve the performance of each process so that it is better than the preceding ones.

It is expected that repeating such a learning process and recording the results accumulated in the entire process would steadily improve the performance of the controlled system.

8.1.2 Reasons for Studying Learning Control

In designing linear controllers, it is usually necessary to assume that the parameters of the controlled system model are reasonably well known. However, many control systems involve problems of uncertainty for the model parameters. These problems may be due to a slow time variation of the parameters (e.g., of ambient air pressure during an aircraft flight), or to an abrupt change in parameters (e.g., in the inertial parameters of a robot when a new object is grasped). A linear controller based on inaccurate or obsolete values of the model parameters may exhibit significant performance degradation or even instability. Nonlinearities can be intentionally introduced into the controller part of a control system so that model uncertainties can be tolerated. Adaptive control and robust control are developed for this purpose.

Adaptive control system can make conditional decision in the environment with uncertainties. With the development of control theory and applications, the controls involve a wider range of problems, and have to make decisions in complex environment with uncertainty. Therefore, a control system is required with more intelligent factors. Learning control is a development and extension of the adaptive control. Learning control can continuously improve knowledge and algorithm according to the experience and lesson, simulate fine behaviors and functions of the human being in advanced reasoning, decision-making and pattern recognition.

Under unknown environment adaptive control makes control decisions conditionally, its control algorithm relies upon a more accurate identification for a mathematical model of the controlled object, and the parameters and structures of the object or environment may abruptly change in a large range. Thereby, the controller is required to have stronger adaptivity, real-time capability and fine control quality. Under these circumstances, the algorithm would be too complex, computational work will be too large, and the real-time control and other requirements will be too hard to be satisfied. As a result, adaptive control only has a limited application field. As the motion of a controlled object possesses repeatability, i.e., the controlled system does the same work every time, then learning control can be applied to this object. In the process of learning control, the actual output signal and expected signal are simply needed to be measured, whereas the complex computation of the dynamic description and parameter estimation for the controlled object can be simplified or omitted. Therefore, learning control has a wide range of application prospects for the repetitive motion of controlled objects such as industrial robots, numerical control machine tool and aircraft flights.

Thus, learning control has become an important area of intelligent control. Learning basic principles and techniques of learning control can significantly enhance the ability of a control engineer to deal with some practical control problems effectively. It also provides a sharper understanding of the real world, which is with uncertainty. In the past thirty years, the application of learning control had been limited by the computational difficulty associated with learning control design and analysis. In recent years, however, advances in intelligence system and computer technology have greatly relieved this problem. Therefore, there is currently considerable enthusiasm for the research and application of learning control. The topic of learning control for a large range of operations has attracted particular attention because, on one hand, the advent of powerful microprocessors has made the implementation of learning control (includes on-line learning control) a relatively simple matter, and, on the other hand, modern technology, such as high-speed high-accuracy robot or high-performance aircraft, is demanding control systems with much more stringent design specifications. Learning control occupies an increasingly conspicuous position in the intelligent control and intelligent automation.

8.1.3 Development of Learning Control

The imagination and research of learning machines began in the 1950's. The learning machine is an automatic device that can simulate memory and conditional reflection of the human. The concept of the learning machine emerged at the same time as cybernetics. Later, the concept of learning machine was beyond the predictor and filter. The chess-game machine was a successful example of a learning machine at the early research stage.

In order to solve a large amount of stochastic problems, adaptive and self-learning methods were developed in the 1960's. Learning control was first used in areas of control, pattern classification and communication of aircraft, then in the control of electric power systems and production processes. Learning systems can be divided into two types: off-line trainable system and on-line self-learning system. In the former, the response of the controlled system is evaluated by the 'rewards and punishments' feedback so the control algorithms can be improved according to the feedback information. In the latter, the system can make various trial-and error, searching, quality-evaluating, decision-making and self-revision of *a priori* knowledge to a certain extent. The on-line self-learning control needs large-capacity and high-speed computers. Due to the limitation of the development level of computers, a hybrid off-line and on-line learning method was applied to implement the early learning control.

Learning control theory was studied on the topic of two-fold control and artificial neural networks beginning in the 1960's, and the control mechanism is based on the pattern recognition method. K.S. Narendra *et al.* proposed a performance feedback-based correction method in 1962. F.W.Smith presented a bang-bang control method by the use of adaptive technique of pattern recognition in 1964. At the meantime, F.B.Smith studied a trainable flight control system, and Butz proposed a learning bang-bang regulator. Mendel used the trainable threshold logic method to control system as a technique of artificial intelligence.

Another way of pattern recognition based learning control is to use linear reinforcement technique to learning control systems. Waltz and Fu proposed a heuristic approach to reinforcement learning control system [72]. Some other works including sub-goal selection and two-level learning control methods were done by others, and one work was used to a precise pose control of satellite.

The third way for studying pattern recognition based learning control is to use Bayes learning estimation that was proposed by K.S.Fu in 1965. Tsypkin, Nikolic and Fu studied a stochastic approximation method and built models of learning systems by using stochastic automata in the 1960's.

Wee and Fu proposed the fuzzy learning control system in 1969, and Saridis *et al.* developed a hierarchically semantic learning method during 1977 to 1982. These learning methods possess higher degrees of intelligence in processing the commands and decisions, thus, can be used to higher the level of hierarchical control systems.

Owing to the limitations and difficulties, pattern recognition based learning control developed slowly. Other two learning control mechanisms, the iterative learning control and repetitive learning control, were proposed and developed in the 1980's, and can be used in engineering with simple learning control laws.

Uneyama first proposed the repetitive control method in 1978 [70], and used the method to control a robot manipulator. Inoue and Nakano developed the repetitive learning control from the point of view of frequency domain [34]. Arimoto *et al.* further developed the Uneyama's preliminary research result (so-called revised algorithm) and proposed a learning control method in time-domain, the iterative learning control [5, 6].

The iterative learning control has a wide range of applications. Kawamura *et al.* studied the inverse system, limited reality, sensibility and optimal regulation of the iterative self-learning control [38-40], and deeply revealed that the iterative learning process is essentially a process that approximates inverse system. Furuta *et al.* proposed an optimal iterative learning control for multi-variable systems in 1986. Gu and Loh studied a multi-step iterative learning control method that improves the system robustness effectively in 1987. In the late 1980's, many topics for iterative learning control were studied, which dealt with discrete-time system, on-line parameter estimation, convergence, optimum design method, 2D-model and nonlinear system, etc.

The origin of learning control is neural network-based learning at the beginning of the study of pattern recognition-based learning control. The early parameter learning control method was developed for simple controlled objects and was done by neurons rather than neural networks; therefore, there exist some disadvantages such as slow speed of convergence, large amounts of memory, complex selection of classificator, difficult selection and extraction of features in implementing this control. As a result, the study of neural networks fell into a difficult position in the 1970's.

Connectionist learning method has transported a new power to learning control since early 1980's. Rumelhart *et al.* proposed an error back-propagation (BP) model [60] that can implement multilayered NN. Meanwhile, Hopfield put forward a feedback interconnected network with the function of associated memory, so-called Hopfield network [31-33]. Up to now, about one hundred models of neural networks have been published. We have introduced typical ones in Section 7.2 and Section 7.3. Owing to the fine properties of ANN described in Sub-section 7.1.2, including parallel distributed processing, nonlinear mapping, learning by training, adaptation and integration, etc., researches on connectionist learning controls have been very active in recent years. The representative works in the domain include NN-based learning control with reinforcement learning [2, 10, 49], NN-based iterative learning control [1, 36, 44, 45, 67], NN-based self-learning control [57, 62], and Rule-based learning control [46, 47], and so on. The successful application examples involve controls for inverted pendulum, robot manipulator and underwater telerobot as well as flight aircraft.

8.2 Machine Learning

Machine learning is one of the important research and application areas of artificial intelligence. A few current computer systems and AI systems possess learning ability, and their learning abilities are very limited. The research of machine learning in theory and application has become more and more active and effective since 1980's. Especially, the research on connectionist learning have brought machine learning to a new stage and given learning control a new motivation. In this section, the basic strategy and structure of machine learning will be presented first; which will be followed by the main modes of machine learning; discussion at the end, application of machine learning in control will be briefly introduced.

8.2.1 Basic Strategies and Structure of Machine Learning

1. **Definition and basic strategies of machine learning**

What is machine learning? It is very difficult to have an exact and unified definition for machine learning. However, in order to do the following: (1) draw an approximate boundary around the concept to provide a perspective on the following discussion, (2) evaluate the advancement of the discipline of machine learning, (3) Compare different levels in different countries, a definition of machine learning is the study of making machines simulate learning activities of human beings. More strictly speaking, machine learning is the study of making machines acquire new knowledge, new skill, and recognize existing knowledge. In the definition, 'machines' are computers, electronic computers now, photon computers or neurocomputers in the future.

Learning is a very complex intellectual activity, and the learning process is closely related to the inference process. According to the level of inference in learning and information provided by the environment, the strategies used for machine learning can be roughly classified into five types, i.e., rote learning, learning through instruction analogue learning and learning from cases(examples) and learning from discovery [64]. More the application of inference in learning, the stronger the capability of the learning system.

Rote learning. The environment provides information exactly at the level of performance task and, thus, no hypotheses are needed. No inference or other transformation of knowledge is required on the part of the learner. The term 'rote learning' is used primarily in learning to memorize given facts and data with no inferences drawn from the incoming information. The earliest game-playing program, Samule's checkers program, learnt to play checkers well enough to beat its creator, and exploited rote learning and parameter adjustment. This program used the **minmax** search procedure to explore checkers game trees. Rote learning recorded large quantities of data to affect the subsequent behavior of the program. Data in this program is organized by indexing board positions with a few important characteristics.

Learning through instruction. Acquiring knowledge from a teacher (supervisor) or other organized source, such as a textbook, requires the learner to transform the knowledge from the input language to an internally usable representation, and the new information will be integrated with prior knowledge of the environment is too abstract or general, and the learning element must hypothesize the missing details. The learner is required to perform some inference.

Learning from cases. Given a set of examples and counterexamples of a concept, the learner induces a general concept description of all the positive examples and none of the counterexamples. Learning from examples(cases) is a method that has been widely investigated in artificial intelligence. The information provided by the environment is too specific, the learning element must hypothesize more general rules.

Learning by analogue. Learning by analogue is acquiring new facts or skills by transforming and augmenting existing knowledge that bears strong similarity to the desired new concept or skill into a effectively useful from in the environment and is relevant only to an analogous performance task. Learning by analogue requires more inference on the part of the learner than the inference in learning through instruction. A fact or skill analogous in relevant parameters must be retrieved from memory, then the retrieved knowledge must be transformed, applied to the new situation, and stored for future use.

Learning from discovery. Learning from discovery is a general form of inductive learning that includes the discovery of systems, theory-information tasks, the creation of classification criteria to form taxonomic hierarchies, and similar tasks without the benefit of an external teacher. This form of unsupervised learning requires the learner to perform more inference than any other approaches thus far discussed.

2. Essential structure of machine learning system

In terms of Simon's view of learning, a simple model of learning system is shown in Figure 8.1. The environment supplies some piece of information to the learning element, and the learning element applies the information to make improvements in an explicit knowledge base, and the performance element uses the knowledge base to perform its task. Finally, the information gained during performing the task can serve as a feedback to the learning element. This structure is a primitive model and many important functions are omitted. This model allows us to classify learning systems according to how they 'fill' these four functional units. In any particular application, the environment, the knowledge base, and the performance task determine the nature of the particular learning problem and the particular functions that the learning element must fulfill. The role of each of these functional units can be examined in the following.

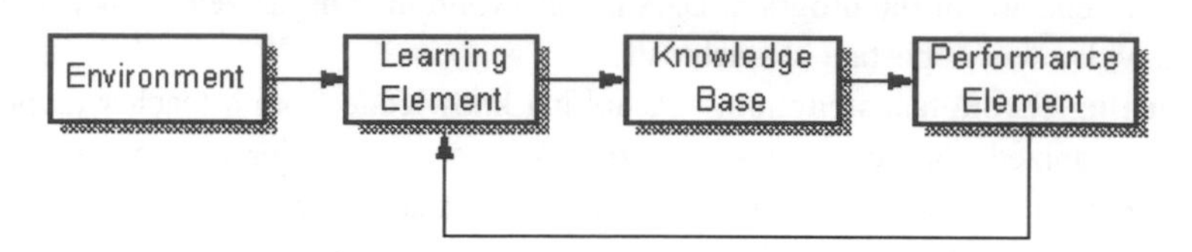

Figure 8.1. Essential structure of learning system.

The Environment. The environment supplies the source information to the learning system. The level and quality of the information will significantly affect the learning strategy. According to the degree of generality the information can be divided into different levels. High-level information is abstract information that is relevant to a broad class of problems. Low-level information is detailed information that is relevant to a single problem. If the learning system is given very abstract advice about its performance task, it must fill in the missing details, so that the performance element can interpret the information in the particular situation. On the contrary, if the system is given very specific information about how to perform in particular situations, the learning element must generalize this information into one or more rules that can be used to guide the performance element in a broader class of situations.

The quality of information has a significant effect on the difficulty of the learning task. In inductive learning, the teacher provides training examples to learning systems. If the training examples are classified without noise, the induction is the easiest one.

The Knowledge Base. The knowledge base is another unit of the learning system and has a significant effect on the design of the learning system. The knowledge base contains predefined concepts, domain constraints, heuristic rules and so on. Up to now, there are many knowledge representation approaches that can be used in learning systems. These approaches include production rules, frame, semantic networks, predicate calculus, vectors and matrices, graphs, formal grammar, and procedural encoding.

A learning system may acquire rules of behavior, descriptions of physical objects, problem-solving heuristics, classification taxonomies over a sample space, and many other types of knowledge useful in the performance of a wide variety of tasks. When designing a learning system, one or more representational forms should be chosen.

The Performance Element. The learning element is trying to improve the action of the performance element. There are three important issues related to the performance element: complexity, feedback, and transparency.

The learning system containing the predefined knowledge depends on the complexity of performance tasks. Complex tasks require more knowledge than simple ones. According to the complexity, the performance tasks can be classified into three classes, the simplest performance task, more complex one, and integration one. When the performance task becomes more complex and the knowledge base grows more in size, the problems of integrating new rules and diagnosing incorrect rules become more complicated.

The performance element should provide feedback to the learning element. All learning systems must have some way of evaluating the hypotheses that have been proposed by the learning element. Some programs have a separate body of knowledge for such evaluation. For the learning element to which credit or blame is assigned to individual rules in the knowledge base, it is useful for the learning element to have access to the internal actions of the performance element.

8.2.2 Paradigms of Machine Learning

There are many modes in machine learning, they are as follows [12, 63, 64]:

- rote learning
- learning by instruction
- learning by recording case
- learning by analogue
- learning by discovering
- analytic learning
- learning by explaining experience
- learning by correcting mistakes
- learning by deduction
- learning by genetic algorithms
- learning by building identification tree
- connectionist learning

and so on.

The above first five models of machine learning have been briefly introduced in Section 8.2.1 from the point of view of strategy. In order to understand how the machine learning is related, the following five paradigms are discussed.

Supervised concept learning.

This is the most mature machine-learning paradigm. A type of inductive learning, supervised concept learning constructs a concept description in some predefined description language based on a collection or training set of examples. Elements of the training set are marked as positive or negative examples; for example, as either members of the target concept description can then be used to predict concept membership of future examples.

Such algorithms differ in several ways. The first is the language for expressing the target concept. Different description languages have different representational power: Concepts that are easily expressed in one language might be inexpressible in another. The application domain can also impose constraints on the representation language.

A second important difference is the inductive bias applied in constructing a concept description. For any set of input examples, a representation language can express multiple (perhaps infinitely many) concepts that accurately classify the entire training set. Selecting a description thus depends on the system's inductive bias. In some systems, the bias is implicit in the algorithm. In other systems, the bias might be explicitly input to the algorithm.

A third difference is whether the algorithm operates incrementally or in batch mode. Traditional inductive algorithms generally operate in batch mode, where all training-set examples are made available to the system at the same time. Once the concept description is constructed, it is used to predict concept membership for future examples. More useful inductive algorithms accept training-set examples incrementally, using the current description for prediction and updating it as inconsistencies are discovered. Incremental algorithms are also more efficient, since they donot have to rederive the entire concept description each time they observe a new example that is inconsistent with the current concept.

Conceptual clustering.

Conceptual clustering systems differ from supervised learning systems in that the training examples are not marked as positive or negative by an outside agent or teacher. These systems must recognize the similarities between examples and group them according to some preestablished notion of similarity. Applications for these systems are readily drawn from the same problems usually addressed by traditional statistical clustering systems, with one important difference: Unlike statistical clustering algorithms where the number of outcome clusters is predetermined, conceptual clusters is predetermined, conceptual clustering algorithms determine the most appropriate number of clusters and then allocate examples to those clusters.

In general, conceptual clustering shares many open problems with supervised concept learning (this is hardly surprising given the close relation between them). Like supervised concept learning, conceptual clustering is a relatively mature paradigm that has been applied to real problems. An officiated success story for such methods is the discovery of a new categorization of stellar spectra that differed from the generally accepted clustering.

Analytic learning

Analytic learning is one of newly developed learning. A chief example of this paradigm is explanation-based learning (EBL) algorithms, which are intended to improve the efficiency of a problem-solving system. While they generally do not change the problems that are in principle solvable by the problem solver (that is, the problem solver's deductive closure), they do bias the problem solver's search space. For this reason, EBL has sometimes been described as speed-up learning. Naturally, given unlimited resources, a problem solver would eventually find a solution to any problem within its deductive closure; thus, EBL only makes sense when used to alter the future performance of a resource-limited problem solver.

For some EBL systems, this bias takes the form of acquired problem-space macro-operators, which alter the search space by compressing generalizations of previously useful solutions into more efficiently applicable idioms. Essentially, EBL integrates redundant problem-space operators with existing operators to bias the exploration of the search space. Acquired macro-operators can lead to quick solutions, but in other circumstances they can delay the discovery of a goal.

Other EBL systems represent acquired bias as explicit search-control heuristics for existing problem-space operators. These heuristics typically alter the ordering of alternative choices by promoting heuristically more promising operators so that they are tried first. Some heuristics reject certain operators outright, while others select a particular operator as especially suitable to the current situation (to the detriment of all other operators). As in the macro-operator systems, while heuristics should contribute to a quicker solution, the time spent evaluating these heuristics can slow down the search.

Several problems within this paradigm remain to be addressed. Speeding up real applications requires controlling performance degradation. No doubt, this problem can be alleviated or avoided altogether through clever indexing techniques coupled with heuristics for managing learned information in some semiprincipled fashion. Perhaps the biggest remaining problem is that, unlike inductive learning systems, EBL systems are domain-knowledge intensive. Thus most EBL systems require complete and correct problem-space descriptions (or domain theories). Recently, the analytic-learning community has begun to address the problem of revising inaccurate or incomplete domain theories on the basis of classified examples. This involves repairing inaccuracies in the domain knowledge that are exposed when examples are handled incorrectly by the original domain theory. Thus domain theory revision is a hybrid problem that shares elements with incremental, supervised, inductive learning problems. Starting from an initial theory (a concept description) that might contain some errors, we patch the theory to account for training examples that were misclassified by the current theory.

Genetic algorithms.

Genetic algorithms are inspired by the Darwinian notion of natural selection. First introduced by Holland [30], these algorithms are ideally suited to solving combinatorial optimization problems, since they efficiently search solution spaces for quasi-optimal solutions. For example, in its simplest form, a genetic algorithm might encode a solution to an optimization problem as a bit vector. By applying mutation operators to the best-performing members of a random pool of solutions, the algorithm essentially performs a parallel hill-climbing search for high-quality solutions. After multiple generations, the pool should contain bit vectors representing near-optimal solutions to the initial problem.

Genetic algorithms are adaptive parallel-search algorithms that can locate global maxima without getting trapped in local maxima. Goldberg describes genetic algorithms as search algorithms based on the mechanics of natural selection and natural genetics [23].

A genetic algorithm includes:

(1) a chromosomal representation of a solution to the problem;

(2) a way to create an initial population of solutions;

(3) an evaluation function that rates solutions in terms of 'fitness'; and

(4) genetic operators that alter the composition of solutions during reproduction.

Starting from an initial population of solutions, the genetic algorithm works with one population of solutions at a time. The algorithm evaluates each solution and assigns it a fitness score. By applying recombination and genetic operators to the old population, the algorithm generates a new population of solutions, which it then explores.

Three genetic operators are commonly used. The reproduction operator duplicates the members of the population(solutions) that will be used to derive new members. The number of copies of each member is proportional to its fitness score. After reproduction, new individuals are generated by selecting two individuals at a time from the resulting population and applying the crossover operator. This exchanges genes between the two selected individuals (parents) to form two different individuals, and it is usually applied with a constant probability p_c. The mutation operator randomly changes some genes in a selected individual. It is applied much less often, at a rate of p_m(where $p_m << 1$). When applying a genetic algorithm, the user assigns values for such parameters as the population size, the number of generations, and the probability of applying crossover and mutation operators.

Since the genetic algorithm works with string structures (analogous to chromosomes in biological systems), the solutions should be encoded and represented in string form. This low-level representation is called a genotype, and the corresponding set of apparent characteristics is called a phenotype. The individual elements of the genotype are called genes, and their possible values are alleles.

The basic genetic algorithm is as follows:

Procedure GA (population size n, maximum number of generation Ng)

begin;

select an initial population of n genotypes $\{g\}$;

no-of-generations = 0;

repeat;

```
    for each member b of the population;
        compute f(g), the fitness measure or each member; /*evaluation*/
  repeat;
    stochastically select a pair of
        genotypes g1, g2 with probability
        increasing with their fitness f; /*reproduction*/
        using the genotype representation of g1 and g2, mutate a random bit
        with probability pu; /*mutation*/
    randomly select a crossover point
        and perform crossover on g1 and g2 to
        give new genotypes g1' and g2'; /*crossover*/
    until the new population is filled with n individuals gi';
    no-of-generations = no-of-generations+1;
    until all the members converge; /*termination*/
    end;
```

Genetic algorithms can be incorporated in a performance system in a variety of ways. Perhaps the simplest is to construct a system (for example, a classification system) in which a genetic algorithm adjusts the system's parameters. The system's overall performance would improve as the parameters attain their quasi-optimal settings. A similar architecture might also be used for a problem-solving system.

Several critical factors affect the success of this approach. There must be both an adequate representation of the solution space and an effective set of genetic operators for that representation. Typical applications encode system parameters as bit vectors and rely on biologically inspired mutation and crossover operators. By encoding nonbinary parameters as consecutive bits in the bit vector, crossover operations can generate new bit vectors that differ in a single parameter. Thus, because of this representational locality, the search for a quasi-optimal solution proceeds by tweaking individual parameters of the performance system. Mutation operations can randomize the search, which helps to avoid coming to rest in local minima of the search space.

The development of genetic algorithms has followed a path largely independent of the mainstream machine-learning community, spawning a specialized conference, the International Conference on Genetic Algorithms. Unlike some of the other learning paradigms, work in this area has been largely application driven.

Connectionist learning.

The ground-breaking work on perceptrons in the late 1950s represents some of the earliest work on learning systems. After a hiatus of some 25 years, neurally inspired, fine-grained, massively parallel systems are once more attracting attention. Learning is an integral part of any neurally inspired system; indeed, the development of the backpropagation learning algorithm has largely spurred the recent activity in this area. Unlike the early perceptron work, this algorithm supports the training of networks with internal layers of units separating input and output units. Such networks avoid many of the pitfalls of earlier systems.

As with genetic algorithms, much of the connectionist work is performed within a specialized community. Nevertheless, the basic problem is exactly the same as that addressed by supervised concept learning. Thus some researchers have evaluated the strengths and weakness of connectionist learning schemes and compared them with supervised concept-learning systems.

There are several recurring themes among the articles in this aspect. Firstly, more and more work blends multiple learning paradigms, using techniques from one paradigm to mask or correct the problems of other paradigms. Secondly, applications of learning algorithms are now drawn from more diverse learning paradigms. Thirdly, more attention has been paid to the integration of learning with large database systems.

The principles, modes and algorithms of machine learning have been used in the control systems since the 1960s.

8.3 Schemes of Learning Control

Researchers have proposed a variety of schemes for learning control since the early 1970's. Basically, learning control has the following schemes:

- pattern recognition-based learning control
- iterative learning control
- repetitive learning control
- connectionist learning control including reinforcement learning control
- rule-based learning control including fuzzy learning control
- anthropomorphic self-learning control
- state learning control

and so on.

The learning control possesses four main functions: searching, recognition, memory, and reasoning. In the early research stage of learning control, research was done mainly on searching and recognition, and seldom on memory and reasoning. Similar to the learning systems, there are two kinds of learning control systems; one is the on-line learning control system, and the other is the off-line learning control system as shown in Figure 8.2(a) and Figure 8.2(b) respectively, in which R stands for the reference input, Y — output response, u — control action, s — switch. When the switch is on, the system is in off-line learning state.

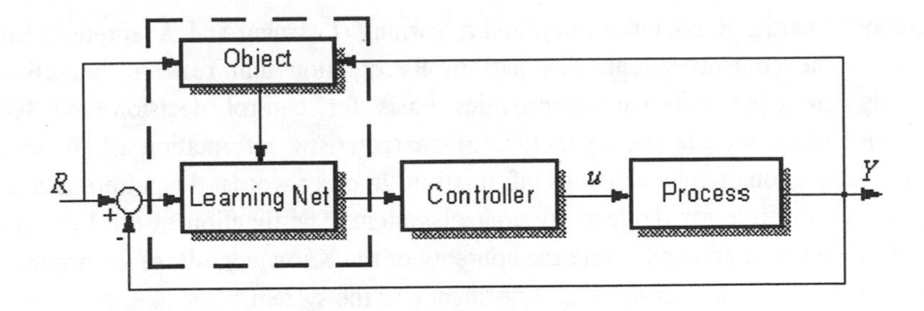

(a) on-line learning control system

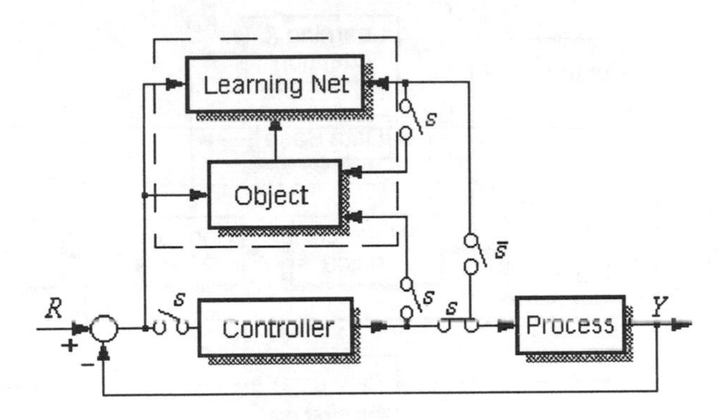

(b) off-line learning control system

Figure 8.2. Block diagrams of learning control systems.

The off-line learning control system is used widely, while the on-line learning control is mainly used in more complex and stochastic environment. The on-line learning control system needs higher speed and larger capability of computers, and spends more time on signal processing. In many cases, the two methods are connected with each other; first, the prior experience is acquired by the off-line method whenever possible, then the on-line learning control is operated.

8.3.1 Pattern Recognition-Based Learning Control

We have introduced the developed process of pattern recognition-based learning control in Section 8.1.3 in some detail. In Section 5.2.2, we proposed a simplified structure of industrial expert controller(see Figure 5.5). In fact, this controller is also a learning controller based on pattern recognition. For the convenience of comparison, the structure of this controller is re-drawn in Figure 8.3, from which it can be seen clearly that a pattern recognition (Feature Recognition) unit and a learning (Learning and Adaptation) unit are included in the control system. The pattern Recognition unit realizes extraction and processing for input information, provides basis for control decision and learning adaptation, which include the extraction of characteristic information of the dynamic process, recognition of characteristics information. In other words, the pattern recognition unit plays a basic role for the learning control system. The function of the Learning and Adaptation unit is to add and revise the contents of the Knowledge Base according to the on-line information, and improve the performance of the system.

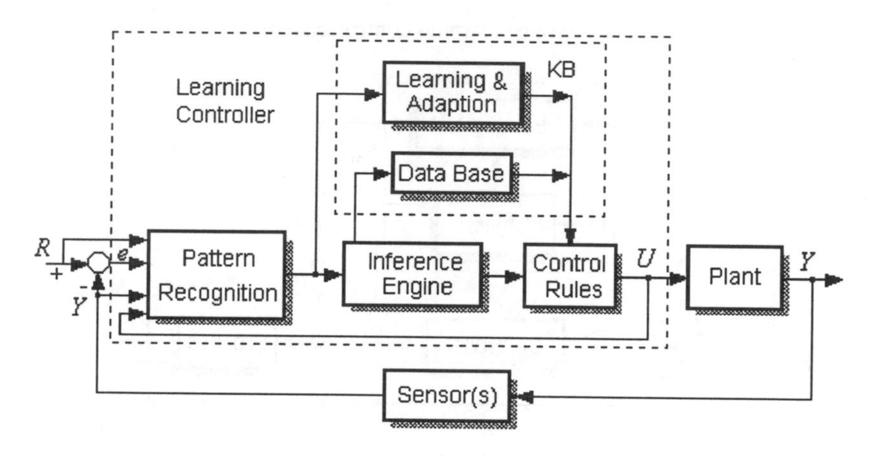

Figure 8.3. A Structure of pattern recognition-based learning control system.

The pattern recognition-based learning control system shown in Figure 8.3 can be generalized to be a hierarchical structure with on-line characteristic identification as depicted in Figure 8.4, from which we know that the control system consists of three levels: the organization level, the self-turning level, and the operation control level. The organization level is implemented by the control rules in the self-learner SL; the self-turning level is used to regulate the controlled parameters of the self-turner ST; and the operation control level is formed by the main controller MC and the coordinator K. The on-line characteristic identifiers CI1 to CI3, the rule bases RB1 to RB3 and the inference engines IE1 to IE3 in MC, ST and SL are placed separately. The common data base CDB is in common use for the three levels so that a close and rapid communication can be made. The processes of information processing and decision-making for every level are described by three triple sequences{A, CM, F}, {B, TM, H} and {C, LM, L} respectively.

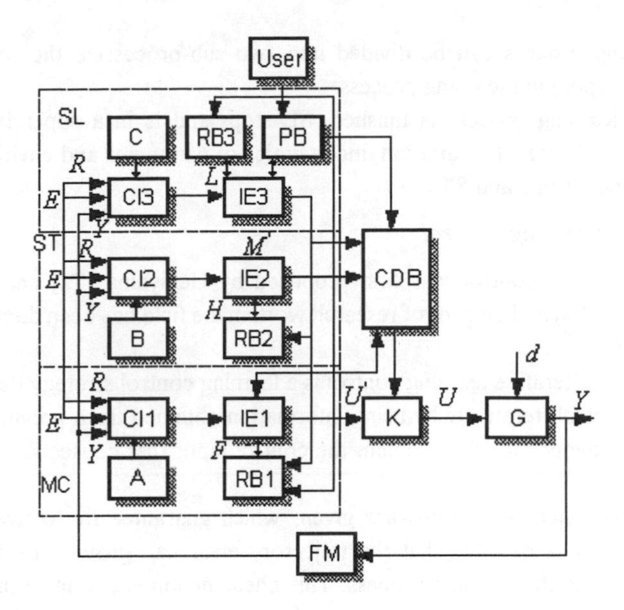

Figure 8.4. A multilevel learning control system.

The on-line information that is command R, system output Y and error E is sent to CI1 of MC and CI2 of ST, compared with the corresponding characteristic model A (system characteristics set of dynamic operations) and model B (system characteristics set of dynamic variations), and identified; the identified information is mapped to the set of control mode CM and the set of parameter correction TM through the set of production rules F and H, then the control output U' and the set of corrected parameters M' are generated. The control output U' enters into the coordinator K and an input vector U of the controlled object G is formed; meanwhile, the M' is input into the common data base CDB to substitute the previous control parameters M.

For operation control level and self-turning level, i.e., for MC and ST, the triple sequences {A, CM, F} and {B, TM, H} are assigned by the designer or the prior knowledge formed in SL, and stored in rule bases RB1, RB2 and CDB. The RB3 in SL is used to store the assembly of the control expertise {C, LM, L} which includes the {A, CM, F}, {B, TM, H}, selecting, modifying and generating rules, and the evaluation of rules. The performance index stored in RB3 involves the set of global index PA and the set of sub-index PB; the PA is given by the user(s); the PB is the decomposition sub-set of the PA and consists of the results of the characteristics identification of CI3; the PB is used as learning basis for the different learning stages and learning types.

The learning process can be divided into two sub-processes, the starting learning process and the operating learning process.

After the learning process is finished, SL stops and is in a supervision state. For constant controlled type, the variation uncertainties of parameter and environment can be rapidly self-tuned by MC and ST.

8.3.2 Iterative Learning Control

The iterative learning control was first proposed by Uchiyama [72] and developed by Arimoto *et al.*[6]. Since then a lot of research work in the field has been done as mentioned in Section 8.1.3.

Definition 8.13. Iterative learning control is a learning control strategy that can improve the quality of control iteratively by using information obtained from previous trials rather than system parameter models to obtain the control input that causes the desired output trajectory finally.

Although sufficient conditions are given, which guarantee the convergence of the learning process, it is possible that the trajectory error can grow quite large before it converges to zero in the learning process. This phenomenon is owing to the fact that the control structure alone does not compensate the output error in each trial. Therefore, the performance in the early stages of learning can be bad for stable plants, and even worse for unstable plants. Employing conventional feedback controllers can be helpful to overcome this kind of a problem in the transient stages of learning since they can compensate the control input to reduce the error [7, 28, 29, 41, 44].

The task of iterative learning control is as follows: giving the current input and current output of the system, determining the next desired input so that the actual output of the system is converged to its desired value. Therefore, in the possible case of parameter uncertainty the best control signal can be obtained through the input-output data obtained from practical operations. The difference between the iterative control and optimal control is that the optimal control computes the optimal input according to the system model, and the iterative control gets the best input from the previous trials. The difference between the iterative control and adaptive control is that the algorithm of the iterative control is implemented off-line after each trial, and the algorithm of the adaptive control is an on-line algorithm and needs a great quantity of computation.

A general block diagram of the iterative learning control system is shown in Figure 8.5, where y_d stands for bounded continuous desired output; u_k – *k*th iterative reference input; u_{k+1} – *(k+1)*th iterative reference input; y_k – *k*th actual iterative output of the closed-loop control system; *k*=1, 2, $\cdots$, *n*.

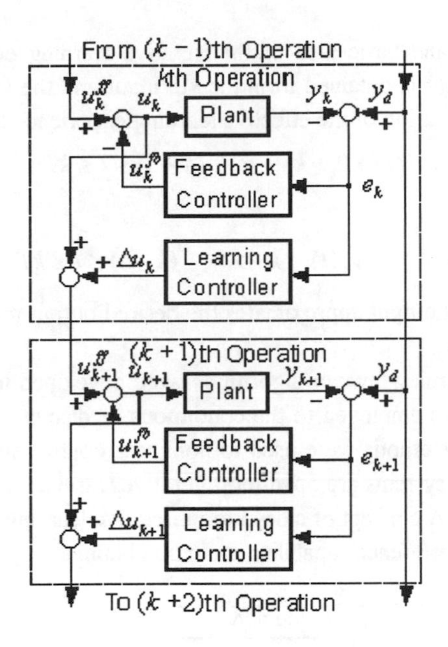

Figure 8.5. Control mechanism for iterative learning control

From Figure 8.5 it can be seen that the total control input consists of two components, one is feedback input u_{k+1}^{fb} generated by the feedback controller (PID or adaptive controller), the other is feedforward input u_{k+1}^{ff} composed by the previous control input u_k and the output of the learning controller Δu_k.

Suppose the controlled object (plant) has the following dynamic process:

$$\begin{cases} \dot{x}_k(t) = f(t, x_k(t), u_k(t)) \\ y_k(t) = g(t, x_k(t), u_k(t)) \end{cases} \tag{8.1}$$

where $x_k \in R^{n\times 1}$, $y_k \in R^{m\times 1}$, $u_k \in R^{r\times 1}$, f, g — vector functions with the corresponding dimension and unknown structure as well as parameter.

Let the error between the desired output $y_d(t)$ and the actual output $y_k(t)$ be:

$$e_k(t) = y_d(t) - y_k(t). \tag{8.2}$$

From Figure 8.5 we can learn that in the *k*th learning the reference input $u_k(t)$ and the revised signal Δu_k are added and stored, then used as *(k+1)*th given input, i.e.,

$$u_{k+1}(t) = L(u_k(t), e_k(t)). \tag{8.3}$$

This gives a clear and basic idea of the iterative learning control, that is, for the $(k+1)$th learning, the input is obtained from the kth input and the kth learning experience; with the iterative accumulation of the effective learning experience the following holds:

$$e_k(t) \to 0, \quad k \to \infty, \quad (k-1) \le t \le kT \tag{8.4}$$

or

$$y_k(t) \to y_d(t), \quad k \to \infty, \quad (k-1) \le t \le kT \tag{8.5}$$

so that the learned actual output approximates the desired output gradually; where T is the learning sampling period.

Several iterative learning laws or algorithms were developed in the middle of 1980's [6, 19, 55, 68]; they were employed to the continuous or discrete linear control systems. Other iterative learning methods were used to nonlinear control systems [10, 21, 22, 28, 44, 45, 58]. All of these systems are open-loop iterative learning control systems in which exists no feedback loop. A concept of closed-loop iterative learning control was proposed, in which a better anti-interference capability could be obtained.

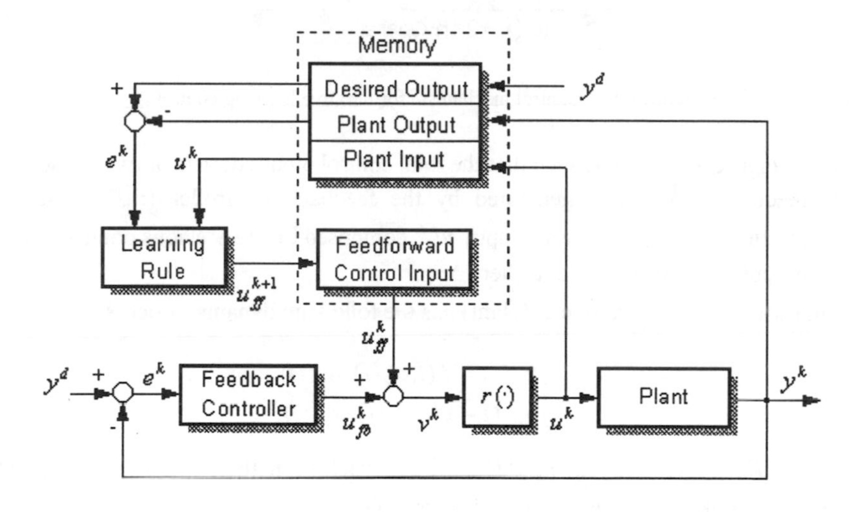

Figure 8.6. Iterative learning control with a feedback controller and an input saturator. (with Jang *et al.* 1995)

Figure 8.6 shows an iterative learning control scheme in feedback systems [36]. This method can achieve precise tracking control of a class of nonlinear systems over a finite time interval, and the learning is done in a feedback configuration, the learning law updates the feedback input from the plant input of the previous trial. By employing plant input saturators the class of nonlinear systems can be extended. A rigorous proof has shown that the convergence condition has no terms reflecting the controller dynamics and thus the feedback controllers have no effect on the convergence condition. The performance of learning can be improved greatly. The learning control scheme is shown in Figure 8.6. The feedforward control input for the next iteration can be updated with the current plant input instead of the current feedforward control input. If a stabilizing controller that gives a reasonably good performance is used and the saturation bounds are set to be sufficiently large, then the feedback control output makes sure that the plant output does not deviate far from the desired output trajectory but stays within a neighborhood of it. Thus, the convergence of the feedforward control input to the desired one can be very fast with the help of the feedback control input. As the feedforward control input makes the actual output follow the desired output trajectory exactly, the output of the feedback controller is zero since the input to the feedback controller is zero.

It is expected that this iterative learning control will be very effective for reducing the error in feedback control systems, and may be used as a tool in nonlinear system theory. The proposed learning control system was applied to the tracking control of a two-link robot manipulator, and a good tracking performance was obtained in the simulation.

8.3.3 Repetitive Learning Control

It is well known that in order to reach a stable tracking under step input or a stable restraining under step disturbance for the servo-controlled system, an integrating compensator has to be introduced into the system. Inversely, as long as an integrating compensator is introduced into the system to make its close-loop system stable, a servosystem without stable error can be realized. According to the internal model principle, for a sinusoidal input with single frequency ω_c of vibration a mechanism with transfer function $1/\left(s^2+\omega_c^2\right)$ is simply set into the closed-loop system as the internal model [15, 16].

If the designed mechanism generates all periodic signal with defined period of L and is set into the closed-loop as the internal model, then arbitrary periodic function with its period of L can be generated by giving an arbitrary initial function corresponding to one period, storing the generated signal, and retrieving the signal repeatedly once for a period L. Therefore, a periodic function generator with period L can be imagined as a time-lag unit shown in Figure 8.7. In fact, let the initial function of the time-lag unit e^{-Ls} be $\varphi(\theta)$, then $\varphi(\theta)$ repeats period by period, and its objective function $r(t)$ can be represented as

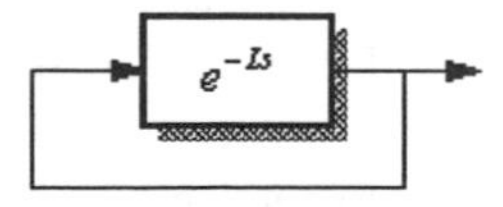

Figure 8.7. Periodic function generator.

$$r(iL+\theta)=\varphi\ (\theta),\quad 0\le\theta\le L, i=0,1,\ldots \tag{8.6}$$

It can be inferred that as the periodic function generator (PFG) is set up in closed-loop as an internal model, then the servosystem without stable error can be constructed for arbitrary objective signals with period L. The periodic function generator is called the repetitive compensator, and the control system that has set up the repetitive compensator is called the repetitive control system as shown in Figure 8.8.

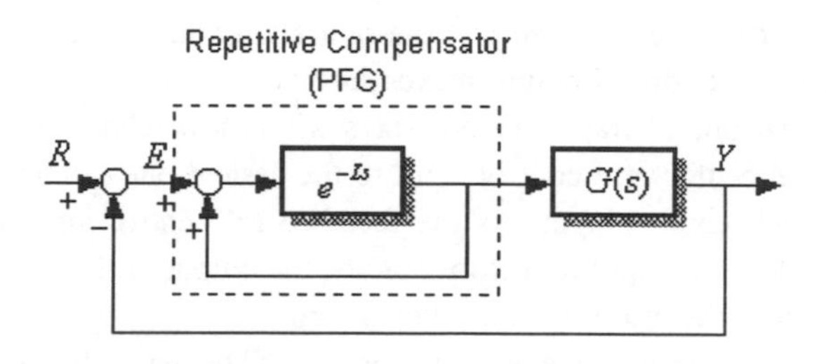

Figure 8.8. Block diagram of repetitive control system.

Repetitive control has a closed relation with the iterative control in their control mode, and use the error function to revive the next input. However, because these two learning controls exist some essential differences as follows.

(1) Repetitive control is constructed in a complete closed-loop system and goes in the continuous model. Inversely, iterative control is independent always, after the trial the initial state of the system is restored. As a result, the stability condition of the system for iterative control is laxer than that for repetitive control.
(2) The convergent conditions for the two controls are different and are defined in different ways.
(3) For iterative control, a deviation of the error is introduced into the revived expression of the control input.
(4) Iterative control can deal with nonlinear systems for which the control input is added linearly.

From the above discussion, we know that iterative control has some advantages. However, iterative control also has its limitation in application [56].

Repetitive control has been used in the servocontrol of a d.c. motor, for the voltage inverter control, and the trajectory control of robot manipulator, etc. [56].

8.3.4 NN-based Learning Control

We have discussed the neurocontrol systems in Chapter 7 in some detail. As a matter of fact, the core of the neurocontrol systems is the neurocontroller (NNC), and the key technique for neurocontrol is the learning (training) algorithms. From the view point of learning, the neurocontrol system is naturally a part of the learning control system. Some may call neurocontrol as connectionist learning control, others may call it neural network-based learning control. We would not add any new content to the neurocontrol system, but one can view Chapter 7 as a section of Chapter 8.

There still are some other mechanisms for learning control. For example, one learning control is based on the fuzzy rule, and a fuzzy-neural learning control system is integrated [27, 37, 43, 48, 52]. Another example is anthropomorphic self-learning control that emulates the human intelligence [13, 50].

8.4 Some Issues for Learning Control

Similar to fuzzy control systems, some properties of learning control systems including dynamic characteristics, stability, convergence, and robustness can be analyzed. Learning control with applications to dynamic systems, such as robot manipulators, has become an interesting and important topic in recent 10 years. A better process and a set of learning control schemes have also been investigated and proposed [4-6, 24]. The motivation for studying properties of learning control can be briefly stated as follows:

(i) Find a relatively simple relation between input and output of dynamic systems;

(ii) Execute each control process updated by learning results of the previous control process;

(iii) Improve the performance of each process so that it is better than the preceding ones.

In this section the learning control modeling for a PID-type learning control system will be discussed first, then the stability and convergence of the CMAC-based on-line and off-line learning control will be analyzed.

8.4.1 Modeling for Learning Control Systems [5]

1. Learning control law for robotic systems

Suppose $\mathbf{u}_k(t)$ and $\mathbf{y}_k(t)$ are the input and output functions of the *k*th control process of a dynamic system respectively, $\mathbf{u}_{k+1}(t)$ is the input function for the (*k*+1)th control process, and $\mathbf{y}_d(t)$ is defined as the output vector of a desired trajectory, then a PID-type learning control law can be written as [5]:

$$\mathbf{u}_{k+1}(t) = \mathbf{u}_k(t) + (\phi + \Gamma d/dt)(\mathbf{y}_d(t) - \mathbf{y}_k(t)) + \psi\int_0^t [\mathbf{y}_d(\tau) - \mathbf{y}_k(\tau)]d\tau \tag{8.7}$$

where $\mathbf{u}_k(t), \mathbf{u}_{k+1}(t) \in \mathbf{R}^n, \mathbf{y}_k(t), \mathbf{y}_d(t) \in \mathbf{R}^r$, and ϕ, ψ and Γ are n by r constant coefficient matrices. The difference $\mathbf{y}_d(t) - \mathbf{y}_k(t)$ is the error in the *k*th process, and will be denoted as

$$\mathbf{e}_k(t) = \mathbf{y}_d(t) - \mathbf{y}_k(t). \tag{8.8}$$

The convergence in the learning control scheme (8.7) requires that

$$\|\dot{\mathbf{e}}_{k+1}(t)\| < \|\dot{\mathbf{e}}_k(t)\| \text{ and } \mathrm{e}_k(0) = 0, \text{ for } k=1, 2, \ldots$$

This implies

$$\mathbf{e}_k(t) \to 0 \text{ as } k \to \infty. \tag{8.9}$$

It may be observed that for any type of learning control schemes, sufficient conditions of convergence are subject to the general state equation of robotic systems. Suppose a robotic system can be modeled by

$$\left.\begin{aligned} \dot{\mathrm{x}}(t) &= \mathrm{f}(\mathrm{x}(t)) + \mathrm{B}(\mathrm{x}(t))\mathrm{u}(t) \\ \mathrm{y}(t) &= \mathrm{h}(\mathrm{x}(t)) \end{aligned}\right\} \tag{8.10}$$

where $\mathbf{x}(t) \in \mathbf{R}^{2n}, \mathbf{u}(t) \in \mathbf{R}^{n}, \mathbf{y}(t) \in \mathbf{R}^{r}$ and $\mathbf{B}(\mathbf{x}(\mathrm{t}))$ is a $2n$ by n matrix. In the sequel, the time dependence of various variables will be suppressed whenever no confusion is caused. Then, for a PD-type of learning control $(\psi = 0)$, the time-derivative of the (k+1)th process error $\dot{\mathbf{e}}_{k+1}$ can be written in terms of $\mathbf{e}_k$ and $\dot{\mathbf{e}}_k$, the kth process error and its time-derivative, i.e.

$$\begin{aligned} \dot{\mathbf{e}}_{k+1} = \dot{\mathbf{y}}_d - \dot{\mathbf{y}}_{k+1} &= (\mathbf{I} - \mathbf{J}_h(\mathbf{x}_{k+1})\mathbf{B}(\mathbf{x}_{k+1})\Gamma)\dot{\mathbf{e}}_k \\ &+ \mathbf{J}_h(\mathbf{x}_{k+1})\left[\mathbf{f}(\mathbf{x}_k) - \mathbf{f}(\mathbf{x}_{k+1})\right] + \left[\mathbf{J}_h(\mathbf{x}_k) - \mathbf{J}_h(\mathbf{x}_{k+1})\right]\dot{\mathbf{x}}_k \\ &+ \mathbf{J}_h(\mathbf{x}_{k+1})\left[\mathbf{B}(\mathbf{x}_k) - \mathbf{B}(\mathbf{x}_{k+1})\right]\mathbf{u}_k - \mathbf{J}_h(\mathbf{x}_{k+1})\mathbf{B}(\mathbf{x}_{k+1})\phi\,\mathbf{e}_k \end{aligned} \tag{8.11}$$

In literature [5], it was derived that $\dot{\mathbf{e}}_{k+1} \to 0$ as $k \to \infty$ if certain conditions, such as $\mathbf{f}(\mathbf{x})$ being Lipschitz continuous, and

$$\|\mathbf{I} - \mathbf{J}_h(\mathbf{x}_{k+1})\mathbf{B}(\mathbf{x}_{k+1})\Gamma\| < 1 \tag{8.12}$$

can be satisfied. It is hard to find a constant matrix Γ such that the condition (8.12) generally holds, because matrices $\mathbf{J}_h$ and $\mathbf{B}$ are the functions of $\mathbf{x}_{k+1}$. If an upper bound of the norm $\|\mathbf{J}_h(\mathbf{x}_{k+1})\mathbf{B}(\mathbf{x}_{k+1})\|$ can be found, or if $\mathbf{J}_h(\mathbf{x}_{k+1})\mathbf{B}(\mathbf{x}_{k+1})$ is a constant matrix and independent of $\mathbf{x}_{k+1}$ in the more special case, then an appropriate Γ can be selected to satisfy the condition (8.12). For the remaining terms of (8.11), it is also not easy to show their influence on the convergence characteristics and the convergence speed of (8.9). However, it can be shown that if a position vector at either the Cartesian level or the joint level is chosen as an output $\mathbf{y}(t)$, then the left side of (8.12) becomes 1 and convergence becomes impossible. Since in robot model (8.10), $\mathbf{B}(\mathbf{x}) = (0 \quad \mathbf{B}_2^T)^T$ for some n by n matrix

$\mathbf{B}_2$, and $\mathbf{y} = \mathbf{h}(\mathbf{x}) = \mathbf{h}(\mathbf{q})$ if some position function $\mathbf{h}(\mathbf{q})$ depending on $\mathbf{q}$ is usually chosen, then

$$\mathbf{J}_h(\mathbf{x}) = \left(\frac{\partial \mathbf{h}}{\partial \mathbf{x}}\right)^T = \left[\left(\frac{\partial \mathbf{h}}{\partial \mathbf{q}}\right)^T \quad \left(\frac{\partial \mathbf{h}}{\partial \dot{\mathbf{q}}}\right)^T\right] = \left[\left(\frac{\partial \mathbf{h}}{\partial \mathbf{q}}\right)^T \quad 0\right] \tag{8.13}$$

so that $\mathbf{J}_h(\mathbf{x})\mathbf{B}(\mathbf{x}) = 0$ and the norm on the left side of (8.12) becomes 1 for all Γ. Thus, the choice of output function has to be limited to velocity or mixture of velocity and position.

Let a state vector $\mathbf{x}^T = (\mathbf{q}^T \quad \mathbf{p}^T) \in \mathbf{R}^{2n}$, where $\mathbf{q}$ is a joint position vector and $\mathbf{p}$ is a generalized momentum defined by

$$\mathbf{p} = \frac{\partial K}{\partial \dot{\mathbf{q}}} = \mathbf{W}\dot{\mathbf{q}} \tag{8.14}$$

in which $\mathbf{W}$ is an n by n total inertial matrix, then the general state equation (8.10) is specified by [24]

$$\mathbf{f}(\mathbf{x}) = \begin{bmatrix} \mathbf{W}^{-1}\mathbf{p} \\ -\frac{1}{2}\left(\mathbf{I} \otimes \mathbf{p}^T\right)\frac{\partial \mathbf{W}^{-1}}{\partial \mathbf{q}}\mathbf{p} \end{bmatrix} \quad \text{and} \quad \mathbf{B}(\mathbf{x}) = \begin{bmatrix} 0 \\ \mathbf{I} \end{bmatrix} \tag{8.15}$$

where $K = \frac{1}{2}\dot{\mathbf{q}}^T \mathbf{W}\dot{\mathbf{q}}$ is kinetic energy, and $\otimes$ denotes the Kronecker product. From (8.15), $\mathbf{B}$ becomes a constant matrix of the simplest form. Also, if output $\mathbf{y} = \mathbf{p}$, then

$$\mathbf{J}_h(\mathbf{x}) = \left(\frac{\partial \mathbf{h}}{\partial \mathbf{x}}\right)^T = \begin{pmatrix} 0 & \mathbf{I} \end{pmatrix} \tag{8.16}$$

and relation (8.11) is thus reduced to

$$\dot{\mathbf{e}}_{k+1} = (\mathbf{I} - \Gamma)\dot{\mathbf{e}}_k + (0 \quad \mathbf{I})\left[\mathbf{f}(\mathbf{x}_k) - \mathbf{f}(\mathbf{x}_{k+1})\right] - \phi\,\mathbf{e}_k\,. \tag{8.17}$$

The second term of (8.17) may be rewritten as a difference of two partial derivatives of kinetic energy based on the well-known Hamiltonian equation, and denoted by

$$\left.\frac{\partial K}{\partial \mathbf{q}}\right|_k - \left.\frac{\partial K}{\partial \mathbf{q}}\right|_{k+1} = \Delta(k, k+1) \tag{8.18}$$

where the partial derivative is taken under fixed $\dot{\mathbf{q}}$ (if under fixed $\mathbf{p}$, as in the case of (8.15), a minus sign should be added to each partial derivative term in (8.18)). Therefore, if we select D-type of learning control law for robotic systems described by (8.10), (8.15) and

(8.16), and let $\Gamma = 1$ for k = 1, 2, ..., then

$$\dot{\mathbf{e}}_{k+1} = \Delta(k, k+1) \tag{8.19}$$

Thus, it follows that $\dot{\mathbf{e}}_{k+1} \to 0$ as $k \to \infty$ if and only if $\Delta(k, k+1) \to 0$ as $k \to \infty$. Furthermore, for the D-type of learning control law, since $\phi = \psi = 0$ and Γ is chosen to be **I**, (8.7) becomes

$$\begin{aligned}\mathbf{u}_{k+1} &= \mathbf{u}_k + \dot{\mathbf{e}}_k = \mathbf{u}_k + \dot{\mathbf{p}}_d - \dot{\mathbf{p}}_k \\ &= \mathbf{u}_k + \left.\frac{\partial K}{\partial \mathbf{q}}\right|_d + \mathbf{u}_d - \left.\frac{\partial K}{\partial \mathbf{q}}\right|_d - \mathbf{u}_k = \mathbf{u}_d + \Delta(d, k)\end{aligned} \tag{8.20}$$

It shows that if $\Delta(d, k) \to 0$ as $k \to \infty$, then $\mathbf{u}_{k+1} \to \mathbf{u}_d$. Therefore, in choosing $\mathbf{y}$ = $\mathbf{p}$ and $\Gamma = 1, \phi = \psi = 0$, the convergence of the learning control process only requires that the function $\partial K / \partial \mathbf{q}$ be Lipschitz continuous.

2. **Realization and simulation in robotic systems**

In robotic systems, the generalized momentum **p** cannot be directly measured. Only after **q** and $\dot{\mathbf{q}}$ have been measured for each step, **p** can be computed by substituting **q** and $\dot{\mathbf{q}}$ into (8.14). On the other hand, the learning control law is now reduced to

$$\mathbf{u}_{k+1} = \mathbf{u}_k + \dot{\mathbf{e}}_k = \mathbf{u}_k + \dot{\mathbf{p}}_d - \dot{\mathbf{p}}_k . \tag{8.21}$$

The time-derivative of **p** requires a number of computation steps in the following symbolic formula

$$\dot{\mathbf{p}} = \mathbf{W}\ddot{\mathbf{q}} + \dot{\mathbf{W}}\dot{\mathbf{q}} . \tag{8.22}$$

This needs a computation of almost the entire dynamic formulation and is time-consuming. In order to guarantee a quick convergence for the learning control process, we retain the computation of **p**, since it is reasonable, and try to modify the computation of (8.22).

For the learning control scheme given by (8.7), an updated input depends only upon an output error between the desired and previous process data at a sampling point. In other words, the learning process just learns one 'point' in the previous process for updating a new 'point'. The human learning process, however, often gathers at least a few 'old points' to acquire the knowledge and decide what to do for the next process. Intuitively, the more points the learning process gathers, quicker and more reliable the convergence it could achieve. On the basis of the human learning manner, we wish to improve the one-point learning control scheme (8.7). An easy and attractive way is to apply a differencing method to the time-derivative of the error $\mathbf{e}_k(t)$, namely let

$$\dot{\mathbf{e}}_k(t) = \frac{1}{h}\sum_{j=-m}^{m} a_j \mathbf{e}_k(i+j-s) \tag{8.23}$$

where i is a point index in the kth process, h is the sampling interval between two adjacent points in the same process, each a_j is a constant interpolation coefficient of difference equation, and the integer s determines where those $2m+1$ learning points are located. The optimal value for each a_j is subject to the number of neighboring points interpolated in the summation of (8.23). For example

$$\begin{aligned} &m=1, \quad (a_{-1}, a_0, a_1) = (s-1/2, -2s, s+1/2); \\ &m=2, \quad (a_{-2}, a_{-1}, a_0, a_1, a_2) = ((2s^3-3s^2-s+1)/12, \\ &\qquad (4s^3-3s^2-8s+4)/6, (2s^3-5s)/2, (4s^3+3s^2-8s-4)/6, \\ &\qquad (2s^3+3s^2-s-1)/12) \end{aligned} \tag{8.24}$$

Thus, the new learning control version becomes

$$\mathrm{u}_{k+1}(i) = \mathrm{u}_k(i) + \frac{1}{h}\sum_{j=-m}^{m} a_j \mathrm{e}_k(i+j-s) \tag{8.25}$$

where $\mathbf{e}_k(i) = \mathbf{p}_d(i) - \mathbf{p}_k(i)$.

Equation (8.25) shows that the new input at each sampling point is a learning result from the old (previous) output data at a number of neighboring points. Obviously, it requires a memory of the output information for each entire process. Once the use of the output information has been completed for updating all new inputs of the next process, the memory can be cleared for the next output storage.

The objective of implementing learning control law (8.25) in robotic systems is to achieve a quick convergence. Although the computational load of the generalized momentum $\mathbf{p}$ for each point in each process may be heavier than that of other kinds of velocity vectors, the desired and first trial generalized momentums $\mathbf{p}_d$ and $\mathbf{p}_1$ can be computed off-line. Furthermore, it has been found that even the second trial $\mathbf{u}_2(t)$ may control the process to produce an output $\mathbf{p}_2$ which is adequately close to the desired $\mathbf{p}_d$.

A simulation study of the new learning control version (8.25) was accomplished for a PUMA 560 robot manipulator, and the procedure of this simulation was defined [25]. From this simulation the following conclusion can be drawn [24]:

(i) The first process has a large output error, because the trial input $\mathbf{u}_1(t) = 0$ is arbitrarily selected. But after the second trial, the output has converged significantly close to the desired one.

(ii) For each process, the initial output must be the same as the desired one, i.e., $\mathbf{e}_k(0) = 0$ for any k = 1, 2, ... If not so, the process will converge to a trajectory parallel to the desired one with a constant distance. This is consistent with the convergence conditions given in Ref. 4.

(iii) Since the convergence of the learning control law (8.25) mainly depends upon $\Delta(k,k+1)$ according to (8.19), if a given desired trajectory has a relatively small variation of $\partial K/\partial\mathbf{q}$, then the convergence of the learning process may become so quick that even the second control process can reach the desired one. But since $K=\frac{1}{2}\dot{\mathbf{q}}^T\mathbf{W}\dot{\mathbf{q}}$ and $\partial K/\partial\mathbf{q}=\frac{1}{2}\left[(\mathbf{I}\otimes\dot{\mathbf{q}}^T)\partial\mathbf{W}/\partial\mathbf{q}\right]\dot{\mathbf{q}}$, when a robot moves fast or is stretching out or pulling back, $\partial K/\partial\mathbf{q}$ becomes significant, and the convergence $\Delta(k,k+1)\rightarrow 0$ may slow down. Nonetheless, the number of factors which have to be considered for quick convergence in learning control has been reduced to one in this paper when compared with any other type of learning control scheme.

8.4.2 Stability & Convergence Analysis of Learning Control

1. **Stability and convergence analysis for off-line learning control system** [42, 51]
We have introduced the CMAC based neurocontrol in Section 7.4.7. For further studying, we analyze the CMAC learning control systems in two respects: off-line control and on-line control.

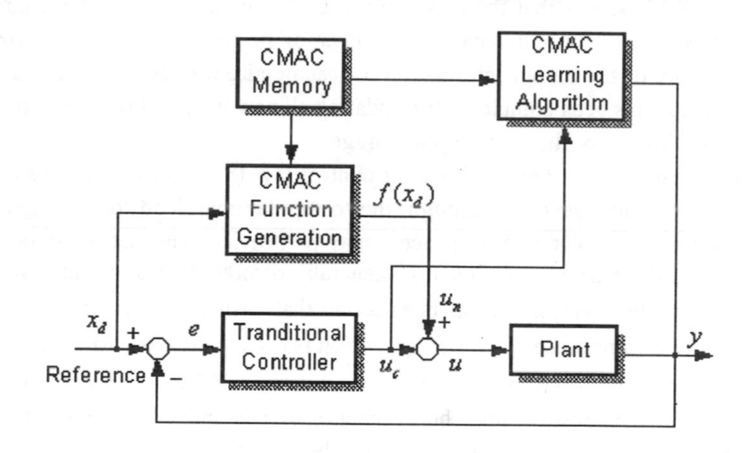

Figure 8.9. CMAC off-line learning control system.

According to the control strategy by Kawato [42], a structure of CMAC off-line learning control system is shown in Figure 8.9 that is similar to Figure 7.25. The control input u of the plant consists of two components:

$$\left.\begin{array}{l} u(t)=u_c(t)+u_n(t) \\ u_n(t)=f(x_d(t)) \end{array}\right\}. \tag{8.26}$$

Consider a class of nonlinear system:

$$\dot{y}(t)=f(y(t),t)+g(y(t),t)u(t) \tag{8.27}$$

where $y^T(t)=\left[y_1^T(t),y_2^T(t)\right]$, $y_1^T(t)\in R^n, y_2^T(t)\in R^n$, $f(y(t),t)=\begin{bmatrix} y_2(t) \\ f_1(y(t),t)\end{bmatrix}$, $g(y(t),t)=\begin{bmatrix} 0 \\ g_1(y(t),t)\end{bmatrix}$, $f_1(y(t),t)\in R^n, g_1(y(t),t)\in R^n$.

Suppose the system satisfies the following conditions:

(i) The system is repeatable in domain $[0, T]$;

(ii) $f_1(\_,\_)$ and $g_1(\_,\_)$ is Lipshitz continuous functions

$$\begin{array}{l} \left|f_1(y_1(t),t)-f_1(y_2(t),t)\right|\le k_f\left|y_1(t)-y_2(t)\right| \\ \left|g_1(y_1(t),t)-g_1(y_2(t),t)\right|\le k_g\left|y_1(t)-y_2(t)\right| \end{array}. \tag{8.28}$$

(iii) $g_1(\_,\_)$ is symmetrical positive definite matrix, and satisfies the following inequality:

$$0<\lambda_1\mathbf{I}\le g_1(y(t),t)\le\lambda_2\mathbf{I}. \tag{8.29}$$

According to Equation (8.26), the control magnitude of the system in the jth iteration can be described as:

$$u^j(t)=u_c^j(t)+u_n^j(t). \tag{8.30}$$

After the jth iterative control the weight of the CMAC net in learning phase can be trained as:

$$\begin{aligned} \mathbf{W}^{j+1}(t) &= \mathbf{W}^j(t)+\frac{\beta}{cq}\cdot\left(u(t)-u_n^j(t)\right)\cdot a_1^j(t) \\ &= \mathbf{W}^j(t)+\frac{\beta}{cq}\cdot u_c^j(t)\cdot a_1^j(t) \end{aligned} \tag{8.31}$$

where $a_1(t)$ is the associative vector of the CMAC and is dependent upon the practical state $y^j(t)$ of the system.

In the $(j+1)$th control, the feed-forward learning control magnitude of the system can be obtained:

$$u_n^{j+1}(t) = (\mathbf{W}^{j+1}(t))^T \cdot a_2(t)$$
$$= \mathbf{W}^{jT}(t) \cdot a_2(t) + \left(\frac{\beta}{cq} u_c^j(t) \cdot a_1^j(t)\right)^T \cdot a_2(t) \tag{8.32}$$
$$= u_n^j(t) + \eta u_c^j(t)$$

where a_2 is the associative vector of the CMAC, and is dependent upon the ideal state $y_d(t)$ of the system and independent of the iterative number j; $\eta = \frac{\beta}{cq} a_1^{jT}(t) a_2(t)$ is a constant.

Theorem 8.1. Apply control (8.32) to the nonlinear system (8.27), there exists a feedback control

$$u_c(t) = \gamma\delta\lambda^{-1}(x_{1d}(t) - y_1(t)) + (\gamma + \delta)\lambda^{-1}(x_{2d}(t) - y_2(t)) \quad (\gamma > 0, \delta > 0) \tag{8.33}$$

so that the state error of the system is bounded

$$|x_d(t) - y^j(t)| \le \sqrt{1 + 4\delta^2} \cdot \frac{\lambda_2 |u_d(t) - u_n^j(t)|}{\delta(\gamma - \theta)} \tag{8.34}$$

where $\theta = \delta(\lambda_1^{-1}\lambda_2 - \mathbf{I}) + (2 + 1/\delta)(k_f + k_g |u_d|_m)$.

The proof can be found in Ref. 26.

Theorem 8.2. Suppose $u_d(t)$ and $\dot{g}_1^{-1}(y^j(t), t)$ are bounded functions:

$$|u_d(t)|_m \le u_0 < \infty$$
$$\|\dot{g}_1^{-1}(y^j(t), t)\| \le g_0 < \infty \tag{8.35}$$

and the feedback gains γ, δ in Theorem 8.1 satisfy the inequality:

$$(\gamma + \delta)(2(\gamma + \delta) - \eta) - \gamma\delta \ge 0. \tag{8.36}$$

Define:

$$m_1 = 2(\lambda + \delta)\lambda_1^{-1} - \eta\lambda_1^{-1} - 2g_0 > 0$$
$$m_2 = 2(\gamma + \delta)\lambda_1^{-1} - \eta\lambda_1^{-1} - g_0 - 2\gamma\delta\lambda_1^{-1} > 0 \tag{8.37}$$
$$m_3 = 1 - \frac{(g_0 + \upsilon/\gamma\delta + \upsilon/(\gamma + \delta))^2}{m_1 m_2} - \frac{4\upsilon}{u} > 0$$

where $u = \min(\gamma\delta m_1, (\gamma + \delta)m_2)$, $\upsilon = k_f + k_g u_0 / \lambda_1$, make the system convergence:

$$\lim_{j \to \infty} y^j(t) = x_d(t). \tag{8.38}$$

$$\lim_{j\to\infty} u_n^j(t) = u_d(t). \tag{8.39}$$

[Prove] From Equation (8.30) and (8.33) we know that the *j*th control input is:

$$\begin{aligned} u^j(t) &= u_c^j(t) + u_n^j(t) \\ &= \gamma\delta\lambda_1^{-1}(x_{1d}(t) - y_1^j(t)) + (\gamma+\delta)\lambda_1^{-1}(x_{2d}(t) - y_2^j(t)) + u_n^j(t). \end{aligned} \tag{8.40}$$

Substitute Equation (8.40) into Equation (8.27), we get

$$\dot{\tilde{y}}^j(t) = \begin{bmatrix} 0 & \mathbf{I}_{n\times n} \\ -\gamma\delta\cdot p^j(t) & -(\gamma+\delta)\cdot p^j(t) \end{bmatrix} \tilde{y}^j(t) + \begin{bmatrix} 0 \\ q^j(t) \end{bmatrix} \tag{8.41}$$

where

$$\begin{aligned} p^j(t) &= \lambda_1^{-1}\cdot g_1(y^j(t),t) \\ q^j(t) &= f_1(x_d(t),t) - f_1(y^j(t),t) + g_1(x_d(t),t)\cdot u_d(t) - g_1(y^j(t),t)\cdot u_n^j(t) \\ &= \left(f_1(x_d(t),t) - f_1(y^j(t),t)\right) + \left(g_1(x_d(t),t) - g_1(y^j(t),t)\right)u_d(t) \\ &\quad + g_1(y^j(t),t)\cdot\left(u_d(t) - u_n^j(t)\right) \end{aligned}$$

in $t \in \left[0, t_f\right]$ we define objective function:

$$V^j(t) = \int_0^t \left(\frac{\lambda_1}{\eta}\cdot \tilde{u}^{jT}(\tau)\cdot\tilde{u}^j(\tau)\right)d\tau \tag{8.42}$$

where $\tilde{u}^j(t) = u_d(t) - u_n^j(t)$.

As $j=1$, $u_n^j(t) = 0$, for $\forall t \in \left[0, t_f\right]$

$$V^j(t) = \int_0^t \left(\frac{\lambda_1}{\eta}\cdot \tilde{u}^{jT}(\tau)\cdot\tilde{u}^j(\tau)\right)d\tau < \infty. \tag{8.43}$$

Let $\omega^j(t) = \gamma\delta\cdot\tilde{x}_1^j(t) + (\gamma+\delta)\cdot\tilde{x}_2^j(t)$,
then the learning law (8.32) is turned into:

$$\begin{aligned} \Delta\tilde{u}^j(t) &= \tilde{u}^{j+1}(t) - \tilde{u}^j(t) \\ &= u_n^j(t) - u_n^{j+1}(t) \\ &= -\eta u_c^j(t) \\ &= -\eta\lambda_1\omega^j(t). \end{aligned} \tag{8.44}$$

From Equation (8.38) we have:

$$
\begin{aligned}
\dot{\omega}^{j}(t) &= \gamma\,\delta\,\dot{\tilde{y}}_1^j(t)+(\gamma+\delta)\dot{\tilde{y}}_2^j(t) \\
&= \gamma\,\delta\,\dot{\tilde{y}}_1^j(t)+(\gamma+\delta)(-\lambda_1^{-1}\cdot g_1(y^j(t),t)\cdot\omega^j(t)+\tilde{f}_1^j(t) \\
&\qquad +\tilde{g}_1^j(t)\cdot u_d(t)+g_1(y^j(t),t)\cdot\tilde{u}^j(t))
\end{aligned} \tag{8.45}
$$

where

$$
\tilde{f}_1^j(t)=f_1(x_d(t),t)-f_1(y^j(t),t)\,. \tag{8.46}
$$

$$
\tilde{g}_1^j(t)=g_1(x_d(t),t)-g_1(y^j(t),t)\,. \tag{8.47}
$$

Define

$$
\Delta V^j(t)=V^{j+1}(t)-V^j(t) \tag{8.48}
$$

then

$$
\begin{aligned}
\Delta V^j(t) &= \int_0^t \frac{\lambda_1}{\eta}\cdot\left(\Delta\tilde{u}^{jT}(\tau)\cdot\Delta\tilde{u}^j(\tau)+2\Delta\tilde{u}^{jT}(\tau)\cdot\tilde{u}^j(\tau)\right)d\tau \\
&= \int_0^t(\eta\lambda_1^{-1}\omega^{jT}(\tau)\cdot\omega^j(\tau)-2\omega^{jT}(\tau)\cdot g_1^{-1}(y^j(\tau),\tau)\cdot(\dot{\omega}^j(\tau) \\
&\quad +(\gamma+\delta)\lambda_1^{-1}\cdot g_1(y^j(\tau),\tau)\cdot\omega^j(t) \\
&\quad -(\gamma+\delta)(\tilde{f}_1^j(\tau)+\tilde{g}_1^j(\tau)\cdot u_d(\tau))-\gamma\,\delta\,\tilde{y}_2^j(\tau)))d\tau \\
&= -\omega^{jT}(t)\cdot g_1^{-1}(y^j(t),t)\cdot\omega^j(t) \\
&\quad -\int_0^t(\omega^{jT}(\tau)\cdot((2(\gamma+\delta)-\eta)\lambda_1^{-1}\mathbf{I}-\dot{g}_1^{-1}(y^j(\tau),\tau))\cdot\omega^j(\tau) \\
&\quad -2\omega^{jT}(\tau)\cdot g_1^{-1}(y^j(\tau),\tau)\cdot((\gamma+\delta)(\tilde{f}_1^j(\tau)+\tilde{g}_1^j(\tau)u_d(\tau))+\gamma\,\delta\cdot\tilde{y}_2^j(\tau)))d\tau \\
&= -\omega^{jT}(t)\cdot g_1^{-1}(y^j(t),t)\cdot\omega^j(t)-\gamma\,\delta\cdot\tilde{y}_1^{jT}(t)\cdot((\gamma+\delta)(2(\gamma+\delta)-\eta)\lambda_1^{-1}\mathbf{I} \\
&\quad -\gamma\,\delta\cdot g_1^{-1}(y^j(t),t))\cdot\tilde{y}_1^j(t)-\int_0^t\Delta W^j(\tau)d\tau
\end{aligned} \tag{8.49}
$$

where

$$
\begin{aligned}
\Delta W^j(t) &= \gamma^2\delta^2\tilde{y}_1^{jT}(t)\cdot\left((2(\gamma+\delta)-\eta)\lambda_1^{-1}\mathbf{I}\text{-}2\dot{g}_1^{-1}(y^j(t),t)\right)\cdot\tilde{y}_1^j(t) \\
&\quad +(\gamma+\delta)^2\cdot\tilde{y}_2^{jT}(t)\cdot\left((2(\gamma+\delta)-\eta)\lambda_1^{-1}\mathbf{I}\text{-}\dot{g}_1^{-1}(y^j(t),t)-2\gamma\,\delta\cdot\dot{g}_1^{-1}(y^j(t),t)\right)\cdot\tilde{y}_2^j(t) \\
&\quad -\gamma\,\delta(\gamma+\delta)\cdot\left(\tilde{y}_1^{jT}(t)\cdot\dot{g}_1^{-1}(y^j(t),t)\cdot\tilde{y}_2^j(t)+\tilde{y}_2^{jT}(t)\cdot\dot{g}_1^{-1}(y^j(t),t)\cdot\tilde{y}_1^j(t)\right) \\
&\quad -2\left(\gamma\,\delta\cdot\tilde{y}_1^j(t)+(\gamma+\delta)\cdot\tilde{y}_2^j(t)\right)^T\cdot g_1^{-1}(y^j(t),t)\cdot(\gamma+\delta)\left(\tilde{f}_1^j(t)+\tilde{g}_1^j(t)u_d(t)\right)
\end{aligned} \tag{8.50}
$$

From the definitions $m_i\,(i=1,2,3)$ and $f_1(y^j(t),t), g_1(y^j(t),t)$ are Lipschitz continuous functions, we have:

$$\Delta W^j(t) \geq \gamma^2\delta^2 m_1 \cdot \left|\tilde{y}_1^j(t)\right|^2 + (\gamma+\delta)^2 m_2 \left|\tilde{y}_2^j(t)\right|^2 - 2\gamma\delta(\gamma+\delta) g_0 \cdot \left|\tilde{y}_1^j(t)\right| \cdot \left|\tilde{y}_2^j(t)\right|$$
$$-2\upsilon \cdot \left((\gamma+\delta+\gamma\delta) \cdot \left|\tilde{y}_1^j(t)\right| \cdot \left|\tilde{y}_2^j(t)\right| + \gamma\delta \cdot \left|\tilde{y}_1^j(t)\right|^2 + (\gamma+\delta) \cdot \left|\tilde{y}_2^j(t)\right|^2\right)$$
$$= \frac{1}{2} m_1 \cdot \left(\gamma\delta \cdot \left|\tilde{y}_1^j(t)\right| - (\gamma+\delta) \frac{g_0 + \frac{\upsilon}{\gamma\delta} + \frac{\upsilon}{\gamma+\delta}}{m_1} \cdot \left|\tilde{y}_2^j(t)\right|\right)^2$$
$$+ \frac{1}{2} m_2 \cdot \left((\gamma+\delta) \cdot \left|\tilde{y}_2^j(t)\right| - \gamma\delta \frac{g_0 + \frac{\upsilon}{\gamma\delta} + \frac{\upsilon}{\gamma+\delta}}{m_2} \cdot \left|\tilde{y}_1^j(t)\right|\right)^2$$
$$+ \frac{1}{2}\gamma^2\delta^2 m_2 \cdot \left(1 - \frac{\left(g_0 + \frac{\upsilon}{\gamma\delta} + \frac{\upsilon}{\gamma+\delta}\right)^2}{m_1 m_2} - \frac{4\upsilon}{\gamma\delta m_1}\right)^2 \cdot \left|\tilde{y}_1^j(t)\right|^2$$
$$+ \frac{1}{2}(\gamma+\delta)^2 m_2 \cdot \left(1 - \frac{\left(g_0 + \frac{\upsilon}{\gamma\delta} + \frac{\upsilon}{\gamma+\delta}\right)^2}{m_1 m_2} - \frac{4\upsilon}{(\gamma+\delta) m_2}\right)^2 \cdot \left|\tilde{y}_2^j(t)\right|^2$$
$$\geq \frac{1}{2} m_3 \cdot \left(\gamma^2\delta^2 m_1 \cdot \left|\tilde{y}_1^j(t)\right|^2 + (\gamma+\delta)^2 m_2 \left|\tilde{y}_2^j(t)\right|^2\right) \geq 0 \tag{8.51}$$

$\Rightarrow$

$$\Delta V^j(t) \leq -\omega^{jT}(t) \cdot g_1^{-1}(y^j(t),t) \cdot \omega^j(t)$$
$$-\gamma\delta \cdot \tilde{y}_1^{jT}(t) \cdot \left((\gamma+\delta)(2(\gamma+\delta)-\eta)\lambda_1^{-1}\mathbf{I} - \gamma\delta \cdot g_1^{-1}(y^j(t),t)\right) \cdot \tilde{y}_1^j(t)$$
$$\leq -\omega^{jT}(t) \cdot g_1^{-1}(y^j(t),t) \cdot \omega^j(t) \tag{8.52}$$
$$-\gamma\delta \cdot \tilde{y}_1^{jT}(t) \cdot \left((\gamma+\delta)(2(\gamma+\delta)-\eta) - \gamma\delta\right) \cdot \lambda_1^{-1} \cdot \tilde{y}_1^j(t).$$

Suppose $g_1^{-1}(y^j(t),t)$ is positive definite matrix, and $(\gamma+\delta)(2(\gamma+\delta)-\eta) - \gamma\delta \geq 0$, then we have:

$$\Delta V^j(t) \le 0. \tag{8.53}$$

From Equation (8.53) we can draw the conclusion that the sequence $\{V^j(t)\}$ is monotone decreasing; because $V^1(t)$ has a boundary, the nonnegative monotone decreasing sequence converges to a nonnegative fixed value; as $j \to \infty$, $V^j(t) \to 0$, i.e.,

$$\forall t \in [0, t_f], \quad W^j(t) \to 0, \tilde{y}_1^j(t) \to 0, \tilde{y}_2^j(t) \to 0.$$

Therefore

$$\lim_{j\to\infty} y^j(t) = x_d(t). \tag{8.54}$$

Finally, $\because W^j(t) \to 0 \Rightarrow \dot{W}^j(t) \to 0$, and from Equation (8.42), we can get:

$$\lim_{j\to\infty} u_n^j(t) = u_d(t).$$

2. **Stability and convergence analysis for on-line learning control system**

In order to distinguish off-line learning control and on-line learning control, we divide the system operation mechanism into two processes, the adaptive process and the learning process.

Definition 8.14. Adaptive process is a process in which the control system trains neural network in real-time in every sampling period. The CMAC off-line learning control belongs to the learning process.

Definition 8.15. Learning process is a process in which the control system trains neural network in off-line after accumulating a series of samples. The CMAC on-line feedback learning control belongs to the adaptive process.

Figure 8.10 shows a block diagram of the CMAC-based on-line learning control system that has a feedback from the system output.

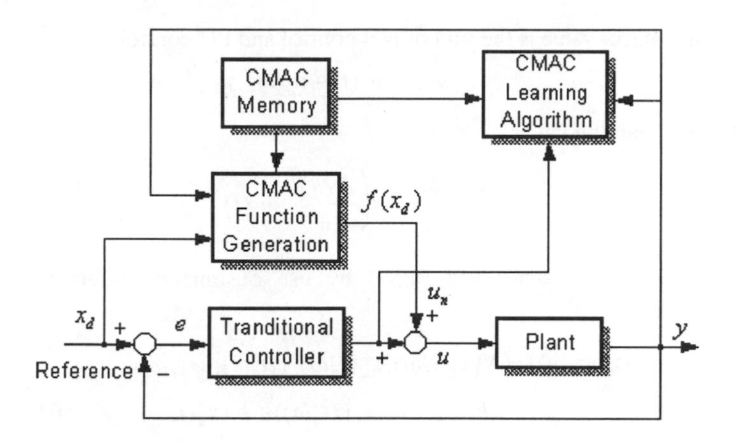

Figure 8.10. CMAC on-line learning control system.

The convergence of the neural network in CMAC can be discussed as follows.

Theorem 8.3. Apply CMAC neural on-line learning control to a class of affine mapping nonlinear system

$$\dot{y} = f(y,t) + g(y,t)u(t) \tag{8.55}$$

where the traditional control uses PD control law, i.e.,

$$u_c(t) = k_2(\dot{x}_d(t) - \dot{y}(t)) + k_1(x_d(t) - y(t)) \tag{8.56}$$

if the system satisfies the following conditions:

(i) $g(y,t)$ is reversible;

(ii) ideal state $\dot{x}_d$ and x_d are in strictly mixed and smooth random process;

(iii) PD control guarantees the system to global approximated stability;

(iv) $g^{-1}(y,t) + k_2$ is positive definite.

Then the algorithm of CMAC neural network is convergent, i.e.,

$$\lim_{t \to \infty} W(t) = W^* \tag{8.57}$$

[Prove] From condition (i) and Equation (8.55), we have

$$u(t) = g^{-1}(y,t)(\dot{y} - f(y,t)) = \varphi(\dot{y}, y, t) \tag{8.58}$$

The inputs of the CMAC neural networks are ideal values of $\dot{x}_d(t), x_d(t)$ and the practical measured value of $y(t)$, so

$$u_n(t) = N(\dot{x}_d(t), x_d(t), y(t), w(t)) \tag{8.59}$$

and the system control value is the sum of NN control and PD control:

$$u(t) = u_n(t) + u_c(t). \tag{8.26}$$

The CMAC learning law is:

$$\frac{dw}{dt} = \beta\left(\frac{\partial N}{\partial w}\right)^T \cdot u_c(t). \tag{8.60}$$

Represent Equation (8.56) and (8.57) by use of random differential equations respectively:

$$\begin{aligned}\varphi(\dot{y}(t,\omega), y(t,\omega)) = N\big(&\dot{x}_d(t,\omega), x_d(t,\omega), y(t,\omega), w(t,\omega) \\ &+ k_2(\dot{x}_d(t,\omega) - \dot{y}(t,\omega)) + k_1(x_d(t,\omega) - y(t,\omega))\big)\end{aligned} \tag{8.61}$$

$$\begin{aligned}\frac{dw(t,\omega)}{dt} = -\beta&\left(\frac{\partial N(\dot{x}_d(t,\omega), x_d(t,\omega), y(t,\omega), w(t,\omega))}{\partial w}\right)^T \\ &\cdot(u_c(\dot{x}_d(t,\omega), x_d(t,\omega), \dot{y}(t,\omega), y(t,\omega)))\end{aligned} \tag{8.62}$$

where ω is the sampling point of stochastic space.

According to Geman Theorem [20], as β is in small value, the solution of Equation (8.62) can be approximated by the following averaging equation:

$$\begin{aligned}\frac{dM}{dt} = \beta(&\left(\frac{\partial N(\dot{x}_d(t,\omega), x_d(t,\omega), y(t,\omega), w(t,\omega))}{\partial w}\right)^T \\ &\cdot(u_c(\dot{x}_d(t,\omega), x_d(t,\omega), \dot{y}(t,\omega), y(t,\omega))))_{\omega=M}\end{aligned} \tag{8.63}$$

Let function P:

$$\begin{aligned}P = \frac{1}{2}&u_c(\dot{x}_d(t,\omega), x_d(t,\omega), \dot{y}(t,\omega), y(t,\omega))^T \\ &\cdot u_c(\dot{x}_d(t,\omega), x_d(t,\omega), \dot{y}(t,\omega), y(t,\omega)).\end{aligned} \tag{8.64}$$

The Liapunov function defining averaging equation is:

$$V(M,t) = E(P(\dot{x}_d(t,\omega), x_d(t,\omega), \dot{y}(t,\omega), y(t,\omega)))_{\omega=M} \geq 0. \tag{8.65}$$

For convenience of writing, every variable expression is simplified as:

$$\frac{dV}{dt} = E\left(u_c^{\,T}(\dot{x}_d, x_d, \dot{y}, y) \cdot (\frac{d}{dt} u_c(\dot{x}_d, x_d, \dot{y}, y)) \right)_{\omega = M}$$
$$= E\left(u_c^{\,T}(\frac{\partial u_c}{\partial \dot{x}_d}\frac{d\dot{x}_d}{dt} + \frac{\partial u_c}{\partial x_d}\frac{dx_d}{dt} + \frac{\partial u_c}{\partial \dot{y}}\frac{d\dot{y}}{dt} + \frac{\partial u_c}{\partial y}\frac{dy}{dt} \right)_{\omega = M} \tag{8.66}$$

From Equation (8.61), $\dot{y}$ can be written as

$$\dot{y}(t,\omega) = \rho(\dot{x}_d(t,\omega), x_d(t,\omega), y(t,\omega), w(t,\omega)) \tag{8.67}$$

Therefore,

$$\frac{dV}{dt} = E\left(u_c^{\,T} \cdot (k_2 \frac{d\dot{x}_d}{dt} + k_1 \frac{dx_d}{dt} - k_2 \frac{d\dot{y}}{dt} - k_1 \frac{dy}{dt} \right)_{\omega = M} \tag{8.68}$$

$$= E\left(u_c^{\,T} \cdot (k_2 \frac{d\dot{x}_d}{dt} + k_1 \frac{dx_d}{dt} - k_1 \frac{dy}{dt} \right.$$

$$\left. - k_2 (\frac{\partial \dot{y}}{\partial \dot{x}_d} \cdot \frac{d\dot{x}_d}{dt} + \frac{\partial \dot{y}}{\partial x_d} \cdot \frac{dx_d}{dt} + \frac{\partial \dot{y}}{\partial y} \cdot \frac{dy}{dt} + \frac{\partial \dot{y}}{\partial w} \cdot \frac{dw}{dt})) \right)_{\omega = M}$$

$$= E\left(u_c^{\,T} \cdot (k_2 \frac{d\dot{x}_d}{dt} + k_1 \frac{dx_d}{dt} - k_1 \frac{dy}{dt} - k_2 (\frac{\partial \dot{y}}{\partial \dot{x}_d} \cdot \frac{d\dot{x}_d}{dt} + \frac{\partial \dot{y}}{\partial x_d} \cdot \frac{dx_d}{dt} + \frac{\partial \dot{y}}{\partial y} \cdot \frac{dy}{dt})) \right)_{\omega = M}$$

$$+ E\left(u_c^{\,T} \cdot (-k_2 \frac{\partial \dot{y}}{\partial w} \cdot \frac{dw}{dt}) \right)_{\omega = M} .$$

The first part of the above Equation can be expressed as:

$$\frac{d}{dt} E\big(P(\dot{x}_d(t,\omega), x_d(t,\omega), \dot{y}(t,\omega), y(t,\omega), M \big) \tag{8.69}$$

From condition (iii) we know that the differential equation of the state y is stable, so $y, \dot{y}$ and P become Baire functions of strictly smooth processes $\dot{x}_d$ and y. Because $E\big(P(\dot{x}_d(t,\omega), x_d(t,\omega), \dot{y}(t,\omega), y(t,\omega), M\big)$ is independent of time, the first part, i.e., Equation (8.69) becomes zero, then Equation (8.68) turns to:

$$\frac{dV}{dt} = E\left(u_c^{\,T} \cdot (-k_2 \frac{\partial \dot{y}}{\partial w}) \cdot \frac{dw}{dt} \right)_{\omega = M} . \tag{8.70}$$

From Equation (8.61) we have:

$$\frac{\partial \varphi}{\partial \dot{y}} \frac{\partial \dot{y}}{\partial w} = \frac{\partial N}{\partial w} - k_2 \frac{\partial \dot{y}}{\partial w}$$

$$\Rightarrow \quad g^{-1}(y(t,\omega)) \frac{\partial \dot{y}}{\partial w} = \frac{\partial N}{\partial w} - k_2 \frac{\partial \dot{y}}{\partial w} \tag{8.71}$$

$$\therefore \quad \frac{\partial \dot{y}}{\partial w} = (g^{-1}(y(t,\omega)) + k_2)^{-1} \frac{\partial N}{\partial w}.$$

Therefore,

$$\frac{dV}{dt} = E\left(u_c^{\,T} \cdot (-k_2 (g^{-1}(y(t,\omega)) + k_2)^{-1} \cdot \frac{\partial N}{\partial w} \cdot \frac{dw}{dt}) \right)_{\omega = M}$$

$$= E\left(u_c^{\,T} \cdot (-k_2 (g^{-1}(y(t,\omega)) + k_2)^{-1} \cdot \frac{\partial N}{\partial w} \cdot \beta (\frac{\partial N}{\partial w})^T \cdot u_c) \right)_{\omega = M}. \tag{8.72}$$

From condition (iv), the matrix $g^{-1}(y(t,\omega)) + k_2$ is positive definite,

$\therefore \quad k_2 (g^{-1}(y(t,\omega)) + k_2)^{-1} \cdot \frac{\partial N}{\partial w} \cdot \beta (\frac{\partial N}{\partial w})^T$ is positive semi-definite, so we have:

$$\frac{dV}{dt} \le 0 \tag{8.73}$$

$\because$ CMAC is global minimum,

$\therefore$ as $w = w^*$, Equation (8.65) and (8.68) hold, then w approaches and converges to w^* gradually.

We have discussed the convergence of NN algorithm by using the averaging equation and the Liapunov method. As the weight value and control value u_n of NN substitute PD and become the desired control inputs, then the system state approximates to its ideal value gradually.

8.5 Example of Learning Control Systems

A learning control system possesses the capability to improve its performance over time by interaction with its environment. The learning control system is designed so that its learning controller has the ability to improve the performance of the closed-loop system by generating command inputs to the plant and utilizing feedback information from the plant. Therefore, the learning control systems, including the fuzzy learning control systems, neural network-based learning control system and self-learning fuzzy neurocontrol system, have been used in real-time industrial fields in recent years [12, 24, 47, 53]. In this section we will introduce a self-learning fuzzy neurocontrol system for the arc welding process

[12]. First of all, the scheme of control system will be discussed, followed by describing the algorithm of the self-learning fuzzy neurocontroller, and then an application of the self-learning fuzzy neurocontrol for arc welding process will be shown.

8.5.1 Models of Self-learning Fuzzy Neurocontrol

Figure 8.11 gives out a schematic of the self-learning fuzzy neurocontrol system with uncertainties, where the fuzzy controller FC maps the regulated error *e*(*t*) into control action *u*(*t*). The output signal *y*(*t*) of the process is detected by the measurement sensor. The NN-based process model is denoted by PMN. The outputs of the process and sensor are denoted by same *y*(*t*) in the sense of neglecting the difference of their transformation coefficients.

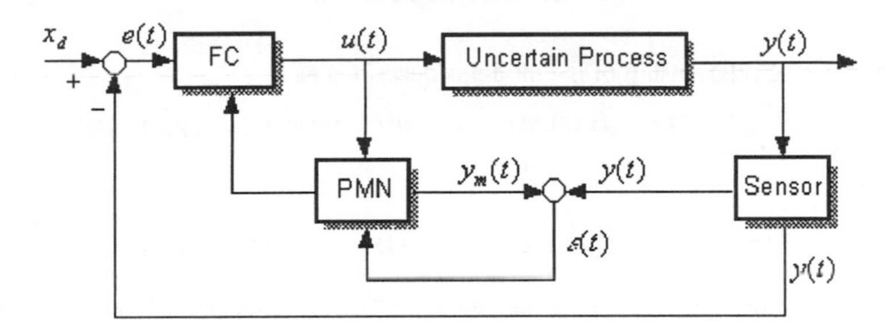

Figure 8.11. Self-learning fuzzy neurocontrol system.

The Models of FC and PMN can be developed as follows:

The fuzzy controller FC is described in an analytical formula instead of common fuzzy rule tables, that is

$$U(t) = \sigma\left[a(t)b(t)E(t) + (1 - a(t)b(t))EC(t) + (1 - b(t))ER(t)\right] \tag{8.74}$$

where $\sigma = \pm 1$ is related to the properties of the controlled process or fuzzy rules, e.g., $\sigma = 1$ is corresponding to $u \propto e, ec$, and $\sigma = -1$ to $u \propto -e, -ec$; *U*, *E*, *EC* and *ER* denote the fuzzy variables corresponding to their crisp variables: control action *u*(*t*), acceleration error *e*(*t*), change in error *ec*(*t*)=*e*(*t*)-*e*(*t*-1), and acceleration error *er*(*t*)=*ec*(*t*)-*ec*(*t*-1); $a(t) \in [0, 1]$, $b(t) \in [0, 1]$. The fuzzy variables and their corresponding crisp variables differ in the transformation factors relating to their universes of discourses. Contrary to common approach, all universes of discourses are considered continuously.

The fuzzy controllers is adaptively regulated by modifying the tuning parameters *a*(*t*) and *b*(*t*) according to the neural network model of the uncertain system.

The model of PMN for the uncertain process and the measurement sensor can be realized by a back-propagation network with four layers as shown in Figure 8.12.

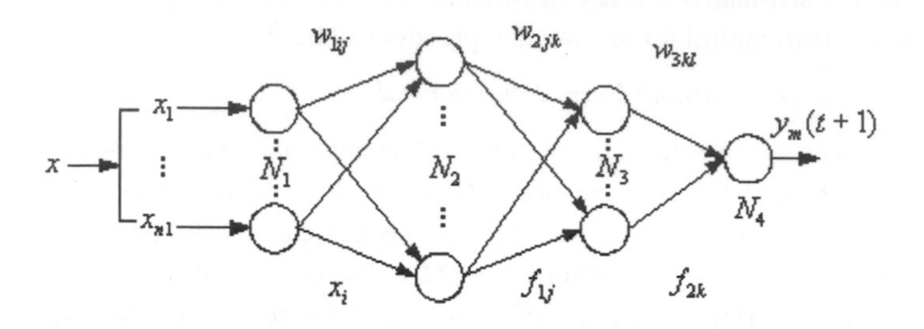

Figure 8.12. Model of PMN.

The mapping relationship of the model is described as

$$y_m(t+1) = f_m(u(t), u(t-1), \ldots, u(t-m); y_m(t), \ldots, y_m(t-n)) \tag{8.75}$$

by defining

$$x^T = [x_1, \ldots, x_{n1}]^T = [u(t), u(t-1), \ldots, u(t-m); y_m(t), \ldots, y_m(t-n)]^T$$

where m and n denote the orders of the uncertain system, which could be roughly estimated by experiments on the system.

The network function of the PMN can be described as follows:

$$f_{1j} = 1 \Big/ \left\{ 1 + \exp\left[-\left(\sum_{i=1}^{N_1} W_{1ij} x_i + q_{1j} \right) \right] \right\}, \quad j = 1, \ldots, N_2 \tag{8.76}$$

$$f_{2k} = 1 \Big/ \left\{ 1 + \exp\left[-\left(\sum_{j=1}^{N_2} W_{2jk} f_{1j} + q_{2k} \right) \right] \right\}, \quad k = 1, \ldots, N_3 \tag{8.77}$$

$$y_m(t+1) = 1 \Big/ \left\{ 1 + \exp\left[-\left(\sum_{k=1}^{N_3} W_{3kl} f_{2k} + q_{3l} \right) \right] \right\} = f_m(f_{2k}(f_{1j}(x))) \tag{8.78}$$

8.5.2 Algorithms of Self-learning Fuzzy Neurocontroller

The self-learning algorithms for the fuzzy controller FC and neural network model PMN are developed as follows:

(i) The index of control error

$$J_e = \sum_{t=1}^{N} \left[x_d - y(t+1) \right]^2 / 2 \tag{8.79}$$

(ii) The index of model error

$$J_\varepsilon = \sum_{t=1}^{N} \varepsilon^2(t+1)/2 = \sum_{t=1}^{N} \left[y(t+1) - y_m(t+1) \right]^2 / 2 \tag{8.80}$$

(iii) Learning algorithms of the model PMN

Off-line and on-line learning algorithms can be used to modify parameters of PMN networks. The initial weights of the PMN can be obtained by batch sample data pairs $\{u(t), y(t+1)\}$. The off-line learning results of the PMN can be used as a reference model of real uncertain controlled process. Using the on-line learning algorithms, network weights of the PMN in real-time can be modified by the index (8.80) and the principle of error gradient descent, that is

$$\Delta W(t) \propto -\partial J_\varepsilon / \partial W(t)$$

$$W(t+1) = W(t) + \Delta W(t). \tag{8.81}$$

The learning algorithms for the PMN are briefly described as below: Defining

$$v_3(t) = (y(t) - y_m(t))(1 - y_m(t)) y_m(t)$$

$$v_{2k}(t) = f_{2k}(t)(1 - f_{2k}(t)) W_{3kl}(t) v_3(t), \quad k = 1, \ldots, N_3$$

$$v_{1j}(t) = f_{1j}(t)(1 - f_{1j}(t)) \sum_{k=1}^{N_3} W_{2jk}(t) v_{2k}(t), \quad j = 1, \ldots, N_2$$

then weights modified as:

$$\Delta W_{3kl}(t) = h_3 v_3(t) f_{2k}(t) + g_3 \Delta W_{3kl}(t-1)$$

$$W_{3kl}(t+1) = W_{3kl}(t) + \Delta W_{3kl}(t) \tag{8.82}$$

$$q_{3l}(t+1) = q_{3l}(t) + h_3 v_3(t) \tag{8.83}$$

$$\Delta W_{2jk}(t) = h_2 v_{2k}(t) f_{1j}(t) + g_2 \Delta W_{2jk}(t-1)$$

$$W_{2jk}(t+1) = W_{2jk}(t) + \Delta W_{2jk}(t) \tag{8.84}$$

$$q_{2k}(t+1) = q_{2k}(t) + h_2 v_{2k}(t) \tag{8.85}$$

$$\Delta W_{1ij}(t) = h_1 v_{1j}(t) x_i + g_1 \Delta W_{1ij}(t-1)$$

$$W_{1ij}(t+1) = W_{1ij}(t) + \Delta W_{1ij}(t) \tag{8.86}$$

$$q_{1j}(t+1) = q_{1j}(t) + h_1 v_{1j}(t) \tag{8.87}$$

where $h_i, g_i \in (0,1)(i = 1,2,3)$ are the learning factors and momentum factors respectively. The equations (8.81)-(8.87) are one-step learning algorithms for the PMN networks in a controlling period.

(iv) Adaptive-modifying of tuning parameters $a(t)$, $b(t)$ of the FC

Assuming the PMN network parameters as known variables which have been obtained by the off-line or the last one-step learning result, we have the following algorithms to modify the tuning parameters $a(t)$, $b(t)$ of the fuzzy controller FC.

$$a(t+1) = a(t) + \Delta a(t) \tag{8.88}$$

$$b(t+1) = b(t) + \Delta b(t) \tag{8.89}$$

$$\Delta a(t) = -h_a(\partial J_e / \partial a(t)) \tag{8.90}$$

$$\Delta b(t) = -h_b(\partial J_e / \partial b(t)) \tag{8.91}$$

Learning factors $h_a, h_b \in (0,1)$.

$$\begin{aligned}\partial J_e / \partial a(t) &\approx \left[x_d - (y_m(t+1) + \varepsilon\right]\left[\partial y_m(t+1) / \partial a(t)\right] \\ &= \left[x_d - y(t+1)\right]\left[\partial y_m(t+1) / \partial a(t)\right], \quad (\ \partial\varepsilon / \partial a \text{ is neglected})\end{aligned} \tag{8.92}$$

$$\partial y_m(t+1) / \partial a(t) = \left[\partial f_m / \partial u(t)\right]\left[\partial u(t) / \partial a(t)\right] \tag{8.93}$$

$$\partial u(t) / \partial a(t) = \sigma b(t)\left[E(t) - EC(t)\right] \tag{8.94}$$

$$\begin{aligned}\partial J_e / \partial b(t) &\approx \left[x_d - (y_m(t+1) + \varepsilon\right]\left[\partial y_m(t+1) / \partial b(t)\right] \\ &= \left[x_d - y(t+1)\right]\left[\partial y_m(t+1) / \partial b(t)\right], \ (\ \partial\varepsilon / \partial b \text{ is neglected})\end{aligned} \tag{8.95}$$

$$\partial y_m(t+1) / \partial b(t) = \left[\partial f_m / \partial u(t)\right]\left[\partial u(t) / \partial b(t)\right] \tag{8.96}$$

$$\partial u(t) / \partial b(t) = \sigma\left[a(t)E(t) + (1 - a(t)) - EC(t) - ER(t)\right]. \tag{8.97}$$

And then

$$\begin{aligned}\partial f_m / \partial u(t) &= \partial f_m / \partial x_1 = \left[\partial f_m / \partial f_{2k}\right]\left[\partial f_{2k} / \partial f_{1j}\right]\left[\partial f_{1j} / \partial x_1\right] \\ &= -\left\{f_m(1 - f_m)\sum_{k=1}^{N_3}\left[W_{3kl} f_{2k}(1 - f_{2k})\sum_{j=1}^{N_2}(W_{2jk} f_{1j}(1 - f_{1j})W_{1ij})\right]\right\}\end{aligned} \tag{8.98}$$

where f, w are related to the states and weights of the PMN.

The equations (8.88)-(8.98) are one-step self-modifying algorithms for the tuning parameters $a(t)$, $b(t)$ of the FC in a controlling period, essentially, which implies to regulate the fuzzy control rules as human operators in real-time.

8.5.3 *Self-learning Fuzzy Neurocontrol System for Arc Welding Process*

A self-learning fuzzy neurocontrol system has been developed for the arc welding process.

1. **Architecture of control system for arc welding**

Figure 8.13 depicts a block diagram of the control system of the pulse TIG welding.

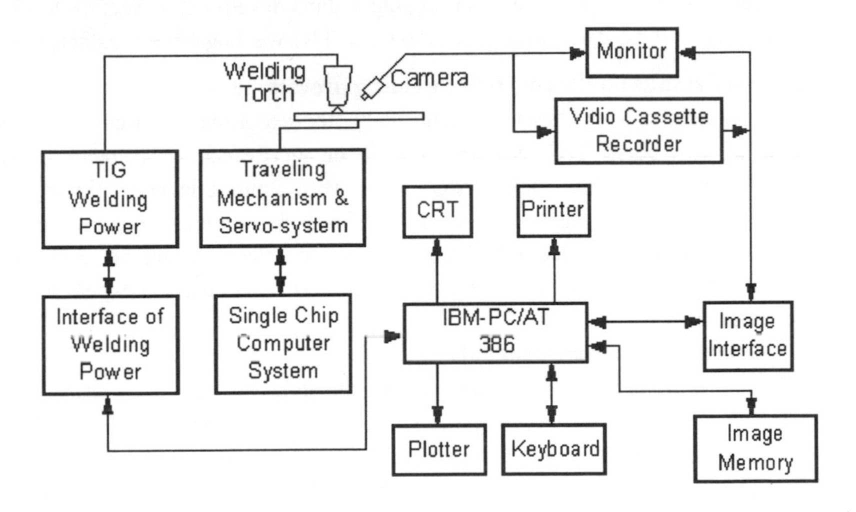

Figure 8.13. Block diagram of arc welding control systems.

The systems consisted of a IBM-PC/AT 386 personal computer for realizing self-learning control and image processing algorithms, a TV camera was used as vision sensor to pick up the front weld pool image, an image interface unit, an image monitor and an alter/direct pulse arc welding power source. The welding current was regulated by the interface of the welding power source, and the regulation of the welding travel speed was realized by a single chip computer system.

2. **Modeling and simulation of the welding process**

By analyzing the technology process of the pulse TIG welding and the test data under the standard conditions, we knew that the main factors influencing welding pool changes are welding current and welding travel speed under the fixed technical normal parameters such as welding pool dynamics for control of the pulse TIG welding was established. The input and output of the model are the plate thickness, joint gap, etc., no loss of its practicality, for simplicity, a SISO model of welding current and the top-bead width of welding pool respectively. Using the batch testing data of input-output pairs and off-line learning algorithm, a neural network model of the process, with nodes N_1, N_2, N_3 and N_4 are 5, 10, 10, and 1 respectively, realizes the following mapping:

$$y_m(t+1) = f_m(u(t), u(t-1), u(t-2), y_m(t), y_m(t-1)) \tag{8.99}$$

$y_m(t+1)$ adds a pseudo-random sequence as the simulation model of the real uncertain process in Figure 8.11. Using the self-learning algorithms developed in Section 8.5.2, the simulation results on the control scheme for the pulse TIG welding were satisfactory [14].

3. **Experiment results on the control of welding process**

Based on the system scheme shown in Figure 8.13, the experiments on control of weld bead width of the pulse TIG welding have been performed. The bead-on-plated experiments were on the low carbon steel plate with 2mm thickness, the dumbbell specimen was used for imitating sudden changes of heat radiation and conduction in the welding process; the tungsten electrode diameter was 3mm; the shielding gas of argon was at 8ml/min flow rate; the constant welding current of 180 A was used in experiments; the arc voltage was 12 to 30 VDC.

The experiment results show that [14]:

(i) The control of the weld specimens in changing heat transfer circumstances demonstrates that the self-learning fuzzy neural control scheme as Figure 8.11 is available to control the welding, speed-welding pool dynamic process of the pulse TIG welding. The controlled results show that regulating actions in the control system is similar to the operating actions or intelligent behaviors of a skilled welder. The improved effects of compensating time delays for the uncertain process is evident.

(ii) The controlled precise results are mainly influenced by the period of completing control algorithm and image processing, which could be improved by realizing parallel processing of neural networks by hardware and enhancing computation speed.

8.6 Summary

Learning is an important behavior and intelligent ability of human beings. Beginning with the definitions of learning and learning control in this chapter, we have studied the concepts of learning, learning control, and the learning system in some details. Various definitions have been discussed. According to the definitions, the mechanism of the learning control has been summarized as: (i) search for and find a relatively simple relation between input and output of dynamic control systems; (ii) execute each control process updated by the learning results of the previous control processes; (iii) improve the performance of each process so that it is better than the preceding ones. Repeating such a learning process and recording the results accumulated in the entire process would steadily improve the performance of the learning control systems.

The learning control systems can deal with processes with uncertainty and nonlinearity and guarantee to keep fine adaptivity, satisfactory stability and speed enough convergence. Therefore, learning control is using widely in recent years. With the development of machine learning, learning control has a new motivation and goes to a new stage.

There are many schemes of learning control such as pattern recognition-based learning control, iterative learning control, repetitive learning control, and connectionist learning control, and so on. We have introduced the mechanisms and structures of the above four schemes of the learning control systems.

In Section 4 of this chapter, we have laid the emphasis of the discussion upon some significant issues for learning control, e.g., the modeling of learning control system, the stability and convergence analysis of learning control for both off-line and on-line systems.

As an application paradigm, a self-learning fuzzy neurocontrol system for the arc welding process has been presented. The models of FC and PMN, the learning algorithms of the fuzzy neurocontroller, and the modeling, architecture, simulation and experiment results of the self-learning fuzzy neurocontrol system for arc welding have been studied. The results have shown that the self-learning fuzzy neurocontrol scheme is available to control the welding speed-welding pool dynamic process of the pulse TIG welding, and the regulating actions in the control system is similar to the operating actions or intelligent behaviors of a skilled welder.

References

1. H. S. Ahn, C.-H.Choi, and K.-B. Kim, "Iterative learning control for a class of nonlinear systems," *Automatica*, **29** (1993) 1575-1578.
2. C. W. Anderson, "Learning to control an inverted pendulum using neural networks," *IEEE Control System Magazine*, **9** (1989) 31-37.
3. S. Arimoto, S. Tamaki, S. Kawamura, and F. Miyazaki, "Convergence of learning control for linear time-varying mechanical system," *System and Control*, **30** (1986) 255-262.
4. S. Arimoto, S. Kawamura, and F. Miyazaki, "Learning control theory for dynamical systems," in: *Proc. 24th IEEE Conf. Decision and Control*, Lauderdale, Florida, (1985) 1375-1380.

5. S. Arimoto, S. Kawamura, and F. Miyazaki, "Convergence stability and robustness of learning control schemes for robot manipulators," *Recent Trends in Robotics, Proc. Int. Symp. Robot Manipulators: Modeling, Control and Education*, Albuquerque, New Mexcio (1986) 307-316.
6. S. Arimoto, S. Kawamura, and F. Miyazaki, "Bettering operation of robots by learning," *J. Robotic Systems*, **1** (1984) 123-140.
7. C. G. Atkeson and J. McIntyre, "Robot trajectory learning through practice," *Proc. IEEE Int. Conf. Robotics and Automation*, San Francisco, 1986, 1737-1742.
8. A. G. Barto, "An approach to learning control surface by connectionist systems," in: *Vision, Brain and Cooperative Computation*, eds. M. Arbib and Hanson (MIT Press, Cambridge, MA, 1988) 665-701.
9. A. G. Barto, "Connectionist learning for control," in: *Neural Network for control*, eds. W. T. Miller, R. S. Sutton, and P. J. Werbos (MIT Press, Cambridge, MA, 1990) 5-58.
10. P. Bondi, G. Casalino, and L. Gambardella, "On the iterative learning control theory for robotic manipulators," *IEEE J. of Robotics and Automation*, **4** (1988) 14-22.
11. Z.-X. Cai, "Learning control systems," in: *Intelligent Control* (Electronic Industry Press, Beijing, 1990) 116-123.
12. Z.-X. Cai, and G.-Y. Xu, *Artificial Intelligence and Its Applications*, Second Edition (Tsinghua University Press, Beijing, 1996).
13. N.-J. Chen and Z.-S. Li, "A novel intelligent controller based on characteristic identification on-line," in: *Advance in Modeling and Simulation*, **14** (1988) 53-63.
14. S.-B. Chen, L. Wu and Q.-L. Wang, "Self-learning fuzzy neural networks for control of the arc welding process," *Proc. Int. Conf. Neural Networks*, Washington, D.C., USA, 1996: 1209-1215.
15. B. A. Francis, "The multivariable servomechanism problem from the input-output viewpoint," *IEEE Trans. Automatic Control*, AC-**22** (1977) 322-328.
16. B. A. Francis, and W. M. Wonham, "The internal model principle for linear multivariable regulators," *Applied Math. & Opt.*, **2** (1975) 170-194.
17. K. S. Fu, "Learning control systems — review and outlook," *IEEE Trans. Automatic Control*, AC-**15** (1970) 210-221.
18. K. S. Fu, "Learning control system and intelligent control system: an intersection of artificial intelligence and automatic control," *IEEE Trans. Automatic Control*, AC-**16** (1971) 70-72.
19. K. Furuta, and M. Yamakita, "The design of a learning control system for multivariable systems," *Proc. IEEE ISIC*, 1987: 371-376.
20. S. Geman, "Some averaging and stability results for random differential equations," *SIAM J. Applied Mathematics*, **36** (1979) 86-105.

21. Z. Geng, R. Carroll, and J. Xie, "Two-dimension model and algorithm analysis for a class of iterative learning control system," *Int. J. Control*, **52** (1990) 833-862.
22. Z. Geng, R. Carroll, and M. Jamshidi, *et al.*, "An adaptive learning control approach," *Proc. Of 30th Conf. On Decision and Control*, Brighten, England, 1991: 1221-1222.
23. D. Goldberg, *Genetic Algorithms in Search, Optimization, and Machine Learning* (Addison-Wesley, Reading, Mass. 1989).
24. Y.-L. Gu and N. K. Loh, "Learning control in robotic systems," *J. Intelligent and Robotic Systems*, **2** (1989) 297-305.
25. Y.-L. Gu, and N. K. Loh, "Imaginary robot: Concept and application to robotics system modeling," *Proc. 26th IEEE Conf. On Decision and Control*, Los Angeles, California, 1987: 1011-1016.
26. S. V. Gusev, "Linear stability of nonlinear systems program motion," *System and Control Letter*, **11** (1988) 409-412.
27. R. Hamid and P. Khedkar, "Learning and turning fuzzy logic controllers through reinforcements," *IEEE Trans. Neural Networks*, **3** (1992) 724-740.
28. J. E. Hauser, "Learning control for a class of nonlinear systems," *Proc. 26th IEEE CDC*, Los Angeles, 1987, 859-860.
29. G. Heinzinger, D. Fenwick, B. Paden and F. Miyazaki, "Stability of learning control with disturbances and uncertain initial conditions," *IEEE Trans. Automatic Control*, AC-**37** (1992) 110-114.
30. J. Holland, *Adaptations in Natural and Artificial Systems* (University of Michigan Press, Ann Arbor, Mich., 1975).
31. J. J. Hopfield, "Neurons with graded response have collective computational properties like those of two-state neurons," *Proc. National Academic Sciences*, USA, **81** (1984) 3088-3092.
32. J. J. Hopfield, and D. W. Tank, "Neural computation of decisions in optimization problems," *Biol. Cybernetics*, **52** (1985) 158-159.
33. J. J. Hopfield, and D. W. Tank, "Computing with neural circuits," *Science*, **233** (1986) 625-633.
34. T. Inoue, Iwai, and M. Nakano, "High accurate control for playback servo system (in Japanese)," *Trans. Electrical Society*, C101 (1981) 89-96.
35. T. Inoue, "High accurate control for iterative operation of portion synchrotron electromagnetical power (in Japanese)" *Trans. Electrical Society*, C100 (1980) 234-240.
36. T.-J. Jang, C.-H. Choi, and H.-S. Ahn, "Iterative learning control in feedback systems," *Automatica*, **31** (1995) 243-248.
37. R. Jyh-shing, "Self-learning fuzzy controllers based on temporal backpropagation," *IEEE Trans. Neural Networks*, **3** (1992) 714-723.

38. S. Kawamura, F. Miyasaki, and S. Arimoto, "Study on system theory in learning control mode (in Japanese)," *Trans. Of Measurement and Automatic Control Society*, **21** (1985) 445-450.
39. S. Kawamura, F. Miyasaki, and S. Arimoto, "Proposal for learning control method betterment process of dynamic system (in Japanese)," *Trans. Of Measurement and Automatic Control Society*, **22** (1986) 56-62.
40. S. Kawamura, F. Miyasaki, and S. Arimoto, "Learning control for motion of robot manipulator," *Trans. Of Measurement and Automatic Control Society*, **22** (1986) 443-450.
41. S. Kawamura, F. Miyasaki, and S. Arimoto, "Realization of robot motion based on a learning method," *IEEE Trans. SMC*-**18** (1988) 126-134.
42. M. Kawato, K. Furukawa, and R. Suzuki, "A hierarchical neural-network model for control and learning of voluntary movement," *Biological Cybernetics*, **57** (1987) 169-185.
43. B. Kosko, *Neural Networks and Fuzzy System* (Prentice-Hall, Englewood Cliffs, 1992).
44. T.-Y. Kuc, J. S. Lee, and K. Nam, "An iterative learning control theory for a class of nonlinear dynamic systems," *Automatica*, **28** (1992) 1215-1221.
45. T.-Y. Kuc, K. Nam, and J. S. Lee, "An iterative learning control of robot manipulators," *IEEE Trans. On Robotics and Automation*, **7** (1991) 835-941.
46. J. R. Layne, and K. M. Passino, "Fuzzy model reference learning control," in: *Proc. 1st IEEE Conference on control Applications*, Dayton, OH, 1992: 686-691.
47. J. R. Layne, K. M. Passino, and S. Yurkovich, "Fuzzy learning control for antisked braking systems," *IEEE Trans. On Control Systems Technology*, **1** (1993) 122-129.
48. C. C. Lee, "A self-learning rule-based controller employing approximate reasoning and neural net concepts," *Int. J. of Intelligent Systems*, **6** (1991) 71-93.
49. C. C. Lee, and H. R. Berenji, "An intelligent controller based on approximate reasoning and reinforcement learning," In: *Proc. IEEE Int. Symposium on Intelligent Control*, 1989, 200-205.
50. Z.-S. Li, Y.-Q. Tu, and Q.-J. Zhou, *et al.*, "Design and application of controllers based on methods for emulating human intelligence," *Proc. IMEKO TC7 Int. Symp. On AIMac*, 1991, 81-86.
51. H. Liu, *The Study of CMAC-Based Learning Control System*, Ph.D. Dissertation, Shanghai JiaoTong University, 1995.
52. X.-Q. Luo, *Learning control and Its Application in Industry*, Ph.D. Dissertation, Zhejiang University, Hangzhou, China, 1994.

53. W. T. Miller, R. P. Hewes, F. H. Glanz, and L. G. Kraft, "Real-time dynamic control of an industrial manipulator using a neural-network-based learning controller," *IEEE Trans. Robotics and Automation*, **6** (1990) 1-9.
54. M. L. Minsky, *The Society of Mind* (Simon and Schuster Inc., New York, 1985).
55. T. Mita, and E. Kato, "Iterative control and its application to motion control of arm — a direct approach to servo-problems," *Proc. 24th CDC*, 1985, 1393-1398.
56. M. Nakano, T. Inoue, Y. Yamamoto, and S. Hara, *Repetitive Control* (Society of Measurement and Automatic Control, Tokyo, Japan, 1989, in Japanese).
57. D. H. Nguyen, and B. Widow, "Neural networks for self-learning control systems," *IEEE Control System Magazine*, **10** (1990) 18-23.
58. S. Oh, Z. Bien, and I. H. Suh, "An iterative learning control method with application for the robot manipulator," *IEEE J. Robotics and Automation*, **4** (1988) 508-514.
59. D. E. Rumelhart, G. E. Hinton, and R. J. Williams, "Learning internal representations by error propagation," In: *Parallel Distributed Processing*, eds. D. E. Rumelhart and J. L. McClelland (MIT Press, Cambridge, 877, MA, 1986).
60. R. M. Sanner, and D. L. Akin, "Neuromorphic pitch attitude regulation of an underwater telerobot," *IEEE Control System Magazine*, **10** (1990) 62-62.
61. G. N. Saridis, *Self-organizing Control of Stochastic Systems* (Marcel Dekker Publisher, New York, 1977).
62. G. N. Saridis, "Architectures for intelligent machines," CIRSSE Technical Report, No.96, RPI, 1991.
63. A. M. Sagre, "Applications of machine learning," *IEEE Expert*, **7** (1992) 30-33.
64. Z.-Z. Shi, *Principles of Machine Learning* (Int. Academic Publishers, Beijing, 1992).
65. H. A. Simon, "Why should machines learning?" In: *Machine Learning: An Artificial Intelligence Approach*, eds. R. S. Michalski, J. G. Garbonell, and T. M. Mitchell (Tioga Publishing Ca, Palo Alto, 1983).
66. R. F. Stengel, "Toward intelligent flight control," *IEEE Trans. SMC*, **23** (1993) 1699-1717.
67. T. Sugie, and T. Ono, "An iterative learning control law for dynamic systems," *Automatica*, **27** (1991) 729-732.
68. M. Togai, O. Yamano, "Analysis and design of an optimal learning control scheme for industrial robots: a discrete system approach," *Proc. 24th CDC*, 1985: 1399-1404.
69. Y. Z. Tsyphin, *Adaptation and Learning in Automatic Systems* (Academic Press, New York, 1971).
70. Uneyama, "Formation of high speed motion of artificial hand with testing (in Japanese)," *Trans. Of Measurement and Automatic Control Society*, **14** (1978) 706-712.

71. H. Waldman, *Dictionary of Robotics* (Macmillan Publishing Company, New York, 1985).
72. M. D. Waltz, and K. S. Fu, "A heuristic approach to reinforcement learning control system," *IEEE Trans. Automatic Control*, AC-**10** (1965).

CHAPTER 9
INTELLIGENT CONTROL SYSTEMS IN APPLICATIONS

In the previous chapters, we have presented many application examples of intelligent control systems. Intelligent control has a very wide range of application areas from laboratories to industries, household appliances to missile guidance, manufacturing to mining, flight vehicle to weapon control, steel rolling mill to mail processing machine, industrial robots to restorative artificial limbs, and so on. If we had selected every successful application example and simulation paradigm of intelligent control, then we might edit and publish a book of selected papers entitled "*999 Application Examples of Intelligent Control*".

Due to the objectives and the limited volume of this book, we can only introduce some typical and representative application examples in this chapter. We believe that the readers and others would create and develop some more paradigms of intelligent control systems after the study of this chapter.

9.1 Intelligent Robotic Planning and Control — Intelligent Control for Multiple Autonomous Undersea Vehicle

There are many members in the robotic family. They are the industrial robotic manipulators, mobile robots, space telerobots, undersea robotic vehicles and restorative limbs etc. In this book, we often take robotic control as an example. For example, in Chapter 4, a hierarchical robotic assembly system has been introduced; in Chapter 6, a fuzzy force control for a biped robot has been presented. All of these examples give the evidence that robot control is an important application field of intelligent control. In this section, we are mainly concerned with a hierarchical control of intelligent machine developed for the space station telerobots and undersea vehicles, i.e., for space robots and undersea robots [1, 22, 53].

One of the major directions that the robot researchers are concerned with is planning and controlling of robots motion. Given a specific task, a motion plan must be made to meet the task requirements; then the plan must be executed by sufficient control for the robot to adequately generate the desired motion. The intelligent control of multiple autonomous vehicles presented below will concentrate on both the planning and the control of the robot motion.

9.1.1 Introduction to Autonomous Vehicles

In order to achieve real-time intelligent control of multiple autonomous vehicle in complex environments, research issues such as distributed control, knowledge-based systems, real-time planning, world modeling, value-driven reasoning, intelligent sensing, intelligent communication, gaming, cooperative problem solving, and learning must be addressed. The types of activities that must be achieved by these autonomous systems include aggression, predation, exploration, stealth, deception, escape, communication and cooperation. These activities are required in order to thrive in a natural and potentially hostile environment.

The goal of this project was to examine some of these issues for Multiple Autonomous Undersea Vehicles (MAUV) in the undersea domain and for space station telerobots in the outside space domain by attempting to achieve intelligent, autonomous, cooperative behavior in multiple vehicles. A control architecture of intelligent control was proposed. A first cut at algorithms and software was downloaded into computer boards mounted on board of the vehicles. A series of demonstration tests was carried out for two undersea vehicles in Lake Winnipesaukee in New Hampshire, USA.

These vehicles were designed and constructed by the Marine Systems Engineering Laboratory at the University of New Hampshire. The vehicle is gravity stabilized in pitch and roll, with thrusters that allow it to be controlled in *x, y, z* and *yaw*. It is battery powered with the batteries stored in cylindrical tanks at the bottom of the vehicle. The vehicle carries three acoustic navigation buoys placed in the water, allowing range and bearing relative to these buoys to be measured. The vehicle also carries a compass, pressure and temperature sensors, and depth and altitude senars. In front, it has an obstacle avoidance sonar. In addition, the vehicle carries both acoustic and radio telemetry systems. All computer boards are mounted in card cages inside the flotation tanks at the upper part of the vehicle.

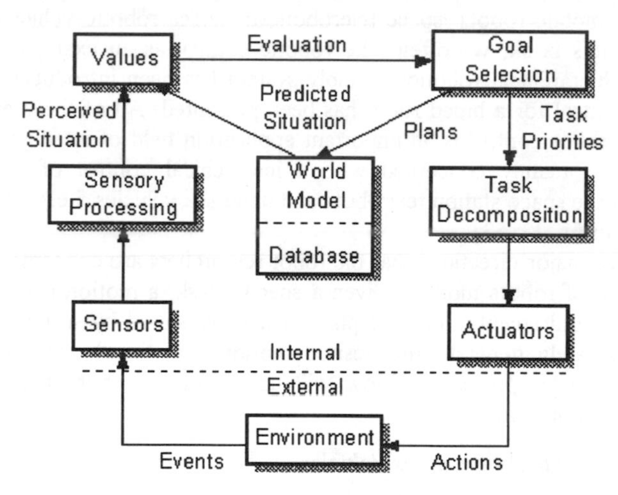

Figure 9.1. An intelligent control architecture for MAUV.

Autonomous vehicles that operate in a complex environment require intelligence. Truly intelligent machines will have complex system architectures in which sensing, acting, sensory processing, world modeling, task decomposition, value judgments, and goal selection are integrated into a system which responds in a timely fashion to stimulation from the environment. Figure 9.1 illustrates a general structure and the basic elements of an intelligent control system.

9.1.2 Control System Architecture of MAUV Vehicle

The control system architecture of the intelligent vehicle is a three-legged hierarchy of computing modules serviced by a communication system and a common memory as shown in Figure 9.2 [1-3].

The three hierarchies are: task decomposition, world modeling, and sensory processing. The task decomposition modules perform real-time planning and task monitoring functions, and decompose task goals both spatially and temporally. The sensory processing modules filter, correlate, detect, and integrate sensory information over both space and time in order to recognize and measure patterns, features, objects, events, and relationships in the external world. The modeling modules answer queries, make predictions, and compute evaluation functions on the state space defined by the information stored in common memory. Common memory is a global database which contains the system's best estimate of the state of the external world. The world modeling modules keep the common memory database current and consistent.

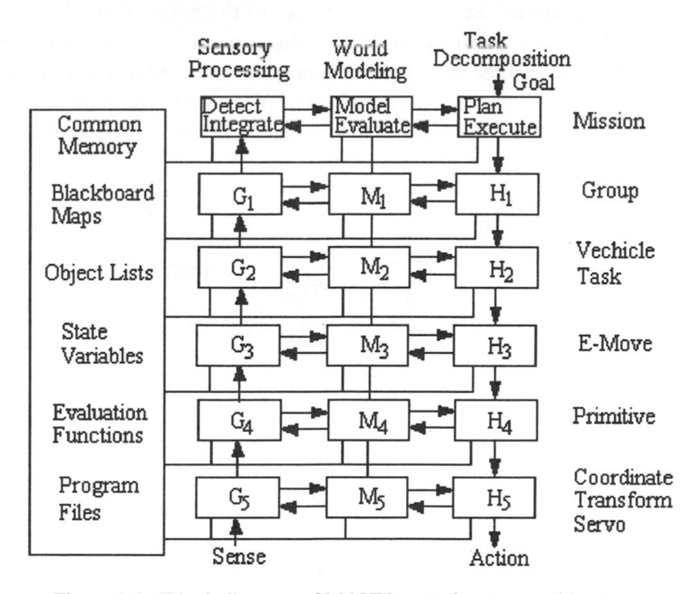

Figure 9.2. Block diagram of MAUV control system architecture.

1. **Task decomposition**

The first leg of the hierarchy consists of task decomposition H modules which plan and execute the decomposition of high level goals into low level actions. Task decomposition involves both a temporal decomposition (into sequential actions along the time line) and a spatial decomposition (into concurrent actions by different subsystems). Each H module at each level consists of a job assignment manager JA, a set of planners PL(*i*), and a set of executors EX(*i*). These decompose the input task into both spatially and temporally distinct subtasks as shown in Figure 9.3.

2. **World modeling**

The second leg of the hierarchy consists of world modeling M modules which model (i.e. remember, estimate, predict) and evaluate the state of the world. The "world model" is the system's best estimation and evaluation of the history, current state, and possible future states of the world, including the states of the controlled system. The "world model" includes both the M modules and a knowledge base stored in a common memory database where state variables, maps, lists of objects and events, and attributes of objects and events are maintained. By this definition, the world model corresponds to a "blackboard". The world model performs the following functions:

(1) Maintain the common memory knowledge base by accepting information from the sensory system.
(2) Provide predictions of expected sensory input to the corresponding G modules, based on the state of the task and estimates of the external world.
(3) Answer "What is?" questions asked by the executors in the corresponding level H modules. The task executor can request the values of any system variables.
(4) Answer "What if?" questions asked by the planners in the corresponding level H modules. The M modules predict the results of hypothesized actions.

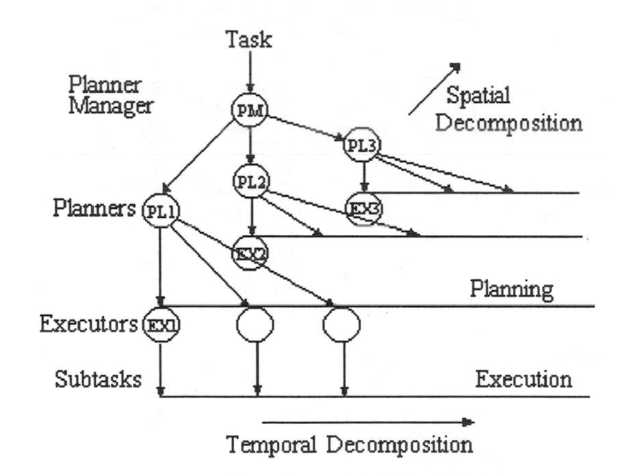

Figure 9.3. Internal structure of the task decomposition modules in MAUV.

3. **Sensory processing**
The third leg of the hierarchy consists of sensory processing G modules. These recognize patterns, detect events, and filter and integrate sensory information over space and time. The G modules at each level compare world model predictions with sensory observations and compute correlation and difference functions. These are integrated over time and space so as to fuse sensory information from multiple sources over extended time intervals. Newly detected or recognized events, objects, and relationships are entered by the M modules into the world model common memory database, and objects or relationships perceived to be no longer in existence are removed. The G modules also contain functions which can compute confidence factors (CF) and probabilities of recognized events, and statistical estimates of stochastic state variable values.

4. **Operator interfaces**
The control architecture defined here has an operator interface at each level in the hierarchy. The operator interface provides a means by which human operators, either in the space station or undersea vehicle or on the ground base, can observe and supervise the vehicle or telerobot. Each level of the task decomposition hierarchy provides an interface where the human operator can assume control. The task commands into any level can be derived either from the higher level H module, or from the operator interface. Using a variety of input devices such as a joystick, mouse, trackball, light pen, keyboard, voice input, etc., a human operator can enter the control hierarchy at any level, at any time of his choosing, to monitor a process, to insert information, to interrupt automatic operation and take control of the task being performed, or to apply human intelligence to sensory processing or world modeling functions.

The operator interfaces include operator control interface, operator monitoring interfaces, and sensory processing/world modeling interfaces.

5. **Common memory**
The common memory is a global database which contain maps, object limits, state variables, evaluation functions and program files, etc.

(1) **Communications.** One of the primary functions of common memory is to facilitate communication between modules. Communication within the control hierarchy is supported by a common memory in which state variables are globally defined. Each module in the sensory processing, world modeling, and task decomposition hierarchies reads inputs from, and writes outputs to, the common memory. Thus each module needs only to know where it should write its output variables in common memory. The data structures in the common memory then define the interfaces between the G, M, and H modules. The operator interfaces also interact with the system through common memory.

(2) **State estimation.** The state variables in common memory are the system's best estimate of the state of the world, including both the external environment and the internal state of the H, M, and G modules. Data in common memory are available to all modules at all levels of the control system. The knowledge base in the common

memory consists of three elements: maps which describe the spatial occupancy of the world, object-attribute linked lists, and state variables.

9.1.3 Levels in Control Hierarchy

Either for the MAUV or for the flight telerobot system, the control system architecture is the same and is described by a six-level hierarchy. Figure 9.3 and Figure 9.4 show the MAUV task decomposition hierarchy. Each module in the task decomposition hierarchy receives input commands from one and only one supervisor, and outputs subcommands to a set of subordinate modules at the next level down in the tree. Output from the bottom level consists of drive signals to motors, actuators, and transducers. Each large box in Figure 9.5 has three levels of small boxes inside it. The top-level box represents the Planner Manager, the middle-level set of boxes represents planners, and the lowest-level set of boxes represents the executors, one associated with each planner. The output of each executor is a subtask command to the next lower level.

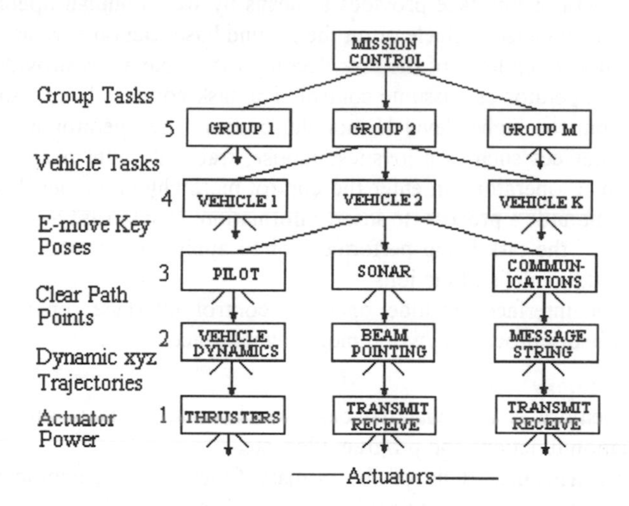

Figure 9.4. MAUV task decomposition with six-level hierarchy.

1. **Mission level**

Missions are typically specified by a list of mission objects, priorities, requirements, and time line constraints. In the implementation of MAUV system, the inputs to mission level are command and a mission value function. The command is a task involving a mission strategy, e.g. SEARCH-AND-DESTORY, SEARCH-AND-REPORT, and MAP. Associated with each command is a list of subtasks that define the command. The mission value function is a function used to score the mission.

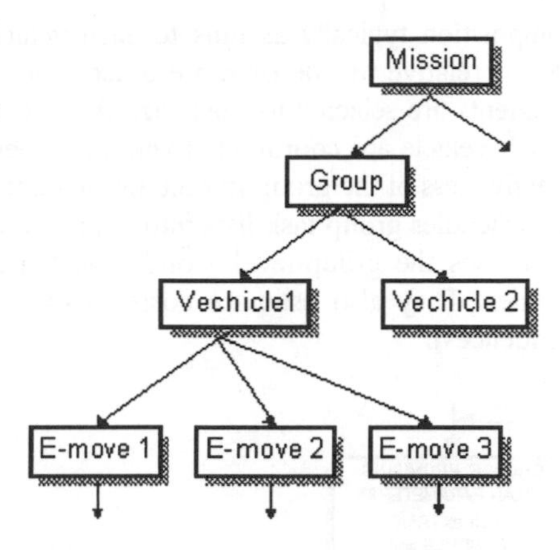

Figure 9.5. MAUV task decomposition architecture.

The function of the mission level is:

(1) Subdivide the vehicles into groups. In this scenario, there is only one group which contains two vehicles.
(2) Determine whether any of the subtasks defining the input mission command should be omitted.
(3) Provide a coarse description of routes and tactics for the mission that are sent to the lower levels.
(4) Determine appropriate priorities to be used by the lower levels in planning the subtasks.

The output of the mission level are the group subtasks and priorities. Priorities are values indicating the importance of the following factors during lower-level planning: time used, energy used, stealth, and vehicle survival.

As indicated in Figure 9.5, the mission level has a Planner Manager, a planner for each group, and an executor for each planner. The Planner Manager assigns vehicles to groups, sets priorities for group actions, and assigns mission objectives to the groups. The planner for each group schedules the activities of the group and sets the priorities mentioned above.

A flow chart for a mission level planner is shown in Figure 9.6. The program attempts to generate an optimal sequence of subtasks as follows. First, a set of promising plan parameters is chosen. These include a specific sequence of subtasks and an estimate of the time and energy priorities. Next, the planner uses outcome calculators to determine the result of choosing these plan parameters.

2. **Group level**

Group task commands define actions to be performed cooperatively by groups of MAUV vehicles on multiple targets. The Planner Manager decomposes group tasks into individual vehicle tasks. This decomposition typically assigns to each vehicle a prioritized list of tasks to be performed on or relative to one or more other vehicles, objects, or targets. Tactics and vehicle assignments are selected to maximize the effectiveness of the group's activity. The actions of each vehicle are coordinated with the other vehicles in the group so as to maximize the effectiveness of the group in accomplishing the group task goal.

Each vehicle planner schedules group task lists into coordinated sequences of vehicle tasks. The vehicle planner uses the group-level world model map to compute vehicle trajectories and transit times. They also estimate costs, risks, and benefits of various vehicle tactics (or task sequences).

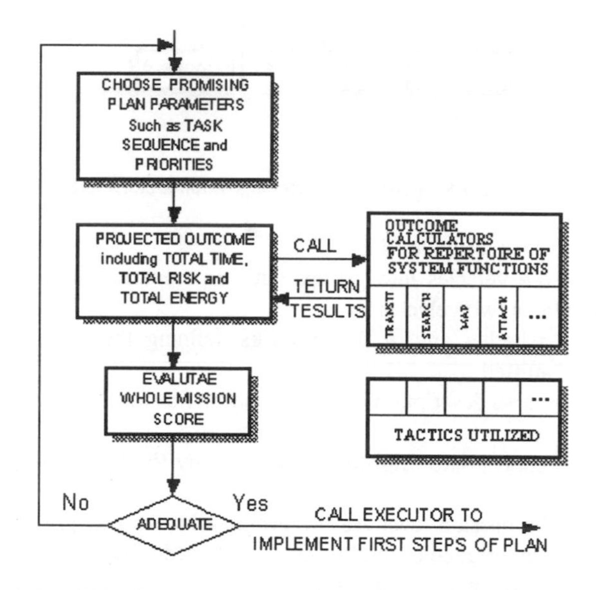

Figure 9.6. Mission-level planner for a single group.

In the implementation, the inputs to the group level are a command and a set of priorities. The command is a task involving multiple vehicles, e.g., TRANSIT, ATTACK, RASTER-SEARCH. The priorities are values indicating the importance of stealth, destruction, time, and energy.

The outputs of the group level are the vehicle tasks and priorities. The group priority values are the same as the input priorities.

3. Vehicle level

The inputs to vehicle level are a command and a set of priorities. The command is a task performed by a single vehicle, e.g. GOPATH, WAIT, RASTER-SEARCH, LOCALIZE-TARGET, RENDEZVOUS. The function of the vehicle level is to decompose the input vehicle task into a sequence of subtasks for each subsystem of the vehicle. These subsystem tasks are called elemental moves or actions (e-moves). These subsystems include the pilot, sensors and communication subsystems.

The output of the vehicle level are the e-move tasks.

4. **E-move level**

The input to the e-move level is a command which is an elemental move or action involving a single subsystem, e.g., GO-STRAIGHT (pilot subsystem), ACTIVATE-ACTIVE-SENSOR (sensor subsystem), SEND-MESSAGE (communications subsystem).

The function of e-move level is to decompose the input e-move command into a sequence of low-level commands to the particular subsystem controller.

The pilot e-move can be defined as a smooth motion of the vehicle designed to achieve some position, orientation, or "key-frame pose" in space or time. The pilot planner computes clearance with obstacles sensed by sonar sensors on-board and generates sequences of intermediate poses that define pathways between key-frame poses. A^* search is used to generate these paths.

The outputs of this level are low-level commands to the subsystem controllers of the MAUV vehicles.

5. **Primitive level**

The primitive level computes inertial dynamics and generates smooth, dynamically efficient trajectory positions, velocities and accelerations. Inputs to this level consist of intermediate trajectory poses which define a path that has been checked for obstacles and is guaranteed free of collisions.

The outputs of this level consist of evenly spaced trajectory points which define a dynamically efficient movement.

6. **Servo level**

The servo level transforms coordinates from a vehicle coordinate frame into actuator coordinates. This level also servos thruster direction and actuator power. There is a planner and executor at this level for every motor and actuator in the vehicle.

Inputs to this level consists of commanded positions, velocities, thrust, power, orientation, and rotation rates of the vehicle. Outputs of this level consist of electrical voltages or currents to motors and actuators.

9.1.4 Real-time Planning

This subsection describes the real-time planning system used at the group level and vehicle level of the hierarchy. Figure 9.7 shows the block diagram of this planning system. An input task command first goes to the Planner Manager, which contains two modules. The first, the Job Assignment Module, divides the input task into several jobs and sends each to a different planner. The different planners then work on these jobs in parallel. The second module, the Plan Coordination Module, coordinates planning among the various planners.

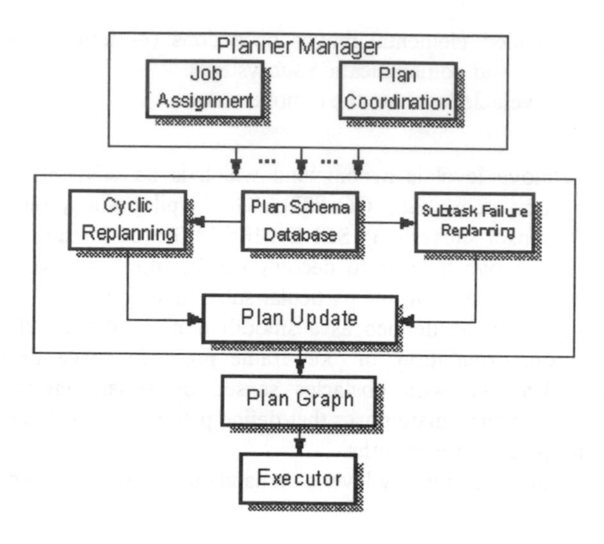

Figure 9.7. Internal structure of planner manager and planners.

After a planner has finished generating a plan in the form of a plan graph, the executor associated with the planner steps through the graph. Each planner contains several modules (see Figure 9.7). The Cyclic Replanning Module accepts an input command (or job) from the Planner Manager and, at regular cycle times, generates a new plan. The primary way in which the system performs replanning is by generating new plans regularly. This approach is based on the notion that the best way to know whether the world has changed in such a way as to require a new plan is to actually run the algorithm that generates the plan, and then to see whether the plan has changed.

One issue that must be considered is real-time planning and how it is handled by the planner. We view a plan as being composed of actions and world events. Execution of the plan by the executor occurs by monitoring for world events and stepping to the appropriate action based on which world events have occurred. Let t_1 be an arbitrary point in time and E be the set of events in the world occurring at t_1. We define real-time planning as the process of generating plans quickly enough so that there is always an action a given to the executor such that (1) Action a is part of a plan p, and (2) Plan p represents an "appropriate" response by the system to events E at time t_1. Let t_1 be as defined above and t_2 be the furthermost point in time at which an action must be executed in order to appropriately respond to the world events E. Then the planning reaction time is defined as the time interval $(t_2 - t_1)$.

The planning reaction time is different at different levels of the hierarchy. At the higher levels, the world representation is course, planned actions occur over large time scales, and world events are coarsely represented. Therefore, the planning reaction time of the system can be relatively slow. At the lower levels, the world representation is detailed, planned actions occur over small time scales, and world events are represented in detail. Therefore, the planning reaction time must be fast.

The cyclic replanning time at each level is determined by the planning reaction time. The cyclic replanning times at the higher levels are longer than that at the lower levels. At the end of a cyclic replanning time interval, the next action to be taken must have already been determined by the planner, for the executor must always have an action to carry out. However, these time intervals will often not be enough for the planners to generate new full plans. Therefore, the planner will pass on to the executor whatever is its best plan at the end of the cycle time, even though the planner may not have finished planning to completion. In this implementation, where A^* search is used, the best plan at any point in time is the path in the search tree from the root to the leaf node with the lowest cost.

When the Cyclic Replanning Module has generated a new plan, the plan is passed to the Plan Update Module which updates the Plan Graph.

If a subtask (i.e. an action) of the current plan is sent by the executor to the level below and the subtask cannot be achieved, then a signal is returned to the current level and the plan is modified by the Subtask Failure Replanning Module. Associated with each subtask command sent to the level below is a set of failure constraints. If these constraints cannot be met, then the subtask fails. Examples of failure constraints are: (1) achieving the subtask within a time window, (2) achieving a goal (e.g. arriving at a given point in space), and (3) not deviating more than a certain amount from a given path.

Both the Cyclic Replanning and the Subtask Failure Replanning modules access the Plan Schema Database to generate plans. A plan schema is used to define a subtask command. It provides all possible sequences of actions that define the command. In order to determine the best sequence in a given situation, it allows the application of a cost function and provides the ability to perform a search which is driven by the plan schema.

9.1.5 Implementation and Experimental Results

1. **Computing systems for implementation**

The control system was implemented on the computing systems shown in Figure 9.8. On the left of this figure is the target hardware for the two MAUV vehicles. In each vehicle, a VME bus supports high bandwidth communication between sensory processing, world modeling, planning, and execution modules at each level of the hierarchy. These modules are partitioned among three separate single-board Ironics computers so as to maximize the use of parallel computation. A two-megabyte common memory board is used for communication between processes, and an 800-megabyte optical disk is used for mass storage. The real-time multiprocessor, multitasking operating system used is PSOS.

On the right of Figure 9.8 is the software development and simulation environment. A variety of computers, including Sun workstations, a VAX 11/785, a micro-VAX, IRIS graphics systems, PCs, Duals, and Ironics development systems are tied into the development environment for code development and simulation. Once the software has been translated to run on the Ironics Unix-based development system, it can be compiled to run under PSOS and downloaded into the 68020 target hardware for real-time execution.

2. Experimental results

Some initial experimental tests on the lake was performed with one of the MAUV vehicles during October 1987. Due to lack of continued funding, the MAUV project was terminated in December 1987. Therefore, these tests have been very limited and preliminary.

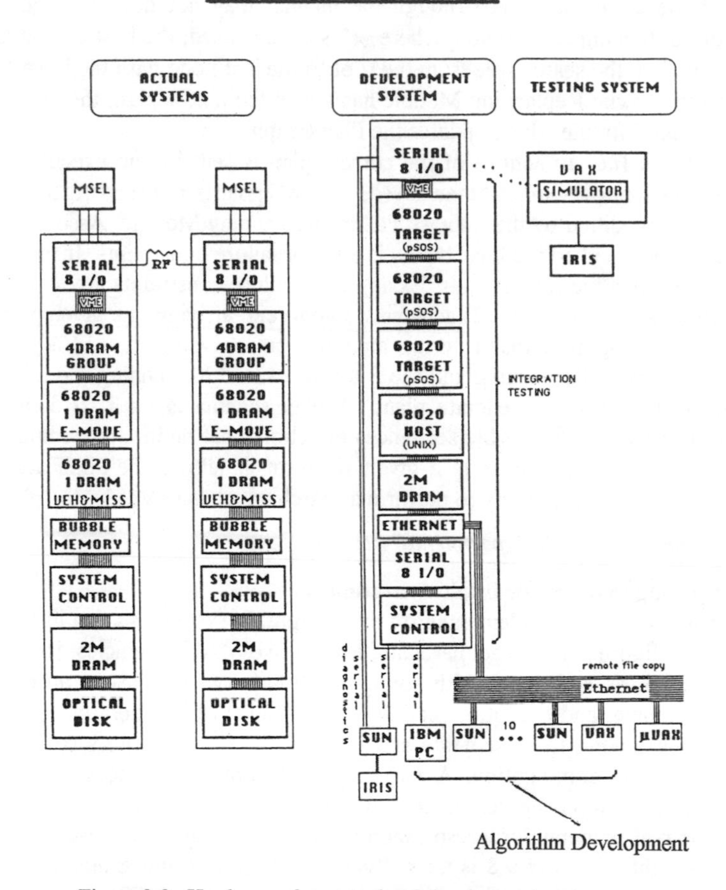

Figure 9.8. Hardwares for control and development systems. (with Herman et al. 1990)

The lake tests were performed at Lake Winnipesaukee and were run using code at the servo, primitive, and e-move levels. The first experiment involved local obstacle avoidance. Figure 9.9 shows the path executed by the vehicle during a test run in which an obstacle was manually entered into the world model map at point *C*, and the vehicle was commanded to go from point *A* to point *B*. The control system successfully planned and executed a path around the obstacle at point *C*.

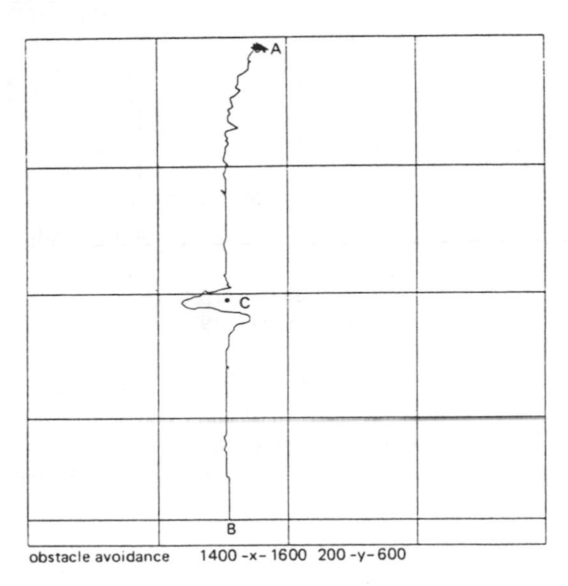

Figure 9.9. Obstacle avoidance test.
(with Herman *et al.*, 1990)

The second experiment involved following along a predefined path. Figure 9.10(a) shows a raster-scan path from point *A* to point *B*. The vehicle determined its *x*, *y* position from on board acoustic navigation transponders which receive signals from navigation buoys placed in the water. The actual path executed by the vehicle during the run is shown in Figure 9.10(b). One of the obvious problems brought about by this run is that the vehicle tends to overshoot when it makes turns. This is a problem with the current low-level control, which allows position control but not velocity control. Because the velocity is at a maximum value when it makes a turn, it will always overshoot. Also, there is considerable error in the position measuring transponders, which largely accounts for the ragged appearance of the pathways.

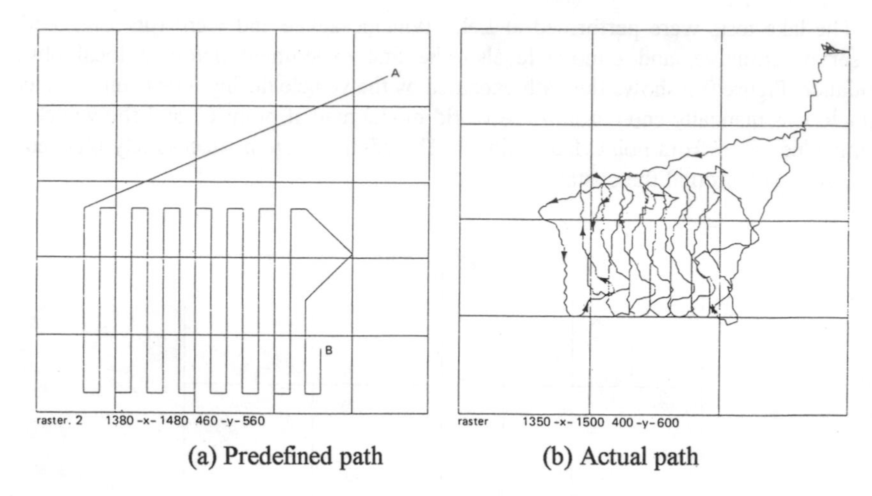

(a) Predefined path (b) Actual path

Figure 9.10. Raster-scan path test.
(with Herman *et al.*, 1990)

The third experiment involved updating the internal model of the lake bottom with altitude information obtained from the downward-looking depth sonar. Figure 9.11 shows three graphs. The top and middle graphs display the x and y positions of the vehicle path respectively. The bottom graph shows the lake depth values obtained from the world model along this path after the world model is updated from the information in the depth sonar.

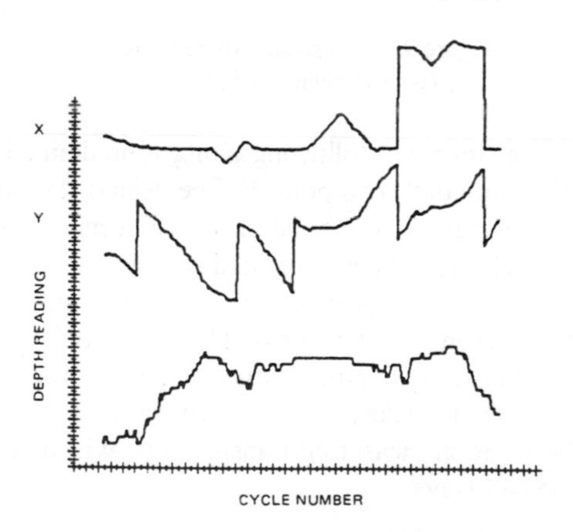

Figure 9.11. Updating world model of lake depth.
(with Herman *et al.*, 1990)

3. Conclusion

The achievement of real-time intelligent control for autonomous vehicles requires a system that integrates AI with control theory, and can be implemented on parallel, possibly special-purpose hardware. This section has presented the basic components of such a system, and has proposed a hierarchical control system architecture that can serve to integrate the various components. A first cut at the algorithms and software for many of these components has been developed. However, many of these algorithms need to be improved to handle more complex scenarios.

A major problem is the achievement of real-time performance. Although the multiprocessor computing system described here can serve as a good basis for achieving real-time performance, it must be augmented with special-purpose hardware such as real-time sensory processing devices and massively parallel devices (e.g. neural networks). Generally, such special-purpose hardware would accomplish the functions of one or two modules in the hierarchical control system architecture, and can thus fit very elegantly into a system that implements this architecture.

Another major problem of intelligent control is that of leaning. Learning and the ability to generalize are very important for autonomous systems that must operate efficiently in a wide variety of situations in a complex real-world environment. The neural network systems can offer promising approaches to this problems. Again, this system can fit very nicely into this hierarchical control system architecture.

9.2 Intelligent Process Supervision and Control — Hierarchically Intelligent Control for Industrial Boiler

Many industrial continuous production lines, e.g. steel making, chemical engineering, oil refining, materials processing, papermaking and nuclear reaction etc. have production processes which require supervision and control with high performances and reliability. In order to maintain physical parameters within a certain change range of precision, and to guarantee high product quality and productivity in these continuous industrial plants or production lines, efficient control modes have to be applied to their process control. As we have mentioned in Chapter 2 that the traditional mathematical model-based controllers of these processes exist great limitation and could only suit to certain range of applications. The intelligent control can imitate human experience and build knowledge-based model or hybrid models. Much research and many developments for using intelligent control to industrial production process have been made and shown superior control performances. For instances, fuzzy control of a rotary cement kiln [24], advanced fuzzy logic control for automotive industry [55], expert control for process monitoring of steel making [36], neural control of steel rolling mill [44], distributed intelligent materials handling system [52], hierarchical intelligent materials processing [65], intelligent *pH* process control [9], intelligent control of synchronous shearing process [70], hierarchically intelligent control for industrial boiler [20], knowledge-based control of nuclear reactors [32, 43], neural-network-based process control for plasma etching and deposition [16], self-learning fuzzy neural networks for control of the arc welding process [10], and others [4, 49].

In this section, we only take the hierarchically intelligent control of industrial boiler [20] as an application example to introduce in detail. A hierarchical structure of intelligent control and algorithms including expert control, multi-mode control and self-tuning PID control will be presented. This hybrid controller adjusts its structure in order to maintain high performance over a wide variation of the plant operating condition. The results of simulation and industrial run will be demonstrated.

9.2.1 Structure of Hierarchical Hybrid Controller

The hierarchical hybrid control structure for industrial boiler plant is shown in Figure 9.12 that is built in a type of learning manner.

At the outset of controller operation, when the system knows little about controlled process or when there is obvious variation of working state, then the expert controller (EC) provides the initial crude control by imitating the actions of the human operator (switch K is turned on to position 1). When there is some ripple of the working condition, i.e. deviation $e(k)$ and incremental deviation $\Delta e(k)$ are slightly big, then multi-mode controller (MC), which control action is strengthened, is selected (switch K is turned on to position 2). As the working condition is smooth and steady, i.e. $e(k)$ and $\Delta e(k)$ are rather small, then the self-tuning PID controller (STPID), which adjusts in a fine-tuning manner, is selected (K is turned on to position 3).

The operation of the hybrid controller is coordinated by the expert system according to the dynamic process. The expert system makes heuristic decisions needed to regulate the control structure as it learns about the controlled process so that the controller switching is governed properly.

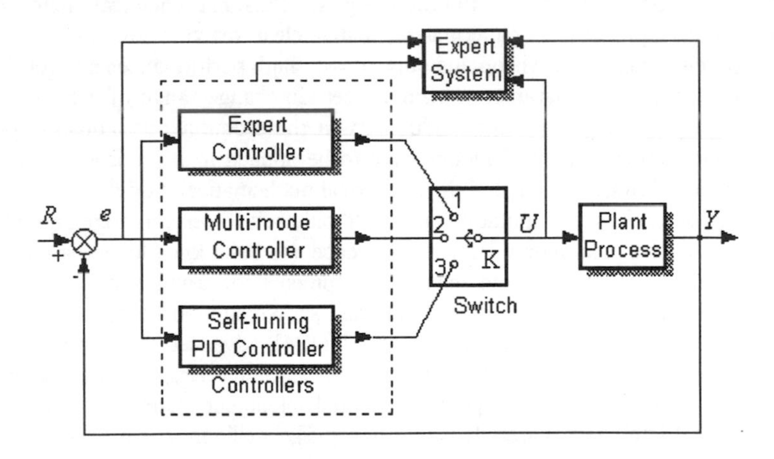

Figure 9.12. Block diagram of hybrid controller.

The expert system is a rule-based production system and the forward chaining is used on its inference engine. The expert system attempts to select the best control method in accordance with its knowledge about the process. The algorithm of the above mentioned hybrid control is designed to provide with hierarchical structure and complied with the principle of increasing precision with decreasing intelligence (IPDI), where expert control is the most intelligent control action, and self-tuning PID control is the most precise control action. It can be presented in the phase plane as shown in Figure 9.13.

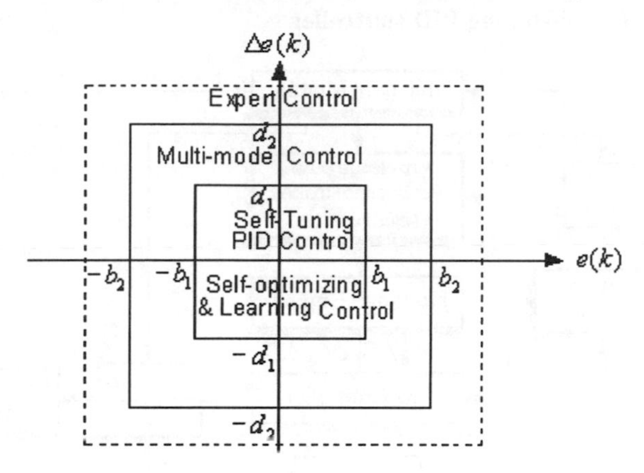

Figure 9.13. Phase plane of IC algorithm.

The internal layer of this hierarchical structure is the module of STPID; the middle layer is the module of MC; and the external layer is module of EC. Under the monitor of expert system, the three control modules can be switched on from one to another at a smooth transition. They may also be expressed with the following production rules:

rule 1 IF $\left[|e(k)| \geq b_2 \cup |\Delta e(k)| \geq d_2\right]$

THEN [EC]

rule 2 IF $\{[(b_1 \leq |e(k)| < b_2) \cap (d_1 \leq |\Delta e(k)| < d_2)]$

$\cup\left[|e(k)| \leq b_1 \cap d_1 \leq |\Delta e(k)| < d_2\right]$

$\cup\left[b_1 \leq |e(k)| < b_2 \cap |\Delta e(k)| \leq d_1\right]\}$

THEN [MC]

rule 3 IF $\left[|e(k)| < b_1 \cap |\Delta e(k)| < d_1\right]$

THEN [STPID]

where b_1 and b_2 denote deviation bounds; d_1 and d_2 indicate threshold values of incremental deviation.

9.2.2 Algorithms in the Control System

1. **Algorithm for self-tuning PID controller**

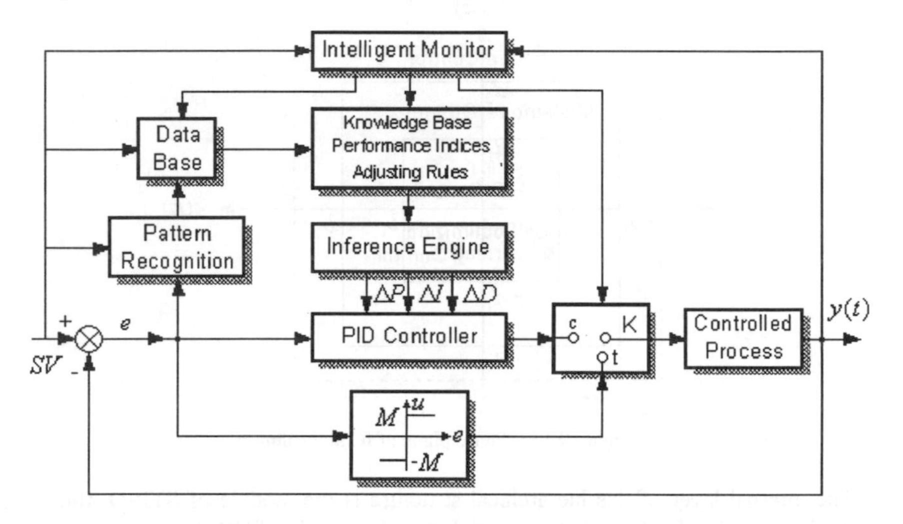

Figure 9.14. Architecture of self-tuning PID controller.

Figure 9.14 shows the architecture for self-tuning PID controller which adopts the method of characteristic recognition and is provided with a expert system. The relay nonlinear unit is used for pretuning and self-tuning control parameters. When switch K is turned to t, the relay unit makes the closed loop to produce an equal amplitude oscillation, thus the characteristic parameters of control system may be obtained. Equivalent amplification coefficient of the system is:

$$K_c = 4M/\pi A_c \tag{9.1}$$

where A_c denotes amplitude; M is characteristic value of the relay, which can be optimized with on-line self-learning algorithm. The critical period of oscillation T_c can be directly detected from output curve of closed loop system. When switch K is turned to c, the system is in a state of PID control. Switch K is supervised by the expert system. The function of pattern recognition module is used to recognize and classify the mode of system transient response. The knowledge base stores control characteristic indices and adjusted rule based on knowledge and experience of control experts. By means of inference engine, real-time adjustment and optimization of control parameters are realized. The self-tuning is suitable for control parameters preselection when the system is put for starting operation or for adjusting parameters when set value or working condition is changed.

2. Algorithm for multi-mode control

Multi-mode control is such a control scheme that adopts various control strategies and modes according to different characteristic states of control process. Using multi-mode control scheme is one of the fundamental principles of simulating human intelligent control (SHIC). It is usually difficult to make account of dynamic and steady state indices by means of single control mode. Multi-mode control makes it possible to adopt appropriate control in the light of different states and periods at control process. Thus, it is able to give consideration to manifold performance indices of control system.

If the input set of multi-mode controller is

$$\varepsilon = (\varepsilon_1, \varepsilon_2, \ldots, \varepsilon_n) \tag{9.2}$$

its output set is

$$U = (U_1, U_2, \ldots, U_n) \tag{9.3}$$

and f is the mapping from ε to U:

$$f: \varepsilon \rightarrow U \tag{9.4}$$

then the established control rule set must satisfy the necessary condition:

$$f(\varepsilon) = U \tag{9.5}$$

that is, in order to avoid possible loss of control, full mapping from ε to U is required.

The substance of multi-mode control is carrying out piecewise control to system transient. The MC algorithm is composed of eight control modes as follows:

(1) Bang-bang control

MC1: IF $e > DL_3$ THEN $U(n) = U_m$

MC2: IF $e < -DL_3$ THEN $U(n) = -U_m$

(2) Output holding

MC3: IF $|e| < b \cap |\Delta e| < d$ THEN $U(n) = U(n-1)$

(3) Enhancing control action $(K_2 > 1)$

MC4: IF $e \cdot \dot{e} < 0 \cap |e| \geq b \cap |\Delta e| < d$

THEN $U(n) = U_c + K_1 K_2 SE_m$

(4) Braking control

MC5: IF $e \cdot \dot{e} < 0 \cap \delta_1 < |e/\Delta e| < \delta_2$

THEN $U(n) = U_c + K_1 SE_m$

MC6: IF $e \cdot \dot{e} < 0 \cap |e| < b \cap |\Delta e| \geq d$

THEN $U(n) = U_c + K_1 SE_m + K_{d1} \Delta e$

(5) others

MC7: IF $e \cdot \dot{e} \geq 0 \cap |e| < b_3$

THEN $U(n) = U(n-1) + K_p\left[(e_n - e_{n-1}) + T_s / T_i(e_n) + T_d / T_s(e_n - 2e_{n-1} + e_{n-2})\right]$

MC8: IF $e \cdot \dot{e} \geq 0 \cap |e| \geq b_3$

THEN $U(n) = U_c + K_b(e_n - e_{n-1})$

where

$$SE_m = \beta \sum_{i=1}^{n-1} e_{i,\max} + e_{n,\max}$$

is known as the memory sum of deviation peak; $\beta < 1$, indicates forgetting factor. In the above rules, U_c is the value position of actuator; K_1, K_2, K_b are proportional amplification coefficients (gains); K_{d1} is differential coefficient; DL_3, b, d, δ_1, δ_2, b_3 are threshold values.

3. Algorithm for expert control of boiler

Owing to the complicated mechanism of the boiler burning process, such as considerable difference of distinct coal quality and great variation of plant characteristics, it is very difficult to establish the mathematical model of boiler combustion process, that should be sufficiently accurate and also easily controlled. Adopting expert control algorithm is an effective way to achieve a practical automatic control for the burning process of industrial coal-fired boiler.

(1) Basic control principles

Taking the chain-grate boiler with coal-ejector of 35T/H as a controlled plant, the simulating human (expert) control algorithms, which regulate air and coal supplies mainly according to the temperature T_b on chamber of furnace, are presented as follows.

- The actual load range of the boiler is the divided into four segments:

(a) Lower than 25% rated load;
(b) Within the scope of 25-50% rated load;
(c) Within the scope of 50-75% rated load;

(d) Higher than 75% rated load.

In each segment of load, different control parameters are used. As the load increase, the adjustment amounts of air supply $U_F(n)$ and fire-coal $U_M(n)$ are increased, the given value of oxygen content $OC(n)$ is decreased, and the velocity of chain-grate $U_p(n)$ is slowed down. Both air supply and oxygen content have their upper and lower limits.

When furnace temperature T_b and steam drum pressure P_b obviously drop during either changing load or constant load, the expert control algorithm adjusts the ratio between air and fire-coal supplies to seek T_b raising as soon as possible. Thus

$$\Delta T_b = T_b(n) - T_b(n-1) > 0 \tag{9.6}$$

is a criterion; if $\Delta T_b > 0$, then continue regulating with adapted control rule, otherwise alternate other rules. Adjusted amounts are related to accumulation $\sum_{i=1}^{m} T_b(i)$ and load range.

(2) Expert control rule set

The expert control rule set of industrial coal-fired boiler includes five modules as follows:

(a) Fault monitoring for sensors
(b) Load variation detecting
(c) Control of constant load condition
(d) Control of increasing load condition
(e) Control of decreasing load condition

This control rule set consists of 22 production rules. The module of fault monitoring for sensors is used to monitor the zirconia transducer of oxygen content and thermocouple in the chamber of furnace whether or not in good condition. For examples:

rule 20 IF $(OC(n) < OC_{\min} \cup OC(n) > OC_{\max})$

THEN (setting mark of oxygen transducer failure, $OCF = 1$)

rule 23 IF $(T_b(n) < T_{b,\min} \cup T_b(n) > T_{b,\max})$

THEN (setting mark of thermocouple failure, $TF = 1$)

where $OC_{\min}$ and $OC_{\max}$ denote normal lower limit and upper limit of oxygen transducer respectively; $T_{b,\min}$ and $T_{b,\max}$ denote normal lower and upper limits of thermocouple respectively; OCF — oxygen transducer failure factor; TF — thermocouple failure factor. According to increment of load value $\Delta D = D(n)\text{-}D(n\text{-}1)$ and its accumulation $\sum_{i=1}^{10} \Delta D(i)$, the module of load variation detecting decides and enters appropriate control program such as

rule 28 IF $|\Delta D(n)| < \delta_1 D_r$

THEN(entering control of constant load condition)

rule 29 IF $\left[|\Delta D(n)| \geq \delta_1 D_r \cap \Delta D(n) \cdot \Delta D(n-1) > 0 \cap \sum_{i=1}^{10} \Delta D(i) > \delta_2 D_0 \right]$

THEN(entering control of increasing load condition)

where D_r is rated load; D_0 is the steam flux when load starts to change; $\delta_1 = 0.005$ and $\delta_2 = 0.15$ are coefficients related to load increment.

When the external load has none or little variation, if internal factors such as distinct coal quality cause variation of pressure in steam drum and furnace temperature, then $U_M(n)$ and $U_F(n)$ are adjusted according to pressure deviation $e_p(n)$ in steam drum:

$$E_p(n) = P_{set} - P_b(n) \tag{9.7}$$

and sum

$$\sum_{i=1}^{10} \Delta T_b(i) \tag{9.8}$$

as well as load range, where P_{set} is the set pressure of the steam drum.

If external load raising causes descent of main steam pressure $P_M(n)$, then $U_F(n)$, $U_M(n)$ are regulated according to earlier changing tendency of furnace temperature and adjusted mark is set after controlling, e.g.,

rule 41 IF $\left[\left(\sum \Delta D > 0 \cap \sum \Delta P_M < 0 \cap \sum \Delta T_b \geq 0 \right) \right.$

$\left. \cup \left(\text{FR41}(n\text{-}1) = 1 \cap \Delta T_b > 0 \cap D(n) \in D_i, i = 1,2,3,4 \right) \right]$

THEN $\left[\Delta U_M(n) = K_{yi} e_p(n) + K_{li}(D(n) - D(n-1)) \right.$

$\cap \Delta U_F(n) = K_{FMi} \Delta U_M(n)$

$\cap U_F(n) = U_F(n-1) + \Delta U_F(n)$

$\cap (\text{IF } U_F(n) > U_{FHi} \quad \text{THEN } U_F(n) = U_{FHi}$

$\cap U_M(n) = U_M(n-1)$

$\cap U_p(n) = U_p(n-1)$

$\left. \cap \text{FR41}(n) = 1 \right]$

When external load reducing causes ascent of $P_M(n)$, then fire-coal and air supplies are decreased until they adapt present load.

9.2.3 Results of Simulation and Industrial Running

1. **Hybrid simulation**

For controlled plant, the transfer function is

$$G_0(s) = 1/[(21s+1)(10s+1)(9s+1)] \tag{9.9}$$

The hybrid simulation data are shown in Table 9.1, while the sampling period $T_s = 2$ sec.

Table 9.1. Hybrid simulation data of self-tuning

No.	M	ε	T_c	K_c	$PB\%$	$1/K_i$	K_d	K_p	T_i	T_d
1	18	0.7	52.49	6.837	24.37	3.199	13.46	4.10	26.25	6.55
2	12	0.2	52.49	6.297	26.46	3.473	12.39	3.78	26.24	6.55
3	15	.05	52.49	7.751	21.50	2.821	15.26	4.65	26.24	6.56
4	15	0.1	52.49	6.879	24.22	3.188	13.54	4.13	26.32	6.56
5	15	0.1	48.28	7.383	22.59	2.727	13.36	4.43	24.14	6.03
6	25	0.1	56.69	8.373	19.90	2.821	17.80	5.02	28.34	7.09
7	25	0.1	50.39	8.054	20.69	2.607	15.22	4.83	25.20	6.30
8	25	0.1	48.29	8.400	19.84	2.396	15.21	5.04	24.15	6.04
9	25	0.1	50.39	6.860	24.29	3.061	12.96	4.12	25.20	6.30

Step response curves of PID control adapting self-tuning parameters and multi-mode control are shown in Figure 9.15(a) and Figure 9.15(b) respectively.

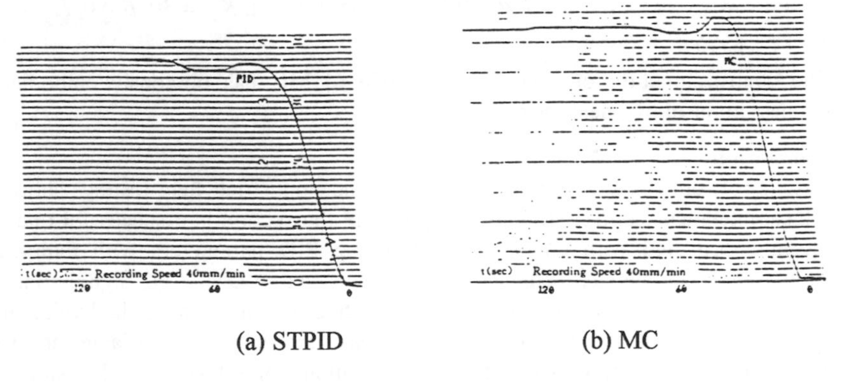

(a) STPID (b) MC

Figure 9.15. Control curves for step response.
(with Guo *et al.* 1992)

2. Industrial worksite running

There are two boilers in the self-supply power plant of a factory in Dalian, China, each has an evaporation flow rate of 35T/H, they are allocated with two deaerators in connection. Four DMC-200 intelligent digital controllers have been installed in the system, separately regulating water level and pressure of the deaerators. Several field tests of self-tuning and multi-mode control were made in above control system. Part of the data from the test are shown as follows:

Given value $SV = 25$; sampling period $T_s = 2\text{sec}$; relay parameter $M = 15$; self-tuning precision $\varepsilon = 0.1$.

(1) Self-tuning

No.1 Deaerator pressure: $T_{c1} = 12.60; K_{c1} = 3.72; PB_1\% = 44.86; 1/K_{i1} = 1.413; K_{d1} = 1.755; K_{p1} = 2.23; T_{i1} = 6.30; T_{d1} = 1.57.$

No.2 Deaerator pressure: $T_{c2} = 10.5; K_{c2} = 4.167; PB_2\% = 40.1; 1/K_{i2} = 1.05; K_{d2} = 1.638; K_{p2} = 2.5; T_{i2} = 5.25; T_{d2} = 1.31.$

When PI control is used $K'_{p2} = 1.875; T'_{i2} = 8.925.$

The above self-tuning parameters have been put into running in corresponding to control loops, the control variables have few ripples, steady-state error accords with the demands of production process. Tracking performance also shows the test is a successful one.

(2) Multi-mode control

Two groups of MC parameters are as follows:

Group 1: $DL_3 = 15; b = 3; d = 0.12; K_1 = 1.05; K_2 = 1.1; K_{d1} = 30; \beta = 0.1; K_p = 2.23; T_i = 6.30; T_d = 1.57.$

Group 2: $DL_3 = 5; b = 7; d = 0.3; K_1 = 1.05; K_2 = 1.1; K_{d1} = 30; \beta = 0; K_p = 1.25; T_i = 5.5; T_d = 1.375.$

The control function of each control mode of MC has obviously embodied its effectiveness in the above field tests.

3. Conclusion

The results of digital and hybrid simulation and industrial field-testing show the hierarchical intelligent control structure proposed in the section is a successful one. The IC algorithm provided with combining auto-tuning PID parameters with multi-mode control as well as on-line self-learning function is achieved in DMC-002 digital controller and applied in industrial spot and proved quite effective. When the controlled value of the system is rather rippled, the MC algorithm can make the controlled variable tend to be stable in short time, hence, it creates favorable conditions for auto-tuning PID parameters under the circumstance of load changing or set value varied. Thus, the robustness of control system is improved. This method is suitable for various control loops of a boiler plant as well as other fields of industrial process control. Expert control for the boiler burning process is applicable on occasions such as load increasing or decreasing or changing kind and quality of fire-coal. Self-optimizing and learning algorithm make it possible to realize economical combustion. The efficiency of the industrial boiler is obviously increased by the expert control system.

9.3 Intelligent Control for Automatic Manufacturing Systems

The Computer Integrated Manufacturing Systems (CIMS) and Flexible Manufacturing Systems (FMS) have been developed rapidly in recent years [6, 8, 35, 41, 57, 69, 70].

In a complicated manufacturing process, several operations under different conditions are needed to generate a final product. The uncertainty of the environment and the system complexity in hardware and software challenge today's engineers in design and implementation of an effective integrated control system. In this section, we would like to introduce two application examples of intelligent manufacturing systems.

9.3.1 A Petri Nets-based Manufacturing Control System

In order to extend the existing Petri net techniques to modern manufacturing systems, a theory which combines machine intelligence techniques and Petri net theory and intelligent discrete event controllers needs to be developed. In this example, a Petri net-based methodology for such intelligent controller design is formulated and applied to such discrete event systems as flexible manufacturing systems [69].

1. **Architecture of intelligent manufacturing control systems**

(1) Hierarchical structure

The Petri net-based intelligent control scheme of three hierarchies is shown in Figure 9.16. Except those manufacturing processes, sensors, and actuators, a real-world manufacturing system includes decision-maker, supervisory controller, local controllers, database, and knowledge base. The third level focuses on local control of processes or individual operations, and the robot and NC-programming languages are used. The second level concentrates on supervisory control or coordination using Petri nets and C^{++} programming language etc. The top level aims at effective decision-making whenever a conflict or an abnormal situation arises, and the programming languages Auto-Lisp, OPSS etc. are employed. More levels should be included when such functions as process planning, long-term and short-term scheduling are considered [15].

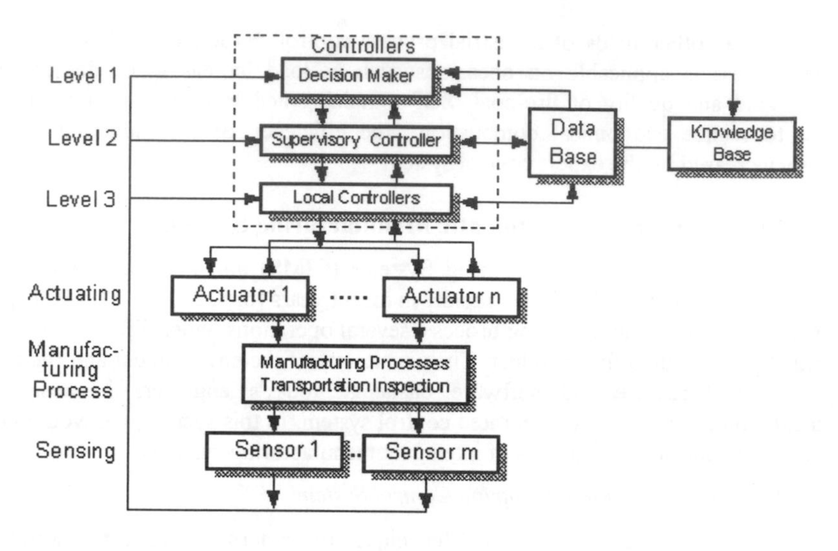

Figure 9.16. Structure of intelligent control for manufacturing systems.

Local controllers can be interpreted as process controllers for manufacturing facilities such as machine tools and robots.

Knowledge base and database are important parts to support all control activities. The following will focus on two components: Petri net based supervisory controller and intelligent decision maker.

(2) Petri net controller

Invented as a graphical tool for modeling and analysis of asynchronous communication systems in the early 60's, Petri nets have been extended to the design of manufacturing control systems since the 70's. A Petri net controller is defined as a control-logic based on a marked Petri net for a discrete event system [71]. The flow of tokens through places is regulated by the firing of transitions in a marked Petri net. It is this flow of tokens which defines the supervisory control actions for a materials manufacturing system. To illustrate the above concepts, a portion of a Petri net is shown in Figure 9.17. It models a part of the manufacturing system. The places represent either the status of resources or operations and transitions start of operations as shown in Figure 9.17.

For this net, initial marking $(1,1,1,0,0,1,1)^T$ implies that all resources including a raw material piece are available. Only one transition, i.e., t_1 is initially enabled since its two input places p_1 and p_2 have tokens. Firing t_1 removes one token from p_1 and p_2 respectively and deposits one token into its output place p_4. Place p_4 being marked implies that the workpiece is being machined by machine 1. The new state is described in

Figure 9.18 and the marking becomes $(0, 0, 1, 1, 0, 1, 1)^T$.

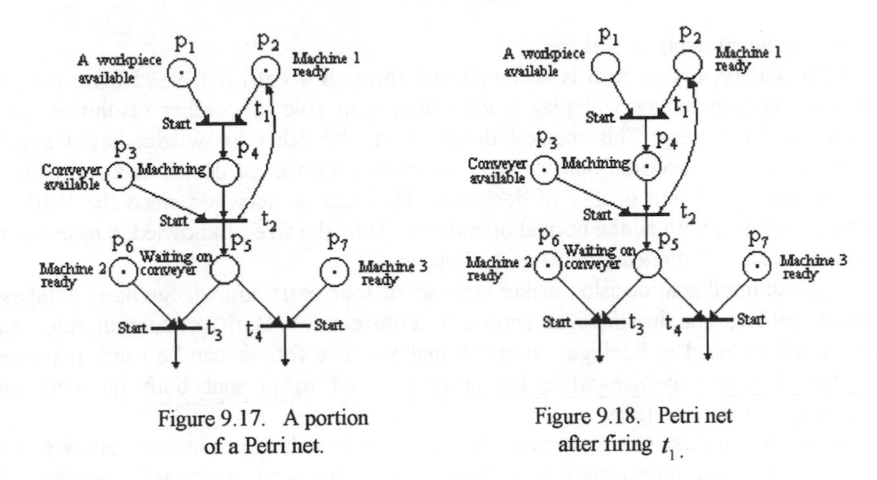

Figure 9.17. A portion of a Petri net.

Figure 9.18. Petri net after firing t_1.

Since only transition t_2 is enabled at $(0, 0, 1, 1, 0, 1, 1)^T$, the supervisor will fire t_2 when the matching process is completed.

Then one token is removed from p_3 and p_4 and deposited into p_2 and p_5 respectively, implying that an intermediate part is on the conveyor for further processing and Machine 1 is released, see Figure 9.19.

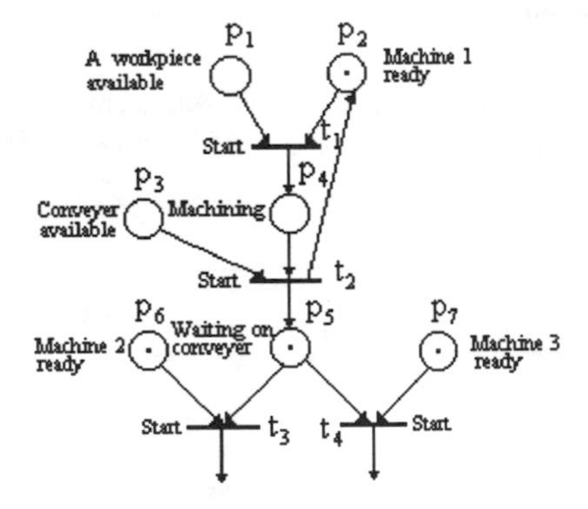

Figure 9.19. Petri net after firing t_1 and t_2.

When both Machine 2 and Machine 3 are available, then t_3 and t_4 are enabled. This implies either operation can be followed. Depending on a strategy implemented such as alternative firing or random firing policies, the next operation will be chosen. If only one of Machines 2 and 3 is available, then only one transition is enabled, and there is no choice raised.

(3) Intelligent decision maker

While supervisory control is accomplished through a Petri net based controller, an intelligent decision maker will play a very important role in conflict resolution, fault diagnosis and recovery. This method differs from the other knowledge-based system methods in that this decision maker makes as small a number of decisions as possible to guarantee the speed and quality of decisions. This can be achieved since the Petri net controller will deal with all the normal operations. Thus the size of knowledge base can be reduced to achieve high decision making efficiency.

A typical intelligent decision maker consists of four parts: knowledge base, database, inference engine, and interface as shown in Figure 9.20. Both production rules and schemes will be used in intelligent decision-makers. The former can be used to present individual domain expertise while the latter is used to present both the data and relationship between the rules.

There are various reasoning methods based on knowledge base and database. The two most common approaches are: forward chaining and backward chaining. An approximate reasoning for imprecise knowledge and model-based reasoning for the case that mathematical models are available have also been applied. The manufacturing systems considered involve mostly the assessment of a situation and formulation of a solution. Thus the forward chaining method is the most suitable. However, the system will be designed the way that various knowledge representations and reasoning methods such as approximate reasoning, model-based reasoning, and evidential reasoning can be integrated in the computer environment.

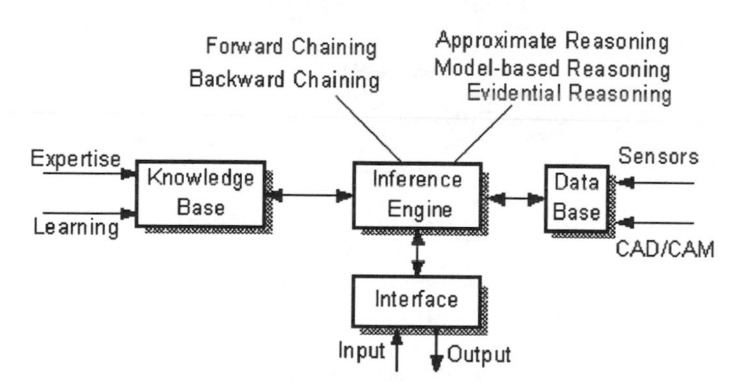

Figure 9.20. Structure of intelligent decision maker.

Most domain knowledge for manufacturing systems can be acquired by interviews with experts. However, to enhance the ability of an intelligent maker, the system should be designed the way new knowledge can be accumulated by various learning methods. The useful approaches for manufacturing systems include learning by instruction or teaching, learning by example, and learning by analogy. The others include learning by observation, learning by experiment, learning by discovery.

The interface between inference engine and the outside world serves as input and output information interpreters which make information accessible and legible to the inference engine, the controllers, and users. Also, the system is designed such that human intervention is possible.

2. Application illustration

(1) FMS cell description

The proposed design theory has been applied for designing an FMS cell controller. The FMS cell is located in the Center for Manufacturing Systems at New Jersey Institute of Technology. The layout of this system is shown in Figure 9.21. The system is designed to manufacture a family of products. Components of the system include a GE P-50 industrial robot, an IBM 7535 SCARA robot, an NASA II CNC miller, an SI Handling Cartrac Conveyor System, an IBM PC XT computer, and a GE Series SIX PLCs.

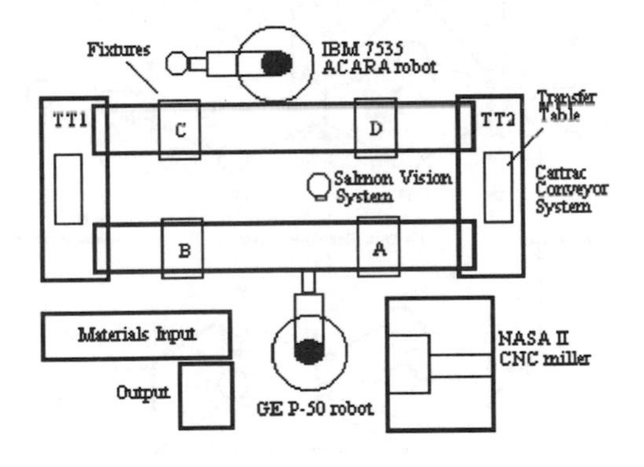

Figure 9.21. A FMS cell layout.

(2) Petri net modeling

This system can be modeled by using the bottom-up procedure. First we decompose the system into several parts as follows:

(a) Cartrac Conveyor System: four carts or fixtures, two transfer tables.

(b) Milling system: GE P-50 robot and NASA II CNC miller.

(c) Drilling system: IBM 7535 SCARA robot and driver.

(d) Vision system for inspection.

By modeling these four subsystems, we can get the final Petri net model by sharing transitions as shown in Figure 9.22. The detailed process can be referred to [69].

(3) Intelligent manufacturing control system

Using the proposed architecture, an implementation strategy for this existing manufacturing system has been formulated. Since there are already PLCs for Cartrac conveyor system, GE P-50 robot, etc., the Petri net model is translated into PLCs and the decision-maker is incorporated to deal with choices and exception handling. In order to enhance the productivity, system maintenance rules and error handling rules are programmed into the knowledge base, thus human intervention can be reduced to the minimum. These rules are currently obtained from experienced engineers on different machines and robots. Later development includes certain knowledge acquisition algorithms, for example, learning by instruction and teaching and learning by example. The CAD software is also integrated into the computer system such that changing CNC programs is easily achieved.

This project is still under way and the final results will be compared against the existing FMS capacity and productivity. The experience gained by working on this system serves to draw many useful conclusions for today's FMS control systems design.

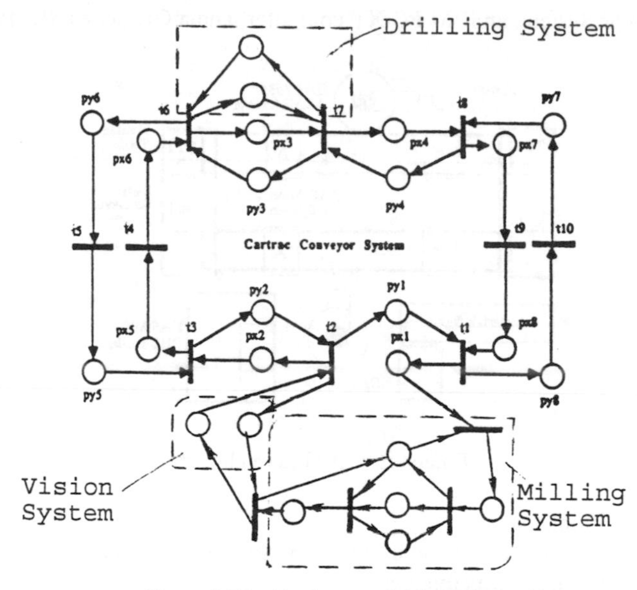

Figure 9.22. Petri net model for a FMS cell.

3. **Future development of application**

Further development of this application includes the real-time implementation of intelligent decision makers based on different reasoning methods and knowledge acquisition methods. The results will be tested based on the actual operation of the system. Different Petri net controller implementation methods will be compared against each other.

Furthermore, the method will be compared against the conventional manufacturing control systems design method as a benchmark test. The application to meaningful industrial FMS will be explored.

9.3.2 An Intelligent Control for Automatic Manufacturing Systems

Let us look at another application example of intelligent control for manufacturing systems [57].

1. **System structure and improved performances**

Manufacturing systems are required to optimize many performances, such as cost performance, productivity, speed, accuracy, due date and energy saving, etc. It is not easy to satisfy these requirements by a centralized control system. A practical solution to this problem is for the multi-control loops in the hierarchy of manufacturing systems to be improved to satisfy their respective performances in different ways.

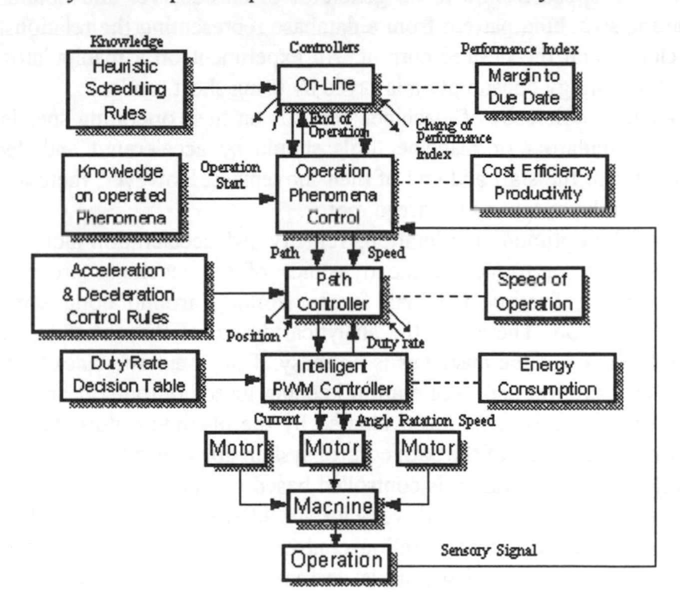

Figure 9.23. An intelligent control system with hierarchy loops. (with Watenabe and Kawata, 1987)

An intelligent hierarchical control system using theoretical/heuristic knowledge based rules with sensory information is presented. Figure 9.23 shows its composition and performances improved at each level. In the proposed system, the energy consumption of Pulse Width Modulation (PWM) control for DC motors is saved by using the knowledge of electromagnetism and information on motor rotation speeds. Acceleration-deceleration along the hand/tool path of machines is adequately adjusted by using heuristic knowledge of servo dynamics and information on motor controller duty rates. The cost performance/productivity of machining is optimized by using theoretical/experimental knowledge of its physical phenomena and sensory observed signals. The margins to the due date of production by a group of intelligent machines are improved by using heuristic rules for scheduling problems and information on their operating conditions.

2. **Intelligent PWM control of motor**

PWM is widely used for DC motor control because it effectively saves on energy consumption by on-off switching of power transistors. The switch-on and switch-off patterns of the transistors and the duty rate necessary to produce the required current are investigated by computer simulation based on the knowledge of electromagnetism and control theories. At every sampling time of motor control, the motor controller obtains the motor rotation speed from a tacho-generator or an observer and obtains the output duty rate and the switching pattern from a database representing the relationship between the motor velocity and the desired current. An experiment on a manipulator shows that nearly 24% of the energy consumption is saved by using short mode.

The productive efficiency of machines depends on their operating speeds. The hands of the robot manipulators or machine tools should be accelerated and decelerated as quickly as possible at the start and end of their movement. However, there are constraints such as the limit in the output of the servo motor.

Planning of the optimum acceleration, velocity and deceleration along the hand/tool path might be easily performed if the dynamics of the machines are clearly known. However, the inertia, friction and load forces working around/along servo axes vary widely during operation. Therefore, the analytical design of the optimum velocity curve based on the dynamics of the machines is not easy. Thus, heuristic rules to automatically determine the acceleration and deceleration keeping motor outputs at an adequate value are adopted. The motor outputs are evaluated by the maximum duty rate among them. This is because the accuracy of the path control system goes over 1.

The acceleration-deceleration is controlled based on heuristic rules that represent the relationship between duty rate and velocity in different working modes. When the knowledge-based acceleration-deceleration control is performed on the manipulator the same way as above, the operating time is decreased by adopting the proposed method [57].

3. **Optimization of machining operation**

Machines in manufacturing often control physical phenomena such as cutting, grinding, welding, electrolysis operation, and so on. Knowledge of the phenomena is effective in improving the performance of the operations. An example of knowledge-based control for machining will be presented next.

Before searching for the optimum speeds of the tool, some constraints in actual operations must be taken into consideration. A strong constraint in machining is that of the cutting force. The cutting force has to be kept lower than its limit to protect from tool breakage. In ordinary research, the cutting force is kept below the limit by PID control or model reference adaptive control. Another type of control using knowledge of cutting, that is, "the cutting force is proportional to tool velocity l/rotational speed v" is proposed. The control algorithm is made as follows [59, 60]:

$$\dot{l}(k+1) = \dot{l}(k)M_r / M(k), \ k = 1, 2, 3 \qquad (9.10)$$

where M_r — limit of cutting force; $M(k)$ — cutting force at sampling time k; $\dot{l}(k)$ — speed of tool at sampling time k.

It is supposed here that the variation in the tool rotational speed v is very slow in comparison with that of $\dot{l}$ due to the spindle speed inetia.

The knowledge gives the system superior characteristics. The system responds more quickly to the parameter change than an ordinary PID controller and can keep stability against it the same way as an adaptive controller [61].

In ordinary machining, the velocity components of the tool such as its rotational speed v and velocity $\dot{l}$ along its path are determined by referring to the database for the given combination of tool and work-material. However, it has been found that the database can not often give the optimum data because the parameters of the machining phenomena vary widely. For example, the hardness of the work-material varies easily due to slight differences in heat treatment. In such a case, sensing of the phenomena gives the necessary information. Figure 9.24 shows an example of how to introduce the shear strength and the level of tool wear α_f from the measured bending moments in the tool by using the knowledge of cutting theory [58]. Furthermore, the speed of tool wear for given velocities of the tool can be calculated from the identified parameters by using knowledge based on heat transfer and wear theories. The tool life T can be calculated from the speed of the tool wear.

A performance index of machining is given as follows:

$$PI = a \cdot b \cdot \dot{l} / \left[K_1 + K_2 + (K_1 \cdot T_c + C \cdot K_3) / T\right] \qquad (9.11)$$

where a, b — width and depth of cut; K_1 — repayment of the machine per time; K_2 — operating cost of the machine per time; K_3 — tool cost; C — constant, $0 \le C \le 1$; T_c — tool exchange time; T — tool life; PI — the cost efficiency for $C = 1$, that is, the cutting volume per unit cost. PI is proportional to productivity for $C = 0$, that is, the cutting volume per unit time.

Velocity $\dot{l}$ and rotational speed v of the tool to maximize the PI can be introduced by using the precise knowledge of machining mentioned above. However, theory-based knowledge of cutting is often not perfect for many combinations of work-material and tools. In such a case, knowledge obtained from the observation of experiments should be used, that is, a simple abstract equation to fit experimental data as follows:

$$T = K_4 \cdot a^{\alpha} \cdot b^{\beta} \cdot \dot{l} \cdot v. \qquad (9.12)$$

The optimum j and v to maximize the *PI* can be solved by using the theoretical or experimental model of cutting, mentioned above, with constraint on the cutting force.

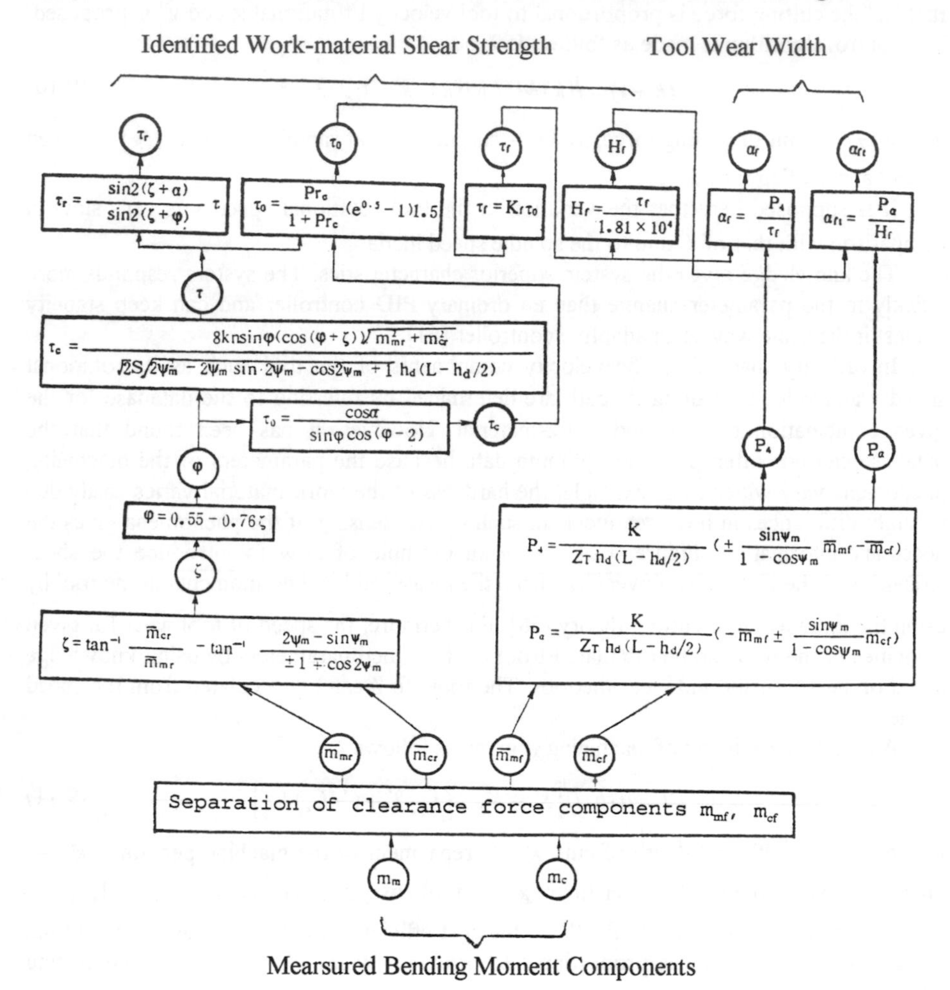

Figure 9.24. Knowledge of cutting process for identifying parameters.

4. On-line scheduling control of intelligent machines

The problem of a factory using intelligent machines, such as that mentioned above, is that they decide their operation speed j by themselves, and so, the pre-determined time schedule of their task operations is disturbed. Better results would be expected when the schedule is changed according to the progress of machine operations than when it is fixed and pre-determined.

Another problem in total control is that of determining what should be selected as the performance index. The cost efficiency of the total system is desirable as the performance index, however, its optimization is a difficult problem. Thus, the lateness in the due dates of the products is selected as the performance index. Optimum scheduling is not easy for batch manufacturing, in which the order of the task operations is different for each product. The theoretical method is not useful for this problem. The branch and bound method used in AI also requires too much computation time to be processed by on-line scheduling.

Therefore, simple heuristic algorithms using priority rules, such as SLACK and Look Ahead, are adopted. The SLACK rule gives first priority to the operation which has the shortest limit time margin in starting the operation without a generation of delay to its due date. The SLACK rule selects the task served first from among those which can be operated. The Look Ahead rule also has the same algorithm, but it selects the shortest margin task A from among the tasks which can be operated at the moment and the shortest margin task B from among those which cannot be operated now but will be possible in the near future, that is, their preceding tasks are being operated now. Both tasks A and B are compared to each other, and one of them is selected as the operated one by using the following rule:

IF [the margin of task A] - $K_5 *$ (the time to wait for the arrival of task B)] > [the margin of task B]

THEN select task B first

ELSE select task A

The on-line scheduling program has operation-databases, blackboards on which current operations, pre-simulated and executed schedules are written, and an inference engine, as shown in Figure 9.25. Each database has data on operations, due dates and part arrival time. Figure 9.26 shows the structure of the data of operations, the numbers of the following and preceding operations, the kind of machine used, the state of the operation shown by a flag if the state is impossible, possible, currently operated, or the completed states of the operation, the simulation of times of the following tasks to the due date, the limit time to start, actual start time and the actual completed time.

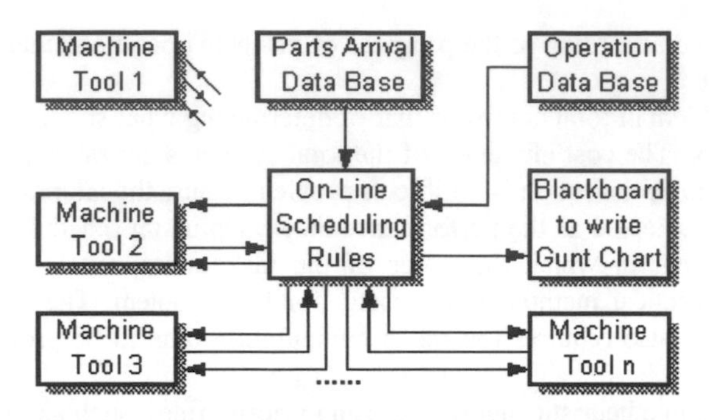

Figure 9.25. Composition of the on-line scheduler.

Task Number	Following Task Numbers	Proceeding Task Numbers
Due Date	Margin to Due Date	Time Limit to Due Date
Actual Starting Time	Actual Stop Time	Prospective Operation Time

Figure 9.26. Components of the database of tasks.

The operation times are calculated so as to maximize the performance index *PI* given by Equation (9.10), in which the parameter *C* is selected so as to adequately satisfy cost efficiency and productivity. The scheduling software, that is an inference engine, starts its activity when a machine has completed its operation. It evaluates the margins to the limit times of the operation tasks. The next processed task is selected by the SLACK or Look Ahead rule.

The effect of on-line scheduling is researched by computer simulation. In general, it is shown that the on-line scheduling is effective when the number of machines and products is large rather when the number is small. Figure 9.27 shows the effect of on-line scheduling. Ten studies for five machines and twenty products are performed. The decrement of the maximum lateness to the due dates is shown in percentages. The horizontal axis is the variation in operating time. While on-line scheduling using the SLACK rule introduces worse results than a fixed one, it shows improvement when the Look Ahead rule is used.

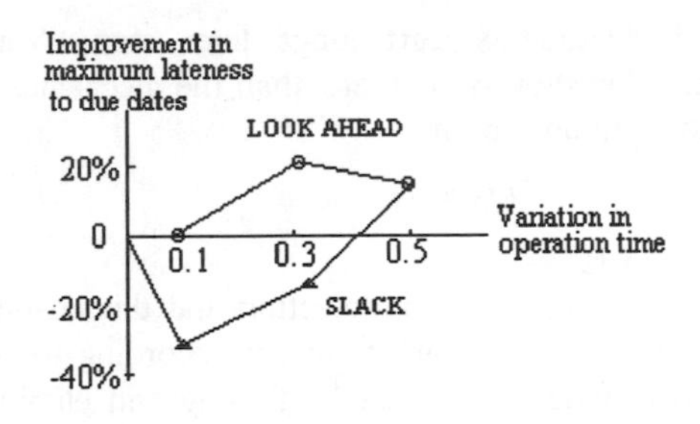

Figure 9.27. The effect of on-line scheduling.

5. Conclusions

Several types of intelligent control are presented. They show adaptability to variations in environmental conditions by using sensor feedback signals and hybrid theoretical/heuristic control rules. Since it is not easy for the present control technology to improve the multiple characteristics of manufacturing systems by a centralized control loop, they are improved at respective control loops in the hierarchy of the manufacturing system. Energy consumption is improved at the motor control level, the operation speed at the path control level, cost efficiency/productivity at the manufacturing operation control level, and the lateness to due dates at the scheduling level by intelligent control in respective ways. It is verified that the proposed methods are significantly effective in improving the performances in comparison with ordinary control systems. Rule-based control algorithms with database and sensor signals are made on languages, such as ASSEMBLER and FORTRAN to attend practical abilities, such as calculation speed and easy interfacing. A rule-based control algorithm makes programming easy even with these conservative languages.

In the proposed method, partial optimization is done. It does not guarantee true total optimization. Furthermore, research on performance indices and group control are necessary and could introduce better results.

9.4 Intelligent Fault Detection and Diagnosis

Fault detection and diagnosis are closely connected with process supervision and control. A highly automatic control system should have the capability of self-fault detection and diagnosis in order to ensure the whole system has worked with high reliability.

The fault diagnostics of very complex systems has gained extreme importance in the past few years after some disasters occurred due to unsatisfactory control or missed diagnoses of failures, for instance, the disaster of the Chernobyl Nuclear Power Plant [7, 18, 25, 39, 46, 51, 72, 73].

Naturally, many advanced medical diagnostic systems are like intelligent fault diagnostic systems too; the difference between them being in the objects that are diagnosed. For the medical diagnostic systems, the diagnosed objects are the organs and systems in the human body such as heart, lungs, liver, stomach and the breath system, blood circulation system, digestion system, etc. than the mechanical, electrical, electronic, thermal and chemical systems and so on.

9.4.1 Tasks and Types of Fault Diagnosis

1. Tasks of fault diagnosis

The common task for all intelligent fault detection and diagnostic (FDD) systems is to infer the fault cause(s) of a system, part or organ according to the observed situation, domain knowledge and experience, so that by finding and eliminating the failure(s) as soon as possible, the reliability of the systems or devices is increased. The intelligent diagnosis systems can understand the characteristics of the system's components and the relationships among the characteristics, find a phenomenon being covered by another phenomenon, provide detected data to user(s), and draw diagnostic conclusion as correctly as possible from the uncertain information. Precisely speaking, the main tasks of the fault detection and diagnosis systems are as follows:

(1) Detect the system to check whether it is in failure condition at a moment or period, and send an alert as soon as a fault is happened.
(2) Rapidly determine the fault position and components (parts), identify the fault cause(s) and degree, fault tendency as well as fault influence to the system as soon as possible.
(3) Provide effective measures and actions to remove the related failure such that to isolate and replace the fault component(s) or part(s), to restrain and remove the fault situation, and make the system work again in the normal state.

2. Types of fault diagnosis

In developing intelligent systems to carry out fault diagnosis on process plants, two quite different types of knowledge can be deployed. These two types of knowledge have been referred to as shallow knowledge and deep knowledge. The shallow knowledge consists of rules which capture conclusions that can be made from the given certain plant condition. In this type of diagnostic system, knowledge about failures in the plant are collected together in the form of fault associations. In such a system, knowledge can be acquired from both plant operator expertise and qualitative simulations of the plant. In the effective system, the knowledge is compiled such that redundancy is built into the rule set. The redundancy ensures that when a failure occurs there is sufficient sensory data to correctly diagnose the fault. The price to be paid for this redundancy in the rule-set is an additional requirement for conflict resolution amongst the rules. In developing such systems all the faults are grouped together to form a diagnostic tree, or decision table. During diagnosis when an abnormal event occurs, the data enters the system via sensors and triggers the rules in the rule base. If the fault lies within the diagnostic tree, then the system will trigger enough rules to diagnose the fault correctly.

Deep knowledge can be also used as an effective way to effect fault diagnosis, but in a different manner from that employed in the shallow knowledge approach. The deep knowledge provides lower level causal, functional and physical quantitative information about the plant. The knowledge mathematically models how the plant works. Therefore, in order to facilitate such diagnosis on process plant, knowledge is needed in the form of a mathematical model describing both the normal behavior and the faulty behavior of the plant.

Faults are detected via two separate methods. The first method looks for discrepancies between the real output from the plant and the behavior of the model. By using powerful pattern recognition techniques, the discrepancy in the two responses can be used to trigger rules which will, in turn, detect the fault. The second method looks at discrepancies between model parameters and normal plant operating parameters and uses this information to trigger fault detection rules in the rule base. The modeling process for such fault diagnostics can be effected by using parameter estimation routines which are now well established.

Both types of fault diagnosis have their good and bad points. In the case of shallow knowledge, the knowledge about faults is brought into the system via the plant sensors. These variables are then checked for upward or downward trends or transgressions of fixed-limits. Such systems are very robust and have proved to be a powerful tool in fault detection. However, such methods are only effective after measurable outputs from the plant have been considerably effected, i.e., some time after the fault has occurred.

In the case of deep knowledge, it is possible to use further algorithms to enable the detection of faults in the process earlier, i.e., as the faults occur in the plant. This involves the use of mathematical models of the process and parameter estimation routines. Furthermore, it may be possible to infer non-measurable quantities from the process model using state estimation methods. However, there are significant problems associated with such identification methods as the model of the process has to be accurate for such a scheme to operate. The scheme assumes the process is linear, has a fixed number of known states, and has low signal-to-noise ratios. Moreover, the identification algorithms are often difficult to operate for long periods whilst retaining their full diagnostic power.

In many cases, a methodology for fault detection and diagnosis are based on the combination of qualitative and quantitative approaches, i.e., based on combination of shallow and deep knowledge.

We are going to present two application examples in the following subsections. One is a fault diagnostics for the space station thermal control system using a hybrid analytical/intelligent approach [51]; another is a consulting system for Electrocardiogram (ECG) diagnosis [18].

9.4.2 Fault Diagnostics for Space Station Thermal Control System

The proposed common module thermal control system (TCS) for the space station is designed to integrate thermal distribution and thermal control functions in order to transport heat and provide environmental temperature control through the common module. When the thermal system is operating in an off-normal state due to component faults, an intelligent controller is called upon to diagnose the fault type, identify the fault location and determine the appropriate control action required to isolate the faulty component. The basic methodology used in this system for fault diagnosis is based on a combination of signal redundancy techniques and fuzzy logic. An expert system utilizes parity space representation and analytic redundancy to derive fault symptoms the aggregate of which is assessed by a multivalued rule-based system.

1. **Design goals and mechanism of the intelligent controller**

A hierarchical controller has been designed to direct the fluid flow so as to remove excess heat and maintain an equilibrium such that the temperatures at certain specified points in the network take on prescribed values. To achieve this objective, the available control inputs are the overall fluid mass flow rate and the relative closures of certain by-pass values. The available outputs consist of flow rate and temperatures measured at several points in the network.

The intelligent controller (IC) interacts and works harmoniously with the hierarchical controller to maintain system integrity. The goals of the IC design are to develop a fault detection and isolation scheme which:

(1) Maximizes the sensitivity to component failure detection, and

(2) Minimizes the rate of failure detection-fault alarms.

Usually, these two design goals involve conflicting criteria and the IC design is called upon to optimize the trade-off.

The block diagram of the intelligent controller is depicted in Figure 9.28. The system or thermal control process is viewed to consist of the TCS and the hierarchical controller with their combined objective directed towards achieving specified control goals under normal operating conditions. Temperature, flow, and pressure sensors provide, through data management system, status information to the IC for all important process variables. Controller command signals are also provided to the IC directly from the hierarchical controller. On the basis of available information, the IC calls upon a triggering module to initiate the data validation/fault detection routines; upon identification of a fault, it produces a status report detailing the fault type, its location, and such pertinent information as the degree of severity of the particular failure incident; it finally decides as to what appropriate measures must be taken to isolate effectively the fault component.

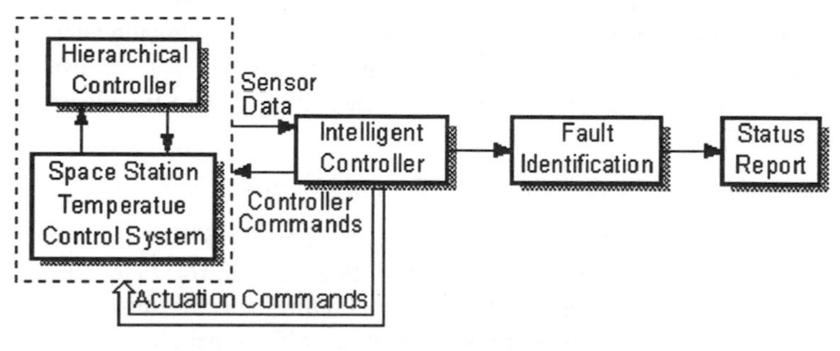

Figure 9.28. Block diagram of intelligent controller.

Figure 9.29 is a schematic representation of a typical thermal control system (TCS) used to demonstrate proof-of-concept feasibility. For purposes of illustration, let us identify three pipe segments and four valves as "components" that may be subjected to a failure mode — a leaky pipe segment and a sticking valve. They represent typical piping and valuing configurations in the TCS network. The detection methodology may be easily extended to include any number of components as failure candidates as long as a sufficient amount of measurements is available at the end nodes of each component.

Furthermore, the approach may include sensors in the candidate set and is designed to accommodate multiple failures.

A functional block diagram of the fault diagnostics approach is shown in Figure 9.30. It consists of the following hardwares and softwares:

(1) Triggering mechanism to initiate the diagnostic algorithm;
(2) Routine for determining a validated vector of direct and analytical measurements;
(3) Fault detection algorithm based on parity space considerations;
(4) Means of identifying the faulty component;
(5) An error trending routine;
(6) Algorithm for isolating the failed component;
(7) Estimate of the "safest" values of the system variables.

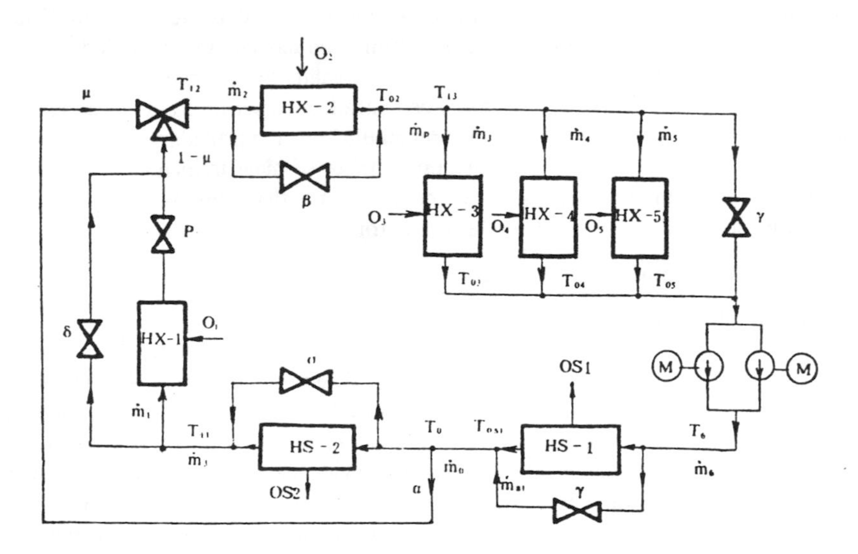

Figure 9.29. Schematic diagram of thermal system test process. (with Vachtsevanos and Davey, 1987)

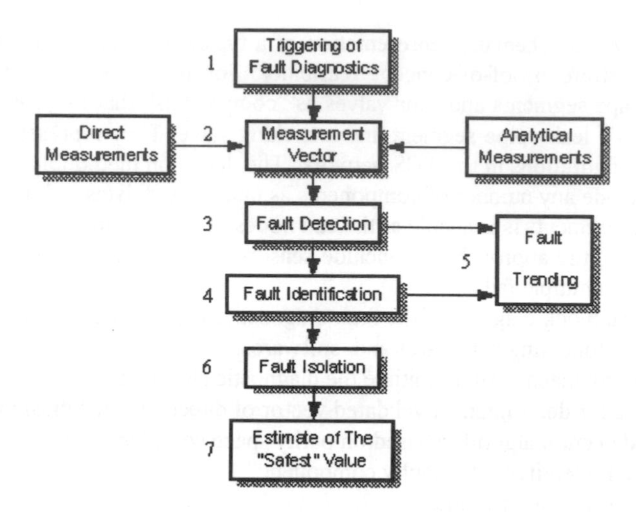

Figure 9.30. Functional diagram of fault diagnostics approach.

Fault triggering may be based on a marked deviation of system performance from its planned trajectory. In the actual TCS, periodical limit checking, disturbances under steady state conditions, or the initiation of a transient state may be used as the trigger mechanism. Sensor data validation, on the other hand, uses signal redundancy in a modified version of the parity space algorithm for signal validation. The validation procedure compares each reading to all other like readings. The "best" estimate from a set of good measurements is defined as that value which gives the minimum of a particular function of the measurements. Where more than the minimum number of measurements is available, the "best" estimate is taken in the least squares sense, i.e., it is that value which minimizes the square of the length of the measurement error vector.

2. Fault detection and identification

(1) Fault detection

Fault detection and identification uses validated sensor and analytic data to analyze the status of the process components. Fault diagnosis involves the incremental accumulation of evidence in the form of symptoms to determine the "health" of the system components. Parity space representation and analytic redundancy, in addition to limit checking, are applied to accumulate the required evidence. Parity space representation transforms an array of redundant measurements into a new array, called a parity vector, in such a way that the true value of the underlying variable is suppressed, leaving only components of the measurement errors as elements of the parity vector. In parity space, the parity vector remains near zero when the redundant measurements are consistent, i.e., when no faults are present. When a fault occurs, the parity vector grows in magnitude, and its direction of growth is uniquely associated with the faulty measurement. The fault detection part of the algorithm capitalizes upon the behavior of the magnitude of the parity vector to detect the presence of a fault.

The space spanned by the independent parity equations is called parity space. Its dimension is equal to the number of measurements minus one for a scalar variable. For three measurements, the parity space is a two-dimensional plane while for four measurements the parity space is a three-dimensional space. If there are l parity equations, then there are only $q=l-1$ independent relations among the l parity equations. The inconsistency values may, therefore, be expressed in terms of q independent variables, say $p_1, p_2, \ldots, p_q$. The vector $p=\left[p_1, p_2, \ldots, p_q\right]^T$ is called the parity vector. All information contained in the C_1^t parity equations is available from the q components of the parity vector. Because the dimension of the parity vector is less than the dimension of the measurement vector m, considerable simplification and insight is provided by expressing the failure detection algorithm in terms of the parity vector. Moreover, the parity vector embodies attributes of error distribution through the system components. The magnitude of the parity vector p is, in general

$$P^T P = \sum_{i=1}^{l} \varepsilon_i^2 - \left(\sum_{i=1}^{l} \varepsilon_i\right)^2 \Big/ l. \tag{9.13}$$

If the sensor noise probability density function is uniform and if the vector in each measurement for an unfailed component is assumed to be bounded, i.e., $|\varepsilon_i| < b_i$, then

$$\delta_t = \max(P^T P) = \sum_{i=1}^{l} b_i^2 - (\min \sum_{i=1}^{l} b_i)^2 / P. \tag{9.14}$$

A search algorithm is used to locate a minimum of $\sum b_i$ and, finally, δ_l is obtained from Equation (9.13).

Fault detection is based upon the relative magnitude of the maximum allowable error bound compared to the length of the parity vector, i.e., $\delta_l / p^T p$. A threshold condition is reached when $p^T p = \delta_l$. In general, both $p^T p$ and δ_l are not known accurately and, therefore, a fuzzy representation of the quantity $\delta_l / p^T p$ is most appropriate.

For fault detection of a leaky component, additional symptoms, such as the flow out of the make-up fluid tank, may enter the rule base.

The rule base and inference scheme are intended to provide a quantitative measure of the severity of the fault depending upon the prevailing process conditions. To illustrate the structure of the proposed rule base, consider the variable ($\delta_l / p^T p$). δ_l is estimated from Equation (9.14). When the magnitude of the parity vector $p^T p$ is in the neighborhood of δ_l, then a fuzzy belief curve or membership function is associated with ($\delta_l / p^T p$). The membership function may be linear or nonlinear (for example, a cosine curve) in this particular range of the universe of discourse. When $\delta_l / p^T p$ is very small, it is certain that $p^T p$ is sufficiently greater than δ_l and, therefore, a fault condition exists. On the other hand, when $\delta_l / p^T p$ is much larger than unity, then the length of the parity vector is too small to allow for a positive detection of a faulty component. The compositional rule of inference [68] is used to infer the degree of severity of the fault condition.

(2) Fault identification

The fault identification part of the program is intended to identify which component is faulted. Following Ref. 42, a technique based on the concurrent checking of the relative consistency of smaller size subsets of measurements is proposed.

Let us consider only scalar measurements, i.e., $n = 1$. If **m** is the $l \times 1$ measurement vector, and $v_1, v_2, \ldots, v_l$ represent the failure directions in the (n-1)-dimensional parity space, then only k measurements are considered, $1 < k < l$. The projection of the (l-1)-dimensional parity vector (generated from all l measurements) on the (k-1)-dimensional subspace orthogonal to the subspace spanned by $v_{k+1}, v_{k+2}, \ldots, v_l$, is the parity vector directly generated from the measurements $m_1, m_2, \ldots, m_k$.

The consistency of large subsets of measurements can be determined in terms of the consistency of all possible smaller subsets of (n+1) = 2 measurements. For **m** measurements, we form t subsets, $s_1, s_2, \ldots, s_t$, of (n+1) = 2 measurements each with all possible combinations of m_i, i = 1, l. For each subset s_i, we calculate corresponding reduced parity vector p_i. For each subset s_i, a threshold value θ_i is computed from the corresponding component normal error bounds.

Finally, p_i is compared with θ_i to determine whether the component associated with the v_i measurement axis has failed or not.

This procedure allows not only the identification of a faulty sensor but also the estimation of the "best" value for the corresponding variable from measurements which remain fairly consistent. Component fault evidence is accumulated by comparing each p_i with the corresponding θ_i value. The rule base and inference scheme are constructed, for the fault identification algorithm, in a similar manner as for the fault detection case.

(3) Isolation of multiple faults

In an attempt to detect and isolate multiple faults, a technique which is a hybrid least squares to fault detection has been developed. The hybrid least squares parity approach to fault isolation is an efficient means of utilizing parity space to isolate multiple faults in large systems. The concept is based on layered functional, each of which is minimized when the most probable fault(s) is isolated. The first functional is simply the square of the difference of pairs of individual parity vector components:

$$F = (P_1 - P_2)^2 + (P_1 - P_3)^2 + (P_2 - P_3)^2 + \ldots = \sum_{ij} (P_i - P_j)^2 ,\ j > i . \tag{9.15}$$

Because each parity component is designed to be zero (or within the measurement tolerance), the negative gradient (with respect to the parity vector) of this functional yields the direction in n dimensional space which most effectively reduces the functional. Thus, the gradient also represents the most likely fault component direction:

$$\nabla F_1 = \partial F / \partial P_1 = (P_1 - P_2) + (P_1 - P_3) + \ldots \tag{9.16}$$

$$\nabla F_2 = \partial F / \partial P_2 = (P_2 - P_1) + (P_2 - P_3) + \ldots \tag{9.17}$$

$$\nabla F_3 = \partial F / \partial P_3 = (P_3 - P_1) + (P_3 - P_2) + \ldots \tag{9.18}$$

The process is developed at the next level by forming a hybrid parity vector equal to the square root of the product of any two parity components (SQR (parity vector * transpose of the parity vector)). Let $A = pp^T$, then

$$u = \begin{cases} \sqrt{A_{12}} \\ \sqrt{A_{13}} \\ \vdots \\ \sqrt{A_{23}} \\ \sqrt{A_{24}} \\ \vdots \\ \sqrt{A_{34}} \end{cases}$$

or

$$u = \sqrt{A_{ij}} ,\ j > i . \tag{9.19}$$

The second functional by analogy to the first is the square of the difference of pairs of individual components of this hybrid vector

$$G = (u_1 - u_2)^2 + (u_1 - u_3)^2 + (u_2 - u_3)^2 + \ldots = \sum_{ij} (u_i - u_j)^2 ,\ j > i . \tag{9.20}$$

Similar to the single fault case, the gradient of this second functional indicates the direction in the second order hybrid space which most effectively minimizes the functional and thus indicates the most probable direction of the double fault:

$$\nabla G_1 \equiv \partial G / \partial u_1 = (u_1 - u_2) + (u_1 - u_3) + \ldots \tag{9.21}$$

$$\nabla G_2 \equiv \partial G / \partial u_2 = (u_2 - u_1) + (u_2 - u_3) + \ldots \tag{9.22}$$

$$\nabla G_3 \equiv \partial G / \partial u_3 = (u_3 - u_1) + (u_3 - u_2) + \ldots \tag{9.23}$$

The process in like manner is extended to any order fault desired. The computational penalty is roughly proportional to the order of the parity vector raised to the power of the fault divided by the factorial of the order of the fault being sought.

3. **Simulation results**

The technique has been tested on a 100-component system and modeled via a Monte Carol technique to simulate a multiplicity fault system. The simulation correctly isolates the faults but does require considerably more computation time as the order of the fault increases.

The intelligent control algorithm has been developed and tested using the thermal system as a prototype case.

Some simulation examples have been presented in details [51]. The simulation results show that the structure, rules and algorithms are effective for detecting and identifying the faults within the pipe segments and the stuck valves of the thermal system. Further tests would be carried out.

9.4.3 Radar Fault Diagnosis Based on Expert System

Automatic test equipment (ATE) software systems have been developed to isolate faulty components and to recommend replacements. ATE software is designed and implemented through the use of flowcharts by the test expert, based on an understanding of the functionality of the unit under test (UUT). When a unit is tested, a fixed series of tests is performed independent of outside knowledge about the fault. Historical data, field symptoms, upgrades, and improvements are difficult to incorporate because of the rigid flowchart implementation. When intermittent faults occur in either the UUT or the test station itself, the flowchart might lead to errors in replacement recommendations. In addition, it is often difficult to follow the reasoning of the program flow. The program's behavior may only be understood by the test software designer, if still available.

An expert system solution for these problems has been developed by Hughes Aircraft Company [7]. This project has progressed through two distinct phases. Phase 1 is focused on a single circuit board from an operational airborne Radar system for which both design and field experience was available. Knowledge acquisition activities were initiated, primarily studying the functionality of the board, investigating failure histories and experience-derived diagnostic heuristics. This information was used to develop an initial architectural concept and select an environment suitable for developing prototyping activities. Phase 2 was concerned with exploring new architectural ideas and exploiting expert system technology. The basic architectural concepts were refined, with emphasis on both the inherent functional hierarchy of the board and the body of heuristic knowledge correlating observed symptoms to faults. Radar Automatic Test Expert System (RATES) features a generic, extensible hierarchical architecture based on the functional composition of the system being diagnosed.

This subsection focuses on the ideas and results of the second phase, and includes the following contents: (1) The RATES hierarchical architecture and operational philosophy. (2) The feasibility demonstration effort. (3) Implementation and performance data. (4) Conclusion and further development.

1. Architecture description of RATES

RATES is based on the expert system technique. Expert systems have much to offer to the problems mentioned above. In particular, the abilities to reason over large hypothesis space, perform diagnosis and prescription tasks, and utilize information of various types are vitally important.

(1) Types of knowledge used in fault diagnosis

From analysis of the problem and through discussions with design experts, it was concluded that there are three basic types of knowledge used in fault diagnosis: (a) problem solving, (b) functional, and (c) heuristic. The first is found in the strategies used by expert trouble-shooters. Experts have developed specialized techniques for tackling fault diagnosis. The hierarchical approach used in the RATES architecture is in part based on this type of problem solving strategy.

Knowledge of how to device functions is the second type to be considered. The abstract and physical models held by the expert are utilized at this point. These models represent, for example, the relationships between inputs and outputs of constituent devices, transfer functions, thermal and electrical properties, and interconnections, both physical and logical. This knowledge is used in conjunction with the problem solving knowledge to isolate a fault.

Although the two previous forms of knowledge are sufficient to isolate a fault in an unit, they are not necessarily the quickest way. Based on years of experience with the involved technology, experts compile a set of heuristics which speed their work. This knowledge includes such items as repair history, pattern of faults, location of best test points, and interpretation of symptoms and measurements. Of the three types, this knowledge is potentially the most powerful in improving the convergence of the diagnostic process. It is the judicial use of all three types of knowledge which makes expert performance superior to novice performance, systematic search or other conventional methods of automated diagnosis.

(2) Hierarchical architecture

Experts performing electronic fault diagnosis use a mental model of the UUT shown in Figure 9.31, often augmented by detailed functional and topological documentation. This model transcends levels of abstraction depending on the amount of knowledge available on the current failure, histories of previous failures, and the current stage of diagnosis. The RATES architecture captures this model through the use of functional hierarchy of knowledge bases. As the expert collects symptoms and performs measurements, the uncertainty in the location of the faulty component is reduced. Certain symptoms may exist which strengthen these suspicions. The expert then performs a set of key measurements, in light of the previous measurements and symptoms, to confirm this hypothesis. Given a confirmation, the expert changes levels to consider the functional partitions of the analog half of the unit, and the cycle is repeated. The division of the unit into analog and digital portions constitutes a model of the unit at one level of abstraction. The knowledge used to isolate on a level is specific to the level of abstraction of the current model. The RATES system performs in exactly this manner.

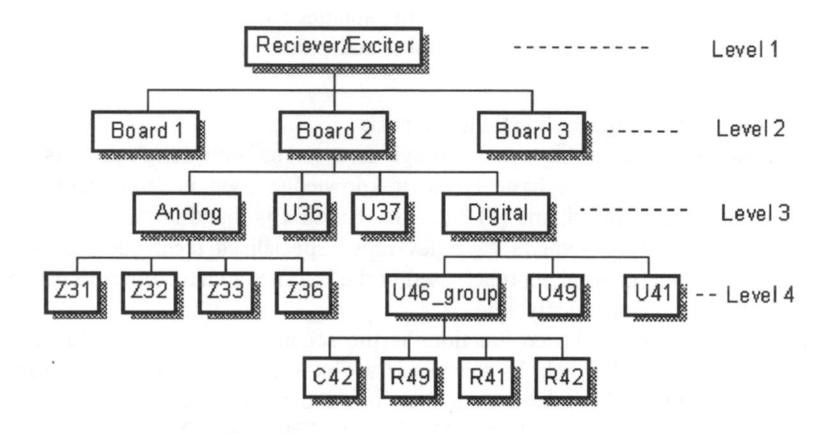

Figure 9.31. Hierarchical architecture of RATES.

For the level structure of RATES, each level of the system contains knowledge specific to the list of possible fault groups. Here, a fault group is defined as either a set of smaller fault groups, such as channel, or individually replaceable components, such as chip. It is this partitioning which represents the functional model of the UUT. Associated with each fault group on the list for a level is a confidence factor (CF) which quantifies the current belief that the group is faulty. The initial values of the CF are preset from a statistical analysis of the repair history of the unit. The application of the diagnostic process discussed above modifies these CF as symptoms, data, and knowledge that are used. A fault group is selected and a level changed when its CF exceeds a threshold determined by the expert.

Each level of abstraction has three distinct knowledge bases which contain the rules on how to modify these CFs. They are: (a) Symptom knowledge, (b) Measurement knowledge, and (c) fault group selection knowledge. Symptom knowledge rules inquire about the presence of certain symptoms in the context of previously collected measurements and symptoms. If found, the symptom knowledge will modify the fault group/CF list to reflect this information. Measurement rules consist of two types, invocation and interpretation. The first type represents the knowledge of when and where key measurements should be made, while the second type uses the results of the measurements to modify the individual CFs. Finally, the fault group for expansion is based on the magnitude of the CFs. This knowledge includes the individual thresholds as mentioned above.

(3) Process control flow

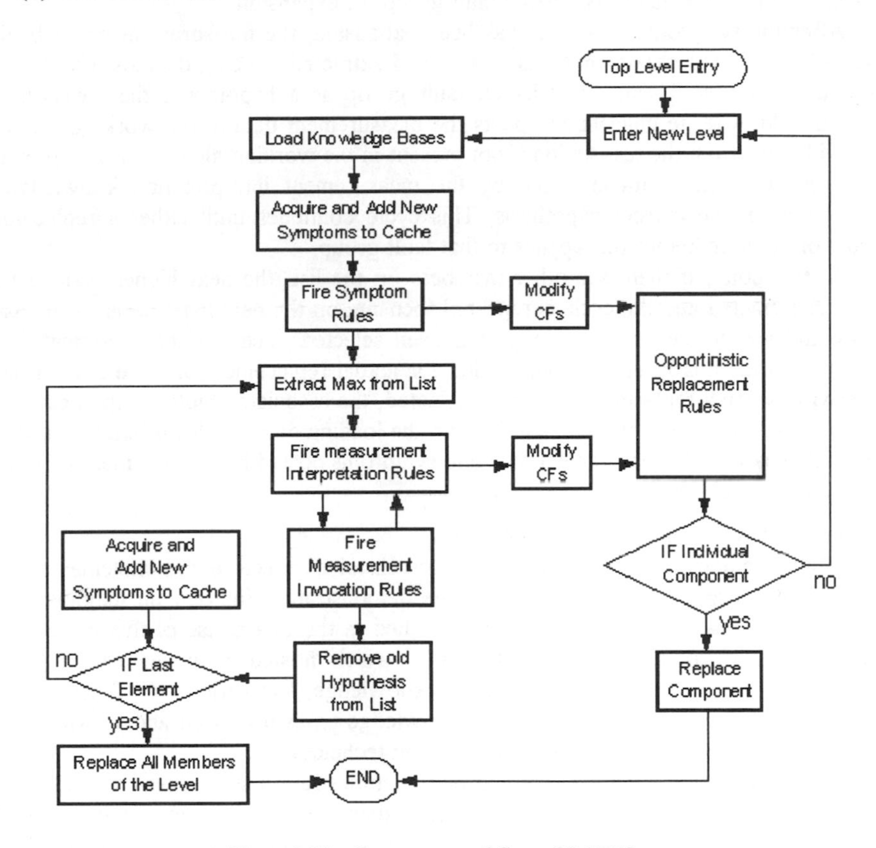

Figure 9.32. Process control flow of RATES.

The relationship between these knowledge bases and system architecture is tied through the process control flow as shown in Figure 9.32. Upon entering the system at an arbitrary level, a fault group list is loaded which covers the partitioning of the unit appropriate for this level of abstraction. Associated with this list are the initial CFs extracted from repair histories, forming a list of ordered pairs. Each of these CFs are initially below the expansion thresholds. At this point, symptoms relevant to this level are gathered either from the user through the use of menus or from working memory posted from higher levels. These newly acquired symptoms are posted in working memory for use by the symptom knowledge, which modifies the CFs in specific fault groups contained on the list. At each modification of a CF, the component replacement knowledge tests for threshold crossing. This opportunistic behavior of the fault group selection or replacement knowledge continues through the reasoning process. When a CF exceeds a threshold, the replacement rule selects the associated fault group for expansion.

After the symptom knowledge has been exhausted, the measurement phase begins. The fault group with maximum CF is located and extracted as being the most profitable to be scrutinized first. Using this selected fault group as a hypothesis, the measurement invocation knowledge collects and posts the measurement data in the working memory. This will come from the test station if not present in the working memory from a previous measurement. These data are used by the measurement interpretation knowledge to modify CFs in the current hypothesis. This cycle continues until either a replacement occurs or no more knowledge applies to that fault group.

At this point, if there are other members on the list, the next highest valued fault group is extracted and the cycle is repeated focusing on the new hypothesis. If the list is empty and no replacement knowledge has been selected, then the entire original list of fault groups for this level is recommended for actual replacement and the system halts. Otherwise, when a replacement has been selected, the associated fault group is expanded into the next lower level of the hierarchy with the loading of a new list of fault groups and knowledge bases. All measurement data and symptoms posted in working memory remain intact during level transitions.

2. Feasibility demonstration of RATES

A feasibility demonstration of the RATES hierarchical architecture was implemented over a period of approximately four months. The demonstration focused on diagnosing the fault in a particular Radar circuit board is studied in the first phase of this project. The development environment was an IBM PC, which hosted both the expert system development tool M.1 (version 1.3) from Teknowledge, and various 'C' functions. The resultant expert system utilized all of the knowledge types discussed above, which were acquired using a variety of knowledge acquisition techniques.

The application for the RATES feasibility demonstration system was a circuit board in the receiver/exciter (R/E) module of an existing Radar system manufactured by Hughess Aircraft Company. The main function of this complex and expensive board is to provide analog-to-digital (A/D) conversion for the R/E. Figure 9.32 depicts the functional hierarchy present in this board, which determined the four-level RATES structure implemented in the demonstration system.

(1) Development environment

The availability of microcomputers and a limited budget made the choice of a microcomputer-based expert system the most viable of the available alternatives. It was decided that an expert system building tool was preferable to a language such as LISP or PROLOG in order to facilitate the rapid prototyping demanded by the project schedule. Relative inexperience in AI technology within the implementation team and budget constraints also influenced this decision. The IBM PC and M.1, a rule based, backward chaining expert system shell especially suited for solving diagnosis problems, were chosen for the first phase activities. In addition, M.1 also provides limited forward chaining capabilities, good explanation facilities, compatibility with other programming language, and confidence factors. Since much of the Phase 1 knowledge would be useful in the RATES demonstration, and the implementation team was already familiar with M.1, it was decided to build the feasibility system on the IBM PC in M.1. Later the system was supplemented with a library of special purpose routines written in 'C' language, to improve executional performance and enhance the used interface.

(2) Knowledge acquisition

Three types of knowledge were incorporated as M.1 rules into the RATES demonstration system. These were diagnostic heuristic knowledge, functional circuit knowledge, and historical failure data.

Heuristic or 'expert' diagnostic knowledge was obtained from discussions and interviews with the expert assigned to the RATES project. In the demonstration system, heuristic knowledge took the form of symptom rules which could take effect of "fire" whenever the user observed a particular symptom in the process of testing the R/E. Knowledge was acquired from the expert using the following methods:

(a) Discuss about the functional characteristics of the various components making up the A/D converter board. This included tolerance levels, input/output characteristics, and the effects of interaction between components.

(b) Propose hypothetical situations, such as the existence of a fault in a certain component, and having the expert explain what conditions, characteristics, and measurement values would manifest themselves in such a situation.

(c) Assume the existence of a symptom at the R/E module level and have the expert explain how he would go about using the symptom to guide his fault detection and diagnosis of the A/D converter board. This provides a method of capturing the techniques by which an expert carries out his task.

(d) Assuming no symptom was available, have the expert predict where a fault might be most likely to appear and also have him specify what measurements he would make to verify the existence of such a fault.

These techniques allowed the knowledge engineer to identify the methods applied by the expert in completing his fault diagnosis. The techniques used by the expert were then incorporated into the knowledge base as symptom or measurement interpretation rules.

The other two types of knowledge were obtained more conventionally. Functional knowledge involves how the board works, testing requirements, and performance tolerances. This information was acquired from analyzing existing test software and test requirement documents, and from discussions with the expert. Historical data concerned the types and frequencies of failures on the A/D converter board. This was obtained from data bases and fault records maintained separately from the actual test station.

(3) Implementation of RATES

The RATES system was implemented as an M.1 control shell using level specific knowledge bases in a microcomputer environment. M.1 provides CFs to indicate the degree of belief as facts and represent them with an integer between 0 and 100. CFs may only increase in certainty and do so by the application of the following formula where CF1 and CF2 represent the current and new evidence CFs:

$$(100 * CF1 + 100 * CF2 - CF1 * CF2)/100 \quad (9.24)$$

The final CF will approach, but never exceed 100 and is never less than either of CF1 and CF2.

Table 9.2 shows examples of the four types of rules which represents the RATES knowledge. The example symptom rule indicates that if on a previous level, it has been observed that only one channel failed, and that channel is the *i*1 channel, then the confidence in parts *z*304 and *z*301 being faulty should be increased by 25 and 20 respectively. The example measurement rule calls for the performance of two measurements (*m-q*1-target and *m-i*2-target) by the test station, and based on the results may increase the confidence in part *z*301 being faulty by 90. Finally, the invocation rule declares that before the measurement freq_*e*1_035 is performed, set up one and two of the dcoffset must first be performed.

Table 9.2. Sample rules of RATES

Category	Example
symptom	IF old_*s*=one_channel_failed and *s* = *i*1_channel_failed and *cf*_update-*z*304-25 and *cf*_update-*z*301-20 THEN *f*=*z*304 *cf* 25 and *f*=*z*301 *cf* 20.
measurement	IF *m-q*1-target=no and *m-i*2-target and *cf*_update-*z*301-90 THEN fault_verified(*z*301) *cf* 90.
invocation	IF dcoffset_setup1 and dcoffset_setup2 THEN freq_*e*1_035-setup.
control	IF do(reset symptoms_processed and update_display(*F*) and symptoms_processed and fault_found is unknown and fault_list_processed(*F*) is sought and check_return(*F*) is sought THEN check_complete(*F*).
	IF fault_found is unknown and th_check(*F*)=*X* and *X*==1 and file_loaded(*F*) and check_complete(*F*) is sought THEN th_check.

The RATES hierarchical approach was implemented as a rule-based control structure which provided a control shell applicable at all levels of the hierarchy. The control rules need only access those knowledge bases dealing with the fault class (level) currently under consideration. The RATES control structure guides the diagnosis session through the hierarchy providing such functions as flow control, symptom accumulation, measurement control, as well as opportunistic threshold checks. The two example control rules given in

Table 9.2 form the basis for control on any level. The first rule controls the consultation logic flow. This rule declares that on any level *F*, the display should be updated, symptoms processed, and if a fault still has not been found, measurements should be performed. Whenever the confidence in a fault is increased, the second rule is fired. This rule checks if the confidence in the fault has exceeded its threshold, and if so, loads the knowledge base for the new level, and calls the first rule to start processing on the new level. This cycle continues to explore the UUT hierarchy in greater detail at each level until a fault is found.

The RATES system uses either a graphical display or an interactive text display during a consultation.

When the knowledge base needs to directly interact with the user, the screen is cleared and the user is asked a question and given a menu of legal answers. Besides entering the answer, the user may also interrogate the system by asking why or viewing selective parts of the knowledge base or cache (memory). At the end of a consultation all members at all levels are displayed graphically and the replacement candidates are made to flash. This final display shows the threshold and current confidence of all faults, those faults which have been measured, as well as the final confidence in the replacement recommendations.

3. System characteristics

(1) Static characteristics

There are 199 unique rules in the RATES knowledge base. The breakdown by type and number appears in Table 9.3. The history and hierarchy facts are level-unique pieces of knowledge used only by the rules for that particular level. An example of such a fact would be the default confidence of a particular component belonging to a level.

Table 9.3. Types of rules in RATES

Rule Set	Number of Rules
Control Structure	63
Symptom Processing	18
Measurement Invocation	36
Measurement Interpretation	26
History and Hierarchy Facts	56

Only a portion of the 199 rules are needed by RATES at any one time. This was implemented using M.1's overlay system to essentially "load on demand" these rules needed at a particular RATES level. Performance was increased by reducing the number of rules that must be searched. Also, writing modular rule sets facilitates debugging and maintenance operations. At any particular level the control structure rules are always present and accessible, but only a subset of the non-control structure rules are required.

The control structure, knowledge base, and measurement data files are all given in

ASCII file size since they must be interpreted by M.1. The measurement data files contain the predefined measurement results used by the various test cases. The M.1 development executable file provides the development environment used to run consultations and create run-time versions of RATES. The M.1 object file contains the code required by M.1 to execute a run-time system. The external 'C' functions object file consists of the code to handle list processing and graphics support in RATES.

To run a consultation in RATES, at any given time, 141K bytes of memory is required to store the two M.1 files, the external C functions object file, and the C libraries and data. In addition the control structure file and each one of the knowledge base and measurement data files will require memory. Therefore, approximately 160K bytes of memory will be needed during the consultation.

(2) Executional characteristics

Table 9.4 compares the number of measurements required by RATES versus the number needed by the existing ATE test software for various test cases used to validate RATES. These test cases reflect "hard to detect" faults which require much time and effort to diagnoze using existing test software. For the purposes of this comparison, a measurement is defined as the need to compare a measured result or value against a known, expected, or fixed value. RATES makes no measurements for Test Case 1 because it involves the use of a symptom which is so indicative that a direct component replacement recommendation can be made without the need for measurements. Note also that the primary reason for the large difference in the number of required measurements is the rigid end-to-end testing format of the existing test software.

Table 9.4. Comparison in number of measurements

Test Case Number	Rates Measurement	Existing Test Software Measurement
1	0	206 - 257
2	5	158 - 159
3	11 - 14	168 - 174
4	7 - 17	128

Table 9.5 indicates the amount of time required to complete a single execution or consultation of RATES versus the existing test software. The measurement for RATES was made using an IBM PC/AT while an HP-1000 was the host computer for the existing test software. The former used Interpreted PROLOG language on IBM PC/AT; the latter used ATLAS (Abbreviated Test Language for All Systems) and FORTRAN on HP-1000. Note that the required execution time for RATES does not include any of the prerequisite or overhead setups and measurements that are included in the existing test software time requirement. In addition the execution time for RATES includes the use of predefined measurement results as opposed to the existing test software which must interact with actual test equipment to obtain these values. A comparison of the two systems can still be made, though, as the setup and measurement times are negligible compared to the total times. The reason for the existing test software's widespread execution time is its end-to-

end test sequence which can diagnoze a fault at any time in that series of tests. On the average, 80 percent of all faults detected by the RATES are found in approximately five minutes. Therefore RATES primary advantage over the existing test software is its ability to diagnoze "hard to detect" faults by making far fewer measurements and therefore execute its task quicker.

Table 9.5. Comparison in execution times

Software Class	Execution Time (minutes)
RATES	4 - 4.5
Existing Test Software	1 - 50

4. Conclusion and further development

An assessment for this project was made which included: (a) knowledge acquisition; (b) expert system development shell; (c) microcomputer development hardware; (d) RATES architecture; and (e) use of expert system techniques [7]. The RATES has much better performances than the existing test software does in radar fault diagnosis.

The research and development of the RATES system has been on-going. The system capabilities are being extended in two distinct directions. First, the RATES demonstration system is being extended and integrated with a production ATE station. Second, the knowledge base is being analyzed to evaluate the feasibility and desirability of separating general from specific knowledge in the next (prototyping) development phase. Alternative approaches of utilization of hybrid knowledge (generic and specific) will be formulated and alternative architecture approaches will be identified.

9.5 Intelligent Control for Aircraft Flight and Landing — A Neural Network Baseline Problem for Control of Aircraft Flare and Touchdown

In the aircraft flight the final approach (flare) and landing are critical and require constant monitoring and control. Intelligent control has been used to aircraft flight control since the middle of 1980s [26, 28, 47, 56].

This section introduces the problem of aircraft flight control first, then describes the design issues of neural network controller for the aircraft flight, and deals with the applications and implementation of the neural network controller finally.

9.5.1 Introduction to Aircraft Flight Control

Most commercial aircrafts currently have available an optional automatic landing system or "autolander". The first automatic landing was made in England on June 10, 1965, using a Trident aircraft operated by the British airline BEA. Since then, these systems have been significantly enhanced and are now used routinely on a variety of planes including the Lockheed L1011s and Boeing 747s. They are most often activated in clear, calm weather but can also be used in fog or rain if winds are calm. When they are used in the place of a pilot it is usually for two reasons. First, the automatic systems give a reliably smoother landing, leading to increased passenger comfort and to a reduction of wear on tires and landing gear. Secondly, their use in calm weather provides an important training function, accustomising pilots to a system that they must understand and trust if it is to be used in adverse conditions in the future.

Automatic landing systems rely on an airport based system, called the Instrument Landing System (ILS), to position an aircraft for the final phases of landing. The ILS has two radio beacons, called the glide-slope and localizer beams, which guide the aircraft into the proper height (elevation) and approach angle (azimuth) relative to the runway threshold. The actual distance to the runway threshold may also be measured. At approximately 50 feet (about 15 meters) above the runway surface, the flare is initiated to elevate the nose of the aircraft, bleed off airspeed, and cause a soft touchdown on the runway surface. ILS signals are dispensed with before the actual flare because ILS signals are too noisy to provide reliable guidance below about 200 feet altitude. Below that altitude, a radio altimeter or visual data is used as the primary reference. From the flare-initiation point until touchdown, the aircraft follows a control program which decreases both vertical velocity and air speed. Because the full three dimensional dynamics of an aircraft are quite complex, the simplified baseline model considers movements only in the longitudinal and vertical axes (i.e., horizontal displacements are not considered). Figure 9.33 illustrates the geometry of the resulting two dimensional flare and landing. This simplification permits the display of aircraft movement and controller errors on a standard personal computer screen without elaborate graphics.

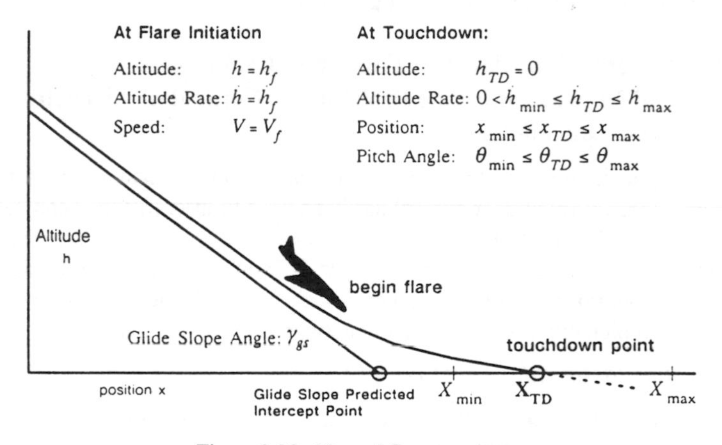

Figure 9.33. Normal flare geometry.
(with Jorgensen and Schley, 1990)

A typical strategy used by a conventional controller is to decrease the vertical velocity in proportion to the altitude above the runway (i.e., altitude decreases exponentially with time or in space) [38, 40].

It should be noted that existing autolanded systems work reliably only within a carefully specified operational safety envelope. Also, the system cannot be operated if strong downdrafts called microbursts are present. Usually, turbulence is not a significant problem for large commercial aircraft, but there are limits. A conventional autoland system may have difficulties in severe turbulence. Thus, to increase the safety and smoothness of landing it would be desirable if new technology could expand the operational envelope to include safe responses under a wider range of conditions. Neural networks may provide one method of exploring alternative control systems capable of dealing with turbulent, possibly nonlinear control conditions. Consequently, it is very useful from a research standpoint to have available a baseline problem within which alternative network configurations can be tried and compared. In this section we mention one such system defined in terms of linearized component equations. First, however, it is useful to elaborate on why a neural controller might be useful and how one might go about exploring the autoland problem.

9.5.2 Design Issues of Neural Network Controller

A trained pilot develops skilled intuitions called flight sense which help him to make appropriate responses in new and unforeseen situations. It is unclear exactly what flight sense is operational except that it appears experienced pilots are able to handle aircraft better than automated controllers in situations outside normal conditions. An example would involve control behaviors during weather related events such as wind shear (varying head or tail winds at different altitudes), unexpected wake vortices, or microbursts (abrupt violent changes in vertical wind speed). One possible use of neural net technology would be to have a network acquire some of these subtle temporal and spatial skills from a human pilot through observation. Current linearized controllers don't do a good job of emulating pilot responses to emergencies or in generating creative solutions to rare phenomena. This is perhaps due to the nonlinear or unrecognized linear relationships that pilots may use which are not captured in a conventional controller. Because a neural net is capable in principle of generating a mapping from one large set of variables (*e.g.*, sensor streams) to another (*e.g.*, operational modes or control actions), it can potentially capture critical behaviors implemented in very complex functions and hence may, if provided with the correct training set, be useful for the construction of such functions.

Another potential application is to let a network adaptively capture variable interrelationships not specified by a design engineer through the use of on-line training. Interactive trainability may be possible because a neural network can operate in a variety of learning modes, one of which is to learn by example. Thus, if presented with example pilot behaviors during selected scenarios, a net may be able to extract critical initiating variables as well as the essence of a correct response without being explicitly told what to do at each step.

Some recent research on neural networks emulating simple classical control problems has shown encouraging results that indicate an extension of such technology to more complicated problems that may well be feasible [5, 19, 66]. However, much research remains to be done before valid comparisons between real world control applications and neural network methods can be made. Figure 9.36 gives a block diagram of some of the most important components of a generic aircraft landing control system. Notice that there are a number of operational modes which in turn depend upon the stage of the landing. After several minutes of altitude hold, ILS capture occurs, followed by ILS tracking, and finally flare. Output from these modes are a series of control commands, the most important of which for us is θ_{cmd} which controls the aircraft elevator servomechanism and consequently the pitch up during landing. Because the aircraft is often flying under reduced power at landing, the throttle and autothrottle have minimum effect. These control variables are then fed into the aircraft dynamic model which in turn is influenced by the environment. Thus, to train a neural network we must first create the mathematical models of a standard controller (simplified and adjusted for our problem), a reduced complexity model of the airframe upon which the controller will act, and some type of environmental disturbance (in our case a wind model which exhibits logarithmic shear or different strengths at different altitudes following a logarithmic variation with altitudes). After these models are generated, a series of experiments can be conducted using variations on neural network architectures, training rules, and performance criteria. The above models have been presented in detail [26].

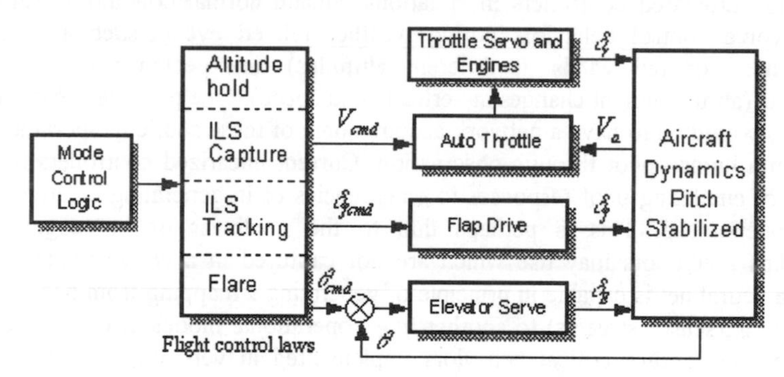

Figure 9.34. Components of aircraft landing system.

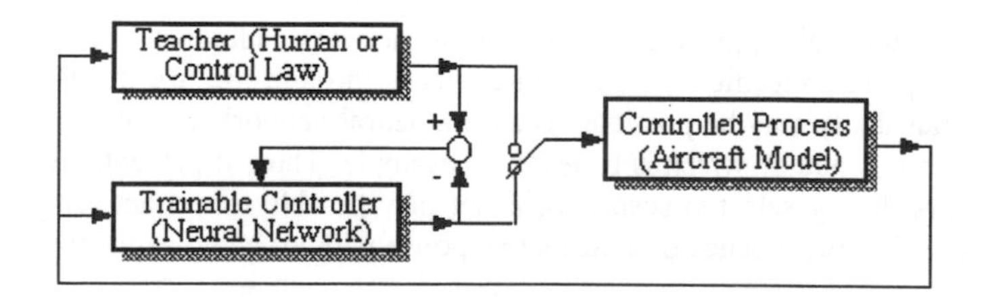

Figure 9.35. Structure of trainable adaptive controller.

A general approach is to apply process control using a trainable controller like that represented in Figure 9.35. It consists of a teacher (a human or a linear controller), a trainable neural network (using a backpropagation learning algorithm), and the process to be controlled (e.g., keeping an aircraft icon on a predetermined flight path by issuing an attitude command θ_{cmd}). The teacher could have used a linear or nonlinear law, linear in the case of the traditional controller and probably nonlinear if humans were to provide landing values. The linear control law was derived [27] by using conventional control theory methods. For a human teacher, we would propose using actual input/output time histories obtained through direct experiments using a simple interface and a mouse or a joystick. Supervised learning in the network is used to provide the greatest amount of control over the mapping so the learning is not taking place in real-time. The true state of the control process was provided by both the teacher and the network. The teacher defined the desired behavior of the controller by providing outputs stored in a data base sufficiently large to provide a detailed sample of the possible input state vectors.

9.5.3 Applications and Implementation of Neural Network Controller

To implement this framework we supplement the generic aircraft model with some supporting structures including a flare controller for the vertical/longitudinal axes [23]. Once the net using nominal conditions (i.e., no disturbances) is trained, its performance on off-nominal conditions such as various wind disturbances is also studied. This is because in order to make the neural net perform robust control, we have first to understand its ability to perform well understood linear control behaviors and then determine how well the network could generalize to new environmental situations.

A simple graphic tracking program is generated to provide human input through the interface of a workstation. The neural controller could learn through observation of a human working with the elevator angle only.

Regarding application to the real world, it is assumed that existing trends in pilot usage of autoland systems will continue. Consequently, a neural controller may best be used only as a supplementary enunciator device monitoring pilot landing behavior (i.e., an ILM enhancement). If this operational mode were selected, the pattern recognition capabilities of a neural net would have to be significantly enhanced. Particular attention should be paid to early detection of hazardous external conditions such as wind shear and the monitoring of ongoing pilot control actions. In the case of abnormal weather events, it would be necessary to consider whether a neural network could function adequately in a real-time predictive mode as well, so as to anticipate emerging problems.

1. **Implementing the baseline aircraft model**

To facilitate the use of the baseline model it is useful to construct a simulation environment on a Symbolic computer which provides three types of experimental environments.

(1) A process control environment to simulate the plant to be controlled and to generate trajectories used for display and neural net learning.

(2) A neural net fabrication environment to easily construct neural net architectures and train them.
(3) A statistical analysis environment to study statistical properties of data to be used for neural net training.

The baseline aircraft simulation is then coded into this environment. Provisions have been made for training data storage and graphical display interfaces to observe performance. Figure 9.36 is a modified screen image of the nominal (no-wind) aircraft response produced by the classical controller during the execution of a simulated landing. The lower graph shows a landing descending from 500 feet after glide-slope capture while the upper graph shows the aircraft vertical velocity ranging between 5 and -15 ft/sec.

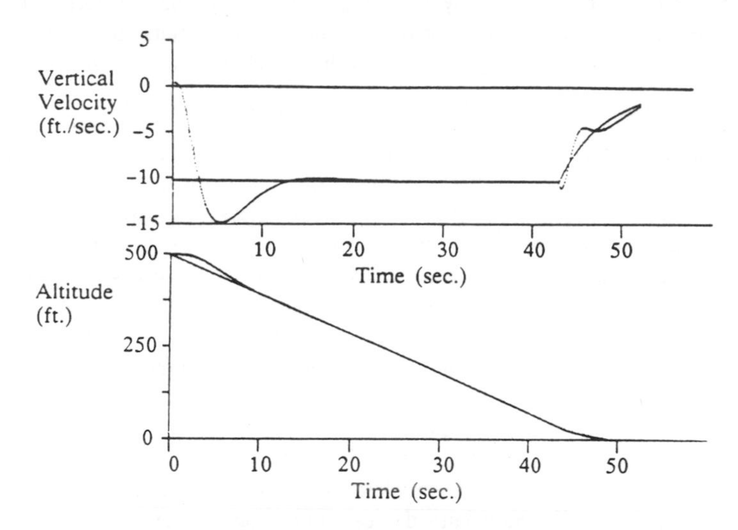

Figure 9.36. Aircraft simulation nominal trajectory in process control environment. (with Jorgensen and Schley, 1990)

At present, alternative neural net input architectures have been explored to address the dynamics inherent in the control problem. Experiments centered around the specification of various input combinations and the issue of including system dynamics. Typically, a control problem requires three elements: past history by means of integration, current values by means of proportional combination, and future predictions by means of differentiation. Consequently, some degree of information recurrence must be used to mimic a conventional control system.

Recurrence within a neural net complicates learning immensely. One approach for recurrent loops is to break them apart and handle them separately. Learning takes place very slowly under these circumstances. Therefore, a scheme was devised to provide the effects of recurrence by defining variables external to the neural net not only current values for aircraft variables (i.e., altitude, altitude commands, *etc.*), but also values at some basis to formulate a model of proportional combination and differentiation of system variables. Additionally, the past value of the neural net output (i.e., itch attitude command) was also back as an input to allow the neural net to represent internally the integration of system variables. This scheme has a beneficial effect of providing recurrence outside of the neural net. Standard backpropagation can then be used for training. Clearly, there are other approaches that could be explored using the current baseline equations.

Figure 9.37 illustrates the construction of one of the neural net structures used in the implementation, where h_{cmd} and h_{est} are the aircraft altitude rate commands and altitude rate estimates respectively. Inputs are considered of current altitude and altitude rate error values along with values at the previous simulation time step (taken as 0.025 here). Also provided is the value of the pitch attitude command at the previous time step, providing exterior (or broken) recurrence.

The architecture of the neural net shown in Figure 9.38 consists of an input layer with nine inputs, a hidden layer with four neurons and an output layer with one neuron. Operation of the neural net is as follows. The input values are scaled so that the input neurons have activity values from -1 to +1. Then, each input neuron is connected to each hidden layer neuron by a weight value adjusted during learning. Hidden layer neuron activity values are computed by summing the input neuron activity values multiplied by the weight value. This is followed by application of a standard nonlinear (sigmoidal) squashing function which provides hidden layer neuron activity value from -1 to +1. Similar weight connections, summation of products and squashing are used to determine the activity value of the output neuron. Finally, the output neuron activity is scaled to provide the requisite range for pitch attitude commands.

2. **Training the neural network**

The proposed neural net has been trained using a version of the backpropagation algorithm under a variety of wind conditions. Trajectories involving no wind, constant wind, and logarithmic wind shear have been generated. Also, turbulent gusts have been added. Figure 9.40 shows the result of controlling the aircraft using a neural net trained in gusty wind conditions. The upper graph of Figure 9.40 shows the altitude rate of change while the lower graph shows the altitude. These initial indications imply that the neural net results provide good comparisons with conventional controllers when exposed to conditions similar to the training set. It is noted that the conventional controller response for these wind conditions is very similar to the neural controller response shown in Figure 9.38.

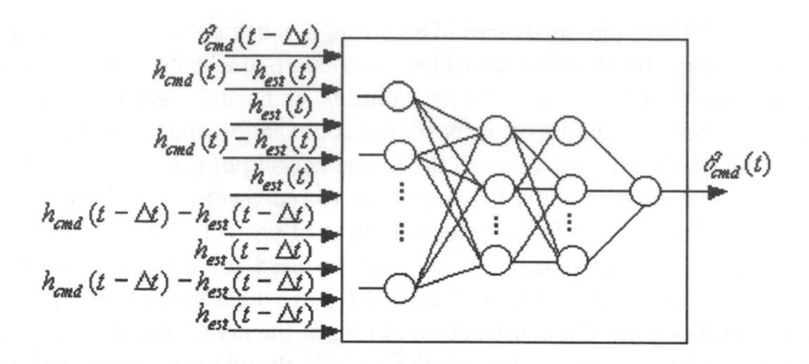

Figure 9.37. Neural net architecture for glide slope and flare control.

Further experimentation with learning is needed to show that this or alternative neural nets can indeed fuse the various modes of flight near landing (*e.g.*, glide-slope, flare). To accomplish this fusion, networks need to be trained with a wide variation of environmental conditions.

3. Training neural network via human interface

Learning experiments can also be conducted using human-derived training data. A simple aircraft navigation interface has been constructed. Currently, the system uses a mouse and does not run in real-time. This system is not intended to be a realistic autolander simulator. Rather, it is intended to experiment with training selected neural networks with other than conventional linear controllers.

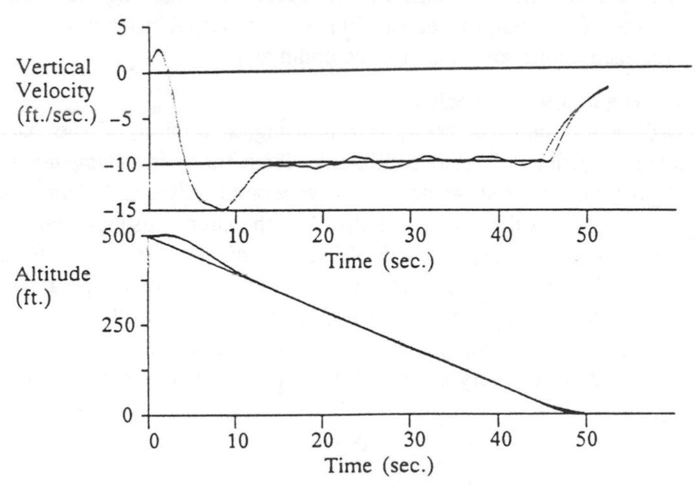

Figure 9.38. Gusty wind trajectory of neural net controlled aircraft. (with Jorgensen and Schley, 1990)

The human interface includes a Flight Manager, similar to that used in real aircraft, which displays the angle of inclination and height of the aircraft along with additional information. A preview indicator also displays a desired flight-path profile. The Flight Manager has the following components:

(1) A plane symbol, whose height above/below the horizon indicates the aircraft's degree of indication from horizontal.
(2) A pair of triangular guides, located just outside the aircraft symbol's wingtips, which approximates the correct angle of inclination at the current moment for proper flight-path navigation.
(3) A circular guide that allows the pilot to direct the aircraft's angle of inclination with the use of a mouse input device.
(4) An altitude bar, which monitors the current height of the aircraft as a rising/falling horizontal line in the bar.
(5) A triangular guide adjacent to the altitude bar indicates the correct altitude at the current time for proper flight-path navigation.
(6) A constant, computed at initiation of the flight manager, relates overall window size to the number of pixels per degree of inclination.
(7) The desired flight path is displayed on an *x-y* Cartesian graph. Height is represented along the vertical axis and distance on the horizontal. The plane must then "fly" the indicated line, with the current position of the plane represented as the leading dot of a series of dots, which mark the plane's actual path.
(8) Upon initiation, the entire flight path is shown. At a predetermined altitude, the flight path is redrawn at a larger scale, to assist the pilot during crucial flare and touchdown phases of the flight.

4. Conclusions and discussions

In this section we have introduced a simplified model of a realistic aircraft which potentially provides a baseline model and controls relationships against which alternative neural network controllers can be evaluated. The detailed model, especially the mathematical equations, were presented in Ref. 26. We have mentioned several ways in which the problem could be approached including some methods currently being explored. These efforts are only in their early stages and the general area of comparison with classical control approaches for aviation problem areas should provide a rich topical source for further work. Of particular interests are the relationships between human and machine controllers, how displays generate conditions that influence human control strategies, and the generation of mappings between classical control laws and the outputs that may be generated by a neural controller as it extracts control regularities from data. As a result of these experiments, we have observed the necessity for refinements in a number of neural network concepts when faced with a complex application. Among them are the ability of a neural net to learn discontinuities, the importance of and potential risks associated with grouping training data around discontinuities, the vital need for faster convergence methods, and the advantages of incorporating *a priori* knowledge into a network to facilitate convergence. Although the networks can land the aircraft in headwinds and tailwinds successfully, we have not found a single network which has been able to capture the performance capabilities of a linear controller throughout the entire domain of its input space. Further research is needed, and present work may interest other researchers in solving this problem for evaluation of their own neural network design and in using neural network controllers to other areas.

9.6 Intelligent Control in Medicine — Fuzzy Logic Control of Blood Pressure during Anesthesia

The techniques of fuzzy logic and expert system have been used in the medical area since middle 1970s [21, 34, 37, 45, 62]. For example, the MYCIN that has been used to diagnoze and consult for antibiotic treatment is well known in AI and medicine worlds.

This section describes an example of intelligent control in medicine, a fuzzy logic controller which controls mean arterial pressure (MAP) that was taken as a parameter for the depth of anesthesia. The fuzzy membership functions and the linguistic rules have been applied to the design and implementation of the controller that was tested in different surgical operations [34]. In the next section, Section 7 of Chapter 9, we will introduce another example, a consulting system for Electrocardiogram diagnosis. As a matter of fact, that would be another example of intelligent control in medicine.

9.6.1 Introduction to Depth Control of Anesthesia

Before referring to the fuzzy logic control system, let us discuss the problem of depth control of anesthesia in order to acquire some preliminary knowledge about this topic. One of the main tasks of the anesthetist during surgery is to control the depth of anesthesia. However, the depth of anesthesia is not readily measurable. In clinical practice, the depth of anesthesia is evaluated by measuring blood pressure, heart rate, and clinical signs such as pupil size, motor activity, etc. A good correlation of blood pressure to the anesthetic was given [13]. EEG signals have been used for the dosage of proposal. However, in clinical routine EEG is difficult to be used to artifacts and large intra- and interindividual variations partly caused by differing anesthetic agents. Continuous electromyographic recording of spontaneous activity of the upper facial muscles has been used, but results in unpredictable values when the patient receives muscle relaxants. Evoked responses have been used for various agents; they show an agent-dependent effect, too.

None of the above-mentioned methods has been established on a routine basis. Anesthesiologists still use blood pressure as the most reliable guide for dosing inhaled anesthetics. An argument heard frequently against this practice is that blood pressure depends on many other factors such as blood volume and cardiovascular function. This is true, of course, but anesthesiologists, when unable to treat rapidly enough the primary cause of hypo- or hypertension (hypovolemia and hypervolemia, cardiac failure, etc.) will always adjust the concentration of the inhaled anesthetic to rapidly bring blood pressure to normal ranges. Thus, control of blood pressure still appears as the most appropriate way of controlling depth of anesthesia.

To obtain adequate depth of anesthesia, MAP has to lie within a predefined range. The main reason for automating the control of depth of anesthesia is to release the anesthetist so that he/she can devote his/her attention to other tasks, such as controlling the fluid balance, ventilation, and drug application, which cannot yet be adequately automated, thus increasing the patient's safety.

Depth of anesthesia is controlled by using a mixture of drugs which are injected intravenously or inhaled as gases. Most of these agents decrease MAP. Among the inhaled gases, isoflurane is widely used, most often in a mixture of 0-2% by volume of isoflurane in oxygen and/or nitrous oxide. The isoflurane concentration in the inspired air is adjusted by the anesthetist depending on the patient's physiological condition, surgery, MAP, and other clinically relevant parameters. To deliver the anesthetic agent to the patient, a semi-closed circle breathing system is used that allows the reuse of the exhaled anesthetic gases.

The first experiments in the automatic control of the depth of anesthesia started in the 1970s with constant gain PID-controllers. Then came controllers with one-step adaptation of the gains, which were unable to cope with the time-varying parameters of the patient during surgery. It was thought that this problem could be solved by using adaptive controllers [54]. Such controllers show robustness with respect to variations in the patient's parameters. More fashionable studies use rule-based and fuzzy controllers have to cope also with the time-varying structure of biological systems [30, 31, 67]. Some reviews of this subject have been made [11, 29].

9.6.2 Controller Design and Simulations

The controller should mimic the control actions of the anesthetist. This way a supervising anesthetist can easily ensure proper functioning of the controller. Furthermore, modeling a biological process like anesthesia is very complex, because it has a nonlinear, time-varying structure with time varying parameters. These facts suggest the use of rule-based controllers like fuzzy controllers which are suitable for the control of such systems.

The control loop studied has the structure shown in Figure 9.39. There are two different kinds of disturbances: (a) system noise due to pain caused by surgery, cardiovascular disease, concomitantly applied drugs, etc. (for example, a skin incision can lead to rapid changes in blood pressure of more than 10 mmHg). (b) Measurement noise in the blood pressure and artifacts caused by calibration, electrocautery, etc.

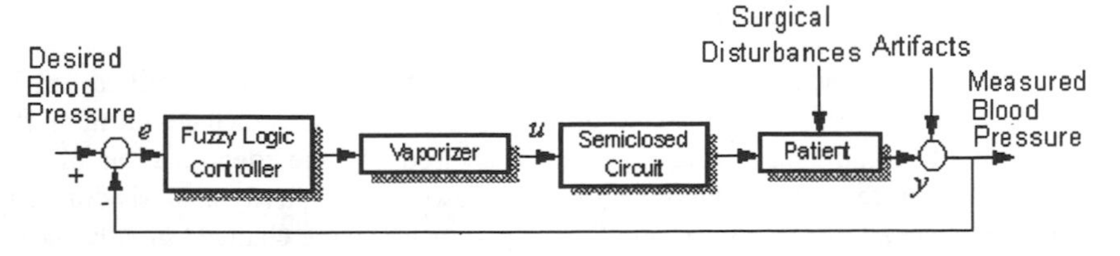

Figure 9.39. Block diagram for depth control of anesthesia.

For simulation and controller design purposes the relationship between inflow concentration of isoflurane $u(t)$ and the resulting blood pressure $y(t)$ is modelled as the sum of two first order terms each with a pure time delay (the model includes the patient and also the semi-closed circuit). The step response $h(t)$ corresponding to a unit step input can be written as follows:

$$
\begin{aligned}
h(t) &= f_1(t) + f_2(t) \\
f_1(t) &= \begin{cases} 0 & \text{for } t < \tau_1 \\ K_1[1-\exp(-\alpha_1(t-\tau_1))] & \text{for } t > \tau_1 \end{cases} \\
f_2(t) &= \begin{cases} 0 & \text{for } t < \tau_2 \\ K_2[1-\exp(-\alpha_2(t-\tau_2))] & \text{for } t > \tau_2 \end{cases}
\end{aligned} \tag{9.25}
$$

where $K_1 = -3, K_2 = -7.3, \tau_1 = 23(s), \tau_2 = 101(s), \alpha_1 = 0.01, \alpha_2 = 0.006.$

For the purpose of this analysis it is assumed that τ_1 and τ_2 are integral multiples of the sampling period T such that $\tau_1 = c_1 T$ and $\tau_2 = c_2 T$. The z-transform of the transfer function of the step response in (9.25) can be expressed as:

$$
\begin{aligned}
Y(z) &= \sum_{i=1}^{2} K_i \alpha_i [1 - z^{-1}\exp(-\alpha_i T)]^{-1} z^{-ci} U(z) \\
&= [(b_1 z^{-c1} + b_2 z^{-1-c1} + b_3 z^{-c2} + b_4 z^{-1-c2}) / (1 + a_1 z^{-1} + a_2 z^{-2})] / U(z)
\end{aligned} \tag{9.26}
$$

where $Y(z)$ and $U(z)$ are z-transforms of the output blood pressure and input isoflurane concentration respectively, and

$$
\begin{aligned}
b_1 = K_1\alpha_1, b_2 = -K_1\alpha_1\exp(-\alpha_2 T), b_3 = K_2\alpha_2, b_4 = -K_2\alpha_2\exp(-\alpha_1 T), \\
a_1 = -\exp(-\alpha_1 T) - \exp(-\alpha_2 T), a_2 = \exp(\alpha_1 T) + \exp(-\alpha_2 T).
\end{aligned}
$$

It is convenient to write the z-transform notation in the recursive form as shown with $T = 10s, a_1 = -1.331, a_2 = 0.335, b_1 = 0.030, b_2 = -0.048, b_3 = 0.017, b_4 = -0.041$ and $b_4 =$ -0.041:

$$
\begin{aligned}
y(KT) = -a_1 Y((k-1)T) - a_2 Y((K-2)T) + b_1 u((K-c_1)T) + b_2 u((K-c_1-1)T) \\
+ b_3 u((K-c_2)T) + b_4 u((k-c_2-1)T).
\end{aligned} \tag{9.27}
$$

The parameters used in the controller design process were identified off-line recursively from the MAP data collected during surgery after step changes in the inflow concentration of isoflurane. A single patient was used for the identification of these parameters. At first, this might sound insufficient. However, sensitivity analysis during the design phase where all or a combination of these parameters were changed greatly showed no major effect in the simulated controller performance. Actually, this very robustness makes a fuzzy controller the appropriate choice for the task at hand.

The first simple linguistic rules that describe the anesthetist's actions were tested and systematically extended in several simulation runs. During the first design phase the error $e(t)$, the change of error $ce(t)$ and the integral of error $ie(t)$ were used for the computation of the control variable $u(t)$. The reason for using the integral part was to eliminate the steady state error; the point in using the derivative part was to speed up the controller. The membership functions and the reference points were chosen by using the data recorded during an actual operation where MAP was controlled by an anesthetist. The bell shaped membership functions used can be described by the following exponential equation:

$$\eta = \exp[-k(\zeta - \lambda)^2] \tag{9.28}$$

where ξ is the input value and λ the shifting of the function in relation to zero; the factor k determines the "width" of the bell. The evaluation of the linguistic rules was performed using the max-min composition and the center of gravity method.

The control characteristics and behavior under different disturbances were tested in simulations with different noise amplitudes and parameters. The best results were achieved with the linguistic rules and parameters of the membership functions shown in Table 9.6 and Figure 9.40. The rule in Table 9.7, for example Rule 7, says that "if the error is around 0 (ze) and if the integral of the error is around -90 (ns), then set the inflow concentration of isoflurane about 3% (pb)". In Figure 9.40, the memberships μ_e for the error e, and μ_u for the inflow concentration u and their reference points; nb negative big, ns — negative small, ze — zero, ps — positive small, pb — positive big, pm — positive medium, pv — positive very big. The value of k in (9.28) was chosen as 5 for all membership functions. The controlled variable was the absolute amount of inflow concentration of isoflurane, not the incremental value. There are three stabilized situations considered, according to Table 9.7 (rules 5, 6 and 7), i.e., a stable output is limited by any combination of ps, pm, and pb. To stabilize at a value different than 2% there must be an integrative error (i.e., it forces a certain trajectory). On the other hand, there will not be an inflow concentration higher than 4% (safety measure). Therefore, rules 3 and 8 are only for transients, Figure 9.41 depicts a simulation run with the reference points and linguistic rules shown in Table 9.6 and Table 9.7. From Figure 9.41 it can be seen that at 200 s the set value is raised from 80-88 mmHg. At 800 and 1400 s two disturbances are applied. The controller is fast and only a small overshoot is observed. At 2100 s the set value is reduced to 84 mmHg.

Table 9.6. The reference points of membership function

	Input		Output	
	e[mmHg]	ie[mmHg]		u[%]
nb	-10	-160	ze	0
ns	- 5	- 90	ps	1
ze	0	0	pm	2
ps	5	90	pb	3
pb	10	160	pv	4

Table 9.7. The Linguistic rules resulted in the best control and noise rejection characteristics

Rule No.	Input		Output
	e	ie	u
1	ns	-	pb
2	ps	-	ps
3	nb	-	pv
4	pb	-	ze
5	ze	ze	pm
6	ze	ps	ps
7	ze	ns	pb
8	-	nb	pv
9	-	pb	ze

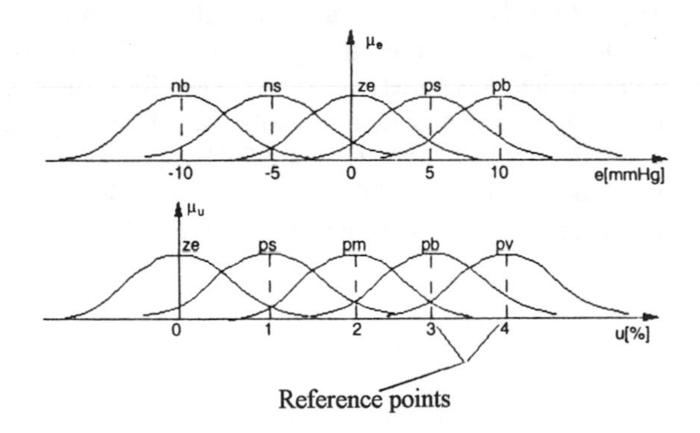

Figure 9.40. Membership functions.

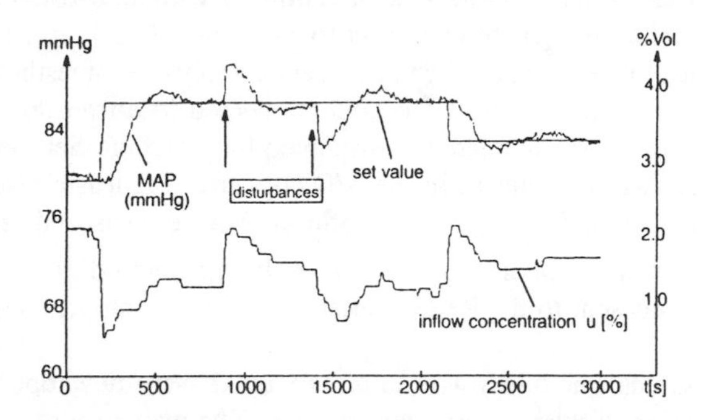

Figure 9.41. A simulation run with linguistic rules and reference points. (with Meier *et al.*, 1992)

Some conclusions drawn from the simulation study are: (a) More rules do not necessarily result in better control characteristics. Conflicting rules can even lead to unstable behavior. Furthermore, computational requirements increase with an increasing number of rules. (b) The contribution of the derivative part is minor. Even after smoothing, relatively fast changes still remain in the signal. Thus, a controller using the derivative part shows a lot of switching of the control variable without improving the quality of control (just as in classical control). (c) The controller is robust with respect to variations in process parameters.

The results of the simulations encouraged to apply the controller during surgery under "real life" operating room conditions.

9.6.3 Controller Implementation and Test Results

(1) Implementation

A IBM compatible personal computer (Toshiba T5100) was used for collection, display, control and storage of data as well as feedback control. The analog input and output data were converted to/from digital data by using a Burr-Brown PCI-20041C-A carrier board in connection with the analog input module PCI-20019M-1A and the analog output module PCI-20003M. Additional digital communication was done via the RS232 serial port. The desired inflow isoflurane concentration was obtained using a Dräger Vapor 19.3 vaporizer driven by an external servo-motor (the servo-motor was driven by a PID-controller-amplifier, which in turn was controlled by a conventional electronic interface). The servo-motor had a switch enabling toggling between manual and automatic control at any time. After determining the curve relating the D/A output voltage (0...10V) to the resulting isoflurane concentration (0...5% by volume), the computer software calculated the D/A gain required for a specified inflow isoflurane concentration, taking into account the actual ambient temperature and barometric pressure. The precise concentration was reassured by a Dräger Irina infrared gas analyzer with an analog output voltage for the A/D converter. A Dräger Sulla ventilator with an 8-ISO circle system was used for ventilation. The fresh gas flow was set to a constant 3 l/min oxygen.

Automatic control was not started before an acceptably deep anesthesia was achieved. A catheter was inserted into a radial artery and connected to a transducer. The obtained electrical signal was then preamplified and displayed by a Hellige Servomed monitor. Its analog output voltage was transmitted to the A/D converter to measure MAP.

The gas concentrations (O_2, CO_2 and isoflurane) were measured at the mouthpiece of the patient by a Datex Cappnomac gas analyzer. It transformed the gas concentrations to a data string that was sent to the RS232 serial port of the computer. Update rate of the data was 0.1 Hz.

Computer programs (modules) written in Modula-2 were developed to perform the tasks of data acquisition, display, control and storage. The main program can be started at any instance. The controller starts working. During automatic control a loop, consisting of MAP-updating (A/D conversion), filtering, control calculations, control signal (i.e., isoflurane concentration) updating (D/A-conversion), data display and data storage on file is repeated every 10 s. The computer displays the actual and the desired blood pressure, as well as the inspiratory and endtidal isoflurane concentrations. The desired MAP may be changed at any time without interrupting automatic control. A filtering procedure for MAP was developed, in order to both smooth the incoming signal and to detect disturbances during data acquisition. Therefore, the incoming MAP signal is processed by a median-filter with a moving window size of 21 samples. Then, the signal is checked for consistency, i.e., it is compared with a number of conditions that have to be satisfied if the data acquisition works properly. Otherwise, the MAP data are not updated and, if the malfunction continues, a warning is displayed. All data of interest are stored in files for later examination.

(2) Test results

The controller was first tested during surgery in the so called "consultative mode", which means that the controller suggested a value for the inflow concentration and the anesthetist could then decide to set this value or not. This way the patient's safety was assured even in the very early stage of development and the controller could be tested under realistic operating room circumstances. Subsequent analysis of data collected during surgery showed that the inflow concentration suggested by the controller was always very close to the actual concentration set by the anesthetist. Therefore, a fully automatic run was completed during which the controller actually set the inflow concentration administered to the patient. Figure 9.42 shows the results of that operation. As adjusted by the controller during the first 2000 s of an abdominal operation it shows that the controller can adjust the blood pressure adequately. The upper curve in Figure 9.42 is the MAP (left scale) and the lower one is the inflow concentration of isoflurane (right scale). The set value of MAP is 85 mmHg. The disturbances at 1000 and 1500 s are due to skin incisions.

In a further operation, MAP was first controlled by a decision based controller (not described here), then by the fuzzy controller and finally by an experienced anesthetist. Figure 9.43 shows the resulting MAP. It can be seen (with a "trained eye"!) that the fuzzy controller does a better job of keeping the blood pressure of the patient within a tight range of the set value. This point is quantified in Figure 9.44. With the fuzzy controller, the variation of the blood pressure was within 5 mmHg of the set value (85 mmHg) 69% of the time and within 15 mmHg of the set value all the time. When the anesthetist controlled the blood pressure, these numbers were 47% and 92% respectively and 8% of the time the patient's blood pressure deviated more than 15 mmHg from its desired value. This shows that the fuzzy controller was measurably more accurate in keeping the blood pressure within a small range of its set value of 85% mmHg.

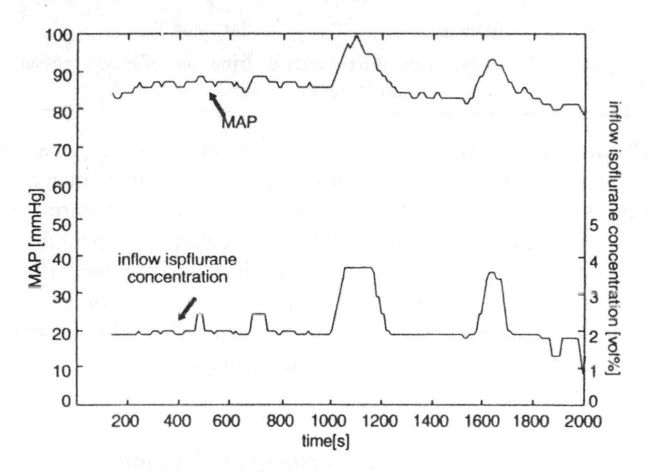

Figure 9.42. Variation of MAP and inflow isoflurane concentration. (with Meier *et al.*, 1992)

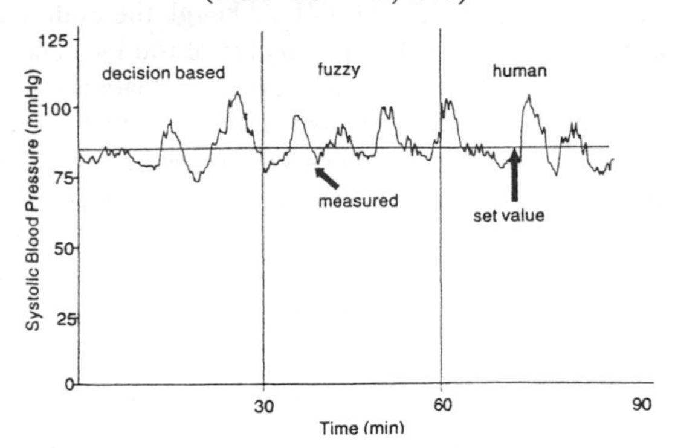

Figure 9.43. Blood pressure control during surgery. (with Meier *et al.*, 1992)

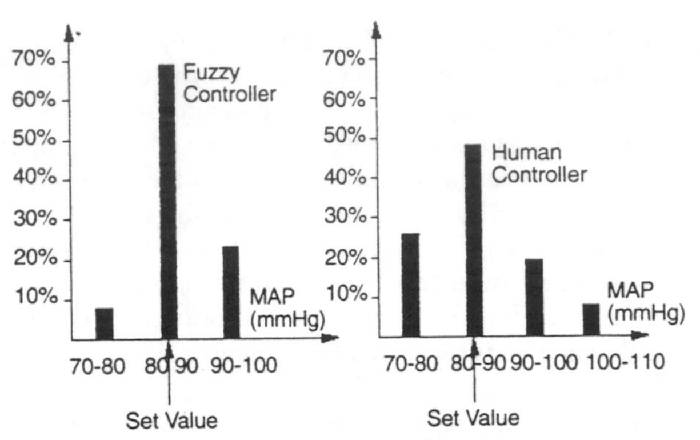

Figure 9.44. Comparison of frequency distributions of blood pressure. (with Meier *et al.*, 1992)

The results in Figure 9.46 relate to a typical patient. However, in all cases, the fuzzy controller also proved to be superior or at least equivalent to the human controller.

In summary, a proportional-integral fuzzy controller which controls the MAP during anesthesia with isoflurane was designed and implemented on a personal computer. The controller was tested in 11 surgical cases. The anesthetists supervising the controller never had to intervene or override it. Furthermore, the quality of control achieved by the fuzzy controller proved to be superior to manual control. Therefore, it is concluded that such a controller can be routinely used during anesthesia with an agent like isoflurane.

9.7 Intellectualized Instruments — Consulting System for Electrocardiogram Diagnosis

During the past ten years a lot of research work has been done in developing systems for electrocardiogram (ECG) diagnosis [18, 63, 64]. Although the evalution of these systems shows that many of them are reliable, their use in clinical routine is limited. This is due to many reasons. There is a lack of an established way to evaluate these systems. Hence the doctor has no criteria to choose the best system, apart from the advertisement of the producers, while it is well-known that trust in a system entails responsibility and is related to medical ethics problems. In most systems the user's interface follows the question-answering approach. The user answers questions posed by the system and eventually the system gives the diagnosis. The final conclusion does not always agree with the user's opinion. The explanation utilities which the systems provide usually are not satisfactory to the user.

9.7.1 Architecture and Mechanism of ECG Diagnosis System

The present consulting system is based on an expert system for electrocardiogram (ECG) diagnosis [17]. The system accepts an ECG signal as input and starts an interactive dialogue with the physician, regarding the measurement parameters of the ECG. It asks questions to the physician and judges his answers, in relation to the results of the automatic ECG analysis which is performed by the system. In case there is a contradiction between the physician's answers and the results of the automatic ECG analysis, the system informs the physician accordingly. The physician can reconsider his estimation or he can force the system to proceed to the diagnosis phase, considering both sets of data, i.e., those of the physician and those that are automatically measured.

The system produces two answers, corresponding to the physician's data and the automatically measured data. These answers are produced by built-in expert system for ECG diagnosis. The physician can ask for an explanation concerning the two answers, or reconsider his estimated data during the measurement phase and require a new diagnosis phase.

The architecture of this expert system of ECG diagnosis is shown in Figure 9.45. Module **A** represents a typical expert system. Module **B** is a system for automatic ECG analysis [12]. After the ECG acquisition and signal processing an array of all useful ECG diagnostic parameters is sent to modules **C** and **D.**

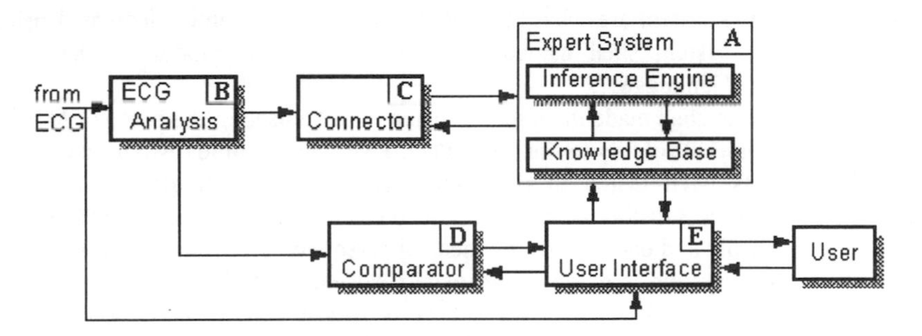

Figure 9.45. Architecture of ECG diagnosis system.

Module **C** provides the communication with module **A**. To each question posed by Module **A** in order to proceed to the diagnosis phase module **C** provides as an answer, the appropriate data based on the array of measured parameters formed by module **B**. The array of parameters created by module **B** is also sent to module **D**. Module **A** poses also the same questions (as with the case of module **C**) to the user via module **E** in a user friendly form. It receives again appropriate answers from the physician which are also received by module **D**. Module **D** forms, from the user's answers, a corresponding array of measured parameters which are compared with those received from module **B**.

It is noted that the communication of the module **E** with the user is accomplished in a graphical form which makes easier to the user the estimation of the measured parameters. Moreover, module **E** provides mechanisms for the correction of some kinds of errors due to the user. For example, if module **A** asks "What is the QRS duration in seconds?" then module **E** presents the whole ECG on the screen and asks the user to mark the onset and offset of the QRSs. After that, module **E** calculates the QRS duration and sends this value back to module **A** and **D** as an answer to the previous question. If some QRSs which have been marked are out of some statistically approvable limits, they are rejected by module **E**. In order to produce the required graphical forms, module **E** has already the ECG data. Module **E** is also a screen signal editor with zoom capabilities.

Additionally module **E** communicates with **D**, in order to receive reports about the differences between the two arrays of measured parameters described above. If there is a mismatch, the user can reconsider some of his estimated parameters.

Finally, with the help of module **A** two diagnoses are presented to the user, one is based on the array parameters produced by module **B** and another is based on the array parameters that the user has estimated. The user may again reconsider his estimated array parameters, in order to have another diagnosis phase.

9.7.2 Implementation and Conclusion of ECG Diagnosis System

This diagnosis system has been partially implemented. More specifically all modules except module **D** have been implemented. The implementation of **D** and the completion of the integration of the system are what remains. Module **A** has been implemented using a commercial expert system shell named M.1. Module **E** and module **C** have been implemented in the C language.

An expert system shell made in Greece is planned to be used in the place of M.1 later on. This will give more flexibility to the system since M.1 poses some restrictions.

Module **A** has been tested in the Alexandra Hospital, Department of Clinical Therapeutics with very satisfactory results. The results show that this approach makes module **A** user friendly and encourages the medical personnel to use this technology.

This consulting system is based on an expert system for ECG diagnosis. The main advantage of the proposed system over others is that it can be used not only as a consulting system, but also as an automated diagnosis tool as well as a training tool for the interpretation of ECG signals. The system can accept ECG signals from patients, or it can use an existing ECG library which has been properly prepared. The last capability is very useful for training new cardiologists. On the other hand, the experienced cardiologist interacts with the system and can improve its diagnostic capabilities in collaboration with the knowledge engineer, by changing the rules of the expert system.

9.8 Summary

Seven different and representative application examples have been presented in this chapter. They are distributed into a wide range of fields in industries, aeronautics and aerospace, ocean engineering, medicine and others.

The first example of application is the hierarchically intelligent control for advanced robots, multiple autonomous undersea vehicle, where robot planning and controlling have been concerned.

The second paradigm is in the area of intelligent process supervision and control, where a hierarchical rule-based and hybrid intelligent control system has been employed to industrial boilers.

The third and fourth examples deal with the intelligent control of automatic manufacturing systems: one is Petri nets-based control system for Flexible Manufacturing Systems (FMS); another is a knowledge-based system with hierarchy loops for optimization of machining operation and on-line scheduling control of intelligent machines.

The fifth and sixth instances are applied to intelligent fault detection and diagnosis: one is fault diagnostics for space station thermal control system; another is radar fault diagnosis system. The former is based on a combination of signal redundancy techniques and fuzzy logic. The latter is based on the expert system technique.

The seventh paradigm of application is involved in aircraft flight control, the neural network controllers for aircraft flare (approach) and touchdown (landing).

The eighth typical case is an intelligent control system in medicine, the fuzzy logic control of blood pressure during anesthesia.

The last example, the ninth example, is the consulting system for Electrocardiogram (ECG) diagnosis, an expert system-based intellectualized instrument.

All of these application examples are the lately developed great achievements in intelligent control and have exhibited superior control performances. As new research and development techniques, meanwhile, further work is still needed to go on in order to improve the controlled performances in some aspects. We expect more and more successful applications of intelligent control systems to be developed and used.

References

1. J. S. Albus, R. Lumia, and H. McCain, "Hierarchical control of intelligent machines applied to space station telerobots," *Proc. IEEE Int. Symp. Intelligent Control*, 1987, 20-26.
2. J. S. Albus, "A control system architecture for intelligent machine systems," *Proc. IEEE Conf. On Systems, Man, and Cybernetics*, Arlington, VA, 1987.
3. J. S. Albus, "System description and design architecture for multiple autonomous undersea vehicles," *NIST Technical Note 1251*, 1988.
4. J. L. Alty, "Knowledge engineering and process control," in: *Knowledge-Based Systems for Industrial Control* ed. J. McGhee (Peter Peregrinus Ltd, UK, 1990) 47-70.
5. A. Barto, R. Sutton, and C. Anderson, "Neuro-like adaptive elements that can solve difficult learning control problems," *IEEE Trans. SMC*, 13 (1983) 834-836.
6. B. A. Catron and B. H. Thomas, "Generic manufacturing controllers," *Proc. IEEE ISIC*, 1988, 742-744.
7. S. K. Chao, T. P. Caudell, and N. Ebeid, "An Application of expert system techniques to radar fault diagnosis," *Proc. IEEE Western conference on Knowledge-Based*

Engineering and Expert Systems, IEEE Computer Society Press, 1986, 127-135.

8. S. S. Chen, "Adaptive neural network control in CIMS," *Proc. IEEE ISIC*, 1988, 719-723.
9. M.-Y. Chen and Q.-J. Zhou, "Intelligent control for PH process," (in Chinese) *Information and Control*, **16** (1987)2:1-6.
10. S. B. Chen, L. Wu and Q. L. Wang, "Self-learning fuzzy neural networks for control of the arc welding process," *Proc. IEEE Int. Conf. On Neural Networks*, 1996: 1209-1214.
11. R. T. Chilcoat, "A review of the control of depth of anesthesia," *Trans. Measurement and Control*, **2** (1980) 38-45.
12. F. Critzali, *Automatic analysis of the ECG signal using mathematical methods*, Ph.D. Dissertation, NTUA Athens, 1988.
13. D. J. Cullen *et al.*, "Clinical signs of anesthesia," *Anesthesiology*, **36** (1972) 21-36.
14. L. K. Daneshmend and H. A. Pak, "Model reference adaptive control of feed force in turning," *Trans. ASME J.DSMC*, **108** (1986) 215-222.
15. A. A. Desrochers, *Modeling and Control of Automated Manufacturing Systems*, (IEEE Press, New York, 1990).
16. G. Erten, A. Charbi, F. Salam, T. Grotjohn, and J. Asmussen, "Using neural networks to control the process of plasma etching and deposition," *Proc. IEEE Int. Conf. On Neural Networks*, 1996: 1091-1096.
17. E. A. Giakoumakis *et al.*, "Formalization of the ECG diagnostic criteria using an expert system shell," *Proc. IEEE Computer in Cardiology*, 1988.
18. E. A. Giakoumakis and G. Papakonstantinou, "A consulting system for Electrocardiogram diagnosis," in: *Engineering Systems with Intelligence: Concepts, Tools and Applications*, *ed.* S. G. Tzafestas (Kluwer Academic Publishers, Netherlands, 1991)133-137.
19. A. Guez and J. Selinstey, "A trainable neuromorphic controller," *J. of Robotic Systems*, **5** (1988)363-388 S.A.Framentec, M.1 Reference Manual, 1987.
20. C. Guo, Z.-Q. Sun, S.-L. Shu and G.-Y. Fu, "Hierarchical intelligent control for industrial boiler," *Proc. IEEE ISIC*, 1992, 116-121.
21. F. Hayes-Roth, D. A. Waterman and D. B. Lenat, *Building Expert Systems*(Addison-Wesley, Reading, MA, 1983).
22. M. Herman, J. S. Albus, and T.-H. Hong, "Intelligent control for multiple autonomous undersea vehicles," in: *Neural Network for Control*, *eds.* W.T.Miller, R.S.Sutton and P.J.Werbos (MIT Press, Cambridge, MA, 1990) 427-474.
23. W. E. Holley, *Wind Modeling and Lateral Control for Antomatic Landing*, Ph.D. Dissertation, Stanford University, Stanford, CA, 1975.
24. L. P. Holmblad and J.-J. Ostergaard, "Control of cement kiln by fuzzy logic," in: *Fuzzy Information and Decision Processes*, eds. M. M. Gupta and E. Sanchez (North-Holland, 1982) 389-399.
25. A. H. Jones, B. Porter and R. N. Fripp, "Qualitative and quantitative approaches to the diagnosis of plant faults," *Proc. IEEE ISIC*, 1988, 87-92.

26. C. C. Jorgenson and C. Schley, "A neural network baseline problem for control of aircraft flare and touchdown," in: *Neural Network for Control, eds.* W. T. Miller, R. S. Sttton and P. J. Werbos (MIT Press, Cambridge, MA, 1990)403-425.
27. H. Kwakernaak and R. Sivan, *Linear Optimal Control Systems* (Wiley, NewYork, 1972).
28. A. A Lambregts, *Integrated system design for flight and propulsion control using total energies principles*, AIAA Paper No.83-2561, New York, 1983.
29. D. A. Linkens, and S. S. Hacisalihzade, "Computer control systems and pharmacological drug administration: a survey," *J.Medical Engineering and Technology*, **14** (1990) 41-54.
30. D. A. Linkens, and S. B. Hasnain, "Self-organising fuzzy logic control and application to muscle relaxant anesthesia," *IEE Proceedings-D*, **138** (1991) 274-284.
31. D. A. Linkens, and M. Mahfouf, "Fuzzy logic knowledge-based control for muscle relaxant anesthesia," *1st IFAC Symp. Modeling and Control of Biomedical Systems*, 1988, 185-1990.
32. J. E. McDonall, and B. V. Koen, "Application of AI techniques to digital computer control of nuclear reactors," *Nuclear Science and Engineering*, **56** (1975) 142-151.
33. R. Meier, J. Nieuwland, S. Hacisalihzade, D. Steck and A. Zbinden, "Fuzzy control of blood pressure during anesthesia with isoflurane," *Proc. IEEE Int. Conf. Fuzzy Systems*, 1992, 981-987.
34. R. Meier, J. Nieuwland, A. M. Zbinden and S. S. Hacisalihzade, "Fuzzy logic control of blood pressure during anesthesia," *IEEE Control Systems*, **12** (1992) 12-17.
35. K. L. Mikhailov, R. Schockenhoff, F. Pautzke, and H. A. Nour Eldin, "Towards building an intelligent flexible manufacturing control system," *Preprints IFAC Int. Symp. on Distributed Intelligence Systems*, 1991, 83-88.
36. R. B. Moore, "Saving gas and saving money — integrating an expert system into the process monitoring for steel making," *Proc. Int. Symp. On Intelligent Control*, 1992: 584-585.
37. J. Nie and D. A. Linkens, *Fuzzy-Neural Control: Principle, Algorithms and Applications* (Prentice-Hall, London, 1995) 37-68.
38. E. H. I Pallett, "Autopilot logic for flare maneuver of STOL aircraft," in: *Automatic Flight Control*, 2nd Edition,(Grenada, London, 1983)
39. K. R. Pattipati and M. G. Alexandridis, "Application of heuristic search and information theory to sequential fault diagnosis," *Proc. IEEE ISIC*, 1988, 291-296.
40. M. Pelegrin, *Encyclopedia of Systems and Control* (Pergamon Press, New York, 1987).
41. Y. Qu, "An intelligent control approach for machining," *Proc. IEEE ISIC*, 1991, 340-346
42. A. Ray, M. Desai and J. J. Deyst, "Sensor validation in a nuclear reactor," *Proc. 12th Annual Conf. On Modeling and Simulation*, 1981.

43. L. A. Rovere, P. J. Otaduy, C. R. Brittain and R. B. Perez, "Hierarchical control of a nuclear reactor using uncertain dynamic techniques," *Proc. IEEE ISIC*, 1988, 713-718.
44. D. Sbarbaro, D. Neumerkel and K. Hunt, "Neural control of a stell rolling mill," *Proc. IEEE ISIC*, 1992, 122-127.
45. E. H. Shortlife, B. G. Buchanan and E. A. Feigenbaum, "Knowledge engineering for medical decision making: A review of computer based clinical decision aids," *Proc. IEEE*, **67**(1979)1207-1224.
46. A. Soumelidis and A. Edelmayer, "Modeling of complex systems for control and fault diagnostics: a knowledge base approach," in: *Engineering Systems with Intelligence: Concepts, Tools and Applications, ed.* S. G. Tzafestas (Kluwer Academic Publishers, Netherlands, 1991)125-132.
47. R. F. Stengel, "Toward intelligent flight control," *IEEE Trans. SMC*, **23** (1993) 1699-1717.
48. M. Tomizuka, J. H. Oh, and D. A. Dornfeld, "Model reference adaptive control of the milling process," in: *Control of Manufacturing Processes and Robotics Systems, eds.* D. E. Hardt and W. J. Book (ASME, New York, 1983) 55-63.
49. S. G. Tzafestas (*ed*), *Knowledge Based Systems: Diagnosis, Supervision and Control* (Plenum Press, NY, 1989)181-216.
50. A. Ulsoy, Y. Koren, and F. Rasmussen, "Principal developments in the adaptive control of machine tool," *Trans. ASME J. DSMC*, **105** (1983)107-112.
51. G. Vachtsevanos and K. Davey, "Fault diagnostics for the space station thermal control system using a hybrid analytic/intelligent approach," *Proc. IEEE ISIC*, 1987, 54-58.
52. K. P. Valavanis, G. Seetharaman, M. A. Bayoumi and M. C. Mulder, "On a distributed intelligent materials handling systems," *Preprints of IFAC DIS*, 1991, 217-222.
53. K. P. Venugopal, A. S. Pandya and R. Sudhakar, "A recurrent neural network controller and learning algorithm for the on-line learning control of autonomous underwater vehicles," *Neural Network*, **7** (1994) 833-846.
54. R. Vishnol, and R. J. Roy, "Adaptive control of closed-circuit anesthesis," *IEEE Trans. on Biomedical Engineering*, **38** (1991) 39-46.
55. C. Von Altrock, B. Krause and T.-J. Zimmermann, "Advanced fuzzy logic control technologies in automotive applications," *Proc IEEE Int. Conf. Fuzzy Systems*, 1992, 835-842.
56. S. M. Wagner and S. W. Rothstein, "Integrated control and avionics for air supertority: Computational aspects of real-time flight management," *AIAA Conference on Guidance, Navigation and Control*, 1989, 321-326.
57. T. Watanabe, and K. Kawata, "Intelligent control in the hierarchy of automatic manufacturing systems," *Proc. IEEE ISIC*, 1987, 42-47.
58. T. Watanabe, "Model based approach to adaptive control optimization for milling," *Trans. ASME J. DSMC*, **108** (1986) 56-64.

59. T. Watanabe, "A control system to improve the accuracy of finished surfaces in milling," *Trans. ASME J. DSMC*, **105** (1983) 192-199.
60. T. Watanabe, "Design of an adaptive control constraint system of a milling machine tool," *Proc. IFAC Int. Symp. CAD of Multivariable Technological Systems*, ed. G. G. Leininger (Pergamon Press, Oxford, UK, 1982) 515-526.
61. T. Watenabe, "Real-time programming of computer numerical control of a machine tool in CAM," *Proc. IFAC/IFIP Workshop on Real-Time Programming*, ed.T.Hasegawa (Pergamon Press, Oxford, 1981) 93-100.
62. S. M. Weiss, and C. A. Kulikowski, *A Practical Guide to Designing Expert Systems* (Rowmand and Allenkeld Publishers, Totowa, NJ, 1984).
63. J. Willems *et al.*, "Assesment of the performance of ECG computer programs: the CSE diagnostic pilot study," *Proc. IEEE Computer in Cardiology*, 1988.
64. J. Willems *et al.*, "Assesment of the use of a reference database," *Circulation*, 71 (1985).
65. D. J. Williams, A. A. West and C. J. Hinde, A hierarchical model of intelligent materials processing, *Proc. IEEE ISIC*, 1991, 330-333.
66. D. Yeaung, "Supervised learning of action probabilities in associative reinforcement learning," in: *Proc. IEEE Int. Conf. on Neural Networks*, San Diego, CA, 1988, 162-171.
67. H. Ying, and L. C. Sheppard, "Real-time expert-system-based fuzzy control of mean arterial presure in pigs with sodium nitroprusside infusion," *Medical Progress through Technology*, **16** (1990) 69-76.
68. L. Zadeh, "Outline of a new approach to the analysis of complex systems and decision processes," *IEEE Trans. Systems, Man, and Cybernetics*, **3** (1973) 28-44.
69. M.-C. Zhou, "Combination of Petri nets and intelligent decision makers for manufacturing systems control," *Proc. IEEE ISIC*, 1991, 146-151.
70. D.-Z. Zhou, N.-R. Yuan, and M. Li, "Intelligent control for synchronous shearing process (in Chinese)," ATCA CINICA, **17** (1991)161-165.
71. M. C. Zhou and F. DiCesare, "Adaptive design of Petri net controller for error recovery in automated manufacturing systems," *IEEE Trans. SMC*, **19** (1989) 963-973.
72. Z.-L. Zhu, "Fault diagnosis expert system of power plant automation control system," *Proc. China-Japan Reliability Symposium*, 1987.
73. Z.-L. Zhu, "Fault detection and diagnosis for water supply system of boiler of power plant (in Chinese)," *Proc. of Chinese Association of Automation*, **1** (1988), 944-947.

CHAPTER 10
PROSPECTIVES OF INTELLGENT CONTROL

It is without a doubt that during the past 10 years intelligent control has been rapidly developing and has gained wide applications. More and more researchers have investigated into the young scientific discipline in various aspects; they believe that intelligent control would be able to admirably devote its new achievements to science, technology, economics, society, and human living. However, intelligent control is a new establishing discipline and is still incomplete both in theory and applications. In this chapter we are going to demonstrate future issues of intelligent control, then acquire a prospective for the development of intelligent control.

10.1 Future Research Issues of Intelligent Control

10.1.1 Intelligent Control Will Play More Significant Roles

Like artificial intelligence and advanced robotics, intelligent control is involved in many areas of science and technology such as artificial intelligence, cybernetics, systematology, informatics, cognitive psychology, cognitive physiology, cognitive engineering, linguistics, logic, bionics, robotics and computer science, etc.

The advancement of related science and technology has given a powerful motivation to the discipline of intelligent control; inversely, the research and development of intelligent control have provided a suitable test-bed and application field for all the mentioned areas. Figure 10.1 proposes a relationship among the related sciences, technology and intelligent control systems. The relationship emphasizes the knowledge and intelligence that are necessary for the development of intelligent science including intelligent control.

In the field of automatic control, intelligent control has also played more significant roles. As we discussed in Chapter 1, the control methods have opened the door to a wide spectrum of complex applications; such complex systems are characterized by uncertainty, nonlinearlity, and so on. The traditional control has encountered many difficulties in its applications, and the degree to which the control system deals successfully with the difficulties depends upon the level of intelligence in the system. Intelligent control has become a more effective way to overcome these difficulties, and has got wide applications in advanced robotics, intelligent planning and scheduling, automatic manufacturing, fault detection and diagnosis, traffic transportation, medical treatment, intellectualized instruments, and so on.

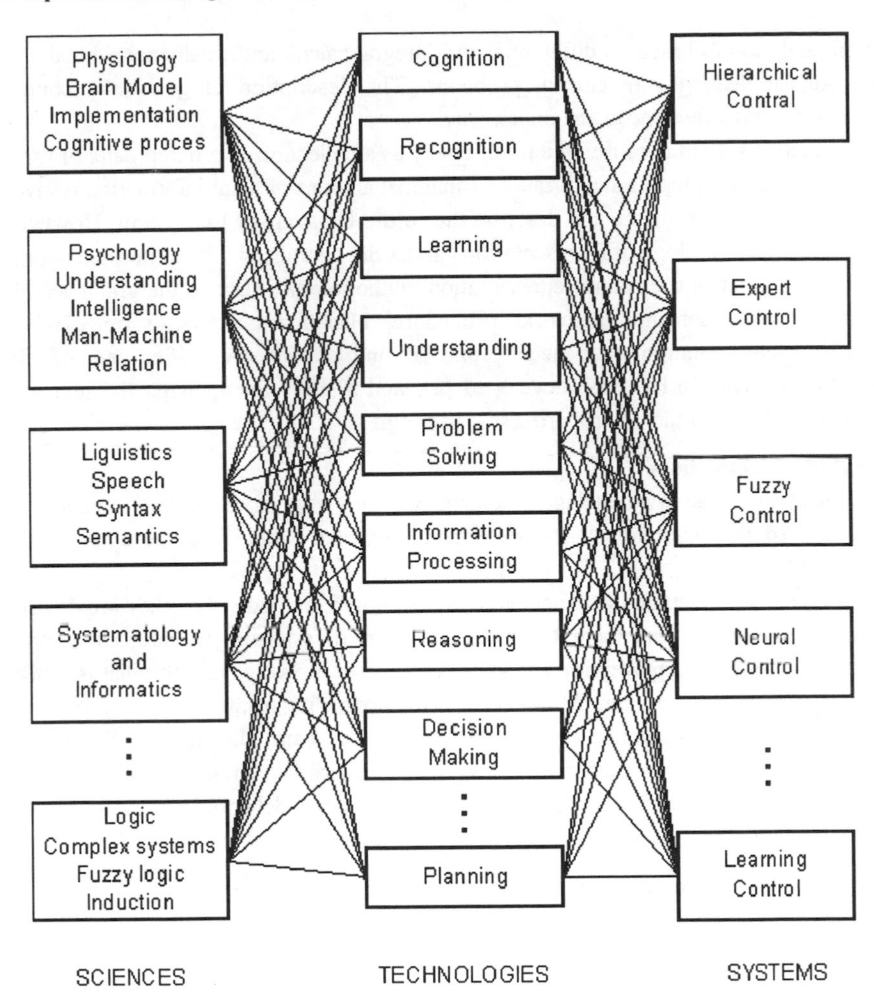

Figure 10.1. Relationship among intelligent control system and related S & T.

10.1.2 Issues for Future Investigation

In order to consolidate the results of intelligent control, several issues for future investigation are considered as follows [1, 3, 6, 7, 10]:

1. Representation

In the past, the problem of world representation of control system was given limited attention, and models based on differential and integral calculi are usually considered as a major tool for dealing with control problems. The description of goals and control concepts are frequently done in an implicit way.

The control system is considered traditionally as a collection of ordered pairs of time-functions representing inputs and outputs. A minimal amount of plant information is given which is required to complete and describe the future behavior of the system. However, the traditional methodology of representation can hardly be utilized within the framework.

Intelligent control uses more representation methodologies such as the graph search, predicate calculus, semantic network, procedure, blackboard, frame, script. Hybrid methods of representation have been proposed and applied. A unified method for representing knowledge has been developed [8], and a study for applying the temporal logic language XYZ to the control process is in progress.

2. Cognitive controllers

As the knowledge base in an intelligent control system is very large and the perception is very diversified the operation of the intelligent controller would become close to the human cognition.

There are many metaphorical models of human cognition. However, they are devised with no orientation to use them for intelligent control systems. Hierarchical control systems with hierarchical perception and knowledge bases are very close to what is similar to the human cognitive activities. Moreover, the understanding to the cognitive process of the human being is still limited and incomplete. In this case, the design of cognitive controllers would certainly encounter many difficulties in the coming future. With the advance of research on cognitive science, a better cognitive model could be developed, and so a better cognitive controller.

3. Suggestion for research on intelligent machines

Several suggestions for possible consideration in the research for intelligent machines are considered [10], these include:

(i) A fuzzy set formulation of the problem for the analytical design of intelligent machines will provide a generalized (and global) methodology for the organizer functions.

(ii) Design of intelligent machines with judgement capabilities will provide greater flexibility to plan organization, formulation and execution.

(iii) A linguistic based approach for the design of intelligent machines is an alternative solution. A comparison with the mathematical approach will provide useful information about possible deficiencies related to both of them.

(iv) A sensitivity analysis due to misinterpretation of the user commands will provide useful information about noise effects to the intelligent machines.

(v) A reliability analysis of the individual functions of the intelligent machines will provide a measure of robustness of the machine and help in the selection of alternate programs.

(vi) More research work is necessary to optimize the Boltzman machine; the modifications of the mathematical and architectural models to account for interactive functions of the different coordinators may be of interest.

4. Control modes

If a control system simply uses sequential control, then it would not be considered as a class of intelligent control. Other control modes such as interactive control, autonomous control, logic control, cognitive control, and direct voice control are considered as intelligent control. The interactive control is mostly utilized and easier to implement. Autonomous control is used to the system or plant that requires high control performance and is usually with uncertain environment. Direct voice control for complex systems is still interesting and a breakthrough in this field is expected. Hybrid modes of control are often used.

10.2 Prospectives for the Development of Intelligent Control

We begin our discussion with some trends in the intelligent control. In order to be provocative, it is generally necessary to predict something new and unexpected. At the same time, anyone who readily accepts the prediction and does not find it "foolish" fends it to be consistent with her /his own current beliefs, and it is thus to some extent already "expected" and not very provocative.

10.2.1 Newer Frames of Theory

Compared with the goal and the definition of intelligent control, the research on intelligent control suffers from several shortcomings that need to be overcome in the future:

(i) Isolation between macro and micro

As the basis science related to intelligent control, from the aspect of macro, researches of physiology, psychology, linguistics, cognitive science and logic are in a level that can be too high and too abstracted. In the micro aspect, the logic symbol, neural network, actionism, hierarchy, and algorithms of intelligent control are all in a low level of research. The isolation between the macro and micro is too large to eliminate; there exist many middle levels that have not been studied until now. A further study is needed into connecting the macro and the micro aspects and in their permeation.

(ii) Isolation between global and local

Like artificial intelligence, the mechanism of intelligent control is completely effective with rich levels and multiple aspects. However, Symbolism simply catches the abstracted thinking behaviour of the human brain; Connectionism only mimicks the thinking in terms of images; Actionism views artificial intelligence as the human intelligent behaviour and its evolutionary process. All of them do exist in evident limitations. It is necessary to have a multilevel, multifactor, multi-dimension and global investigation for intelligent control in order to overcome the mentioned limitations.

(iii) Distance theory from application

We already know something about the practical work of the human brain; however human intelligence is different in thousands of ways, ever changing, and too complex to understand. From the micro point of view, we learn less about the working mechanism of the human cerebrum, and some ideas are apparently right but actually wrong so that it is hard to find laws. In this case, the various proposed intelligent theories are only subjective guesses by someone, and present "intelligence" is limited. For reasons of technology, this distance might be kept for a long time.

From the above discussion we know that the structure and function of the human brain would be much more complex than we have imagined, the difficulties encountered in AI research would be much larger than that we have estimated, the tasks of research in AI and intelligent control would be much more arduous than that we have discussed. These tell us that a newer frame of AI and intelligent control is needed to set up so that the foundation for further development of intelligent control can be established.

Whatever be the new theories on intelligent control, we should deeply investigate on the basic theories and concepts, look for new theories that have not been found up to now, and set up a new mechanism for intelligent control. For instances, to form a unified description (representation) for the control knowledge and the control system, to study stability, robustness and dynamic properties of the intelligent control systems completely and systematically, to build a new generation of the expert control system based on models, to develop a new control mechanism based on bionics and anthropomorphism, and so on, all are very important issues in building new frames of intelligent control theory.

10.2.2 Better Technology of Integration

Similar to artificial intelligence, intelligent control has a developing trend of integrating AI techniques with other information-processing technology, especially, informatics, systematology, cybernetics and cognitive engineering, etc. From the view point of disciplinary structure, different ideas have been proposed; among them the Four-element Intersection Structure of Intelligent Control is a representative one. Within the intelligent control, different control schemes have been integrated; for example, the fuzzy self-learning neurocontrol integrates the techniques of fuzzy control, learning control and neurocontrol.

Achieving such integration presents many challenges. Some of these are involved with creating standard forms for control knowledge representation and transmittal and getting these standards accepted. Others are to understand the useful interactions among the various kinds of control subsystems to develop the representation method for coupling numerical models with non-numerical knowledge, to integrate the quantitative models and qualitative models, in order to support qualitative reasoning at acceptable speeds.

Information technologies that need to be integrated are diverse. Besides the above-mentioned methods, such technologies include computer networks, database, computer graphics, speech and other acoustical techniques, robotics, process control, parallel computing, optical computing, and possibly biological information-processing technology. For future intelligent control systems, the integration would also involve cognitive science, physiology, psychology, linguistics, sociology, anthropology, systematology and philosophy, etc. Figure 10.2 presents a block diagram of disciplines involved in the intelligent control system, where computing technology involves system architecture, object-oriented languages and programming, software engineering, human-machine interaction, database and knowledge bases, reasoning systems, and so on [1, 5, 11].

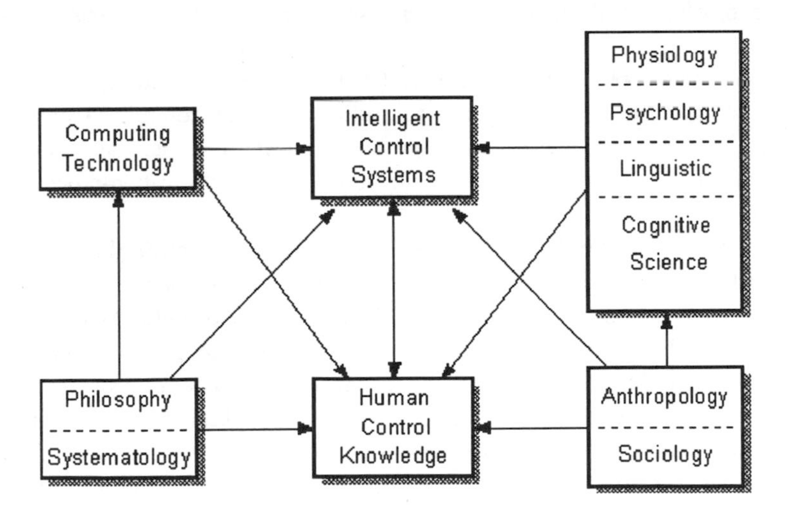

Figure 10.2. Disciplines for integrating intelligent control system.

Intelligent control will develop towards higher technical levels with multi-control and multi-variables, nonlinear, large time-delay, rapid-response, distributed parameter and large-scale systems.

10.2.3 More Mature Methodologies of Application

In order to implement intelligent control, both the new hardware and software have to be developed. Although the implementation of intelligent control needs the support of hardware, software should be the core of intelligent control systems. The discipline of software engineering arose in response to the need to develop complex software systems for many applications. The control software provides standard procedures for producing solutions to certain classes of control problems. Because of the complexity and often open-ended scope of intelligent control applications, traditional software design methods are not adequate. One problem is that the specifications of what function the intelligent control software is to perform are much more likely to change as the control system is developed. The technology for intelligent control must support the experimental aspect of intelligent control development and allow the control system to grow organically from a small kernel prototype to the complete application system.

The direct voice control system that uses the voice of the human operator as input and can understand and respond to the content of the speech will continue to develop and improve, and will get wide applications in industry, management and commerce, say, in robot control.

There are many topics in present research for the application of intelligent control. For examples, to increase operation speed, to realize real-time control, to enhance the capabilities for sensing and interpreting the environment, to design new modularized sensors and interfaces, and to improve the functions of information-processing and recognition.

In the aspect of control softwares, research topics may include the following: to develop task-oriented, independent and general purpose programming languages, to consider the requirements of CAD, CAM, CIMS and CIPS (Computer-Aided Production Systems), to be able to describe various control processes, to process slowly varying signals and faults, and to realize optimization for the intelligent control systems.

With more mature application methodologies, the application fields of intelligent control would be more and more wide. Besides the control of advanced robots (including space robot, undersea robot and mobile robot), process intelligent control and intelligent fault diagnosis, the following would become new application fields: traffic control such as high-speed trains, bus transportation and aircraft flight, automatic manufacturing control for CAD/CAM, CIMS and CIPS, control for medical processes, communication, agriculture, culture-education and entertainment, etc. [2, 4, 5, 9].

It is expected that there will be very general and useful development of intelligent control technologies that are available in the future. In applying intelligent control, new ways are being discovered to classify and solve complex control problems. By developing new tools and algorithms, methodologies are evolving that eventually will allow the successful application of intelligent control to many more fields than have been tracked so far.

We can predict that the artificial intelligence, intelligent machines, intelligent systems, including intelligent control and intelligent control systems, will have a wider scope of application fields than traditional computers have ever had. We also believe that intelligent control would certainly have a magnificent prospect, although those glorious days may still be far from today, and harder work and more expenses cost would be devoted, continuous efforts of researchers generation by generation might also be needed.

References

1. Z.-X. Cai, *Intelligent Control* (Electronics Industry Press, Beijing, 1990) 179-180.
2. Z.-X. Cai, "Applications of intelligent process control," *J. Central South University of Technology*, **26** (1995) 297-300.
3. Z.-X. Cai and G.-Y. Xu, *Artificial Intelligence: Principles and Applications*, Second Edition (Tsinghua University Press, Beijing, 1996) 399-403.
4. Z.-X. Cai and Z.-J. Zhang, "Some issues of intelligent control," *Pattern Recognition and Artificial Intelligence*, **1** (1988) 41-46.
5. B. R. Gaines, "Intelligent systems as a stage in the evolution of information technology," In: Z. W. Ras and M. Zemankova eds. *Intelligent Systems, State of the Art and Future Directions* (Ellis Horwood Limited, New York, 1990) 440-452.
6. A. Meytel, "Intelligent control: issues and perspectives," *Proc. IEEE Symposium on Intelligent Control*, (1985) 1-15.
7. Z.-Z. Shi, *Principles of Machine Learning* (International Academic Publishers, Beijing, 1992) 354-358.
8. C. S. Tang and C. Zhao, *Introduction to the Temporal Logic Language XYZ/E* (Academic Press, Beijing, 1996).
9. S. L. Tanimoto, *The Elements of Artificial Intelligence* (Computer Society Press, New York, 1987) 435-483.
10. K. P. Valavanis and G. N. Saridis, *Intelligent Robotic Systems: Theory, Design and Applications* (Kluwer Academic Publishers, Boston, 1992) 217-218.
11. Z.-J. Zhang and Z.-X. Cai, "Intelligent control and intelligent control systems," *Information and Control*, **18** (1989) 30-39.

BIOGRAPHY

Zi-Xing Cai was born in the Fujian Province of China. He got a diploma and graduated from the Department of Electrical Engineering, Jiao Tong University at Xi'an, China in 1962. He has been doing teaching and research work on automatic control, artificial intelligence, and robotics in the Department of Automatic Control Engineering, College of Information Engineering, Central South University of Technology, Changsha, China since 1962.

During May to December of 1983, he visited the Center of Robotics, Department of Electrical Engineering and Computer Science, University of Nevada at Reno(UNR), Nevada, USA. Then he visited the Advanced Automation Research Laboratory(AARL), School of Electrical Engineering, Purdue University, West Lafayette, Indiana, USA during December 1983 to June 1985. In the two universities, as an visiting exchange scholar he did research and teaching work on AI, expert system, robotics and high-level robot planning. During October 1988 to August 1989, he was a senior research scientist in the National Laboratory of Pattern Recognition, Institute of Automation, Chinese Academy of Sciences. During October 1989 to July 1990 he visited the National Laboratory of Machine Perception, Center of Information, Beijing (Peking) University as a visiting research professor. During September 1992 to March 1993, he visited the Center for Intelligent Robotic Systems for Space Exploration (CIRSSE), Department of Electrical, Computer and System Engineering, Rensselaer Polytechnic Institute (RPI), Troy, New York, USA as a Visiting Professor. Since February 1989, he has become an UN (United Nations) Expert granted by UNIDO, the United Nations Industry Development Organization. His name has been listed into the "*Who's Who of UN Experts*" (1989), "*The Excellent Books and Authors of S&T in China*"(1990), "*Chinese Grand Encyclopedia*"(1991), "*Who's Who of S &T in China*" (1992), "*Who's Who of Chinese Experts*" (1994), "*Who's Who of Modern Scientists and Inventors in China*" (1995), "*Dictionary of Chinese Modern Notables*"(1996), and "*Who's Who in Australasia and The Pacific Nations*" (1996/1997), and "*Who's Who of International Intellectuals*" (1997), ect.

Professor Cai's research interests include Artificial Intelligence, Robotics, Industry Automation and Intelligent Control. Current projects include intelligent control for the operation process of high-speed trains, robustness of intelligent control systems, structure and properties of intelligent control, applications of neurocontrol in industry process, intelligent control process in CIPS, etc. Over 260 papers, articles and book chapters have been published since 1986. He is the author or co-author of 12 books/textbooks such as *Artificial Intelligence: Principles and Applications* (1987; Second Edition 1996),

Robotics: Principles and Applications (1988), and *Intelligent Control* (1990). A special mention should be given to the fact that he used the technique of expert system to high-level robot planning successfully in 1985 for the first time in the world. In 1986, Professor Cai proposed the idea of Four-element Structural Theory of Intelligent Control (AI, Cybernetics, Informatics and Operation Research) for the first time. Over 12 awards of S&T have been received.

Professor Cai is currently the Director of the Research Center for Intelligent Control, and a doctor's supervisor at the Central South University of Technology, the Member of the Board of Directors or Standing Directors for the followings: Asia-Pacific Federation of Intelligent Automation (APFIA), Chinese Association of Artificial Intelligence (CAAI), Chinese Society of Computer Vision and Intelligent Control (CVIC, CAAI), Society of Intelligent Robots (SIR, CAAI, as President), Society of Process Control and Society of Intelligent Automation, Chinese Association of Automation (CAA), Society of Artificial Intelligence and Pattern Recognition, Chinese Federation of Computer (CFC), Sub-Society of Automation, Chinese Association of Computer Users, Chinese University Forum on Automation and Robotics. He is also a Member of the Editorial Board of the National Journal *Robot*, and so Journal *Computer Technology and Automation*. He has been a member of eight national and four international societies/associations/federations.

SUBJECT INDEX

S

T

U

V

W